Biochemistry of
the Lanthanides

BIOCHEMISTRY OF THE ELEMENTS

Series Editor: Earl Frieden
Florida State University
Tallahassee, Florida

A Continuation Order Plan is available for this series. A continuation order will bring delivery of each new volume immediately upon publication. Volumes are billed only upon actual shipment. For further information please contact the publisher.

Biochemistry of the Lanthanides

C. H. Evans

The Ferguson Laboratory
University of Pittsburgh
Pittsburgh, Pennsylvania

PLENUM PRESS • NEW YORK AND LONDON

Library of Congress Cataloging in Publication Data

Evans, C. H., 1950–
    Biochemistry of the lanthanides / C. H. Evans.
        p.   cm. — (Biochemistry of the elements; v. 8)
    Includes bibliographical references and index.
    ISBN 0-306-43176-9
    1. Rare earth metals—physiological effect. 2. Rare earth metals—Metabolism.
    I. Title. II. Series.
QP532.E93 1989   89-8703
574.19′214—dc20   CIP

© 1990 Plenum Press, New York
A Division of Plenum Publishing Corporation
233 Spring Street, New York, N.Y. 10013

Printed in the United States of America

For Lillian Rose Bioletti

Preface

By a happy coincidence, the completion of this text coincided with the 200th anniversary of the discovery of gadolinite, the mineral with which the lanthanide story begins. For a group of elements which occur in only trace amounts biologically, and which have no known metabolic role, the lanthanides have spawned a surprisingly large biochemical literature. Serious interest in the biochemical properties of these elements can be traced to concerns about the safety of radioactive lanthanides toward the end of World War II. As recent events at Chernobyl indicate, this concern remains topical. However, the literature on lanthanide biochemistry predates the atomic era, beginning with sporadic, medically motivated studies in the latter part of the 19th century. Much of the present biochemical activity involving the lanthanides centers around their ability to provide important information on the interactions of Ca^{2+} with macromolecules and with eukaryotic cells. With the increasing industrial use of the lanthanides, their toxicological properties will need to be examined more closely. Rare earth pneumonoconiosis has already been identified as a disease produced by industrial exposure to lanthanides. Several of the biochemical properties of the lanthanides are of relevance to modern medicine. Already cerium-based ointments are used to treat burn wounds, while paramagnetic lanthanides find application in nuclear magnetic resonance imaging.

This book is an attempt to collate and to present in reasonable detail existing knowledge of lanthanide biochemistry before the literature becomes unmanageable. The information it contains should be of value to those engaged in inquiry at the postgraduate level into all aspects of lanthanide biochemistry. Because the subject matter encompasses such a wide range of subdisciplines, it has not been possible to explicate the underlying principles of each topic. For those seeking further information, I have included an appendix listing all the review articles I could find on the various aspects of lanthanide biochemistry. The attempt to be comprehensive will make certain sections of the book tedious for those whose main interest lies elsewhere. However, this volume is not intended as bedtime reading to be read from start to finish; rather it should serve as a repository to be dipped into for the purposes of retrieving specific information.

The problems associated with covering such an array of different disciplines in one volume have been many. *Rare earth metals* did not appear as a key phrase on the MEDLINE information retrieval system until 1975. Since much of the older literature on the lanthanides remains pertinent today, I have had to fall back on traditional methods of literature searching for the years before this. Much of the early work on the metabolism of the lanthanides was published in the reports of private laboratories and government agencies, rather than in the normal scientific journals, and some of these reports have been hard to obtain. I apologize in advance for any omissions, and would welcome any additional information. A second challenge has been the broad span of topics covered, ranging from inorganic chemistry, via molecular and cellular biochemistry, to physiology, toxicology, and medicine. In covering this spread, I have had to trespass into areas of knowledge with which I ordinarily do not deal. I trust that the specialists in these fields will forgive my intrusions and look leniently upon any inaccuracies. Again, I would gladly receive any criticisms or supplemental information.

ACKNOWLEDGMENTS. So many have contributed to the writing of this book that this risks being the longest section of the entire volume. Without the respect for and desire to search for knowledge instilled in me by my parents, Mr. and Mrs. Idwal Evans, such an undertaking would not even have been contemplated. Their constant support and concern have been invaluable and are deeply appreciated. My wife Mindy has been a "book widow" now for many months, a circumstance she has endured with good humor and understanding. Had it not been so, this book would still be uncompleted.

I owe my introduction to the lanthanides to my good friend Dr. Roderic Bowen, someone whose generosity far exceeds that normally attributed to a *Cardi: Diolch yn fawr iawn, 'was; 'r wy'n mewn dyled mawr iti.* Mr. Vernon Westcott provided my first opportunity to work with the lanthanides, and has remained a valued friend, colleague, and advisor. My intellectual life of the last eight years or so has been greatly enriched through the friendship of Dr. Tony Russell, one of the most creative of individuals. His collaborations with me on lanthanide biochemistry have always been stimulating, productive, and, above all, enjoyable.

Mrs. Diana Montgomery could not have predicted the enormity of the task that lay before her upon agreeing to type the several drafts of the manuscript. Even with a word processor, it turned out to be a monstrous undertaking. I am very grateful for her perseverance. Much of the reference section owes its existence to Mr. David Obi, a tireless, good-natured postgraduate student with remarkable stamina. Our bibliographic efforts would have been far less fruitful without the efficient yet cheerful assistance of Mrs. Margaret Norden and her colleagues at the Falk Library, whose staff must have had cause to put in overtime during our periods of peak activity. The writing of this book was greatly expedited when Dr. Norman Curthoys permitted me to colonize a vacant laboratory in the Biochemistry Department. My "book room" became an indispensable island of retreat. Several chapters were improved immensely through critical review by Dr. Tony Russell, Dr. Jimmy Collins, and Dr. Rex Shepherd. Any remaining flaws are my own responsibility.

One of the biggest contributions to the successful completion of the book was the patience and understanding of the members of the Ferguson Laboratory for Orthopaedic Research, which has been running on auto-pilot for so long now. These individuals, too numerous to mention by name, have been wonderful. My expressions of gratitude to Dr. Albert B. Ferguson have been left until last not owing to lack of importance but, on the contrary, because his support, encouragement, and friendly guidance have sustained the past decade of my research in his department. Without "Ferg" this book would not exist.

C. H. Evans

Note Added in Proof

Well over a thousand publications on lanthanide biochemistry have appeared since completion of the text. While it is impossible to incorporate all such new information into this book, wherever possible some of the more important new findings have been added at the proofing stage. However, none of the recent data alter the underlying concepts of lanthanide biochemistry described in this volume.

Contents

4. The Interaction of Lanthanides with Amino Acids and Proteins   85

5. Interactions of Lanthanides with Other Molecules of Biochemical Interest 173

6. Interactions of Lanthanides with Tissues, Cells, and Cellular Organelles 211

Historical Introduction

1.1 Opening Remarks

In the summer of 1787, Karl Arrhenius, a lieutenant in the Swedish army, chanced upon a new mineral, which he named "ytterbite" after the nearby Swedish town of Ytterby. This book is a descendent of that discovery, which foreshadowed the identification of a new group of elements, the rare earths or lanthanides. (For a discussion of nomenclature, see Section 2.1.) The intervening 200 years have been colorful ones. Owing to the close chemical similarities between the members of the lanthanide series, they resisted easy purification and separation from one another. Numerous misidentifications, false claims, and counterclaims are scattered through the pages of this chapter of chemical history. For a number of years, the existence of the lanthanides challenged the accuracy of Mendeleev's periodic table of the elements. Taken together, there is enough material here for an historian of science to write an instructive book on the identification of the lanthanides in its own right. For reasons to be discussed below, the biochemical properties of the lanthanides have received increasing attention over the last three decades or so. The brief historical orientation presented below sketches the intellectual route from Arrhenius's new mineral (Arrhenius, 1788) to this book.

1.2 The Identification of the Lanthanides

The first chemical analysis of ytterbite was undertaken by the Finnish chemist Johann Gadolin. He determined that the mineral contained the oxides of beryllium, silicon, and iron, in addition to a new earth which Gadolin christened "ytterbia" (Gadolin, 1794). Ekeberg (1797) subsequently shortened its name to "yttria." The choice of the word "earth"

1

was unfortunate, as it is more properly applied to the oxide of a metal, rather than the metal itself. In Gadolin's time, "earth" was taken to include "all substances which possessed the properties of alkalis, did not float and did not change on heating, were almost insoluble in water and evolved gas bubbles during reaction with alkalis" (Trifonov, 1963). Actually, the French chemist Vauquelin had concluded at the beginning of the nineteenth century that cerium was a metal and not an earth, but this finding was ignored. The true nature of "earths" was discovered in 1808 by Sir Humphrey Davy. As we shall see, the term rare earth is a double misnomer, as not only are these elements not earths, but they are also not particularly rare (Section 7.1). Indeed, some of the commoner lanthanides are more abundant on earth than, for example, lead, tin, zinc, mercury, and gold. They are also present in the sun's atmosphere, in meteorites, and on the moon, where they are relatively abundant. Nomenclature has historically proved unfortunate for the lanthanides, with misnaming, conflicts over names, and changes of name frequently leading to confusion. We have already seen that Gadolin's "ytterbia" was soon changed to Ekeberg's "yttria." The parent mineral ytterbite was subsequently renamed gadolinite in honor of Gadolin. In view of its important chemical legacy, the Finnish mineralogist Flink is said to have written that gadolinite "perhaps played a greater role in the history of inorganic chemistry than any other mineral" (Trifonov, 1963).

Much of the subsequent hundred years was spent in isolating new lanthanides from their ores, often demonstrating in the process that preparations previously considered pure actually contained a mixture of more than one lanthanide. Most of these separations involved tedious rounds of fractional crystallization. The chemistries of these elements are so close that eons of geochemical activity had done little in the way of natural fractionization, beyond providing ores enriched in heavy or light lanthanides. Most mineral sources thus contain mixtures of several different members of the series. Some idea of the extent of the preparative problems can be gauged from the fact that, before the development of ion-exchange techniques in the mid-twentieth century, some of the rarer lanthanides required as many as 40,000 fractional crystallizations before they were really pure (Spedding, 1951).

Through the work of Berzelius and Hisinger in Sweden and, independently, Klaproth in France, cerium was the first lanthanide to be processed to a reasonably high degree of purity. It was named after an asteroid, Ceres, that had recently been discovered (Table 1-1). Thus, until 1839, two rare earths, yttrium and cerium, were known. In this year, Karl Mosander, a student of Berzelius, discovered lanthanum. Shortly thereafter, Mosander separated a new earth from lanthanum oxide. Because

Table 1-1. Discovery of the Lanthanides[a]

Lanthanide	Year of identification	Discoverer	Origin of name
Lanthanum	1839	Mosander	Lanthanein; Greek for "to lie hidden"
Cerium	1803	1. Berzelius and Hisinger 2. Klaproth	Ceres, an asteroid discovered in 1801
Praseodymium	1885	Von Welsbach	From Greek: *Prasios* = green; *dymium* = twin
Neodymium	1885	Von Welsbach	From Greek: *Neo* = new; *dymium* = twin
Promethium	1947	Marinsky Glendenin Coryell	Prometheus, the Greek god who stole fire from heaven for men's use
Samarium	1879	De Boisbaudran	From its ore, samarskite, named after the Russian engineer Samarski
Europium	1889	Crookes	Europe
Gadolinium	1880	Marignac	After the Finnish chemist Gadolin
Terbium[b]	1843	Mosander	After the town of Ytterby in Sweden
Dysprosium	1886	De Boisbaudran	From Greek: *Dysprositos* = hard to get at
Holmium	1879	1. Soret 2. Cleve	*Holmia,* Latinized version of Stockholm
Erbium[b]	1843	Mosander	After the town of Ytterby in Sweden
Thulium	1878	Cleve	After *Thule,* the Roman name for the northernmost region of the inhabitable world
Ytterbium	1878	Marignac	After the town of Ytterby in Sweden
Lutetium[c]	1908 1907	1. Von Welsbach[c] 2. Urbain	*Lutetia,* Latin for Paris
Yttrium	1794	Gadolin	After the town of Ytterby in Sweden

[a] Difficulties exist in determining the precise year of identification and in deciding who should be given credit for the discovery of several of the lanthanides. Thus, although Gadolin identified yttrium in 1794, he did not realize that his preparation contained a mixture of at least three lanthanides. Yttrium was not purified, as its oxide yttria, until the work of Mosander in 1843. Similarly, what Marignac called ytterbium in 1878 was shown by Urbain in 1907 to be a mixture of lutetium and "neoytterbium" or ytterbium, as it is now called. In constructing this table, I have tried to list as the discoverer the chemist who first identified the element, regardless of the state of purity then obtained. Cases of simultaneous, independent discovery are indicated in the appropriate places. Most of these elements were first identified as their oxides, in which the ending -ium is replaced by -a, e.g. La_2O_3, lanthana; Ce_2O_3, ceria. The discoverer of each element did not necessarily give it its modern name.

[b] Terbium was originally designated element 68 and erbium, element 65. The names were reversed in 1877.

[c] Lutetium is the name given by Urbain. Von Welsbach called it cassiopeium.

of its resemblance to lanthanum, he called the new substance "didymium," meaning twin. It was not until 1885 that Auer Von Welsbach demonstrated that didymium was itself a mixture of two lanthanides, praseodymium (green twin) and neodymium (new twin). Soon after this discovery, Von Welsbach invented an illuminating device, known as the Welsbach gas candle, and a lighter flint, both of which used lanthanides. Indeed, his discovery of an alloy of cerium and iron, which emitted sparks when struck, started the flint industry. Fame and fortune followed. As Spedding (1951) noted, Von Welsbach is one of the few scientists for whom a statue has been erected in a public square.

The identification of new lanthanides was aided immensely by the introduction of spectroscopic analysis in the second half of the nineteenth century (e.g., Gladstone, 1858). This method was partly responsible for the discovery of, or confirmation of, terbium, ytterbium, holmium, thulium, and samarium. However, as with many sensitive new techniques, spectroscopy also produced its share of false leads. This was partly due to uncritical use of the equipment and a failure to realize that even small amounts of an impurity could greatly alter the recorded spectrum. Sir William Crookes, despite being a pioneer spectroscopist, was misled in this way into proposing the existence of meta-elements. In so doing, he wished to explain the closeness of the properties of the rare earths by postulating a single core element which existed in varieties of different atomic weight. Nevertheless, as Table 1-1 shows, by the end of the nineteenth century, all the lanthanides except promethium and lutetium had been identified. Several eponymic references to "hidden" or "difficult" reflect the nature of their discovery. Among the alleged new earths which turned out not to exist were some with imaginative names, including cosmium, neocosmium, decipium, austrium, celtium, demonium, damarium, glaucodymium, lutium, incognitium, phillipium, and victorium. Between 1843 and 1939, the identification of over 70 different new lanthanides was claimed. Element number 71 was discovered independently in 1907–08 by Von Welsbach in Germany and Urbain in France. Urbain named it lutetium for his native city of Paris, while Von Welsbach named it cassiopeium. These names remained in competition for several decades, with cassiopeium being in common use in Germany and Austria until the 1950s.

An early problem facing Mendeleev's periodic table of 1869 was the placement of the lanthanides. It is interesting to note that since the time of Berzelius, the lanthanides had been considered to be bivalent, although Vauquelin made the important discovery that cerium had two oxidation states. Mendeleev invoked his periodic law to conclude, correctly, that they are trivalent. In a similar manner, he was also able to correct the atomic weights of the lanthanides. However, the correct placing of the rare earths proved a thornier problem, eliciting a number of ingenious modifications to the table. According to Trifonov (1963), "most of the modifications of the periodic system were produced by an attempt to solve the problem of placing the rare earths." Its resolution came with Moseley's discovery in 1913 that the atomic number of an element could be determined from analysis of its X-ray spectrum. This clearly demonstrated the position of the lanthanide elements and revealed that one space, that of element number 61, remained to be filled. Its discovery, as the new element illinium in honor of the state of Illinois, was prematurely announced in 1926. An Italian group from the University of Florence claimed to have discovered the element, under the name florentium, two years earlier. Both claims were wrong; element 61 turned out to be unstable, with a half-life of only 2.7 years. Finally produced by Pool and Quill in 1937 by bombarding neodymium with deuterons, its first successful separation and characterization was reported by Marinsky *et al.* (1947). Element 61 was later named promethium, after the god Prometheus, who stole fire from heaven for use by men. As punishment, Zeus chained Prometheus to a rock, where vultures would come every day to torment him. According to the discoverers, "the name not only symbolizes the dramatic way in which the new element was obtained in appreciable quantities, thanks to the harnessing of nuclear energy, but also warns men of the threatening danger of punishment by the vulture of war."

More detailed reviews of the historical aspects of the identification and separation of the lanthanides are to be found in Weeks (1948), Vickery (1953), and Trifonov (1963). Less comprehensive, but more recent accounts have been provided by Gschneidner and Capellen (1987) and Evans (1989).

1.3 Biochemical Studies on the Lanthanides

Four major motives have lain behind research into the biochemical properties of the lanthanides. In roughly chronological order, these are: the search for medicinal applications (Chapter 9), toxicological concerns (Chapter 8), intellectual curiosity, and their use as informative probes in biological and biochemical research (Chapters 3–6).

In the latter half of the last century, cerium oxalate was widely prescribed as an antiemetic of particular use in the sickness that accompanies pregnancy. It is not clear how effective it was, although it may have acted as an inert coating to the wall of the stomach, which resisted the acidity of the stomach, in the same way as other insoluble minerals, such as kaolin. Studies of its effects in dogs (Baehr and Wessler, 1909) are among the earliest biological experiments using the lanthanides. Shortly before this, their *in vitro* antimicrobial properties had been discovered (Drossbach, 1897). This led to clinical trials of lanthanides in the treatment of tuberculosis and leprosy. Although encouraging results were published, these studies did not have any lasting clinical impact. Radioactive species of various lanthanide elements became generally available in highly purified form after the Second World War. These have found use in irradiation therapy for tumors. Lanthanides have a certain ability to localize preferentially in tumors, a property leading to their experimental use as injectable antitumor drugs and tumor scanning agents. Again, the clinical results have not lived up to their early promise.

From the 1920s to the 1950s, much serious research was devoted to attempting to harness the anticoagulant properties of the lanthanides for clinical use. However, their potency as antithrombotic agents was offset by problems of toxicity. Before this disadvantage could be overcome, highly purified heparin became readily available as a cheap anticoagulant for routine use. However, lanthanides are presently finding medicinal use as contrast-enhancing agents in nuclear magnetic resonance (NMR) imaging. In addition, they are being applied to burn wounds as topical antiseptic agents which may also prevent the immunosuppressive sequelae of severe burns (Chapter 9).

A review of the pharmacologic properties of the lanthanides (Chapter 8) reveals that they lower blood pressure, lower serum cholesterol and glucose levels, reduce appetite, inhibit blood coagulation, and prevent atherosclerosis in experimental animals. Here, it would seem, we have an instant cure for the country club diseases of the Western world! The potential clinical benefits of substances with such pharmacologic properties should surely not remain uninvestigated. In addition, certain complexes of the lanthanides have anti-inflammatory activity. One such compound, "phlogodym" (Fig. 9-5), is presently used as a topical anti-inflammatory agent in Hungary. The medical future of the lanthanides could indeed turn out to be a very interesting one. Their present status in this respect is reviewed in Chapter 9.

Although workers had been exposed to high levels of lanthanides since the days of Auer Von Welsbach's flint industry, toxicological concerns were not voiced until it was realized that radioactive lanthanides were an important product of nuclear reactions. At least 16 radioactive species of lanthanides and yttrium are produced during fission of ^{235}U. Prominent among these are ^{90}Y, a daughter of ^{90}Sr and ^{144}Ce. Because of its 290-day half-life and relatively high yield, ^{144}Ce often predominates in the radioactive fallout one to four years after fission. The recent Chernobyl accident released radioactive lanthanides into the atmosphere, reminding us of the continued topicality of this area of lanthanide biochemistry. As a result of its relevance to nuclear matters, research into the lanthanides became incorporated into the Manhattan Project. Part of this involved the application of ion-exchange chromatography to the separation of fission products from nuclear reactors. The impact was tremendous. Ion-exchange chromatography provided, in days, what had previously taken many months of repeated fractional crystallization and precipitation to prepare. Relatively large amounts of highly purified lanthanide elements now became available. This development, in turn, fostered the third motive, that of intellectual curiosity. Metabolic and toxicological aspects of the lanthanides are discussed in Chapters 7 and 8, respectively. As the industrial use of the lanthanides and the health consciousness of our society continue to increase, these aspects will receive greater attention in the future.

The fourth and final motive rests upon the strongest theoretical foundation. Although the effects of lanthanides upon the physiology of muscle had been documented early in the twentieth century (Mines, 1910; Hober and Spaeth, 1914), Lettvin et $al.$ (1964) first penetrated beyond the mere phenomenology in drawing attention to similarities in the ionic radii of Ca^{2+} and La^{3+}. They concluded that La^{3+} could be profitably used to study the interactions of Ca^{2+} with nerves. The richness of this prediction is amply demonstrated by the size of Chapter 6, in which the effects of lanthanide ions on cellular metabolism and physiology are discussed. The idea that lanthanide ions could replace Ca^{2+} in biochemically interesting ways has been confirmed experimentally by many workers. However, it is to R. J. P. Williams of Oxford University that we owe the sound theoretical basis upon which these studies rest (e.g., Williams, 1970). The similarities between calcium and lanthanide ions are not restricted to their ionic radii but include important aspects of their coordination chemistries and binding behavior (Section 2.5; Table 2-11). This would not be so biochemically important had not nature selected Ca^{2+}, an experimentally uninformative ion, to play such a crucial role in the structure and function of living things. By replacing Ca^{2+} ions with lanthanide ions, the investigator can bring a range of spectroscopic techniques (Chapter 3) to bear upon their biochemical behavior. The experimental work of Williams and his collaborators pioneered the use of lanthanides in NMR spectroscopic studies of metal-binding sites on molecules of biochemical interest and of the conformations of such molecules in solution (Barry et $al.$, 1971). At about this time, Darnall and his colleagues demonstrated the usefulness of lanthanides as functional replacements for Ca^{2+} in enzymic reactions (Darnall and Birnbaum, 1970). It is as biochemical probes of metal-binding sites that Ln^{3+} ions have really come into their own and given prominence to the field of lanthanide biochemistry, as the length of Chapters 4 and 5 confirms. The use of lanthanides as isomorphous replacements for Ca^{2+} is now a standard experimental approach to the study of appropriate biochemical materials. In addition, La^{3+} is routinely used as an electron-dense stain in electron microscopy, where it permits good imaging of cell membranes and the identification of gap junctions. As the rate of increase of the literature on these topics shows no sign of decreasing, much future progress can be expected.

References

Arrhenius, K., 1788. *Svenska Akad. Handl.* 9:217.

Baehr, G., and Wessler, H., 1909. The use of cerium oxalate for the relief of vomiting: an experimental study of the effects of some salts of cerium, lanthanum, praseodymium, neodymium and thorium, *Arch. Intern. Med.* 2:517–531.

Barry, C. D., North, A. C. T., Glasel, J. A., Williams, R. J. P., and Xavier, A. V., 1971. Quantitative determination of mononucleotide conformations in solution using lanthanide ion shift and broadening NMR probes, *Nature* 232:236–245.

Darnall, D. W., and Birnbaum, E. R., 1970. Rare earth metal ions as probes of calcium ion binding sites in proteins, *J. Biol. Chem.* 245:6484–6488.

Drossbach, G. P., 1897. Über den Einfluss der Elemente der Cerund Zircon gruppe auf das Wachstrum von Bakterien, *Zentralbl. Bakteriol. Parsitenkd. Abt. 1. Orig.* 21:57–58.

Ekeberg, G., 1797. *Svenska Akad. Handl.* 18:156.

Evans, C. H., 1989. Two hundred and one years of rare earth elements, *Chem. Brit.* (In Press).

Gadolin, J., 1794. *Kgl. Svenska Vetenskapsakad. Handl.* 15:137.

Gladstone, J. H., 1858. On an optical test for didymium, *J. Chem. Soc.* 10:219–221.

Gschneidner, K. A., and Capellen, J., Eds., 1987. *1787–1987. Two hundred years of rare earths,* North-Holland, Amsterdam.

Hober, R., and Spaeth, R. A., 1914. Über den Einfluss seltener Erden auf die Konträktilitat des Muskels, *Arch. Ges. Physiol.* 159:433–456.

Lettvin, J. Y., Pickard, W. F., McGulloch, W. F., and Pitts, W. S., 1964. A theory of passive ion flux through axon membranes, *Nature* 202:1338–1339.

Marinsky, J. A., Glendenin, L. E., and Coryell, C. D., 1947. The chemical identification of radioisotopes of neodymium and of element 61, *J. Am. Chem. Soc.* 69:2781–2785.

Mines, G. R., 1910. The action of beryllium, lanthanum, yttrium and cerium on the frog's heart, *J. Physiol.* 40:327–345.

Spedding, F. H., 1951. The rare earths, *Sci. Am.* 1951(Nov.):89–101.

Trifonov, D. N., 1963. *The Rare-Earth Elements* (translated by P. Basu; R. C. Vickery, ed.), Pergamon Press, New York.

Vickery, R. C., 1953. *Chemistry of the Lanthanons,* Academic Press, New York.

Weeks, M. E., 1948. *Discovery of the Elements,* Mack Printing Co., Easton, Pa.

Williams, R. J. P., 1970. Cation and proton interactions with proteins and membranes, *Biochem. Soc. Trans.* 7:481–509.

Chemical Properties of Biochemical Relevance

2

2.1 Introduction

The chemistry of the lanthanides, unlike their biochemistry, has been extensively reviewed. Several books have been written on this subject (e.g., Vickery, 1953; Spedding and Daane, 1961), and numerous reviews exist (e.g., Moeller, 1973; Morris, 1976). In addition, many standard textbooks on inorganic chemistry have good sections on the lanthanides (e.g., Phillips and Williams, 1966; Cotton and Wilkinson, 1980). It is neither possible nor necessary to discuss all aspects of lanthanide chemistry in this chapter. Instead, those properties of the lanthanides which are most pertinent to their biochemical interactions will be reviewed. Particular attention will be paid to their similarities to Ca^{2+}, as this forms the basis for many of the biochemical studies which employ lanthanides. The strategy of substituting Ca^{2+} by Ln^{3+} ions in biological systems owes much of its theoretical basis and development to R. J. P. Williams (Vallee and Williams, 1968; Williams, 1970; Williams, 1979). Such an approach has proved valuable because of the irony whereby Ca^{2+}, biologically one of the most important metal ions, is chemically one of the least informative. The lanthanides, on the other hand, lend themselves to a number of analytical, investigative techniques (Chapter 3). The approach has proved an extremely successful one and is largely responsible for the sustained interest in lanthanide biochemistry.

Although not nominally a lanthanide, yttrium has sufficient chemical and biochemical similarity to the lanthanides to be included in the subject matter of this book. However, scandium, the other group 3 member, is sufficiently different to be excluded. While on the subject of demarcation, it is worth drawing attention to some of the different nomenclatures that are found in the literature. Strictly speaking, the lanthanides are the 14 elements that follow lanthanum in the periodic table. However, the clas-

9

sification is usually taken to include the element lanthanum itself. Other designations include lanthanons, lanthans, and lanthanoids. These elements were originally called the rare earths (Section 1.1), a term that is still commonly encountered. Some confusion exists in the older literature, where the members of the actinide series are sometimes referred to as the "actinide rare earths." As the lanthanides are not particularly rare and are not earths (oxides), some have suggested that the classification "rare earths" be abandoned. Others have argued that it should be retained to designate elements 57–71 plus Y and Sc, in contradistinction to the title of lanthanides (or lanthanons, lanthanoids, etc.), which includes only elements 57–71. Ignoring such semantics, this book will use the word "lanthanide," abbreviated as Ln, to designate the fifteen elements 57–71 (La–Lu) only. Whenever yttrium (Y) is discussed, it will appear under its own name.

Attempts have also been made to split the lanthanides into various subgroups. One of these, based upon their occurrence in different minerals, distinguishes between the "cerium earths" (La–Gd) and the "yttrium earths" (Tb–Lu, Y). Alternative names for these same groups are the "light" and "heavy" lanthanides. Others have divided the series into three, making reference to the cerium group (La–Sm), the terbium or transitional group (Eu, Gd, Tb), and the yttrium group (Dy–Lu, Y). Such distinctions are largely avoided in the present work as, from the biochemical point of view, one of the advantages of the lanthanides is the relatively smooth and progressive nature of the changes in their chemical properties throughout the series.

2.2 Electronic Configurations and Their Consequences

The outer electronic configurations of the lanthanides, yttrium, and calcium and their most common ions are given in Table 2-1. In progressing from La, the lightest lanthanide, to Lu, the heaviest, there is progressive filling of the $4f$ orbitals. As we shall see, it is these $4f$ electrons which endow the lanthanides with the special properties of such importance to structural biochemical analysis. Of great significance is the electronic configuration whereby the $4f$ electrons are not in the outermost orbitals but are shielded by the electrons of higher orbitals (Table 2-1). All the lanthanides have filled $5s$, $5p$, and $6s$ orbitals, while La, Gd, and Lu each have $5d$ electrons. As the outermost electrons are the valence electrons (Table 2-1), the most important properties resulting from the inner, $4f$ electronic behavior are not lost upon ionization. Furthermore, upon coordination with ligands, the $4f$ electrons remain sufficiently shielded that

Table 2-1. Some Pertinent Chemical Properties of the Lanthanides, Calcium, and Yttrium[a]

Element	Symbol	Atomic number	Atomic weight	Outer electronic configuration — Atomic (Ln⁰)								Outer electronic configuration — Ionic (Ln³⁺)			Ionic radius (Å)				ΣI[b] (eV)
				$4s$	$4p$	$4d$	$4f$	$5s$	$5p$	$5d$	$6s$	$4f$	$5s$	$5p$	CN6	CN7	CN8	CN9	
Calcium	Ca	20	40	[c]											1.00	1.06	1.12	1.18	—
Lanthanum	La	57	139	2	6	10		2	6	1	2		2	6	1.03	1.10	1.16	1.22	36.2
Cerium	Ce	58	140	2	6	10	2	2	6		2	1	2	6	1.01	1.07	1.14	1.20	36.4
Praseodymium	Pr	59	141	2	6	10	3	2	6		2	2	2	6	0.99	—	1.13	1.18	37.55
Neodymium	Nd	60	144	2	6	10	4	2	6		2	3	2	6	0.98	—	1.11	1.16	38.4
Promethium	Pm	61	147	2	6	10	5	2	6		2	4	2	6	—	—	—	—	—
Samarium	Sm	62	150	2	6	10	6	2	6		2	5	2	6	0.96	1.02	1.08	1.13	40.4
Europium	Eu	63	152	2	6	10	7	2	6		2	6	2	6	0.95	1.01	1.07	1.12	41.8
Gadolinium	Gd	64	157	2	6	10	7	2	6	1	2	7	2	6	0.94	1.00	1.05	1.11	38.8
Terbium	Tb	65	159	2	6	10	9	2	6		2	8	2	6	0.92	0.98	1.04	1.10	39.3
Dysprosium	Dy	66	162.5	2	6	10	10	2	6		2	9	2	6	0.91	0.97	1.03	1.08	40.4
Holmium	Ho	67	165	2	6	10	11	2	6		2	10	2	6	0.90	—	1.02	1.07	40.8
Erbium	Er	68	167	2	6	10	12	2	6		2	11	2	6	0.89	0.95	1.00	1.06	40.5
Thulium	Tm	69	169	2	6	10	13	2	6		2	12	2	6	0.88	—	0.99	1.05	41.85
Ytterbium	Yb	70	173	2	6	10	14	2	6		2	13	2	6	0.87	0.93	0.99	1.04	43.5
Lutetium	Lu	71	175	2	6	10	14	2	6	1	2	14	2	6	0.86	—	0.98	1.03	40.4
Yttrium	Y	39	89	2	6	1		2				[d]			0.90	0.96	1.02	1.08	—

[a] From Faktor and Hanks (1969); ionic radii from Shannon (1976).
[b] ΣI = Sum of the first three ionization potentials.
[c] Outer electronic configuration: $3s^2 3p^6 3d^0 4s^2$.
[d] Outer electronic configuration: $4s^2 4p^6$.

complexing groups have only minor effects upon their behavior. Thus, the magnetic and spectroscopic properties (Section 2.2) of the lanthanides are not lost upon ionization or upon binding.

The sum of the first three ionization potentials is shown in Table 2-1. Screening of the $4f$ electrons is such that the $6s$ and $5d$ orbitals change little in energy across the series, as indicated by the relatively constant values of the ionization potentials. The lanthanides are highly electropositive, and their compounds are essentially ionic in nature. The predominant ionic form existing under the conditions appropriate to biochemical investigation is the trivalent cation, Ln^{3+}. Tetravalent states are known for Ce, Pr, Nd, Tb, Dy, and Ho, while Sm, Eu, Er, Tm, and Yb have divalent forms. Of these, only Ce^{4+} and Eu^{2+} are stable enough to exist for any length of time in aqueous solution, and neither of these is as stable as its respective trivalent form under physiological conditions. Thus, throughout this book, we shall be dealing almost exclusively with the biochemistry of the trivalent lanthanide cations.

Ionic radii are also given in Table 2-1. There are two points to note. The first is the progressive decline in ionic radius with increasing atomic number, a phenomenon known as the lanthanide contraction. It occurs because electrons are progressively added to an inner orbital, in which one $4f$ electron incompletely shields another $4f$ electron from the attractive force of the nucleus. As the atomic number, and hence protonic charge, increases, the nucleus is able to exert a greater electrostatic influence over its electrons, thereby contracting the entire ionic structure. The degree of contraction is such that the size of Ho^{3+} (atomic number 67) is approximately equal to that of Y^{3+} (atomic number 39). The second important feature of the ionic radii is that they span the radius of Ca^{2+} at most coordination numbers (Section 2.5; Table 2-1). Three important properties are derived from the electronic energy states of the lanthanides: magnetic susceptibility, radiant energy absorption, and luminescence.

In general terms, the magnetic moments of the lanthanides arise from the presence of $4f$ electrons with unpaired spins. However, there is more to it than this, as Gd^{3+}, which has the highest number of unpaired electrons, does not have the highest magnetic moment (Table 2-2). This occurs because the $4f$ electrons are sufficiently well screened that both their spins and their orbital motions about their nuclei contribute to the magnetic moment. As a result, the lanthanide series contains two magnetic maxima (Table 2-2). Tb^{3+}, Dy^{3+}, Ho^{3+}, and Er^{3+} are among the strongest paramagnetic ions known and, as such, are of great practical use, especially with regard to nuclear magnetic resonance (NMR) spectroscopy (Section 3.2).

Electronic transitions in the $4f$ orbitals give rise to characteristic

Table 2-2. Magnetic Moments and Colors of Yttrium and the
Lanthanide Ions

Ion	Unpaired $4f$ electrons	Magnetic moment (Bohr magnetons)	Color
Y^{3+}	0	0	Colorless
La^{3+}	0	0	Colorless
Ce^{3+}	1	2.39	Colorless
Pr^{3+}	2	3.47	Yellow-green
Nd^{3+}	3	3.62	Reddish
Pm^{3+}	4	2.83	Pink; yellow
Sm^{3+}	5	1.54	Yellow
Eu^{3+}	6	3.61	Nearly colorless
Gd^{3+}	7	7.95	Colorless
Tb^{3+}	6	9.6	Nearly colorless
Dy^{3+}	5	10.5	Yellow
Ho^{3+}	4	10.5	Pink; yellow
Er^{3+}	3	9.55	Reddish
Tm^{3+}	2	7.5	Pale green
Yb^{3+}	1	4.4	Colorless
Lu^{3+}	0	0	Colorless

absorption spectra, some of which produce beautiful colors (Table 2-2). Y^{3+}, La^{3+}, and Lu^{3+} have closed-shell configurations and do not absorb at UV or visible wavelengths (200–1000 nm). All other lanthanides have characteristic absorption bands in this region, and their spectra are shown in Fig. 2-1. The sharpness of the peaks reflects the shielding of the $4f$ electrons. Ce^{3+}, Eu^{3+}, Gd^{3+}, and Tb^{3+} absorb predominantly in the UV region and are colorless. Sm^{3+}, Dy^{3+}, Ho^{3+}, Er^{3+}, and Tm^{3+} absorb in the visible region and are colored (Table 2-2). The molar absorptivities of the lanthanides at their most prominent absorption peaks are shown in Table 2-3. It can be seen that their absorptivities are quite low, with only the UV absorption of Tb^{3+} and Ce^{3+} having a molar absorptivity much greater than 10^2. Absorption spectroscopy is thus of limited sensitivity for analytical use. Furthermore, because of the way in which the $4f$ electrons are shielded, the absorption spectra do not change dramatically upon complex formation. However, certain absorption peaks, known as "hypersensitive" peaks, do respond to the imposition of fields of differing strength around the Ln^{3+} ion upon binding to a ligand. Three types of spectral change are produced. Shifting of the peak may occur. These shifts are usually, but not always, red shifts. Secondly, the peak may split into several smaller maxima. Finally, the molar absorptivity of the peak may alter. The magnitude of these changes is usually proportional to the

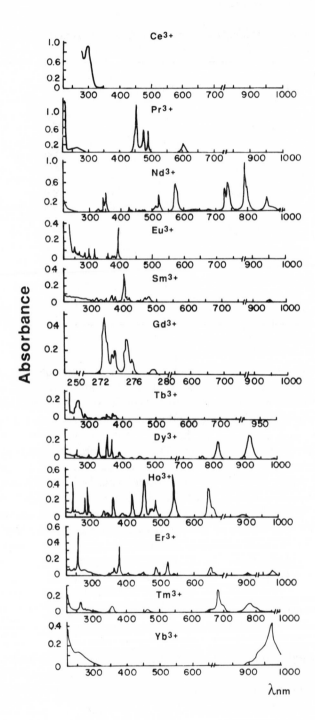

λnm

strength of the interaction between the Ln^{3+} ion and the ligand. As the molar absorptivities of the lanthanides are so low, absorption spectral studies have not found wide experimental use.

Electronic energy transformations also give rise to the luminescent behavior of the lanthanides. This is most prominent with Tb^{3+} and Eu^{3+} (see Fig. 3-2), whose luminescent properties are discussed in more detail in Section 3.3. Depending upon the lanthanide, luminescence can follow excitation by UV irradiation, X rays, fast electrons, neutrons, and certain chemical and mechanical means. The luminescent behavior of the lanthanides has given rise to their industrial use as phosphors in, among other things, color television screens. The emission spectra of the lanthanides are quite varied and can be used for the detection and quantitation of these elements (Section 2.7). Luminescence spectra in aqueous solution are observed for Ce^{3+}, Pr^{3+}, Nd^{3+}, Sm^{3+}, Eu^{3+}, Gd^{3+}, Tb^{3+}, and Dy^{3+}. However, the intensities are usually quite weak, because of low molar absorptivity. One way around this is to excite with a powerful laser source. In addition, luminescence of Ln^{3+} ions in certain complexes can be enhanced by indirect excitation via a suitable chromophore donor in the complex. In this way, luminescence enhancement by a factor of 10^5 has been claimed for Tb^{3+} when it binds to certain proteins. These aspects are described in greater detail in Section 3.3.

2.3 Bonding and Coordination Chemistry

This subject has been reviewed by Karraker (1970). Lanthanide bonding is essentially ionic. Whatever slight degree of covalency exists increases down the series from La^{3+} to Lu^{3+}. The theoretical and experimental evidence of ionicity is very strong. For example, Moeller et al. (1965) have pointed out that even in the most stable complexes, the bond strength is of the same order of magnitude as the Ln^{3+}–water dipole interaction. Measurements of bond distances have also confirmed the highly ionic nature of lanthanide bonding.

In aqueous solutions, Ln^{3+} ions attract around them a hydration shell of water molecules. The exact hydration number has been a source of controversy. Although theoretical considerations (Choppin and Graffeo, 1965) suggested values increasing from 12.8 ± 0.1 for La^{3+} to 13.9 ± 0.1 for Yb^{3+}, the more generally agreed values are 9 for La^{3+}–Nd^{3+} and 8 for Tb^{3+}–Lu^{3+}, with Sm^{3+}, Eu^{3+}, and Gd^{3+} existing as mixed populations

Figure 2-1. Absorption spectra of the lanthanides in 1 M $HClO_4$. Y^{3+}, La^{3+}, and Lu^{3+} do not absorb at UV or visible wavelengths. Redrawn from Banks and Klingman (1956).

Table 2-3. Molar Absorptivity of Rare Earth Ions at Specific Wavelengths[a]

Wavelength (nm)	Molar absorptivity (liter/mol·cm)											
	Ce^{3+}	Pr^{3+}	Nd^{3+}	Sm^{3+}	Eu^{3+}	Gd^{3+}	Tb^{3+}	Dy^{3+}	Ho^{3+}	Er^{3+}	Tm^{3+}	Yb^{3+}
219.8	775	—	—	—	—	—	320	—	—	—	—	—
253.6	—	1.13	0.210	0.166	0.335	4.20	—	0.215	0.227	0.698	0.685	0.365
272.8	—	0.511	1.185	—	0.132	—	—	0.140	3.59	0.316	0.521	0.266
287.0	—	—	2.60	0.052	—	—	—	2.54	0.057	0.096	0.145	—
350.4	—	—	5.20	0.087	—	—	—	0.737	0.047	0.382	0.454	—
354.0	—	—	0.042	0.453	0.053	—	—	0.271	2.34	0.296	0.801	—
361.1	—	—	0.025	0.244	0.062	—	—	2.10	0.331	1.94	0.154	—
365.0	—	—	0.067	0.052	0.299	—	—	0.252	0.047	7.18	—	—
379.6	—	—	0.025	0.148	—	—	—	0.233	0.123	0.096	—	—
394.2	—	—	—	—	—	—	—	0.187	0.038	0.086	—	—
401.5	—	—	—	3.31	3.06	—	—	0.047	0.454	0.306	—	—
444.2	—	10.49	0.025	0.044	0.097	—	—	—	4.16	0.497	—	—
450.8	—	1.29	4.41	—	—	—	—	0.261	0.047	2.10	—	—
521.6	—	—	1.68	—	—	—	—	—	0.076	3.20	—	—
523.5	—	—	—	—	—	—	—	—	5.16	0.076	—	—
537.0	—	—	6.93	—	—	—	—	—	—	—	—	—
575.5	—	0.102	—	—	—	—	—	—	—	—	—	—
640.4	—	—	—	—	—	—	—	—	3.53	0.153	—	—
683.0	—	—	0.336	—	—	—	—	—	—	—	2.56	—
739.5	—	—	7.20	—	—	—	—	—	—	—	0.058	—
794.0	—	—	11.78	—	—	—	—	0.084	0.113	0.143	0.579	—
908.0	—	—	—	—	—	—	—	2.46	—	1.29	—	0.089
974.0	—	—	—	0.052	—	—	—	0.047	—	—	—	2.12

[a] All solutions contained 20 mg of oxide in 1 ml of 1 M $HClO_4$ (except in the case of Er^{3+}, 10 mg/ml). Figures underlined refer to the coefficients at the wavelengths recommended for determining the respective element. From Banks and Klingman (1956). The unstable lanthanide, Pm^{3+}, has absorption peaks at 494.5, 548.5, 568.0, 685.5, and 735.5 nm (Boyd, 1959).

displaying both hydration numbers (Spedding *et al.*, 1966). However, luminescence lifetime measurements have suggested values of 10 for $La^{3+}-Nd^{3+}$ and 9 for $Tb^{3+}-Lu^{3+}$ (Horrocks and Sudnick, 1979). The rate constants for Ln^{3+} hydration are approximately 8×10^7 s^{-1} for $La^{3+}-Eu^{3+}$, 4×10^7 s^{-1} for Gd^{3+}, and around 1×10^7 s^{-1} for $Dy^{3+}-Lu^{3+}$. Part of the changes in hydration rate across the series can be explained by the lower hydration number of the heavier Ln^{3+} ions. These species have more energy involved in the bond to each coordinated water molecule, and thus their reaction rate is decreased.

Water is a very strong ligand for Ln^{3+} ions. The difficulty that competing ligands experience in attempting to dislodge water molecules from the coordination sphere severely restricts the types of ligands with which Ln^{3+} ions can interact in aqueous solution. These ligands are those predicted from the classification of lanthanides as "class A" or "hard" acceptors. One characteristic of such acceptors is their high degree of ionic bonding. Another is their preference for donor atoms in the order O > N > S and F > Cl. Lanthanides are no exception and show a very strong preference for O donor atoms. There are few data to suggest that Ln^{3+} ions can coordinate to S atoms under any but the most unusual of conditions. Coordination with N donors, such as those in *o*-phenanthroline, has been reported (see Sinha, 1966), but the interaction is weak and easily disrupted by the introduction of competing O donors. Under physiological conditions, substances which provide only N donor atoms probably do not complex Ln^{3+} ions, as any such complexes would be hydrolyzed by water. However, $N-Ln^{3+}$ coordination has been reported in polydentate chelation where the other ligands are provided by O donors. Biochemical examples include coordination to the amino groups of amino acids (Section 4.2) and N of purine bases in single-stranded polynucleotides (Section 5.3). However, the overwhelmingly predominant donor in all aqueous systems is the O atom. In most biological instances, this is supplied by the carboxyl groups of proteins or the phosphate groups of nucleotides and nucleic acids. The carboxyl group is likely to act as a monodentate ligand for Ln^{3+} ions (Rhee *et al.*, 1984). Supplementary coordination through the carbonyl and hydroxyl oxygens or the bridging oxygens of diester linkages also occurs. Uncharged groups of this type constitute the sole donor ligands in the Ln^{3+} complexes of certain sugars and nucleosides (Sections 5.1 and 5.4). Measurements on the binding of Ln^{3+} ions to sulfonated sugars (Section 5.4) and the lack of interaction between Ln^{3+} ions and sulfonated buffers such as MOPS, HEPES, and PIPES (Section 2.8) suggest that organic sulfates ($-O-SO_3^-$) are poor ligands for lanthanides. The formation of stable complexes requires coordination through at least two donor groups; complexes with monodentate O ligands tend to dissociate in aqueous solution. Chelation is the predominant form of lanthanide complex formation. Complexes formed in nonchelated structures are weaker than those formed by the same donor atoms when present in chelates.

Coordination numbers of 6 to 12 have been reported for lanthanides. For biological molecules, their coordination numbers are often 8 or 9. It has been suggested that the coordination number changes across the lanthanide series. Precise details of the coordination state of Ln^{3+} ions in several proteins and tRNA are now available from X-ray diffraction studies. There are discussed in Chapters 4 and 5, respectively. There is little or no directionality in Ln^{3+}–ligand interactions. This means that there is little stereochemical preference within the first coordination sphere. Thus, the coordination numbers and geometries are primarily determined by such ligand characteristics as conformation, donor atoms, sizes, and number and by solvation effects. Lanthanide complexes can hence have a variety of geometries. This geometrical flexibility is an important characteristic with regard to the occupation of biological Ca^{2+}-binding sites (Section 2.5). In addition, the low polarizing ability of Ln^{3+} ions and the ionic nature of their interactions means that they do not produce significant changes in the electronic charge distributions at their binding sites. Crystal-field stabilization occurs where a partially filled subshell assumes different energies when the central atom orbitals are placed in a crystal field. This effect is small for Ln^{3+} ions, because of the shielding of the $4f$ electrons. Much attention has been devoted to the complexes that lanthanides form with small organic ligands. Such compounds have been important in the preparation and purification of lanthanides and in the titrimetric determination of Ln^{3+} ions. A number of these molecules are of biochemical interest, and these will be considered here.

Lanthanides form 1:1 complexes with the polyamine polycarboxylic acid chelators EDTA (ethylenediamine-N,N,N',N'-tetraacetic acid), HEDTA (N'-hydroxyethylethylenediamine-N,N,N'-triacetic acid), DTPA (diethylenetriamine-N,N,N',N',N''-pentaacetic acid), and DCTA (1,2-diaminocyclohexane-N,N,N',N'-tetraacetic acid). In addition, NTA (nitrilotriacetic acid) forms 1:1 and 1:2 complexes with Ln^{3+} ions; a species $[Eu_2(NTA)_3]^{3-}$ has also been claimed. The formation constants of these chelates are shown in Table 2-4. The crystal structure of La-EDTA shows it to have the form $[La(EDTA)(H_2O)_3]^-$ with the La^{3+} ion being nine-coordinate. Bond lengths of 2.54 Å (La—O of EDTA), 2.60 Å (La—O of water), and 2.86 Å (La—N) were measured (Lind *et al.*, 1965). The relatively long La—N bond again demonstrates the preference that La^{3+} has for O donors. The high stability of the Ln-DTPA complexes ensures that, following parenteral administration, they pass through the body with-

Table 2-4. Stabilities of Polyamine Polycarboxylic Acid Chelates[a]

Ln^{3+}	Log stability constant				
	NTA	EDTA	DTPA	DCTA	HEDTA
La^{3+}	10.36	15.50	19.48	19.15	14.49
Ce^{3+}	10.83	15.98	20.5	16.26	13.22
Pr^{3+}	11.07	16.40	21.07	16.76	14.08
Nd^{3+}	11.26	16.61	21.60	17.31	14.39
Sm^{3+}	11.53	17.14	22.34	17.68	14.71
Eu^{3+}	11.52	17.35	22.39	18.38	15.15
Gd^{3+}	11.54	17.37	22.46	18.62	15.21
Tb^{3+}	11.59	17.93	22.71	18.77	15.10
Dy^{3+}	11.74	18.30	22.82	19.50	15.10
Ho^{3+}	11.90	—	22.78	19.69	15.08
Er^{3+}	12.03	18.85	22.74	—	15.06
Tm^{3+}	12.22	19.32	22.72	20.68	15.17
Yb^{3+}	12.40	19.51	22.62	20.96	15.38
Lu^{3+}	12.49	19.83	22.44	21.12	15.64
Y^{3+}	11.48	18.09	22.05	21.51	15.79

[a] Data from Moeller (1959) and Moeller (1973), with permission.

out exchange of the Ln^{3+} ions with physiological ligands (Section 7.5.3). This means that they have low toxicity, a property which is exploited medically in the use of Gd-DTPA as a contrast enhancer in the NMR imaging of various organs (Section 9.6.4). By a similar token, DTPA is used to remove radioactive lanthanides from the body following accidental exposure. Weaker chelators, such as EDTA and NTA, permit some degree of physiological exchange, and their metabolism reflects this (Section 7.5.3).

Several lanthanide complexes of mono-, di-, and tricarboxylic acids are of biochemical interest. Acetate forms complexes in which at least three, and possibly four, acetate groups are attached to a single Ln^{3+} ion. The same is true for lactate. The formation constants of these complexes are given in Table 2-5. From Table 2-4, it can be seen that the stability of complexes formed with chelators such as EDTA increases progressively with decreasing size of the Ln^{3+} ion. This may be explained by the increased charge-to-volume ratio that occurs in progressing from La^{3+} to Lu^{3+}, producing greater electrostatic interaction. However, the stability of Ln^{3+}–acetate complexes is greatest around the middle of the series (Table 2-5). A similar effect is seen with diglycollate (not shown). This phenomenon has been called the "gadolinium break," but its explanation is incompletely known. Among the suggested explanations are a

Table 2-5. Formation Constants for Complexes between
Lanthanides and Monocarboxylic Acids[a]

| | Ligand | | | | | |
| | Acetate | | | Lactate | | |
Ln^{3+}	$\log K_1$	$\log K_2$	$\log K_3$	$\log K_1$	$\log K_2$	$\log K_3$
La^{3+}	2.02	1.24	0.50	2.63	1.74	1.30
Ce^{3+}	2.09	1.44	0.44	2.76	1.97	1.23
Pr^{3+}	2.18	1.45	0.47	2.85	2.05	1.20
Nd^{3+}	2.22	1.54	0.45	2.87	2.10	1.42
Sm^{3+}	2.30	1.58	0.59	2.88	2.21	1.26
Eu^{3+}	2.31	1.60	0.60	2.95	2.23	1.26
Gd^{3+}	2.16	1.60	0.82	2.89	2.15	1.20
Tb^{3+}	2.07	1.59	—	2.90	2.31	1.15
Dy^{3+}	2.03	1.61	0.82	3.01	2.34	1.32
Ho^{3+}	2.00	1.59	0.89	3.02	2.40	1.42
Er^{3+}	2.01	1.59	0.82	3.16	2.46	1.58
Tm^{3+}	2.02	1.59	—	3.19	2.52	1.72
Yb^{3+}	2.03	1.64	0.71	3.23	2.59	1.76
Lu^{3+}	2.05	1.64	—	3.27	2.61	1.90
Y^{3+}	1.97	1.63	0.73	2.83	2.10	1.84

[a] Data from Moeller *et al.* (1965).

change in hydration number at this point in the lanthanide series, ligand-field effects, steric factors, a change in the degree of covalency relative to the hydrated form of the ion, and a change in the geometry of the complex. Nieboer (1975) has discussed the matter in some detail and suggested that the characteristic variation in complex stability across the lanthanide series may be of diagnostic value in studying the binding sites of the lanthanides.

As might be expected, Ln^{3+} complexes with dicarboxylic acids are stabler than those formed with monocarboxylic acids. Some representative dissociation constants are shown in Table 2-6. Values for oxaloacetate and malonate are given in Table 2-7. Citrate, a tricarboxylic acid, forms predominantly a $[Ln(citrate)_2]^{3-}$ species between pH 2 and pH 5, while at low pH this is joined by $[HLn(citrate)_2]^{2-}$, $[H_2Ln(citrate)_2]^-$, and $[H_3Ln(citrate)_2]^0$. Stability constants for the reaction $Ln^{3+} + 2\ citrate^{3-} \rightleftharpoons [Ln(citrate)_2]^{3-}$ are shown in Table 2-8. Ln^{3+}–citrate complexes have been widely used in metabolic studies as a means of introducing lanthanides in a form which is stable enough to prevent precipitation at the site of injection, yet labile enough to release Ln^{3+} ions to suitable physiological ligands (Section 7.5.3). More information on the nature of lanthanide

Table 2-6. Dissociation Constants of Some
La^{3+} Dicarboxylates

Dicarboxylic acid	Dissociation constant ($M \times 10^4$)
Malonate	0.13
Succinate	1.10
Glutarate	1.50
Adipate	0.80
Phthalate	0.18
Maleate	0.28
Fumarate	9.80

Table 2-7. Association Constants for
Ln^{3+}-Oxaloacetate and -Malonate

Ln^{3+}	Association constants ($M^{-1} \times 10^5$)	
	Oxaloacetate	Malonate
La^{3+}	1.8	1.00
Gd^{3+}	3.5	2.45
Dy^{3+}	4.6	—
Lu^{3+}	7.5	5.25
Y^{3+}	4.3	—

Table 2-8. Stability Constants
for the Formation of
[Ln(citrate)$_2$]$^{3-a}$

Ln^{3+}	Stability constant
La^{3+}	2.8×10^9
Ce^{3+}	4.5×10^9
Nd^{3+}	5.0×10^9
Pm^{3+}	5.6×10^9
Eu^{3+}	6.3×10^9

^{a} From Sinha (1966), with permission.

Table 2-9. Solubilities of
Lanthanide Sulfates,
$Ln_2(SO_4)_3 \cdot 8H_2O$, in Water[a]

Ln	Solubility (g/100 g H_2O)	
	20°C	40°C
La	3.8	1.5
Ce	23.8	10.3
Pr	12.74	7.64
Nd	7.00	4.51
Sm	2.67	1.99
Eu	2.56	1.93
Gd	2.89	2.19
Tb	3.56	2.51
Dy	5.07	3.34
Ho	8.18	4.52
Er	16.00	6.53
Yb	34.78	22.9
Lu	47.27	16.93
Y	9.76	4.9

[a] Data of Jackson and Reinacker (1930).

complexes can be found in Moeller *et al.* (1965) and Sinha (1966). The complexes of lanthanides with amino acids, nucleotides, and other biological ligands are discussed in Chapters 4 and 5.

In common with the larger, trivalent group 3 ions, the lanthanides form insoluble hydroxides, carbonates, phosphates, fluorides, and oxalates, but soluble chlorides, nitrates, and perchlorates. Their sulfates are sparingly soluble. For salts of weak-acid anions, such as OH^-, solubility generally decreases down the series, whereas the opposite trend is observed for salts of strong-acid anions, such as SO_4^{2-}. Lanthanide sulfates, $Ln_2(SO_4)_3$, have the unusual property of being more soluble when cold (Table 2-9). Of relevance to biochemical studies involving lanthanides is the presence of phosphate and an equilibrium mixture of carbonate and bicarbonate in body fluids. In addition, body fluids usually have a pH of around 7.4. Lanthanide phosphates and carbonates are insoluble, while aqueous solutions containing Ln^{3+} ions hydrolyze above neutral pH to form insoluble precipitates of $Ln(OH)_3$ (Table 2-10). Failure to recognize these properties has unwittingly confused the interpretation of a number of biochemical studies. Thus, the many advantages that lanthanides lend

Table 2-10. Solubility Characteristics of Lanthanide Hydroxides[a]

Ln	$-\log K_{sp}$	pH, $[Ln^{3+}]$ = 0.1 M	pH, $[Ln^{3+}]$ = 0.01 M	Solubility (mol/liter $\times$ 10^6)
La	18.7	8.1	8.43	8.0
Ce	19.5	7.8	8.17	4.0
Pr	19.4	7.89	8.20	4.2
Nd	20.4	7.53	7.87	1.8
Sm	20.9	7.37	7.70	1.1
Eu	21.1	7.30	7.63	0.8
Gd	21.3	7.23	7.57	0.9
Tb	21.4	7.20	7.53	—
Dy	21.5	7.17	7.50	—
Ho	22.	7.00	7.33	—
Er	22.5	6.83	7.17	0.7
Tm	23.2	6.60	6.93	0.6
Yb	23.3	6.57	6.90	0.5
Lu	23.4	6.53	6.87	0.6

[a] Data obtained from Moeller and Kremers (1944) and Glasel (1973). pH values are the upper limits at which a stable, aqueous Ln^{3+} solution can be formed at the indicated concentration of 0.1 M or 0.01 M.

to inorganic biochemical investigation are to some degree compromised by the difficulties of working with them under physiological conditions.

Whereas the use of phosphate or bicarbonate buffering systems can be avoided in biochemical studies, the problem of hydrolysis is less easily countered. Hydrolysis occurs according to the reaction

$$Ln(H_2O)_n^{3+} + H_2O \rightleftharpoons Ln(OH)(H_2O)_{n-1}^{2+} + H_3O^+$$

As the pH is increased, further hydrolysis occurs, until an insoluble precipitate of $Ln(OH)_3$ is formed. Because the tendency to hydrolyze increases with decreasing ionic radius, the pH at which stable aqueous solutions can be formed declines progressively from La^{3+} to Lu^{3+} (Table 2-10). For Ce^{3+}, only about 1% of the metal ion can be hydrolyzed before a precipitate forms. The main reaction appears to be

$$3Ce^{3+} + 5H_2O \rightleftharpoons [Ce_3(OH)_5]^{4+} + 5H^+$$

In addition to hydroxide formation, hydrolysis generates various other species whose composition and solubility are poorly characterized. For example, polynuclear forms such as $Ln[Ln(OH)_3]^{3+}$ may exist (Withey, 1969). Values for hydrolysis constants vary from 10^{-10} for La^{3+} to 10^{-9} for Y^{3+} at 1 mM concentration. It can be seen from Table 2-10 that saturated hydroxide solutions contain approximately micromolar concentrations of free Ln^{3+} ions. The study of hydroxide precipitation is made difficult by such phenomena as supersaturation, colloid formation, dependence of precipitation upon particle size, and rapid partial crystallization.

2.4 Radiocolloid Behavior

Hydrolysis is not the only process complicating the properties of aqueous solutions which contain Ln^{3+} ions. Under certain conditions, lanthanide solutions form colloidal particles, known as "radiocolloids." This name is misleading, as it has nothing to do with radioactivity, but reflects the history of the phenomenon, which was first detected for the ions of radioactive elements. It was discovered by Paneth (1913) for polonium.

The formation of radiocolloids is signaled by a number of changes in the physical properties of the solution. The metal ions are no longer dialyzable, but become filterable, and their diffusion coefficients decrease. Radiocolloids can be recovered by centrifuging. Following the evaporation and autoradiography of solutions in which radiocolloids have formed, discrete points of radioactivity are discernible. Radiocolloids also show anomalous behavior on ion-exchange chromatography. They adsorb onto finely divided materials, filters, walls of containers, and other surfaces. It is of interest that the formation of radiocolloids, unlike ordinary colloids, does not require the solubility product to be exceeded. Radiocolloid behavior has been shown for [86]Y, [90]Y (Schweitzer et al., 1953), [140]La (Schubert and Conn, 1949; Schweitzer and Jackson, 1952b), and [139]Ce (Kurbatov and Pool, 1944). It is likely that all lanthanides exhibit this phenomenon. Schweitzer and Jackson (1952a) have reviewed radiocolloids.

Solutions of $La(NO_3)_3$ below pH 4 and at a concentration of $10^{-8} M$ are not colloidal. As the pH is raised from pH 4 to pH 6, the amount of $La(NO_3)_3$ present as radiocolloids increases to 100% (Schubert and Conn, 1949; Schweitzer and Jackson, 1952b) (Fig. 2-2). Similar behavior is found for [90]Y (Schweitzer et al., 1953). The tendency to form radiocolloids is *inversely* related to concentration and thus cannot be explained simply in terms of exceeding the solubility constant of some salt. As the concentration of Y^{3+} is increased from $10^{-8} M$ to $10^{-2} M$, the pH value at which 50% of the Y^{3+} can be removed by centrifuging increases from around pH 4 to around pH 7 (Schweitzer and Scott, 1955). Radiocolloid formation is inhibited by increasing the concentration of electrolyte or by adding

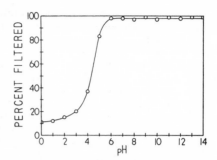

Figure 2-2. Filterability of solutions of ^{140}La at different pH values. From Schweitzer and Jackson (1952b), with permission.

complexing agents (Schweitzer and Jackson, 1952a). Radiocolloids also form slowly in solutions as they age (Chamie and Haissinsky, 1934), a process which can be retarded by storage in vessels lined with paraffin wax (King and Romer, 1933).

The mechanism of radiocolloid formation is incompletely understood. Two main explanations have been suggested. According to one school of thought, the ions adsorb to particulate impurities within the solvent. According to the other, true colloidal particles are formed. The matter remains unresolved, if not ignored. In trying to interpret biochemical data, few, if any, authors consider the intervention of radiocolloids.

2.5 Ca^{2+} and Ln^{3+} Ions Compared

As much of the justification for using Ln^{3+} ions in biochemical investigations turns on their ability to replace Ca^{2+} in a specific, isomorphous manner, it is worth comparing the properties of Ln^{3+} and Ca^{2+} in greater detail. Good reviews of the biologically important chemical properties of Ca^{2+} have been provided by Levine and Williams (1982) and Einspahr and Bugg (1984). Martin (1983) has recently reviewed the structural chemistry of calcium and the lanthanides.

Table 2-11 compares the properties of Ca^{2+} and Ln^{3+} which are of greatest biochemical relevance. In their sizes, bonding, coordination geometry, and donor atom preference, these ions are remarkably similar. It is these similarities which permit Ln^{3+} ions to replace Ca^{2+} so specifically. Nature has tailored many Ca^{2+}-binding sites to exclude competing metal ions, especially Mg^{2+}. This is particularly important intracellularly, where certain proteins can selectively bind Ca^{2+}, which is present at submicromolar concentrations, while immersed in fluids containing milli-

Table 2-11. Salient Properties of Ca^{2+} and Ln^{3+}

Property	Ca^{2+}	Ln^{3+}
Coordination number	6–12 reported	6–12 reported
	6 or 7 favored	8 or 9 favored
Coordination geometry	Highly flexible	Highly flexible
Donor atom preference	O $\gg$ N $\gg$ S	O $\gg$ N $\gg$ S
Ionic radius (Å)[a]	1.00–1.18 (CN 6–9)	0.86–1.22 (CN 6–9),
		depending on species
Type of bonding	Ionic	Ionic
Hydration number[b]	6	8 or 9
Water exchange rate constant (s^{-1})	~5 $\times$ 10^8	~5 $\times$ 10^7
Diffusion coefficient (cm^2/s $\times$ 10^5)	1.34	La^{3+}, 1.30
Crystal-field stabilization	None	Negligible

[a] See Table 2-1.
[b] See discussion p. 15.

molar concentrations of Mg^{2+}. To obtain such selectivity, use is made of the small ionic radius (0.72 Å; CN = 6), inflexible coordination geometry, preference for a coordination number of 6, and the affinity for N donor atoms of Mg^{2+}. Where such constraints do not exist, Mg^{2+} is often the intracellular counterion. Research with tRNA (Section 5.3) has confirmed that Ln^{3+} ions can occupy Mg^{2+} sites in a specific manner. This is also true for certain Mg^{2+}-requiring enzymes. There are also instances of the replacement of Fe^{3+}, Fe^{2+}, and Mn^{2+} by Ln^{3+} ions (Chapter 4). The lanthanides also bear certain chemical similarities to the actinides. For this reason, they are finding use as nutritional markers in studies of actinide metabolism (Chapter 4).

More than one author has commented that Ca^{2+} sites cannot be designed to exclude Ln^{3+} ions. However, although there are indeed numerous examples of the specific replacement of Ca^{2+} by Ln^{3+} ions, exceptions exist. There are several biochemical examples of Ca^{2+}-binding sites that do not accept Ln^{3+} ions, and vice versa: Ln^{3+} ions cannot replace Ca^{2+} in scallop myosin, for example (Chantler, 1983), and are excluded from the Ca^{2+} site on concanavalin A (Richardson and Behnke, 1978). Furthermore, high concentrations of Ca^{2+} cannot displace Gd^{3+} from IgG (Dower et al., 1975) or Tb^{3+} from ferritin (Stefanini et al., 1983). Both the acetylcholine and insulin receptors have two classes of Tb^{3+}-binding site, only one of which accepts Ca^{2+} (Rübsamen et al., 1976; Williams and Turtle, 1984). These proteins are discussed in more detail in Chapter 4. Of interest is the observation that although Ln^{3+} ions are excluded from the Ca^{2+} site of concanavalin A, Eu^{2+} is able to bind here and to restore biological activity to the apoprotein (Homer and Mortimer, 1978). The authors state that in the solubilities and structures of its salts, Eu^{2+} bears a closer resemblance to Ca^{2+} than do Ln^{3+}. Like certain trivalent lanthanide ions, Eu^{2+} can be used in conjunction with Mössbauer spectroscopy, electron paramagnetic resonance (EPR) spectroscopy, magnetic circular dichroism (MCD) spectroscopy, and proton relaxation studies. The chief disadvantages of Eu^{2+} are its powerful reducing properties and the need to work anaerobically (Homer and Mortimer, 1978).

Silber (1974) has suggested that if a Ca^{2+}-binding site is spatially restricted so as to exclude larger ions, and is surrounded by a hydrophobic environment, Ln^{3+} ions will be excluded. The basis for this conclusion is the observation that when the local dielectric constant falls below that of water, Ln^{3+} ions, but not Ca^{2+} ions, bind nonspecifically to counterions. Under these conditions, the Ln^{3+}–counterion adduct would be too large to occupy the Ca^{2+} site. It is not clear whether this phenomenon exists for biological ligands. If such sites do exist biochemically, extracellular proteins may provide the most productive hunting ground for examples of this kind. Levine and Williams (1982) have contrasted the flexibilities of intracellular Ca^{2+}-binding proteins with the rigidity of extracellular Ca^{2+}-dependent enzymes, whose selectivity for Ca^{2+} can be produced by the strict matching of an almost rigid hole with the radius of the Ca^{2+} ion.

There have been various studies of the effect that ionic radius has on the biochemical activity of lanthanides. Such studies are facilitated by the gradual, quantitative nature of the changes in the properties of these ions in progressing from La^{3+} to Lu^{3+}. In some instances, there is no simple relationship between ionic radius and activity. However, in other cases, three trends have been discernible. In many instances, the measured effect increases in passing from La^{3+} to Lu^{3+}, as expected on the grounds of simple electrostatic attraction. According to Smolka *et al.* (1971), the activation of α-amylase by Ln^{3+} ions is an example of this (Section 4.13). In other cases, the effect decreases from La^{3+} to Lu^{3+}; the inhibition of Ca^{2+}/Mg^{2+}-ATPase of skeletal muscle sarcoplasmic reticulum exhibits this trend (Dos Remedios, 1977; Section 4.15). The explanation is based upon size considerations. As the larger Ln^{3+} ions are closer in ionic radius to the Ca^{2+} ions they replace, they are better able to interfere with Ca^{2+}-dependent functions. In other molecules, size specificity appears to have been carried to an extreme degree, with a peak of activity appearing within the lanthanide series at the ionic radius closest to that of Ca^{2+}. Examples include competition with one class of Ca^{2+} site on mitochondria (Tew, 1977; Section 6.5.1; Fig. 6-10), inhibition of trans-

Table 2-12. The Lanthanides Whose Ionic Radii
Equal or Are Closest to That of Calcium at
Different Coordination Numbers[a]

Coordination number of calcium	Coordination number of lanthanide			
	6	7	8	9
6	Ce–Pr	Sm–Eu	Er	<Lu[b]
7	>La[c]	Ce–Pr	Eu–Gd	Er
8	>La	>La	Pr–Nd	Eu
9	>La	>La	>La	Pr

[a] Based upon the ionic radii of Shannon (1976).
[b] <Lu: Ca^{2+} smaller than all Ln^{3+} at these coordination numbers.
[c] >La: Ca^{2+} bigger than all Ln^{3+} at these coordination numbers.

mitter release from the presynaptic terminal of neuromuscular junctions (Bowen, 1972; Section 6.4.5), inhibition of smooth muscle contraction (Triggle and Triggle, 1976; Section 6.4.3), and inhibition of clostridiopeptidase A (Evans, 1981; Section 4.11.1).

Tew (1977) has addressed the theory behind such observations in greater detail. The Ln^{3+} ion preferred by a given binding site in aqueous solution is that whose binding causes the greatest decrease in free energy. Three energy factors are at play here: hydration energy, the free energy of interaction between the Ln^{3+} ion and the negative site, and the energy involved in possible deformations of the binding site necessary to achieve binding. Ignoring the last of these energy factors, binding sites with very low field strength will not be able to overcome the hydration energy, and the larger, more weakly hydrated Ln^{3+} ions will preferentially bind. With a strong-field binding site, the reverse holds true. Combining now the geometry of the binding site with these considerations, Tew (1977) proposed that, under spatially restricted conditions, large ions would be sterically excluded while small ions would not be able to coordinate to all ligands in the binding site. Thus, maximum binding would occur at an optimum ionic radius. With Ca^{2+}-specific binding sites, this would be the Ln^{3+} ion whose radius is closest to that of Ca^{2+} (Table 2-12).

Nevertheless, the interpretation of Ln^{3+} specificity data requires caution, as ionic radii are difficult to determine with precision. Not only will they depend upon how the observed bond lengths are divided up among the contributing ions, but the most commonly quoted radii are crystal radii. The effective radius of Ca^{2+} or a Ln^{3+} ion in solution may be quite different from that in a crystal. Furthermore, the flexibility of their coordination geometries gives these ions the potential to adjust their radii according to the binding site. One way in which Ln^{3+} ions do this is to alter their coordination number. By increasing their coordination number, the smaller Ln^{3+} ions can expand their radii so that they are closer to that of the Ca^{2+} ion. Using the values of Shannon (1976) for the crystal ionic radius, we see that as the coordination number of Ca^{2+} and Ln^{3+} ions increases from 6 to 9, the identity of the Ln^{3+} ion whose ionic radius equals that of Ca^{2+} changes (Table 2-12). Under certain conditions, denoted by $>La^{3+}$ in Table 2-12, the Ca^{2+} ion is bigger than all the Ln^{3+} ions; six-coordinate Ca^{2+} is smaller than all nine-coordinate Ln^{3+} ions. However, Ca^{2+} seems to favor a coordination number of 6 or 7, while Ln^{3+} prefers coordination numbers of 8 or 9. Thus, it is probably still true that in most instances the ionic radii spanned by members of the lanthanide series include that of Ca^{2+}. However, in view of the paucity of detailed information yet available for biological ligands, the possibility of exceptions should be kept in mind.

The most striking chemical difference between Ca^{2+} and the Ln^{3+} ions is the spectroscopic silence of the former and the abundance of spectroscopic signals provided by the latter. From the point of view of research, it is quite unfortunate that nature has chosen one of the least informative metals to be one of the physiologically most important. Whereas the lanthanide series contains members which are colored, highly paramagnetic, and luminescent, Ca^{2+} is colorless, diamagnetic, and nonluminescent. Indeed, apart from its X-ray absorption properties, Ca^{2+} is spectrally inert. Thus, substitution of Ca^{2+} ions by Ln^{3+} ions is a valuable way to study its ligand interactions. Evidence that this substitution often retains the original properties of the liganding molecule has been provided by both structural and functional assays (Chapters 4 and 5).

The major difference between Ca^{2+} and Ln^{3+} complexes is the greater stability of the latter. This is to be expected from the higher charge-to-volume ratios of the Ln^{3+} ions and their greater coordination numbers. These differences are well illustrated by the compilation of Horrocks (1982) comparing the K_d values of 1:1 Nd^{3+} and Pr^{3+} complexes with the K_d values of the corresponding Ca^{2+} complexes (Fig. 2-3). With one exception, the K_d values for Ln^{3+} ions are smaller. In the 1 mM to 1 nM range there is a linear correlation, with the Ln^{3+} ion binding being tighter by a factor of 10^4 to 10^5. Horrocks (1982) pointed out that the K_d values generally correlate with the denticities of the ligands. However, for ligands with potential denticities above 6, the Ln^{3+} complexes were more stable than expected, while the opposite was true for the weakest complexes. Comparison of the K_d values of Ln^{3+} and Ca^{2+} or the K_i values of Ln^{3+}

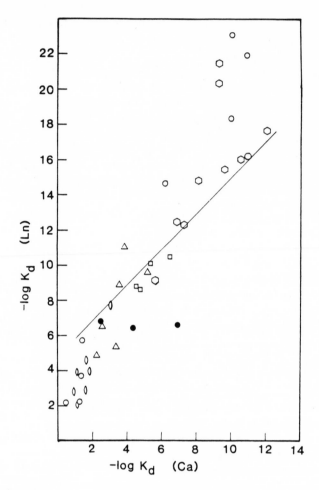

Figure 2-3. Comparison of the affinities of lanthanide ions and calcium ions for a variety of ligands. Ligand denticities: ○, 1; ◇, 2; △, 3; □, 4; ⬡, 6; ●, cryptate ligands. From Horrocks (1982), with permission.

ions in enzyme inhibition (Chapters 4 and 5) generally, but not always, reflects the greater affinity of lanthanides.

The second main difference between Ln^{3+} and Ca^{2+} ions is the ligand exchange rate. Water molecules exchange about 10 times faster around the Ca^{2+} ion (Table 2-11). This is presumably due to the higher charge-to-volume ratio of the Ln^{3+} ions which probably lowers the off rate of complex dissociation.

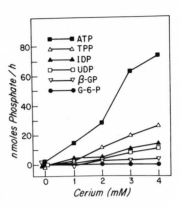

Figure 2-4. Effects of cerium ions on nonenzymic hydrolysis of various phosphorylated substrates. Substrate (1 mM) was incubated at 37°C for 30 min with Ce^{3+}, and the release of inorganic phosphate was measured. ATP, adenosine triphosphate; TPP, thiamine pyrophosphate; IDP, inosine diphosphate; UDP, uridine diphosphate; β-GP, β-glycerophosphate; G-6-P, glucose 6-phosphate. From Robinson and Karnovsky (1983), with permission.

2.6 Catalytic Properties of Lanthanides

Lanthanides depolymerize RNA in a reaction taking several days at 37°C (Bamann *et al.*, 1954) or two hours at 64°C (Eichhorn and Butzow, 1965). The rate of reaction of poly(U), poly(C), and poly(A) with 0.2 mM La^{3+} at pH 7 exceeds that of poly(I). Unlike Bamann *et al.* (1954), Eichhorn and Butzow (1965) found Ce^{3+} to be a more efficient catalyst than La^{3+}. The latter ion cleaved polyribonucleotides randomly at the phosphodiester bonds linking the 5'-phosphate position, to form oligonucleotides and mononucleotides bearing a 3'-phosphate group. The ability of Ln^{3+} ions to cleave phosphodiester bonds has been used experimentally to distinguish these bonds from other types of phosphate linkage (Section 3.10.3).

Cleavage of yeast tRNA at 50°C occurred at positions adjacent to those known, from independent evidence, to be Eu^{3+}-binding sites (Rordorf and Kearns, 1976; Section 5.3). The rate of cleavage increased upon the addition of the first two Eu^{3+} ions per tRNA molecule and remained constant as the next two were added before increasing sharply after addition of a fifth Eu^{3+} ion. The reaction rate then slowly increased as the stoichiometry progressed to 15 Eu^{3+} ions per tRNA. Paradoxically, precipitation of RNA with lanthanum acetate protects it from the action of ribonuclease (Opie and Lavin, 1946). DNA resists degradation by lanthanides.

After prolonged incubation, 3'-dephosphorylation of nucleotides also occurs (Eichhorn and Butzow, 1965). Millimolar concentrations of Ce^{3+} at 37°C exhibit phosphatase activity in cleaving phosphate groups from ATP and certain other phosphorylated compounds (Fig. 2-4), including pyrophosphate. AMP resists dephosphorylation by Ce^{3+} (Robinson and

Karnovsky, 1983). Withey (1969) found La^{3+} to increase the rate of hydrolysis of *p*-nitrophenyl hydrogen methyl phosphonate by a factor of 3.6×10^4. However, hydrolysis of diesters of methylphosphoric acid was unaffected. Lanthanum hydroxide gels also have catalytic activity (Butcher and Westheimer, 1955).

Lanthanides also increase the rate of amide proton exchange in aspartyl-phenylalanine (Bleich and Glasel, 1975) and catalyze a range of organic reactions (Long, 1985).

2.7 Methods of Quantitative Analysis

2.7.1 Introduction

This section is intended as a brief survey of the techniques presently available for the quantitative determination of lanthanides. It is not designed to provide a detailed account of the theory or methods involved. Such detail can be obtained from the book *Analytical Chemistry of Yttrium and the Lanthanide Elements* by Ryabchikov and Ryabukhin (1970). Several comprehensive reviews on the subject can be found in Gschneidner and Eyring (1979).

2.7.2 Spectrophotometry

As shown in Fig. 2-1, most lanthanides absorb UV or visible light at specific wavelengths. These absorptions can be used to determine the concentrations of individual lanthanides in solution. In this respect, the narrowness of the absorption peaks aids specificity. However, errors can be introduced by the ability of various counterions or chelators to shift the absorption peaks. This means that the standard curves used for calibration should contain the same population of chelators and counterions as the sample.

The main obstacle to spectrophotometric determinations remains the low molar absorptivities of the lanthanides (Table 2-3). Greatest absorptivity is possessed by Ce^{3+} and Tb^{3+}. However, this occurs in the UV region, where many other substances also absorb very strongly. Furthermore, the absorptivity of Ce^{3+} varies with ionic strength and temperature (Banks and Klingman, 1956). The lower limits of spectrophotometric detection of lanthanides in the visible region range from about $70 \, \mu g/ml$ for Pr^{3+} to $330 \, \mu g/ml$ for Tm^{3+} (Ryabchikov and Ryabukhin, 1970).

Greater sensitivity is achieved by using chromogenic reagents. The

Table 2-13. Some Chromogenic Reagents for the Determination
of Lanthanides[a]

Reagent	Approximate molar absorptivity [(liter/mol·cm) × 10^{-4}]	Approximate sensitivity (μg/ml)
8-Hydroxyquinoline	0.5–0.6	1
Alizarin red S	1	0.06
Pyrocatechol violet	2.6	~1
Bromopyrogallol red	5	0.5
Naphthazarin	1	1–6
Aluminon	—	6
Arseno I	1	0.04
Arseno III	5	0.01
Chlorophosphonazo III[b]	4–7	—
Eriochrome cyanide R.C.[c]	1.1–1.5	0.1

[a] Abstracted from Ryabchikov and Ryabukhin (1970).
[b] O'Laughlin and Jensen (1970).
[c] Munshi et al. (1968).

principle of such methods involves a major change in the absorption
spectrum of the chromogenic substance upon binding Ln^{3+} ions. This
usually involves a color change, which can be monitored spectrophoto-
metrically to provide quantitative data. The increased sensitivity of these
methods over direct spectrophotometric determination derives from the
far greater molar absorptivities of the Ln^{3+}–chromogenitor complex. Some
chromogenic reagents are listed in Table 2-13. Murexide, alizarin red S,
and arseno III are among the most commonly used reagents in this type
of assay. According to Fernandez–Gavarron et al. (1987) arseno III can
be used at the appropriate pH, to measure La^{3+} in the presence of high
concentrations of Ca^{2+} and phosphate.

   Although many of these reagents are quite sensitive, they are susceptible to interference from other metal ions and usually cannot distinguish between one lanthanide and another. However, certain other chromogenic reagents, such as stilbazo and hematoxylin, can be used to
determine independently heavy and light lanthanides or Y^{3+} (Ryabchikov
and Ryabukhin, 1970). Cerium is the only lanthanide that can be readily
identified individually, with chromogenic reagents, by methods exploiting
its ability to undergo oxidation to Ce^{4+}.

2.7.3 Titrimetry

   As an alternative to measuring the spectral changes in chromogenic
reagents to determine lanthanide concentrations, these substances may

be used as indicators in titrimetric determinations. With this method, increments of the lanthanide solution are added to a standard solution of a strong chelator in the presence of the indicator. When the concentration of lanthanide just exceeds that of the chelator, there will be free Ln^{3+} ions to form a complex with the indicator and produce a color change. This method requires the chelator to form complexes with lanthanides that are sufficiently stable to exist at low concentration; the affinity of Ln^{3+} ions for the chelator must also exceed that for the indicator by at least two orders of magnitude. The most widely used chelator in such determinations is EDTA, which forms 1:1 complexes with Ln^{3+} ions having the stability constants shown in Table 2-4. Other strong chelators, such as DTPA, could also be employed. Commonly used indicators are murexide, eriochrome black T, alizarin red, arseno I, and xylenol orange.

The end point of titrimetric measurements can also be determined in the absence of an indicator with a mercury indicator electrode (Reilley *et al.*, 1958). Alternatively, the protons released during complex formation can be measured with a pH electrode. Dowex-50 ion-exchange resin is useful in this regard, as three protons are displaced from the protonated form of the resin for each Ln^{3+} ion which binds. Lanthanides can also be measured by precipitation titrations with a number of substances such as oxalic acid. As lanthanides form extremely insoluble salts with oxalate, the amount of oxalate remaining in solution after mixing with the test solution provides a measure of lanthanide concentration. None of these three variations are as convenient as the EDTA titrations in the presence of a visual indicator.

2.7.4 Luminescence Techniques

Although Ln^{3+} ions alone luminesce too weakly for analytical purposes, the luminescence of Tb^{3+} and Eu^{3+} is markedly enhanced when bound to suitable chromophores (Section 3.3). Barela and Sherry (1976) employed dipicolinic acid (DPA) to enhance Tb^{3+} luminescence and were able to detect nanomolar concentrations of Tb^{3+} by measuring the emission at 545 nm or 490 nm (see Fig. 3-2b), following excitation at 280 nm. Of several metal ions tested, only other Ln^{3+} ions and Ca^{2+} interfered with Tb^{3+} luminescence. Tb^{3+}, Eu^{3+}, and Dy^{3+} are the only Ln^{3+} ions that show detectable luminescence in DPA solutions (Miller and Senkfor, 1982). Although the relative intensity of the emission of Eu^{3+}-DPA at 614 nm approximately equals that of Tb^{3+}-DPA at 545 nm, that of Dy^{3+}-DPA at 477 nm is only about 10% of this value. According to Miller and Senkfor (1982), the quantitative determination is linear between 50 mM and 5 nM Tb^{3+} or Eu^{3+}. For Dy^{3+} the assay is linear between 50 mM and

$50\,nM$. As the formation constants of all the Ln^{3+}-DPA and Ca^{2+}-DPA complexes are known, concentrations of other Ln^{3+} ions and Ca^{2+} can be luminescently detected with Tb^{3+}-DPA or Eu^{3+}-DPA by a competition assay. Sensitivities down to about $1\,nM$ have been suggested. Dakubu and Ekins (1985) have used Eu^{3+}-β-diketone complexes in the same way.

A variation of the luminescence method is to use X rays to excite lanthanide luminescence. Limits of detection well below the ppm level are obtained (D'Silva and Fassel, 1979). When samples are irradiated with sufficiently energetic X rays, fluorescent X rays are emitted which permit the detection and quantitation of lanthanides. X-ray fluorescence spectrometry has the advantage of requiring minimal sample preparation and yielding relatively simple spectra. One of the disadvantages results from the proximity of several of the lines from different elements, leading to internal absorption and enhancement effects. Methods exist to circumvent these difficulties. The lower limits of detection are in the 50–100 ppm range (DeKalb and Fassel, 1979). Lanthanides can also be identified individually during electron microscopy by measuring the energies of the X rays emitted in response to excitation by the electron beam. This method (energy dispersive analysis of X rays; EDAX) is not particularly sensitive but is useful biologically, as it can be employed in conjunction with electron microscopy, where it has been used to map the cellular and pericellular distribution of lanthanides (Section 6.2.3). This method is helped by the distinctive X-ray emissions of Ln^{3+}, which we easily distinguished from those of elements occurring naturally.

2.7.5 Atomic Absorption Spectroscopy

In determinations of the lanthanides, atomic absorption spectroscopy lacks the precision that it brings to the determination of many other metals, because the absorptivities of the lanthanides are so low (Van Loon *et al.*, 1971). Some detection limits are given in Table 2-14. However, of the sophisticated analytical instruments used in trace element determinations, the atomic absorption spectrometer is most likely to be available to the biochemist. As submicromolar concentrations of several lanthanides can be measured with this technique, it is probably of practical use in many circumstances.

2.7.6 Atomic Emission Techniques

The principle of atomic emission techniques is to excite the metals thermally and to monitor the resulting emission spectra. Many elements

Table 2-14. A Comparison of the Sensitivity of
Various Techniques for Measuring Lanthanides[a]

Ln	Detection limit (μg/ml)				
	FES[b]	AAS[b]	AFP[b]	ICP[b]	NAA[c]
La	0.01	2	—	0.003	0.00001
Ce	10	—	0.5	0.007	0.005
Pr	0.07	4	1	0.06	0.0001
Nd	0.7	2	2	0.05	0.005
Sm	0.2	0.6	0.15	0.02	0.00003
Eu	0.0005	0.04	0.02	0.001	0.0000015
Gd	2	—	0.8	0.007	0.001
Tb	0.03	2	0.5	0.2	0.0002
Dy	0.05	0.2	0.3	0.004	0.0000015
Ho	0.02	0.1	0.15	0.01	0.00002
Er	0.04	0.1	0.5	0.001	0.001
Tm	0.02	0.08	0.1	0.007	0.0001
Yb	0.002	0.02	0.01	0.0009	0.0001
Lu	1	3	3	0.008	0.000015
Y	0.03	0.3	—	0.0002	0.0005

[a] FES: flame emission spectroscopy; AAS: atomic absorption spectroscopy; AFS: atomic fluorescence spectroscopy; ICP: induction coupled plasma; NAA: neutron activation analysis.
[b] From DeKalb and Fassel (1979).
[c] From Ryabchikov and Ryabukhin (1970).

emit at characteristic wavelengths, which can be analyzed to identify and to quantitate the metal concerned. The quantitative determination of lanthanides by flame emission spectroscopy (FES), in which the elements are excited by a flame, has been used for many decades. FES is a general method for lanthanide determinations and can be used to measure the concentration of each member in a mixture. The limits of detection vary from one lanthanide to another (Table 2-14).

Much greater sensitivity can be achieved with induction coupled plasma (ICP) spectroscopy (Table 2-14). Another advantage of ICP spectroscopy is the absence of interelement effects, which means that calibration curves are usually valid for complex mixtures, provided there is no spectral line overlap. In addition, calibration curves are usually linear over a four or five orders of magnitude change in concentration.

2.7.7 Neutron Activation Analysis

Neutron activation analysis is the most sensitive method available for the detection and measurement of lanthanides. It has the additional advantages of low interference from other elements and of requiring minimal sample preparation. However, it does have the minor drawback of requiring a nuclear reactor as a source of neutrons with which to irradiate the sample. As a result of this irradiation, distinctive radioactive species are formed which can be identified and measured.

The sensitivity of the technique is directly proportional to the neutron flux. Detection limits for a neutron flux of 10^{13} cm^{-2}·s^{-1} are given in Table 2-14. The neutron activation analysis of lanthanides has been reviewed by Boynton (1979).

2.7.8 Other Methods

Other methods include spark source mass spectrometry (Taylor, 1979; Conzemius, 1979) and polarography (Ryabchikov and Ryabukhin, 1970). Mass spectrometry is sensitive enough to permit the measurement of lanthanides in the ashes of mammalian organs, without the need for concentration (Sihvonen, 1972). The polarographic method is based upon the reduction of Ln^{3+} to Ln^0 or, where a stable divalent state exists, to Ln^{2+}. Polarography is not as sensitive as spectrometric methods and is rarely used now (O'Laughlin, 1979).

2.8 Practical Considerations

The chemical properties of the lanthanides which have just been discussed impose restrictions upon the experimental conditions that can be used in biochemical investigations. To begin with, many of the soluble lanthanide salts, such as the chlorides, nitrates, and acetates, are hygroscopic. Thus, the concentration of any stock solution made with these salts needs to be independently measured by one of the methods indicated in Section 2.7. Titrimetric determinations with EGTA [ethyleneglycol-bis(β-aminoethyl ether)-N,N,N',N'-tetraacetic acid] or DTPA, using an indicator such as murexide, are a simple, inexpensive way to do this. Stock solutions can be made up with greater accuracy by starting with the oxides, Ln_2O_3. These should first be heated for 2 h at 900°C, to drive off water and to break down carbonates. They are then cooled and dissolved in a small amount of concentrated HCl. In some cases, it is necessary to warm the mixture before dissolution is complete. The concentrate can then be diluted to the required volume.

Storage of stock solutions has its pitfalls. Aqueous solutions of lanthanides turn cloudy when kept at room temperature for several weeks.

Precipitates of $Ln_2(CO_3)_3$, which form through reaction with dissolved CO_2 may be responsible for this. Hydrolysis (Section 2.3; Table 2-10) occurs more quickly and results in the formation of precipitates of $Ln(OH)_3$. This can be minimized by keeping the pH well below that at which hydolysis occurs, and by keeping the concentration of lanthanides on the low side, as indicated in Table 2-10. However, reducing the concentration of Ln^{3+} ions favors the formation of radiocolloids and also increases the effect that adsorption on container walls will have on the bulk concentration. Loss of Ln^{3+} ions to the internal surfaces of containers is not a trivial concern. Not only can it affect the concentration of the Ln^{3+} ions in solution but, by displacing Ca^{2+}, it can introduce competing ions. This has been examined in detail by Ellis and Morrison (1975), who stored aqueous, unbuffered solutions of $2\,\mu M$ $^{152}EuCl_3$ at pH values of 4.0, 4.7, and 7.2, or in $0.1\,M$ N-ethylmorpholine–HCl buffer, pH 8.0 in containers made of Pyrex glass, soda glass, polyallomer, or cellulose nitrate. As shown in Fig. 2-5, best results were obtained with the lowest pH. At pH values above neutral, the containers differed remarkably in their suitability. The effects of adsorption could be minimized by increasing the concentration of Eu^{3+} (Table 2-15). When stored in Pyrex glass at pH 7.4 for 7 days, there was no measurable change in the concentration of a $1\,mM$ solution of Eu^{3+}. Pyrex glass, plastic, or acid-washed soda glass was also recommended by Jones and Dwek (1974), who found that up to 50% of the Gd^{3+} present in a $1\,mM$ solution could be lost through contact with soda-glass Pasteur pipettes. Due to the nature of their experiments, these authors were able to add excess EDTA to the stock solution to prevent adsorption losses. Bearing all these issues in mind, it is probably reasonable to suggest that stock solutions of lanthanides be kept fairly concentrated (10^{-3}–$10^{-2}\,M$), under acid conditions (pH 4–6) in Pyrex glass or polyethylene containers, with regular monitoring of the actual concentration of Ln^{3+} ions. As the lanthanide salts are cheap to buy, it is better to make up fresh stock solutions every week or so than to run the risk of introducing experimental variables through the slow formation of hydroxides, carbonates, or other undesirable species. Stock solutions can be diluted and adjusted to the pH required for particular experimental purposes. Freshly prepared, aqueous solutions of Ce^{3+}, Y^{3+}, Yb^{3+}, and Tm^{3+} at pH 7.3–7.4 (Kanapilly, 1980) and of La^{3+} at pH 7.4–7.7 (Schatzki and Newsome, 1975) are apparently free from polymers and precipitates. However, micromolar concentrations of Tb^{3+} in $0.1\,M$ Tris at pH 8 are not dialyzable, even though there is no visible formation of precipitates (Sherry et al., 1978). Although precipitation of $Ln(OH)_3$ occurs at the pH values given in Table 2-10, it is important to realize that the other hydrolysis products, $Ln(OH)^{2+}$ and $Ln(OH)_2^{+}$, are much more soluble than

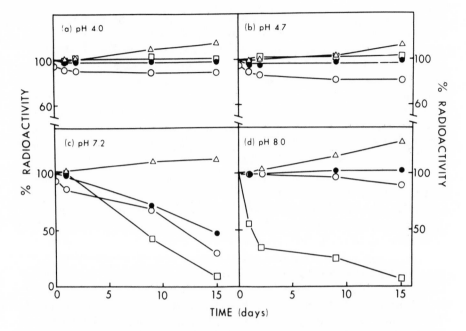

Figure 2-5. Changes in the concentration of $EuCl_3$ as a function of time and pH in different containers. ●, Pyrex glass containers; ○, soda-glass containers; □, polyallomer containers; △, cellulose nitrate containers. In (a)–(c), Eu^{3+} was in aqueous solution. In (d), solutions were buffered with 0.1 M N-ethylmorpholine. Concentration of $^{152}EuCl_3 = 2\,\mu M$. From Ellis and Morrison (1975), with permission.

Table 2-15. Effects of Total Eu^{3+} Concentration
on Adsorption Losses during Storage[a,b]

Eu^{3+} concentration (M)	Percent loss of radioactivity (after 7 days)
10^{-3}	0
10^{-4}	1
10^{-5}	10
2×10^{-6}	46

[a] From Ellis and Morrison (1975), with permission.
[b] Aqueous solution of $^{152}EuCl_3$, pH 7.4; Pyrex glass container.

$Ln(OH)_3$ and form at lower pH values. Thus, just because a lanthanide solution is filterable and free from obvious precipitates, it does not necessarily contain a homogeneous population of unhydrolyzed, $Ln(H_2O)_n^{3+}$ ionic species. In fact, under many conditions of biochemical assay, one would predict the existence of mixed populations of unhydrolyzed and partially hydrolyzed species. Possible contributions of the latter to the experimental findings are rarely, if ever, considered. One exception is the work of Dower *et al.* (1975), who, during their investigations of the interactions of Gd^{3+} with IgG, noted extensive hydroxide complex formation, even at pH 6.

The strong complexing behavior of Ln^{3+} ions requires careful selection of buffers. Many commonly used buffers are disqualified by interacting too strongly with lanthanides. The obvious examples are carboxylic acid or amino acid buffers such as acetate, lactate, citrate, glycine, glycylglycine, tartrate, maleate, and bicine. Failure to recognize this led to confusion over whether lanthanides could or could not activate α-amylase (Section 4.13). Phosphate buffers are, of course, to be avoided at all costs. Of the remaining buffers which are commonly used in biochemistry, Tris forms fluorescent adducts with Tb^{3+} (Short and Osmand, 1983), although it has negligible interaction with Ca^{2+} (Good *et al.*, 1966), and is probably best avoided. EPR data (Stephens and Grisham, 1979) also indicate binding of Gd^{3+} to N-tris(hydroxymethyl)methyl-2-aminoethanesulfonic acid (TES) at pH 7. However, satisfactory results have been reported for N-2-hydroxyethylpiperazine-N'-2-ethanesulfonic acid (HEPES), which has a pK_a of 7.55 at 20°C, piperazine-N,N'-bis(2-ethanesulfonic acid) (PIPES; pK_a = 6.80), imidazole buffer (pK_a = 7.00), and N-morpholinopropanesulfonic acid (MOPS; pK_a = 7.2). This presumably reflects the weak affinity that Ln^{3+} ions have for nitrogen donors and organic sulfate groups (Section 2.3). The successful use of chelators as lanthanide buffers to obtain controlled, standard concentrations of free Ln^{3+} ions has not been reported. However, Cartmill and Dos Remedios (1980) have employed EGTA to achieve stable, low concentrations of Gd^{3+}.

When working with live cells or tissues, the usual Krebs or Ringer solution, which contains a phosphate buffer, is usually replaced by one which is Tris buffered. Although there is evidence that Tris and Ln^{3+} ions interact, no precipitate forms and any inaccuracy thus introduced is probably a minor quantitative one. However, perhaps HEPES would be a better choice in the future. Such simple buffered salt solutions work well for fast-reacting tissues and cells, such as nerves, muscles, platelets, and polymorphonuclear leukocytes whose biochemical responses are measured in seconds or minutes (Chapter 6). However, many types of response such as cell division, enzyme induction, or differentiation may

require hours or days. To study the effects of lanthanides on these processes presents major experimental obstacles. Indeed, some such experiments may be impossible to perform "cleanly." For a start, the main-
tenance of cell viability and responsiveness for such a time period requires
the presence of more than a simple salt solution. Serum, and a complex
mixture of vitamins and other factors, is usually needed. This opens up
all sorts of possibilities for ill-defined interactions between Ln^{3+} ions and
media components. In addition, tissue culture media contain phosphate
and are buffered by a bicarbonate system in equilibrium with the CO_2
environment in the incubator. The latter can be replaced by HEPES, but
phosphate-depleted media are not healthy. One approach may be to introduce the lanthanides as a complex with a chelator which is strong
enough to prevent exchange of Ln^{3+} ions with ligands in the culture
medium, but weak enough to give them up to high-affinity binding sites
on the cell surface. So far as I know, this has never been attempted.

2.9 Summary

The lanthanide series of elements comprises the 15 elements between
lanthanum (La, atomic number 57) and lutetium (Lu, atomic number 71)
in the periodic table. Although not officially a lanthanide, yttrium (Y,
atomic number 39) has sufficiently similar chemical properties for its inclusion in this book. Promethium (Pm) is unstable and consequently not
well characterized.

These elements form trivalent cations (Ln^{3+}). Although some members of the series have divalent and tetravalent states, these are, with the
possible exception of Eu^{2+}, of little biochemical relevance. Unpaired $4f$
electrons endow several lanthanides with valuable magnetic, spectroscopic, and luminescent properties. Furthermore, as these electrons are
well shielded, the spectral characteristics are retained upon binding to
ligands. The absorption spectra of most lanthanides possess a number of
sharp peaks, with only Y^{3+}, La^{3+}, and Lu^{3+} failing to absorb in the
100–200 nm range. Although many lanthanides are beautifully colored
(Table 2-2), their absorptivities (Table 2-3) are too low to be of much
general use in spectroscopy. Low absorptivities also limit the luminescent
emissions of aqueous solutions of lanthanides. However, direct laser excitation or indirect, chromophore-mediated excitation of Tb^{3+} and Eu^{3+}
leads to very strong emissions at characteristic wavelengths (Fig. 3-2).

Lanthanide binding is almost exclusively ionic. The lanthanides are
"hard" acceptors, with an overwhelming preference for oxygen donor
atoms. Thus, the most common biological ligands are carboxyl and phos-

phate groups. Additional ligands can be supplied by carbonyl and hydroxyl groups. Rarely, nitrogen will serve as a donor, but only in multidentate complexes where oxygen atoms act as the other ligands. Because of their high charge-to-volume ratio, Ln^{3+} ions can form very tight complexes with multidentate, oxygen-donor ligands (Tables 2-4–2-8). Water is a strong ligand for Ln^{3+} ions, whose hydration number probably varies from 9 or 10 for La^{3+} to 8 or 9 for Lu^{3+}. Coordination numbers of 6–12 have been reported, with 8 or 9 favored in biological systems. Ionic radii decrease down the series and increase with coordination number (Table 2-1). The former phenomenon is known as the lanthanide contraction.

Although lanthanides form soluble chlorides, nitrates, and perchlorates, their hydroxides, carbonates, phosphates, and fluorides are insoluble. Hydroxide formation occurs in aqueous solutions of lanthanides at approximately neutral pH values (Table 2-10), through hydrolysis reactions. This is a major inconvenience for experimental biochemical studies. Solutions of lanthanides also spontaneously form colloidal particles called "radiocolloids," the chemistry of which is unknown.

Much of the interest in lanthanide biochemistry flows from the ability of Ln^{3+} ions to replace Ca^{2+} in a specific and often isomorphous manner. They do so because of marked similarities between Ca^{2+} and Ln^{3+} ions in their ionic sizes, bonding, coordination geometry, and donor atom preference (Table 2-11). Whereas Ca^{2+} is chemically silent, the Ln^{3+} ions provide many informative spectroscopic signals. In addition, functional changes in a molecule upon substituting Ca^{2+} by Ln^{3+} help in ascribing roles to Ca^{2+} in the native structure (Chapter 3). There is evidence that Eu^{2+} might also be a useful probe of the properties of Ca^{2+}. Under certain conditions, Ln^{3+} ions can also substitute biochemically for Mg^{2+}, Fe^{3+}, Fe^{2+}, and Mn^{2+}. They also share some of the chemical properties of the actinide elements. Structural and functional analyses are aided by the small, gradual, and quantitative nature of the changes in chemical properties throughout the lanthanide series. A number of titrimetric, luminescent, and spectroscopic techniques are available for lanthanide determination (Tables 2-13 and 2-14).

References

Bamann, E., Trapmann, H., and Fischler, F., 1954. Behavior and specificity of cerium and lanthanum as phosphatase models towards nucleic acids and mononucleotides, *Biochem. Z.* 328:89–96.

Banks, C. V., and Klingman, D. W., 1956. Spectrophotometric determinations of rare earth mixtures, *Anal. Chim. Acta* 15:356–363.

Barela, T. P., and Sherry, A. D., 1976. A simple, one-step fluorometric method for determination of nanomolar concentrations of terbium, *Anal. Biochem.* 71:351–357.

Bleich, H. E., and Glasel, J. A., 1975. Catalysis of amide proton exchange by lanthanum ions, *J. Am. Chem. Soc.* 97:6585–6586.

Bowen, J. M., 1972. Effects of rare earths and yttrium on striated muscle at the neuromuscular junction, *Can. J. Physiol. Pharmacol.* 50:603–611.

Boyd, G. E., 1959. Technetium and promethium, *J. Chem. Educ.* 36:3–14.

Boynton, W. V., 1979. Neutron activation analysis, in *Handbook on the Physics and Chemistry of Rare Earths* (K. A. Gschneidner and L. Eyring, eds.), Vol. 4, North-Holland, Amsterdam, pp. 457–470.

Butcher, W. W., and Westheimer, F. H., 1955. La(OH)$_3$ gel-promoted hydrolysis of phosphate esters, *J. Am. Chem. Soc.* 77:2420–2424.

Cartmill, J. A., and Dos Remedios, C. G., 1980. Ionic radius specificity of cardiac muscle, *J. Mol. Cell. Cardiol.* 12:119–123.

Chamie, C., and Haissinsky, M., 1934. Effects of age and Po content on centrifugation of Po solutions, *C. R. Acad. Sci. Paris* 198:1229–1231.

Chantler, P. D., 1983. Lanthanides do not function as calcium analogues in scallop myosin, *J. Biol. Chem.* 258:4702–4705.

Choppin, G. R., and Graffeo, A. J., 1965. Complexes of trivalent lanthanide and actinide ions. II. Inner-sphere complexes, *J. Inorg. Chem.* 4:1254–1261.

Conzemius, R. J., 1979. Analysis of rare earth matrices by spark source mass spectrometry, in *Handbook on the Physics and Chemistry of Rare Earths* (K. A. Gschneidner and L. Eyring, eds.), Vol. 4, North-Holland, Amsterdam, pp. 377–404.

Cotton, F. A., and Wilkinson, G., 1980. *Advanced Inorganic Chemistry,* 4th ed., Wiley, New York, Chapter 23.

Dakubu, S., and Ekins, R. P., 1985. The fluorometric determination of europium ion concentration as used in time-resolved fluoroimmunoassay, *Anal. Biochem.* 144:20–26.

DeKalb, E. L., and Fassel, V. A., 1979. Optical atomic emission and absorption methods, in *Handbook on the Physics and Chemistry of Rare Earths* (K. A. Gschneidner and L. Eyring, eds.), Vol. 4, North-Holland, Amsterdam, pp. 405–440.

Dos Remedios, C. G., 1977. Ion radius selectivity of skeletal muscle membranes, *Nature* 270:750–751.

Dower, S. K., Dwek, R. A., McLaughlin, A. C., Mole, L. E., Press, E. M., and Sunderland, C. A., 1975. The binding of lanthanides to non-immune rabbit immunoglobulin G and its components, *Biochem. J.* 149:73–82.

D'Silva, A. P., and Fassel, V. A., 1979. X-ray excited optical luminescence of the rare earths, in *Handbook of the Physics and Chemistry of Rare Earths* (K. A. Gschneidner and L. Eyring, eds.), Vol. 4, North-Holland, Amsterdam, pp. 441–456.

Eichhorn, G. L., and Butzow, J. J., 1965. Interactions of metal ions with polynucleotides and related compounds. III. Degradation of polyribonucleotides by lanthanum ions, *Biopolymers* 3:79–91.

Einspahr, H., and Bugg, C. E., 1984. Crystal structure studies of calcium complexes and implications for biological systems, in *Metal Ions in Biological Systems* (H. Sigel, ed.), Vol. 17, Marcel Dekker, New York, pp. 51–97.

Ellis, K. J., and Morrison, J. F., 1975. A problem encountered in a study of the effects of lanthanide ions on enzyme-catalyzed reactions, *Anal. Biochem.* 68:429–435.

Evans, C. H., 1981. Interactions of tervalent lanthanide ions with bacterial collagenase (clostridiopeptidase A), *Biochem. J.* 195:677–684.

Faktor, M. M., and Hanks, R., 1969. Calculation of the third ionization potentials of the lanthanons, *J. Inorg. Nucl. Chem.* 31:1649–1659.

Fernandez–Gavarron, F., Brandt, J. G., and Rabinowitz, J. L., 1987. A simple spectrophotometric assay for micromolar amounts of lanthanum in the presence of calcium and
phosphate, *J. Bone Min. Res.* 2:421–425.

Glasel, J. A., 1973. Lanthanide ions as nuclear magnetic resonance chemical shift probes
in biological systems, in *Current Research Topics in Bioinorganic Chemistry* (S. J.
Lippard, ed.), Vol. 18, Wiley, New York, pp. 383–413.

Good, N. E., Winget, G. D., Winter, W., Connolly, I., Izawa, S., and Singh, R. M. M.,
1966. Hydrogen ion buffers for biological research, *Biochemistry* 5:467–477.

Gschneidner, K. A., and Eyring, L. (eds.), 1979. *Handbook on the Physics and Chemistry
of the Rare Earths,* Vol. 4, North-Holland, Amsterdam.

Homer, R. B., and Mortimer, B. D., 1978. Europium II as a replacement for calcium II in
conconavalin A. A precipitation assay and magnetic circular dichroism study, *FEBS
Lett.* 87:69–72.

Horrocks, W. DeW., 1982. Lanthanide ion probes of biomolecular structure, in *Advances
in Inorganic Biochemistry* (G. L. Eichhorn and L. G. Marzilli, eds.), Vol. 4, Elsevier,
New York, pp. 201–261.

Horrocks, W. DeW., and Sudnick, D. R., 1979. Lanthanide ion probes of structure in biology. Laser-induced luminescence decay constants provide a direct measure of the
number of metal-coordinated water molecules, *J. Am. Chem. Soc.* 101:334–340.

Jackson, K. S., and Reinacker, G., 1930. Solubilities of the octahydrates of the rare-earth
sulfates, *J. Chem. Soc.* 1930:1687–1691.

Jones, R., and Dwek, R. A., 1974. The mechanism of water-proton relaxation in enzyme
paramagnetic-ion complexes. I. The Gd^{3+}-lysozyme complex, *Eur. J. Biochem.* 47:
271–283.

Kanapilly, G. M., 1980. *In vitro* precipitation behaviour of trivalent lanthanides, *Health
Phys.* 39:343–346.

Karraker, D. G., 1970. Coordination of trivalent lanthanide ions, *J. Chem. Educ.* 47:424–430.

King, J. F., and Romer, A., 1933. Adsorption of thorium B and thorium C from solution,
J. Phys. Chem. 37:663–673.

Kurbatov, J. D., and Pool, M. L., 1944. Isolation of pure radioactive Ce, *Phys. Rev.* 65:
61–65.

Levine, B. A., and Williams, R. J. P., 1982. The chemistry of calcium ion and its biological
relevance, in *The Role of Calcium in Biological Systems* (L. J. Anghileri and A. M.
Tuffet-Anghileri, eds.), Vol. 1, CRC Press, Boca Raton, FL, pp. 3–26.

Lind, M. D., Lee, B., and Hoard, J. L., 1965. Structure and bonding in a ten-coordinate
lanthanum(III) chelate of ethylenediaminetetraacetic acid, *J. Am. Chem. Soc.* 87:
1611–1612.

Long, J. R., 1985. Lanthanides in organic synthesis, *Aldrich. Acta* 18:87–93.

Martin, R. B., 1983. Structural chemistry of calcium: lanthanides as probes, in *Calcium in
Biology* (T. G. Spiro, ed.), Wiley, New York, pp. 237–270.

Miller, T. L., and Senkfor, S. I., 1982. Spectrofluorometric determination of calcium and
lanthanide elements in dilute solution, *Anal. Chem.* 54:2022–2025.

Moeller, T., 1973. The lanthanides, in *Comprehensive Inorganic Chemistry* (J. C. Bailar,
H. J. Emeleus, R. Nyholm, and A. F. Trotman-Dickenson, eds.), Vol. 4, Pergamon
Press, Oxford, pp. 1–97.

Moeller, T., and Kremers, H. E., 1944. Electrometric study of the precipitation of trivalent
hydrous rare earth oxides or hydroxides, *J. Phys. Chem.* 48:395–406.

Moeller, T., Martin, D. F., Thompson, L. C., Ferrus, R., Feistel, G. R., and Randall,
W. J., 1965. The coordination chemistry of yttrium and the rare earth metal ions, *Chem.
Rev.* 65:1–50.

Morris, L. R., 1976. Thermochemical properties of yttrium, lanthanum and the lanthanide elements and ions, *Chem. Rev.* 76:827–841.

Munshi, K., Chrivastava, S., and Dey, A., 1968. Chromogenic behavior of eriochrome cyanide R.C. in the spectrophotometric determination of scandium, yttrium and lanthanum, *J. Ind. Chem. Soc.* 45:817–820.

Nieboer, E., 1975. The lanthanide ions as probes in biological systems, *Struct. Bonding* 22: 1–47.

O'Laughlin, J. W., 1979. Chemical spectrophotometric and polarographic metals, in *Handbook on the Physics and Chemistry of Rare Earths* (K. A. Gschneidner and L. Eyring, eds.), Vol. 4, North-Holland, Amsterdam, pp. 341–358.

O'Laughlin, J. W., and Jensen, D. F., 1970. Spectrophotometric determination of the lanthanides with chlorophosphoazo III, *Talanta* 17:329–332.

Opie, E. L., and Lavin, G. I., 1946. Localization of ribonucleic acid in the cytoplasm of liver cells, *J. Exp. Med.* 84:107–112.

Paneth, F., 1913. Colloidal structure of radioactive substances, *Kolloid Z.* 13:297–305.

Phillips, C. S. G., and Williams, R. J. P., 1966. *Inorganic Chemistry,* Vol. 2, Oxford University Press, Oxford, Chapter 21.

Reilley, C. N., Schmid, R. W., and Lamson, D. W., 1958. Chelometric titrations of metal ions with potentiometric endpoint detection with ethylenedinitrilotetraacetic acid, *Anal. Chem.* 30:953–957.

Rhee, M. J., Horrocks, W. DeW., and Kosow, D. P., 1984. Laser-induced lanthanide luminescence as a probe of metal ion-binding sites of human factor Xa, *J. Biol. Chem.* 259:7404–7408.

Richardson, C. E., and Behnke, W. D., 1978. Physical studies of the lanthanide binding to concanavalin A, *Biochim. Biophys. Acta* 534:267–274.

Robinson, J. M., and Karnovsky, M. J., 1983. Ultrastructural localization of 5'-nucleotidase in guinea pig neutrophils based upon the use of cerium as capturing agent, *J. Histochem. Cytochem.* 31:1190–1196.

Rordorf, B. F., and Kearns, D. R., 1976. Effect of europium(III) on the thermal denaturation and cleavage of transfer ribonucleic acids, *Biopolymers* 15:1491–1504.

Rübsamen, H., Hess, G. P., Eldefrawi, A. T., and Eldefrawi, M. E., 1976. Interaction between calcium and ligand-binding sites of the purified acetylcholine receptor studied by use of a fluorescent lanthanide, *Biochem. Biophys. Res. Commun.* 68:56–62.

Ryabchikov, D. I., and Ryabukhin, V. A., 1970. *Analytical Chemistry of Yttrium and the Lanthanide Elements* (translated by A. Aladjen), Ann Arbor–Humphrey Science Publishers, Ann Arbor, MI, and London.

Schatzki, P. F., and Newsome, A., 1975. Neutralized lanthanum solution: a largely non-colloidal ultrastructural tracer, *Stain Technol.* 50:171–178.

Schubert, J., and Conn, E. E., 1949. Radiocolloidal behavior of some fission products, *Nucleonics* 4:2–11.

Schweitzer, G. K., and Jackson, M., 1952a. Radiocolloids, *J. Chem. Educ.* 29:513–522.

Schweitzer, G. K., and Jackson, M., 1952b. Studies in low concentration chemistry. I. The radiocolloidal properties of lanthanum-140, *J. Am. Chem. Soc.* 74:4178–4182.

Schweitzer, G. K., and Scott, H. E., 1955. Low concentration chemistry X. Further observations on yttrium, *J. Am. Chem. Soc.* 77:2753–2757.

Schweitzer, G. K., Stein, B. R., and Jackson, W. M., 1953. Studies in low concentration chemistry. III. The radiocolloidal properties of yttrium-90, *J. Am. Chem. Soc.* 75: 793–795.

Shannon, R. D., 1976. Revised effective ionic radii and systematic studies of interatomic distances in halides and chalcogenides, *Acta Crystallogr., Sect. A* 32:751–767.

Sherry, A. D., Au-Young, S., and Cottam, G. L., 1978. Fluorescence properties of terbium-alkaline phosphatase, *Arch. Biochem. Biophys.* 189:277–282.

Short, M. T., and Osmand, A. P., 1983. Luminescence energy transfer studies of C-reactive protein. Binding of terbium(III) ions in C-reactive protein, *Immunol. Commun.* 12: 291–300.

Sihvonen, M. L., 1972. Accumulation of yttrium and lanthanoids in human and rat tissues as shown by mass spectrometric analysis and some experiments with rats, *Ann. Acad. Sci. Fenn.* 168A:1–56.

Silber, H. B., 1974. A model to describe binding differences between calcium and the lanthanides in biological systems, *FEBS Lett.* 41:303–306.

Sinha, S. P., 1966. *Complexes of the Rare Earths,* Pergamon Press, New York.

Smolka, G. E., Birnbaum, E. R., and Darnall, D. W., 1971. Rare earth metal ions as substitutes for the calcium ion in *Bacillus subtilis* α-amylase, *Biochemistry* 10:4556–4561.

Spedding, F. H., and Daane, A. H., 1961. *The Rare Earths,* Wiley, New York.

Spedding, F. H., Pikal, M. J., and Ayers, B. O., 1966. Apparent molal volumes of some aqueous rare earth chloride and nitrate solutions at 25°, *J. Phys. Chem.* 70:2440–2449.

Stefanini, S., Chiancone, E., Antonini, E., and Finazzi-Agro, A., 1983. Binding of terbium to apoferritin: a fluorescence study, *Arch. Biochem. Biophys.* 222:430–434.

Stephens, E. N., and Grishan, C. M., 1979. Lithium-7 nuclear magnetic resonance, water proton nuclear magnetic resonance and gadolinium electron paramagnetic resonance studies of the sarcoplasmic reticulum calcium ion transport adenosine triphosphatase, *Biochemistry* 18:4876–4885.

Taylor, S. R., 1979. Trace element analysis of rare earth elements by spark source mass spectrometry, in *Handbook on the Physics and Chemistry of Rare Earths* (K. A. Gschneidner and L. Eyring, eds.), Vol. 4, North-Holland, Amsterdam, pp. 359–376.

Tew, W. P., 1977. Use of coulombic interactions of the lanthanide series to identify two classes of Ca^{2+} binding sites in mitochondria, *Biochem. Biophys. Res. Commun.* 78: 624–630.

Triggle, C. R., and Triggle, D. J., 1976. An analysis of the action of cations of the lanthanide series on the mechanical responses of guinea pig ileal longitudinal muscle, *J. Physiol.* 254:39–54.

Vallee, B. L., and Williams, R. J. P., 1968. Enzyme action: views derived from metalloenzyme studies, *Chem. Brit.* 4:397–402.

Van Loon, J. C., Galbraith, J. H., and Aarden, H. M., 1971. The determination of yttrium, europium, terbium, dysprosium, holmium, erbium, thulium, ytterbium and lutetium in minerals by atomic absorption spectrophotometry, *Analyst* 96:47–50.

Vickery, R. C., 1953. *Chemistry of the Lanthanons,* Butterworths, London.

Williams, P. F., and Turtle, J. R., 1984. Terbium, a fluorescent probe for insulin receptor binding. Evidence for a conformational change in the receptor protein due to insulin binding, *Diabetes* 33:1106–1111.

Williams, R. J. P., 1970. The biochemistry of sodium, potassium, magnesium and calcium, *Quart. Rev.* 24:331–365.

Williams, R. J. P., 1979. Cation and proton interactions with proteins and membranes, *Biochem. Soc. Trans.* 7:481–509.

Withey, R. J., 1969. Lanthanum ion catalysis of nucleophilic displacement reactions of monoesters of methyl phosphonic acid, *Can. J. Chem.* 47:4383–4387.

Biochemical Techniques Which Employ Lanthanides

3

3.1 Introduction

Only trace amounts of the lanthanides are normally present in living systems (Section 7.2), and they are not known to play any metabolic role. Consequently, they have attracted the attention of biochemists not because of their physiological functions, but because of their several extremely valuable experimental uses. The most important of these exploit their ability to serve as informative spectroscopic probes of the interactions of Ca^{2+} and, to a lesser degree, other metal ions with molecules of biochemical interest. However, there also exists a large literature on their use in electron microscopy, in enzyme assay, and in a number of other types of biochemical investigation. These applications of the lanthanides rest upon their chemical properties, which were described in Chapter 2. In the present chapter, the ways in which these properties are put to experimental use are described.

3.2 Nuclear Magnetic Resonance Spectroscopy

Nuclear magnetic resonance (NMR) spectroscopy is based upon the ability of certain nuclei in a magnetic field to absorb electromagnetic radiation at specific radio frequencies. When these frequencies are plotted against absorption peak intensities, an NMR spectrum is produced. One of the requirements for a nucleus to generate an NMR signal is that its nucleus have a nonintegral spin number. Of biological relevance is that ^1H, ^{13}C, and ^{31}P fulfill this condition. Whereas ^{13}C-NMR spectroscopy usually relies upon the small amount of this isotope that occurs naturally, ^1H and ^{31}P are the predominant natural isotopes of these elements and are thus well suited to NMR spectroscopy of biological materials. In addition, specific extraneous resonating nuclei can be introduced to label particular parts of a molecule. For example, ^{19}F-labeled trifluoroacetyl-trialanine, a competitive inhibitor of elastase, has been employed in conjunction with ^{19}F-NMR spectroscopy to measure the distance between the Gd^{3+} site and the active site of elastase (Dimicoli and Bieth, 1977; Section 4-10).

After the nucleus has absorbed energy and become excited to a higher-energy state, it loses energy to the environment and reverts to the ground state. This mechanism is known as a spin–lattice or longitudinal relaxation process, during which energy is transferred to the molecular lattice. When the time taken for this transfer is shortened, the peak broadens. Thus, the linewidth is inversely proportional to the lifetime of the excited state. Measurement of proton relaxation rates is a valuable aspect of NMR studies with lanthanides. The longitudinal relaxation rates of the protons of water are much greater in the presence of Gd^{3+}, their magnitude being directly proportional to the number of water molecules attached to the cation. Because aqueous solutions contain approximately $55.6 M$ water, these effects are of great relevance. As net enhancement of the water proton relaxation rate occurs upon binding of aquo-Gd^{3+} to macromolecules, hydration numbers and the extent of dehydration upon binding can be determined. Furthermore, the enhancement of protons at the Gd^{3+}-binding site is sensitive to conformational changes and other environmental perturbations. The effects of Gd^{3+} upon spin–lattice relaxation are compared to those of other Ln^{3+} ions and certain other metal ions in Fig. 9-2.

Certain members of the lanthanide series have very high paramagnetic moments (Chapter 2; Table 2-2). When such a Ln^{3+} ion approaches a resonating nucleus, it interferes with the local magnetic environment and alters the resonance peak of the adjacent nuclei. As the effect occurs over a short distance, it will selectively alter the peaks of only those nuclei close to the lanthanide. In this way, particular peaks attributable to nuclei at or close to the lanthanide binding site can be identified, even in a complex molecule giving a complicated NMR spectrum. If the molecule is simpler and specific nuclei can be assigned specific peaks, detailed structural information can be obtained. This is because the magnitude of the change is a precise function of the distance and the angle separating the Ln^{3+} ion and the resonating nucleus. The degree of shifting induced by lanthanides through pseudocontact interaction has a theoretical $(3 \cos^2 \theta - 1)r^{-3}$ dependence, where θ is the angle between the principal axis of symmetry and the metal–nucleus vector, and r is the length of this vector. However, the degree of broadening induced by Gd^{3+} is dependent on the reciprocal of the sixth power of the distance between Gd^{3+} and the resonating nucleus (r^{-6}) and has no angular dependence (LaMar and Faller, 1973). Williams and his colleagues have taken full advantage of this circumstance in developing a method for determining the conformations of small molecules (Section 5.1). In this method, the measured distances and angles are compared by computer to a series of possible values obtained from different permissible conformations, and the best fit is selected. Nonparamagnetic lanthanides serve as valuable controls in such procedures. Although the Ln-NMR method alone cannot provide a complete conformational description of a given molecule (Geraldes and Williams, 1978), it remains of considerable utility.

NMR spectroscopy can also be used to determine binding constants, to obtain pK_a values for the liganding functional groups, and to calculate heats and entropies of binding. Some idea of the stability of a Ln^{3+}–ligand interaction can also be gauged from its lifetime in comparison to the frequency separation of the lines resulting from the complexed and uncomplexed states.

The notion of shift reagents was introduced to NMR spectroscopy in its present form by Hinckley (1969). The action of shift reagents is usually attributed to a through-space dipolar (pseudocontact) interaction, although through-bond contact interactions have been reported in some instances. Horrocks and Sipe (1971) examined a variety of lanthanides as shifting reagents. They employed the three substrate ligands 4-vinylpyridine, 4-picoline-N-oxide, and n-hexanol, in conjunction with Ln(dipivaloylmethanato)$_3$ complexes. Of interest (Fig. 3-1) are the upfield isotropic shifts induced by Pr^{3+}, Nd^{3+}, Tb^{3+}, Dy^{3+}, and Ho^{3+} and the downfield shifts produced by Eu^{3+}, Er^{3+}, Tm^{3+}, and Yb^{3+}. The shifts in response to Pr^{3+} and Eu^{3+} were accompanied by negligible line broadening. Greater shifting occurred with Dy^{3+}, Ho^{3+}, and Yb^{3+}, but this involved some degree of broadening. La^{3+} and Lu^{3+}, which are diamagnetic (Chapter 2), produced no shifts. Gd^{3+} is not a shifting agent but produces extensive peak broadening. Indeed, Horrocks and Sipe (1971) reported that in most cases the spectra were so broadened as to be unobservable. The lack of shift with the Gd^{3+} complexes was interpreted as evidence in favor of a dipolar origin for the shifts in lanthanide systems; as each f orbital in Gd^{3+} has only one electron, there could be no orbital symmetry restriction on electron spin delocalization. Thus, a contact shift, if present in any of the lanthanide complexes, would have been expected to have been observed here. This distinction is especially important when using lanthanides in conformation studies (e.g., Chapter 5). Contact interaction involves electron delocalization through chemical bonds and gives little structural information. A "through-space," dipolar interaction depends on the orientation of the resonating nucleus with respect to the

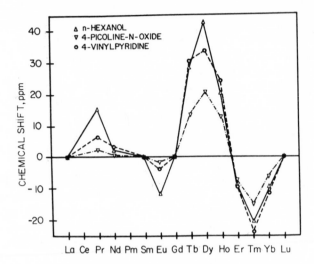

Figure 3-1. Observed isotropic shifts for the most shifted resonances of three chelators in the presence of Ln(dipivaloylmethanato)₃. From Horrocks and Sipe (1971), with permission.

lanthanide and is thus much more informative. In addition, the ability of certain Ln^{3+} ions to produce shifts without line broadening permits spin-decoupling assignments to be made easily. Overall, Yb^{3+} possesses the best characteristics for dipolar chemical shift studies. It produces reasonably large shifts with tolerable line broadening and a high ratio of dipolar to scalar contributions to the shifts (Reuben, 1973; Martin, 1983).

Proton NMR shifts produced by lanthanides are nearly always pseudocontact in origin and thus of much use in structural studies. In this respect, the lanthanides are superior to transition metals. Relative contact shifts for lanthanide perturbations of different nuclei with the same fractional spin occupancy in the same molecule have been estimated (Reuben, 1973). Values relative to [1]H-1 are ^{13}C-9, [14]N-15, [31]P-18, [17]O-24, and [19]F-36. Thus, for common biological nuclei, protons are best suited for structural studies, while ^{31}P shifts often contain considerable contributions from contact perturbations which need to be separated and calculated before conformations can be deduced (Dobson *et al.*, 1973). The theory of the use of lanthanides as NMR shift probes in these sorts of studies has been discussed by a number of authors, including Barry *et al.* (1971), Bleaney (1972), Glasel (1973), Barry *et al.* (1974), Reuben (1979), and Reuben and Elgavish (1979). A recent book on this subject has appeared (Wenzel, 1987).

According to Bradbury *et al.* (1974), peptides can be sequenced with lanthanides by ^{1}H-NMR methods. Addition of Gd^{3+} to a solution of a peptide will broaden the resonances of protons attached to the α-carbon atoms in the peptide chain. As Gd^{3+} attaches to the C-terminal carboxyl group (Chapter 4.2) and as the degree of broadening depends upon a function (r^{-6}) of the distance between Gd^{3+} and the resonating nucleus, it is theoretically possible to establish from the resonances associated with specific amino acids, the sequence of a peptide starting from the C-terminus. This complements standard methods of sequencing, which start from the N-terminus. At a frequency of 100 MHz, peptides containing at least six amino acids could be sequenced in this way (Bradbury *et al.*, 1974).

In addition to its employment in structural studies, NMR spectroscopy has been used in conjunction with lanthanides to study transport across membranes (Chapter 6). In this context, NMR spectroscopy distinguishes internal from external polar head groups in a membrane (Bystrov *et al.*, 1971). One approach has been to complex a paramagnetic Ln^{3+} ion with an ionophore to study transport by monitoring the shifts in the ^{1}H- or ^{31}P-NMR spectra of phospholipids in the membrane. Another method has been to trap specific substances within vesicles and to monitor changes in their NMR spectra as Ln^{3+} ions cross into the intravesicular space. In a variation of this approach, a nontransportable Dy^{3+} complex was introduced to suspensions of erythrocytes, thus producing a strong field gradient in the extracellular space. Using a spin-echo technique, which is sensitive to changes in local field gradients, it was possible to measure the transport of alanine and lactate into the erythrocytes (Brindle *et al.*, 1979). NMR investigations of cellular Na^+ fluxes have been made by employing Dy^{3+} tripolyphosphate, which shifts its NMR peak upon binding [23]Na^+ (Pike and Springer, 1982; Gupta and Gupta, 1982; Balschi *et al.*, 1982).

One disadvantage of NMR spectroscopy is its need for relatively high concentrations of lanthanides and ligands, often in the millimolar range. Many biological materials are simply not that soluble. A particularly readable, nonmathematical review of the applications of NMR spectroscopy to biochemical investigations, which includes mention of the lanthanides, has been written by Moore *et al.* (1983).

3.3 Luminescence Spectroscopy

Luminescence occurs when electrons, having absorbed energy to enter excited states, fall back to their ground states, losing energy by emitting light. Luminescence is a general term for all forms of light emission other than the incandescence which results from high temperatures. The main types of luminescence are fluorescence and phosphorescence. Light emitted as a result of electrons falling from excited levels to the ground energy level is called fluorescence and normally takes 10^{-9}–10^{-6} s. Phosphorescence, however, involves intersystem crossing of electrons from the singlet state to the triplet state, where the excited electron has a reversed spin and the molecule has two independent electrons of the same spin in different orbitals. Quantum theory allows such a molecule to exist in three forms of slightly differing energy known as the triplet state. Decay of excited electrons through phosphorescence takes much longer than through fluorescence, requiring at least 10^{-4} s and sometimes as long as 10^2 s. Both terms are found in the literature to describe the emission of light by lanthanides. The more general word, luminescence, is used in this book.

Excitation of the f electrons in lanthanides can be accomplished directly or indirectly. Although most lanthanides do absorb UV or visible light (Section 2.2; Fig. 2-1), the molar absorptivities are low, and luminescence is correspondingly weak. This limitation can be circumvented in two ways. In one method, whose development owes much to Richardson, Martin, and their colleagues at the University of Virginia, Ln^{3+} ions are excited indirectly by energy transfer from an adjacent chromophore. Alternatively, the Ln^{3+} ion can be directly excited with a powerful laser source, a technique pioneered by Horrocks and his co-workers at Pennsylvania State University. One advantage of luminescence is that it permits study of lanthanides at concentrations in the micromolar range. NMR spectroscopy, in contrast, often requires millimolar concentrations. The emission spectra of Tb^{3+} and Eu^{3+} are shown in Fig. 3-2.

Direct excitation of Tb^{3+} and, in particular, Eu^{3+} has become a widely used technique. The energy levels of Tb^{3+} and Eu^{3+} are compared in Fig. 3-3. Excitation with laser light at 579 nm for Eu^{3+} or 488 nm for Tb^{3+} excites $^7F_0 \rightarrow {}^5D_0$ transitions in Eu^{3+} and $^7F_6 \rightarrow {}^5D_4$ transitions in Tb^{3+}. As emissions are largest at 612 nm for Eu^{3+} or 545 nm for Tb^{3+} (Fig. 3-2), these are usually the wavelengths monitored. The exact wavelength of peak excitation is modified according to the precise configuration of the Ln^{3+}-binding site. Thus, examination of the excitation spectrum permits determination of the number of different classes of binding sites, with each peak corresponding to a discrete binding environment. With special line narrowing techniques, it is possible to make use of these differences in selectively exciting specific Ln^{3+} ions on a protein (Valentini and Wright, 1983). The fine details of the Tb^{3+} emission spectrum are sensitive to the coordination environment of the binding site and may provide a characteristic signature of particular binding domains (Sudnick and Horrocks, 1979).

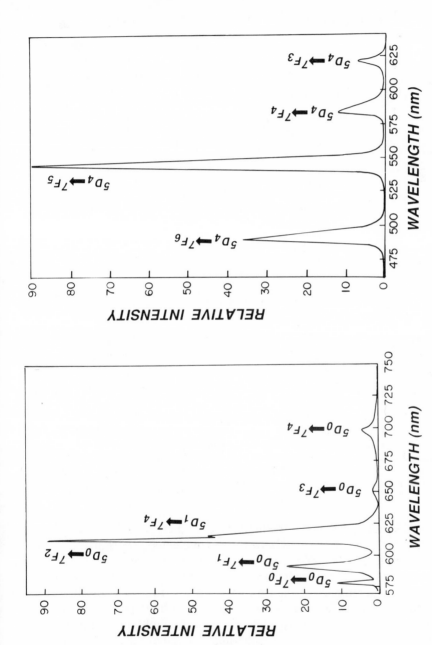

Figure 3-2. Emission spectra of (a) Eu^{3+} and (b) Tb^{3+} in acetic acid sensitized by 4-methylbenzophenone. From Heller and Wasserman (1965), with permission.

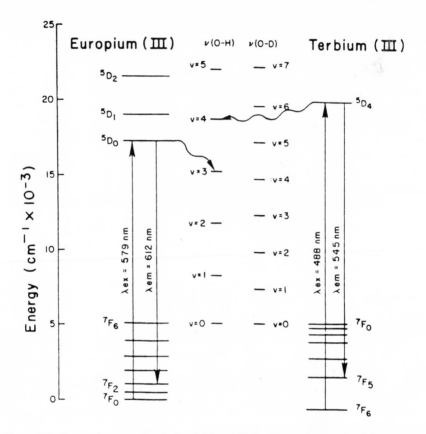

Figure 3-3. Electronic energy levels for Eu^{3+} and Tb^{3+}. The two upward-pointing arrows show the transitions which occur upon laser excitation at the indicated wavelengths. The two downward-pointing arrows label the most intense emissive transitions of the two ions. Radiationless energy transfer competes with the radiative process through coupling of the emissive states to the O—H vibrational overtones of coordinated H_2O molecules. From Horrocks and Sudnick (1981), with permission.

A distinct advantage of the luminescence method is that it allows the direct calculation of the number of water molecules coordinated to the Ln^{3+} ion. It can be seen from Fig. 3-3 that the energy levels of the O—H bonds of water compete much more effectively than O—D bonds of D_2O for the energies of the excited electrons. Energy transfer to H_2O is radiationless and decreases the excited state lifetime. The reciprocal excited state lifetime increases linearly with the mole fraction of H_2O in mixtures of H_2O and D_2O. By comparing the excited state lifetime in H_2O and D_2O, the number of OH oscillators in the first coordination sphere can be calculated. As dehydration occurs upon binding of aquo-Ln^{3+} ions to complexing agents, this method can be used to calculate the number of ligands provided by the molecule to which the Ln^{3+} ion binds.

Lanthanides may also be excited indirectly by transfer of energy from an adjacent chromophore. Suitable chromophores include the aromatic amino acids of proteins and the nucleotide bases of nucleic acids (Chapter 4 and 5). When aromatic amino acid residues act as donors, by far the strongest luminescence is found with Tb^{3+}. In such studies, the Ln^{3+}–ligand complex is excited at a wavelength within the absorption spectrum of the chromophore. If the wavelength of the chromophore's fluorescent emission overlaps the absorption wavelengths of the Ln^{3+} ion (Fig. 2-1), energy is transferred nonradiatively from the chromophore to the lanthanide, which then luminesces at its characteristic wavelengths.

One application of this method is to determine, from the excitation spectrum, the identity of groups laying close to a bound Tb^{3+} ion. For proteins, the excitation peaks of interest are 259 nm (phenylalanine), 280 nm (tyrosine), and 295 nm (tryptophan). The ionized tyrosine phenolate group also absorbs near 295 nm but is unlikely to exist at physiological pH. However, this theoretical simplicity can be confused by several factors. Although the absorption maxima of these three amino acids are different, their UV absorption spectra overlap considerably. Thus, if more than one of these amino acids is contributing to Tb^{3+} luminescence, their identities may be difficult to determine from the excitation spectrum. Some idea of the problem can be gained from examining Fig. 3-4. In addition, the molar extinction coefficients of these amino acids are different, with that of Trp being the highest and that of Phe the lowest. A contribution from Phe may thus be hard to detect in the presence of Tyr or Trp. This condition is exacerbated by the more favorable spectral overlap of Trp-Tb^{3+} relative to Phe-Tb^{3+} and Tyr-Tb^{3+}. Due to this combination of factors, in cases where there is more than one type of donor, Trp will tend to dominate the excitation spectrum. Possibly as a result of this, in the 32 proteins identified by Brittain et al. (1976) as enhancing Tb^{3+} luminescence, Trp appeared as the donor 23 times (Table 3-1. Another potential complicating factor is energy transfer among chromophores before donation to Tb^{3+}. This transfer can occur from Phe $\rightarrow$ Phe, Phe $\rightarrow$ Tyr, Phe $\rightarrow$ Trp, or Tyr $\rightarrow$ Trp. The excitation spectrum only identifies the absorbing chromophore; it cannot identify a secondary donor.

As a result of energy transfer to the Ln^{3+} ions, the intrinsic fluorescence of the protein is often quenched. Ricci and Kilichowski (1974) showed the order of quenching efficiency to be $Eu^{3+} > Yb^{3+} > Sm^{3+} > Tb^{3+} > Gd^{3+} > Ho^{3+}$. This does not reflect the abilities of these ions to

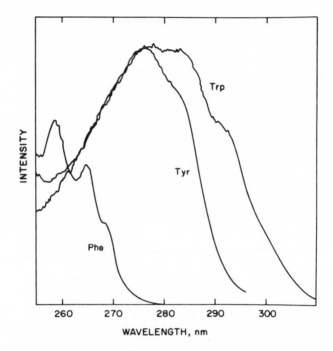

Figure 3-4. Excitation spectra of Tb^{3+}-protein complexes. Emission was monitored at 544 nm. Parvalbumin, calmodulin, and elastase were chosen as examples of proteins which excited Tb^{3+} via energy transfer from Phe, Tyr, and Trp residues, respectively. From DeJersey *et al.* (1981), with permission.

luminesce when bound to proteins. The reason is that, in several cases, the mechanism of energy capture by the Ln^{3+} ion is not one which leads to that energy being radiatively dissipated. With Eu^{3+}, a charge transfer mechanism means that greatly enhanced luminescence occurs only when the donor emits light at a wavelength greater than 400 nm. This is why the luminescence of Eu^{3+} is generally low when it accepts energy from protein donors, but high when accepting energy from 4-thiouridine residues on tRNA (Section 5.3) which fluoresce in the 460–570 nm range. For nucleic acids (Chapter 5), the principle of indirect lanthanide excitation is the same as for proteins. Tb^{3+} luminescence occurs through energy transfer from nucleotide bases. As these bases have different absorption maxima, the excitation spectrum of Tb^{3+} luminescence helps identify the donor base. In many instances this is a guanosyl or deoxyguanosyl residue. In all cases, it should be emphasized that the identification of a specific chromophore as an energy donor for Tb^{3+} luminescence does not

Table 3-1. Abilities of Various Proteins to Enhance the Luminescence of Tb^{3+}[a,b]

Protein	Equiv. of Tb^{3+}	Excitation	TE[c,d]	CPE[d,e]
Parvalbumin, carp	1	Phe	s	m
Troponin C, rabbit skeletal	1	Tyr	s	m
Troponin C, bovine cardiac	1	Tyr	s	m
Trypsinogen	1	Trp	m	n
Trypsin	1	Trp	m	n
Chymotrypsinogen	1	Trp	m	n
α-Chymotrypsin	1	Trp	m	n
δ-Chymotrypsin		Trp	m	n
Subtilisin Carlsberg	1	Tyr	m	n
Subtilisin BPN	2	Tyr	m	n
Pronase		Trp	s	w
Elastase, porcine	1	Trp	vs	s
Collagenase, bacterial	~20	Tyr + Trp	s	s
Prothrombin, bovine	11	Trp + Tyr	s	n
Prothrombin, fragment I	7	Trp	m	n
S-100, calf brain	2	Trp	w	n
Prophospholipase A, porcine	1	Tyr	m	n
Phospholipase A, porcine and equine	1	Tyr	m	n
Phospholipase A, rattlesnake	1	Tyr	m	n
Phospholipase A, honeybee	1	Trp	m	n
Vitamin D Ca-binding protein, porcine	1	Tyr	m	n
α-Amylase, bacterial	2	Trp	s	m
α-Amylase, porcine	2	Trp	s	w
Lysozyme, hen egg	1	Trp	w	n
Lactalbumin, bovine	1	Trp	w	n
Deoxyribonuclease, bovine	2	Trp + Tyr	s	n
Pyruvate kinase, rabbit	4	Trp	w	n
Creatine phosphokinase, rabbit		Trp	w	n
Pyrophosphatase, yeast	2	Trp	w	n
Ca^{2+}-ATPase, rabbit	2	Trp	m	n
Thermolysin	1	Trp	s	w
Concanavalin A		Trp	m	n

[a] From Brittain *et al.* (1976), with permission.
[b] β-Amylase, phosvitin, thrombin, rabbit myosin light chains, soybean trypsin inhibitor, and chicken ovomucoid trypsin inhibitor were inactive.
[c] TE: Terbium emission.
[d] vs, very strong; s, strong; m, medium; w, weak; n, none.
[e] CPE: Circularly polarized emission.

imply any form of complex formation between the two. It merely means that they lie close enough for energy transfer to occur. Despite some of these potential complications, Tb^{3+} luminescence has been very extensively used to very good effect in biochemical studies. Numerous examples are to be found in Chapters 4 and 5.

In addition to identifying adjacent chromophores, Tb^{3+} luminescence permits the distance separating the energy donor from the acceptor Tb^{3+} ion to be measured. The transfer of energy occurs by a Förster dipole–dipole resonance transfer mechanism (Förster, 1959), with the efficiency of transfer depending on the inverse of the sixth power of the distance between Tb^{3+} and the donor (r^{-6}). Based upon this r^{-6} dependence, Martin and Richardson (1979) calculated that a 50% probability of energy transfer occurs at a distance of 5–10 Å for Tyr-Tb^{3+} and Trp-Tb^{3+} pairs. The experimental data of Horrocks and co-workers (Horrocks and Snyder, 1981; Horrocks and Collier, 1981) suggested that 50% energy transfer from Trp-Tb^{3+} in thermolysin and parvalbumin occurs at a separation of 3.4 Å. Although the efficiency of transfer from Trp-Eu^{3+} was greater, there was little enhancement of Eu^{3+} luminescence, as a charge transfer transition mechanism of energy capture by Eu^{3+} was involved.

A modification of this approach permits measurement of the separation of a lanthanide from a chromophore buried within a molecule or a membrane. For this purpose, the Ln^{3+} ion is presented as a chelate, whose distance of maximum approach to the chromophore will normally correspond to the depth to which the chromophore is buried within its parent structure. When the mixture is excited at the excitation wavelength of the chromophore, transfer of fluorescent energy will occur to the Ln^{3+} ion within the complex at the periphery of the parent molecule or membrane. As the efficiency of transfer obeys the r^{-6} rule, the degree of excitation will be a function of the separation of the chromophore from the Ln^{3+} ion. This, in turn, provides a measure of the distance separating the chromophore from the surface. This method, known as "diffusion-enhanced fluorescent energy transfer," has been successfully applied to transferrin (Section 4.8), RNA polymerase (Section 4.25), retinal disk membranes (Stryer et al., 1982), and cytochromes and ferredoxin (Section 4.25). For a review, see Stryer et al. (1982).

Another application of the luminescence method involves quenching by chromophores whose absorption spectra overlap the emission spectrum of Tb^{3+} (Fig. 3-2). This has been exploited in certain Ca^{2+}-requiring zinc metalloproteinases, where Tb^{3+} has been substituted for Ca^{2+}, and Co^{2+} for Zn^{2+} (e.g., thermolysin, Section 4.3). The degree of quenching by Co^{2+} permits calculation of the distance between Co^{2+} and Tb^{3+} on the enzyme. This value should be close to the Ca–Zn separation in the native enzyme. In an analogous manner, it has proved possible to insert Tb^{3+} into one Ca^{2+}-binding site on a protein, and a quenching Ln^{3+} ion, such as Pr^{3+}, Nd^{3+}, Ho^{3+}, or Er^{3+}, into another. Calculations based upon the change in luminescence permit determination of the interlanthanide distance, which presumably reflects the intercalcium separation in the native protein. In a further modification of this technique, absorbing lanthanides have been employed to quench the fluorescence of proflavin, a competitive inhibitor of trypsin and chymotrypsin (Birnbaum *et al.*, 1977; Darnall *et al.*, 1976). The use of Pr^{3+}, Nd^{3+}, and Ho^{3+} in this way has permitted calculation of the distance separating the active site of the enzyme from its Ca^{2+}-binding site.

When the ligand environment is chiral, or contains chiral centers, the luminescence emitted by Tb^{3+} or Eu^{3+} may contain a circularly polarized component (Brittain *et al.*, 1976; Table 3-1). As Martin and Richardson (1979) pointed out, circularly polarized luminescence (CPL) is the emission analogue of circular dichroism (CD) in absorption spectra and is sensitive to details of geometry. CPL has been observed in Tb^{3+} complexes of carboxylates which contain asymmetric carbon atoms in the α- or β-positions (Luk and Richardson, 1975; Brittain and Richardson, 1977). In more than a quarter of 40 proteins screened by Brittain *et al.* (1976), the Tb^{3+} luminescence exhibited CPL. According to Richardson (1982), circularly polarized luminescence enhancement has proved the most sensitive probe of the interaction of Ln^{3+} ions with weak ligands, such as carbonyl, hydroxyl, amide, and amino groups. One example of this is the complex formed between Tb^{3+} and neutral sugars (Davis and Richardson, 1980; Section 5.4) and amino acids (Luk and Richardson, 1975; Section 4.2).

Good reviews of the use of Tb^{3+} and Eu^{3+} luminescence in biochemical studies have been published by Martin and Richardson (1979) and Richardson (1982).

3.4 Other Spectroscopic Methods

3.4.1 Spectrophotometry

Each of the lanthanides, apart from La^{3+} and Lu^{3+}, absorbs light through intraconfigurational f–f transitions (Fig. 2-1). Because the extinction coefficients are very low and because shielding of the $4f$ electrons minimizes spectral changes in response to the ligand environment (Section 2.2), spectrophotometric analysis using lanthanides has received relatively little attention.

One of the highest extinction coefficients in the visible region belongs to Nd^{3+} (Table 2-3), whose interactions with amino acids, bovine serum, glutamine synthetase, staphylococcal nuclease, albumin, trypsin, and trypsinogen have been monitored spectroscopically (Birnbaum *et al.*, 1970; Birnbaum and Darnall, 1973). As subtle changes in the spectrum of Nd^{3+} at 520 nm and 580 nm upon binding to bovine serum albumin (BSA) (Section 4.21; Fig. 4-12) were similar to those produced by standard complexes involving coordination to carboxyl groups, the authors concluded that binding to BSA was mediated through carboxyl groups. Ca^{2+} reversed these changes, indicating possible competition with Nd^{3+} for the same binding site(s) (Section 4.21). Changes in the absorption spectrum of Tb^{3+} upon binding to γ-carboxyglumate have also been investigated (Sperling *et al.*, 1978).

3.4.2 Electron Paramagnetic Resonance Spectroscopy

Gd^{3+} and Eu^{3+} are the only lanthanides which exhibit an electron paramagnetic resonance (EPR) spectrum at room temperature. EPR spectroscopy of lanthanides has not been widely employed in biochemical studies, although it has been used to monitor the interaction of Gd^{3+} with Ca^{2+}-ATPase, parvalbumin (Stephens and Grisham, 1979), immunoglobulin G (Dower *et al.*, 1975), and bovine serum albumin (Reuben, 1971). Considerable broadening of the EPR spectrum of Gd^{3+} occurs upon binding, but limited information has been forthcoming from this technique.

Lanthanides enhance the EPR signals from free radicals by relieving microwave power saturation, thus permitting the use of higher power for better signals (Sarna *et al.*, 1976). In other studies (Case *et al.*, 1976; Blum *et al.*, 1978), the effects of Gd^{3+} and Dy^{3+} on the EPR lines of iron within cytochromes have been measured to determine the location of redox centers within the mitochondrial membrane. Spin labelling of calmodulin (Section 4.4.3) permits the application of EPR to analysis of lanthanide binding (Buccigross and Nelson, 1986).

3.4.3 Mössbauer Spectroscopy

Europium-151, which has a natural abundance of 47.8%, can be used in conjunction with Mössbauer spectroscopy. Shifts are much larger for Eu^{2+} than Eu^{3+} and are sensitive to the local environment of the ion. One limitation to its biochemical use is the tendency of Eu^{2+} to oxidize

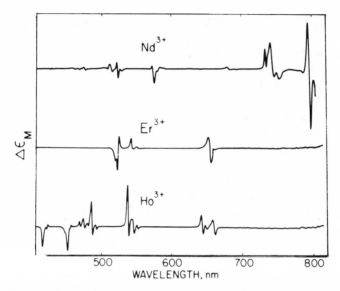

Figure 3-5. Magnetic circular dichroic spectra of the aquo ions of Nd^{3+}, Er^{3+}, and Ho^{3+}. From Holmquist and Vallee (1978), with permission.

under aerobic conditions. Mössbauer studies of Eu^{3+}-transferrin (Spartalian and Oosterhuis, 1973) and of tumor metabolism with ^{153}Sm were unrewarding (Friedman *et al.*, 1976).

3.4.4 Magnetic Circular Dichroism

Although the circular dichroism of lanthanides is weak, several exhibit strong magnetic circular dichroism (MCD). Their MCD spectra are sensitive to the local enviroment of the Ln^{3+} ion, and individual bands in the spectrum undergo changes in shape, sign, and intensity accordingly (Holmquist and Vallee, 1978). The MCD of Nd^{3+} at around 800 nm is particularly powerful (Fig. 3-5) and has been used to study the binding of Nd^{3+} to thermolysin (Section 4.3) (Holmquist and Horrocks, 1975) and parvalbumin (Section 4.4.1) (Donato and Martin, 1974). MCD spectral analysis of Eu^{2+}- and Pr^{3+}-substituted concanavalin A (Section 4.20) has also been reported (Homer and Mortimer, 1978). Parameters describing the MCD spectrum of Pr^{3+} have been derived by Göller-Walrand *et al.* (1980).

3.5 Luminescence Immunoassay and Enzyme Assay

Most immunoassays and enzyme assays employ radioactively labeled substrates. While these methods are accurate and sensitive, they may be expensive and entail concerns about safety and waste disposal. Luminescence labeling has been closely examined as an alternative to radiolabeling. Recently, the lanthanides have received attention in this regard. One of their major potential advantages is the relatively long lifetime of their emissions. This opens the possibility of using time-resolved luminescence to obviate problems due to background fluorescence. Whereas the lifetime of the background fluorescence of a diluted serum sample is typically 20–50 ns, emissions from excited Ln^{3+} ions usually have lifetimes in the millisecond range. The scientific problem to be overcome in such assays has been to attach the Ln^{3+} ion to the desired substrate or antigen in such a way that binding is strong enough to prevent dissociation, while its luminescence is maximized. Both Tb^{3+} and Eu^{3+} have been employed.

Hemmila *et al.* (1984) attached Eu^{3+} to sheep anti-rabbit IgG by means of either diazophenyl-EDTA or isothiocyanatophenyl-EDTA. This was then added to rabbit IgG in a solid-phase assay. Measurement of the amount of bound antibody was not directly possible as the protein–EDTA–Eu^{3+} adduct luminesced only weakly. It was thus necessary to elute the Eu^{3+} and to form a complex with β-diketone, whose luminescence was then mesured. The authors claimed that Eu^{3+} concentrations as low as $5 \times 10^{-14} M$ and IgG concentrations as low as 25 pg/ml could be detected by this method.

Such a method, although apparently accurate and quite sensitive, has drawbacks. According to Bailey *et al.* (1984), β-diketone chelates of Eu^{3+} rapidly lose their luminescent properties at concentrations below about $10^{-6} M$ and are quenched by phosphate. In addition, the need to elute and to measure separately the luminescence of Eu^{3+} is tedious. Various chelating substances have been synthesized which permit extremely strong attachment of Tb^{3+} to proteins, while permitting luminescence to be measured directly (Bailey *et al.*, 1984; 1985). The transferrin complex of Tb^{3+} has also been linked to gentamycin (Wilmott *et al.*, 1984). Human serum albumin has been labeled with Tb^{3+} through a reagent prepared from the *bis*-cyclic anhydride of DTPA (Chapter 2) and *p*-aminosalicylic acid (pAs). The protein–DTPA–pAs–Tb^{3+} complex had a luminescent lifetime of 1.56 s, appeared to be stable at low dilution, was not greatly affected by changes in pH, and was not quenched by phosphate. As little as 5 pmol of Tb^{3+}-labeled albumin could be detected, and successful immunoassay was reported. Labeling of specific antibody with lanthanides through covalent attachment of a chelator based on EDTA has been successfully used in the immunoassay of hepatitis B antigen (Siitari *et al.*, 1983), rubella antibodies (Meurman *et al.*, 1982), phospholipase A (Eskola *et al.*, 1983), and choriogonadotropin (Pettersson *et al.*, 1983). Either Eu^{3+} or Tb^{3+} could be employed in this way with the different emission maxima of Eu^{3+} and Tb^{3+} (Fig. 3-2) providing the prospect of double-labeling luminescence immunoassays. This approach has also been explored as a nonradioactive way to monitor DNA hybridization (Syvänen *et al.*, 1986).

The foregoing principles have been applied to the assay of certain proteinases (Karp *et al.*, 1983). Casein was coupled to Eu^{3+} via *p*-isothiocyanate-phenyl-EDTA, attached to a solid support, and used as a substrate for α-chymotrypsin, trypsin, or subtilisin. At the end of the reaction, Eu^{3+} was dissociated from the degradation products, converted to the β-diketone chelate, and its luminescence measured. Sensitivities down to 1 ng of proteinase were reported.

Time-resolved immunohistochemical staining of cells can be accomplished with Eu^{3+}-labelled antibodies (Soini *et al.*, 1988).

3.6 Cytochemical Detection Techniques

The insolubility of certain lanthanide salts has been used to improve the ultrastructural localization of several enzymes by cytochemistry. Cerium has become the lanthanide of choice in these techniques. The use of Ce^{3+} in cytochemistry was introduced by Briggs *et al.* (1975) in a new method for localizing NADH oxidase in human polymorphonuclear leukocytes. The reaction catalyzed by this enzyme produces H_2O_2, which forms an insoluble precipitate of cerium perhydroxide [$Ce(OH)_2OOH$ and $Ce(OH)_3OOH$] with Ce^{3+} ions. As cerium is of high electron density, these precipitates show up well by electron microscopy. They thus stain sites of H_2O_2 production in stimulated polymorphs. Cerium has also been used to locate the sites of action of other H_2O_2-producing enzymes, including *D*-amino acid oxidase (Robinson *et al.*, 1978), peroxisomal oxidases (Arnold *et al.*, 1979), and mitochondrial monoamine oxidase (Fujimoto *et al.*, 1982).

In an analogous manner, the insolubility of cerium phosphate has been exploited in the cytochemical localization of adenylate cyclase and a variety of phosphatases. Lead had previously been used in such assays. These suffered from the formation of nonspecific deposits, diffusion artifacts, enzyme inhibition, and, in some cases, nonenzymic breakdown of the substrate. Robinson and Karnovsky (1983a) developed an improved method for the ultrastructural localization of 5'-nucleotidase based upon

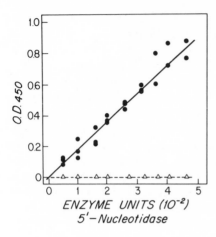

Figure 3-6. Turbidimetric assay of 5'-nucleo-
tidase by precipitation with Ce^{3+}. From
Robinson and Karnovsky (1983a), with permission. Experimental values: ●; control values: △.

Ce^{3+} instead of Pb^{2+} as the capturing agent. In a preliminary soluble
assay, they showed that Ce^{3+} precipitation of the phosphate released
from the hydrolysis of AMP by 5'-nucleotidase could function as a quantitative turbidimetric assay (Fig. 3-6). The possible effect of AMP–Ce
(Section 5.1; Table 5-2) complex formation upon the rate of reaction was
not discussed.

In a subsequent paper (Robinson and Karnovsky, 1983b), the method
was extended to a number of different polymorph phosphatases, including
acid phosphatase, alkaline phosphatase, glucose-6-phosphatase, inosine
diphosphatase, uridine diphosphatase, and thiamine pyrophosphatase. Ce^{3+}
has also been used to localize acid phosphatase, alkaline phosphatase,
glucose-6-phosphatase, and adenosine triphosphatase in rat liver (Hulstaert et al., 1983), thiamine pyrophosphatase in rat hepatocytes (Anger-
muller and Fahimi, 1984), acid phosphatase, alkaline phosphatase, glu-
cose-6-phosphatase, fructose-1,6-biphosphatase, and Na^+/K^+-ATPase in
yeast and other tissues (Veenhuis et al., 1980), and 5'-nucleotidase in
macrophages (Blok et al., 1982). Rechardt and Hervonen (1985) were able
to modify the method to permit the cytochemical detection of adenylate
cyclase in stimulated fat cells of rats.

While providing good cytochemical resolution, the Ce^{3+}-based methods had drawbacks, although these were not insurmountable. Certain
substrates were slowly hydrolyzed by Ce^{3+}. These included p-nitrophen-
ylphosphate, ATP, thiamine pyrophosphate, and UDP. However, rates
of nonenzymic hydrolysis were low in comparison with rates of enzymic
hydrolysis. Glucose-6-phosphate, β-glycerophosphate, and AMP were
stable (Section 2.6; Fig. 2-4). The poor penetration of tissues by Ce^{3+}

has been improved by permeabilization with detergent (Robinson, 1985).

Although cerium is a very good electron-dense marker for electron microscopy, deposits of cerium phosphate are not easily seen by light microscopy. This limitation has recently been removed with a technique where the cerium phosphate is converted to lead phosphate with lead citrate and a final, optically opaque precipitate of lead sulfide is formed by treatment with ammonium sulfide (Zimmerman and Halbhuber, 1985). Both acid and alkaline phosphatase have been localized by the lanthanum–lead method (Halbhuber and Zimmerman, 1986).

Lanthanum nitrate, used in conjunction with hydrogen peroxide, suppresses the nonspecific reduction of Ag^+ in certain histochemical enhancement procedures (Gallyas and Wolff, 1986).

3.7 Electron Microscopy

The use of lanthanides in electron microscopy (EM) dates back to the work of Calvet $et\,al.$ (1948), who used La^{3+} to precipitate and to stain the chromatin of resting lymphocytes. However, their more recent use as electron-dense stains in EM stems from the observation that lanthanum permits very good imaging of eukaryotic cell membranes (Doggenweiler and Frenk, 1965). An example is shown in Fig. 6-1. It also stains bacterial cell walls (Cassone and Garaci, 1974).

Two properties of La^{3+} make it especially useful in EM work. The first is its size. It has been pointed out that the lanthanides are among the smallest available electron-dense tracers. This permits the examination of structures which are otherwise difficult to study at the ultrastructural level. Such an advantage has really come into its own in the investigation of gap junctions. Gap junctions, or nexuses, occur where the plasma membranes of adjacent cells lie parallel to each other, with a gap of 20–40 Å separating the two. Such junctions may play a role in cell–cell communication, electrical coupling, and other activities of this kind. They are to be contrasted with tight junctions, which are regions of membrane fusion. La^{3+} ions can penetrate gap junctions, but not tight junctions, and are thus used to distinguish the two types of junction at the ultrastructural level. They have also been useful in distinguishing between incomplete tight junctions and complete, circumferential tight junctions. In addition to identifying such junctions, La^{3+} staining has permitted their characterization, demonstrating, in particular, the distinctive pattern of hexagonal subunits first reported by Revel and Karnovsky (1967). Where junctional permeability has been assessed both by La^{3+} staining and electrophysiologically, the results have been in good agreement (Tisher and

Yarger, 1975). The second property of importance to EM studies is the inability of La^{3+} ions normally to penetrate the surface of healthy cells. As discussed in Section 6.2, both physiological and EM studies agree with this conclusion. Where La^{3+} has been reported intracellularly, it is usually in endocytic vesicles or in damaged cells. In view of this, La^{3+} has been used as an ultrastructural probe for cell damage in, for example, ischemic heart cells (Chapter 6) and as a vital tracer of permeability *in vivo*.

The exact state of the La^{3+} in solutions used in EM work is unclear. Most such solutions are neutral or slightly alkaline, and thus uncomfortably close to the pH values at which precipitates of $La(OH)_3$ form (Section 2.3; Table 2-10). Indeed, some authors have reported a slight opalescence of these preparations, especially when a fixative such as osmium tetroxide is included. The possible formation of "radiocolloids" (Section 2.4) is unlikely, as the concentrations of La^{3+} used in staining are usually several millimolar. When solutions of La^{3+} are injected intravenously into living animals prior to fixation, the analysis of their chemical form becomes more complicated, as ill-defined interactions with the constituents of blood occur. These are discussed in Section 7.5.2. Although lanthanide deposits do not have sufficiently strong optical contrast for staining in conjunction with light microscopy, their use in fluorescence microscopy has been reported (Scaff *et al.*, 1969). Leif *et al.* (1977) described an automated blood-cell sorter which used Tb^{3+} and Eu^{3+} to label luminescently certain classes of lymphocyte.

The anatomical details revealed by lanthanum staining techniques are not pertinent to the main subject of this book. However, representative examples of the use of lanthanum as a stain are included in Tables 3-2 and 3-3. For a review of lanthanum staining, see Shaklai and Tavassoli (1982).

3.8 Separation and Purification Techniques

3.8.1 Methods Involving Extraction or Precipitation

Several decades ago, biochemists were interested in exploiting the precipitating properties of lanthanides as aids to the preparation and analysis of biological materials (e.g., Neuberg and Grauer, 1953). The literature of this period (e.g., Davidson and Waymouth, 1944; Chargaff *et al.*, 1949) contains references to the use of La^{3+} as precipitating agent in the purification of DNA and RNA. One advantage of this approach was that the La^{3+} salt of RNA resists the action of ribonuclease and can thus be extracted in a more complete form (Opie and Lavin, 1946). These methods

Table 3-2. Examples of the Perfusion of La into Living *In Vivo* or *Ex Vivo*
Tissues as an EM Marker of Permeability and the Extracellular Space

Tissue	Observations	Reference
Rodent pineal gland capillaries	Differences in permeability between rat, gerbil, and hamster	Heuring and Bergmann, 1985
Mouse neuromuscular junctions	Direct connections between the "T" tubules and subneural apparatus	Zacks and Saito, 1970
Blood/brain barrier of pigs and monkeys	Staining of extracellular space of choroid plexus, subependymal white matter, intracellular space of subependymal capillary	Milhorat *et al.*, 1975
Rat mesentery	No penetration of basement membranes by La^{3+}	König *et al.*, 1981
Rat intracerebral arterioles	Passage of La into arteriolar walls and extracellular compartment of surrounding brain. Accelerated passage of La in hypotensive animals.	Nag *et al.*, 1982
Cockroach central nervous system	La penetration of septate dermosomes and gap junctions	Lane and Treherne, 1972
Vascular permeability of rat incisor pulp	Most permeable capillaries lie between the odontoblasts	Bishop, 1985
Rodent enamel organs	Identification of gap junctions	Garant, 1972
Rat lens	Staining of extracellular space and intercellular gaps	Rae and Stacey, 1979
Mouse epidermis	Intercellular penetration	Elias and Friend, 1975
Rat brain	Permeation of the choroid plexus but not cerebral capillaries	Bouldin and Krigman, 1975
Cat brain	Penetration of choroid plexus	Castel *et al.*, 1974
Oral mucosa of rat palate	Intercellular staining confirmed by dispersive X-ray analysis	Squier and Edie, 1983
Rat brain	Hyperosmoticity opens up tight junctions	Dorovini-Zis *et al.*, 1983
Rat testis	La surrounded spermatogonia and spermatocytes. Excluded from tight junctions.	Russell, 1978

are no longer used. However, La^{3+} has been recently used as a precipitant in ribonuclease assays (Jung, 1986).

More recently, concentrated solutions of $LaCl_3$ have been used to extract proteoglycans from cartilage (Mason and Mayes, 1973; Roughley *et al.*, 1978). After 20-h incubation with a $0.5\,M$ solution, approximately 80% of the total hexuronate and hexose was extracted from slices of bovine nasal cartilage. The extraction occurred through dissociation of

Table 3-3. Examples of the Use of La as a Stain of Fixed Tissue in EM

Tissue	Observations	Reference
Mouse heart and liver	Hexagonal arrays of subunits in intercellular junctions	Revel and Karnovsky, 1967
Rat bone marrow	La^{3+} remained extracellular apart from endocytic vesicles and a novel type of stromal cell	Tavassoli *et al.*, 1980
Rat mandibular cartilage	Staining of chondrocyte membranes and extracellular proteoglycans	Oi and Utsumi, 1980
Chick embryonic heart, retina, limb bud	Staining of pericellular layer, removed by phospholipase C, but not trypsin, pronase, neuraminidase, α-amylase	Lesseps, 1967
Rat liver, kidney, small intestine	Surface staining, enhanced by alcian blue or cetylpyridinium chloride	Shea, 1971
HeLa cells	Surface staining, removed by neuraminidase	Boyd *et al.*, 1972
Isolated lymphocyte	Surface staining, removed by α-glucosidase, but not sulfatase, sialidase, β-glucuronidase	Anteunis and Vial, 1975; Bona and Anteunis, 1973
Isolated tobacco protoplasts	Stains plasma membranes	Taylor and Hall, 1979
Rat molar tooth germ	La used to assess permeability of ameloblast layer	Matsuo *et al.*, 1984
Chick cartilage and reaggregating chondrocytes	Staining of atrix proteoglycan	Khan and Overton, 1970
Human gingival epithelium	Gap junctions identified	Barnett and Szabo, 1973
Human epidermis	Penetration of basal lamina, intercellular spaces. Identification of gap junctions	Hashimoto, 1971
Crayfish axon	No penetration of axons. Heavy staining of membrane	Payton *et al.*, 1969
Lizard ovaries	Demonstration of an extracellular link between blood and follicular epithelium	Neaves, 1972

the high-molecular-weight proteoglycan aggregates into their monomeric units, which were free to diffuse from the tissue. Proteoglycans extracted in this way could be precipitated by reducing the concentration of La^{3+} through dialysis or dilution. A variation of this method has been used to extract hyaluronic acid from skin (Munakata and Yosizawa, 1980). Selective fractionation by precipitation with La^{3+} has also been used to monitor the degradation of proteoglycans (Doganges and Schubert, 1964).

3.8.2 Lanthanide Enhanced Affinity Chromatography

Certain enzymes which require Ca^{2+} for substrate attachment are inhibited by Ln^{3+} ions through the formation of an abortive enzyme–Ln^{3+}–substrate ternary complex. In such complexes a Ln^{3+} ion competitively replaces Ca^{2+}, promotes interaction between the enzyme and its substrate, but inhibits the formation of reaction products. Despite inhibition of the enzymic reaction, the affinity of the enzyme for its substrate is usually greater with Ln^{3+} than with Ca^{2+}. Such properties have led to the use of lanthanides as adjuncts to the purification of certain enzymes by affinity chromatography.

The feasibility of this approach was demonstrated with staphylococcal nuclease (Furie *et al.*, 1973) (Section 4.14). Deoxythymidine diphosphate, a substrate analogue, was covalently attached to Sepharose and used as the affinity ligand. In the presence of 1 mM or 5 mM $NdCl_3$, 93% of the nuclease bound to the column and could be eluted subsequently with EDTA. The first application of this approach to enzyme purification resulted from the studies of Furie and Furie (1975). Activation of factor X by snake venom anticoagulant protein was inhibited by several lanthanides through the formation of a nonproductive enzyme–factor X–Ln^{3+} ternary complex (Section 4.5.2). To apply this finding to the purification of the enzyme, factor X was covalently attached to Sepharose and used in an affinity column. Crude snake venom was applied to the column in the presence of 10 mM $NdCl_3$. Because of the high affinity of the activating enzyme for its substrate in the presence of Nd^{3+}, 0.5 M NaCl could be included to minimize nonspecific binding. A highly purified preparation of the activating enzyme could be recovered by eluting with 10 mM EDTA. A yield of 40% was obtained with Nd^{3+}, Tb^{3+}, Gd^{3+}, or La^{3+} as the lanthanide cofactor.

Clostridial collagenase is also amenable to this purification procedure. Its inhibition by Ln^{3+} ions was noted to be uncompetitive with regard to the substrate, but competitive with regard to Ca^{2+} (Evans, 1981; Section 4.11.1). Collagen was attached to polyacrylamide and used as the affinity ligand (Evans, 1985). Crude collagenase was applied to the column in the presence of 400 μM Sm^{3+}, and a purified fraction of high specific activity was eluted with 1 mM EGTA. Although Sm^{3+}, Er^{3+}, and Lu^{3+} could be used in this procedure, La^{3+} was a much poorer cofactor, a property reflected in its weaker ability to inhibit collagenase.

3.8.3 Magnetic Separation Techniques

As observed in Section 2.2, several of the lanthanides have high paramagnetic susceptibilities (Table 2-2). Westcott and his co-workers (Bowen and Westcott, 1980; Evans and Tew, 1981; Russell *et al.*, 1983) have demonstrated that certain materials sequester sufficient Er^{3+} to attain susceptibilities permitting their magnetic capture. This finding has opened the door to improved magnetic separation techniques based upon differential affinities for Er^{3+} ions.

These techniques are presently being applied to the magnetic recovery, separation and isolation of particles. Such an approach has already been tested in the ferrographic analysis of synovial fluids, where the wear particles of human joints are magnetized with Er^{3+} and magnetically arranged along a glass substrate (Evans *et al.*, 1980; 1982). Various types of bacterial and eukaryotic cell have also been recovered in this way (Russell *et al.*, 1983). More recently, the feasibility of resolving mixtures of Er^{3+}-treated particles by "paramagnetic chromatography" has been demonstrated (Evans *et al.*, 1986).

One of the most useful applications of this approach is thought to be in the separation of discrete populations of eukaryotic cells from heterologous mixtures. This should provide a useful alternative to traditional methods of cell separation, as the discriminating factor will be the affinities of different cells for Er^{3+} ions. Since these affinities reflect cellular chemistries, the discriminating parameters are chemical. In most other cell separation techniques, they are physical or physiological. In living cells, which exclude Er^{3+}, the important determinant will be cell surface chemistry. In dead cells, which are permeable to Er^{3+}, parameters such as RNA content and ploidy may be important. One technical complication in these studies is the low magnetic susceptibilities of the particles. Using the isomagnetic method, Russell *et al.* (1987) have calculated the susceptibilities shown in Table 3-4. The use of a high-gradient magnetic separator in this work is presently being investigated.

This field has been reviewed recently by Evans *et al.* (1989).

3.9 Cellular Ca^{2+} Fluxes: The Lanthanum Method

As discussed in Chapter 6, the measurement of Ca^{2+} fluxes is of crucial importance to studies of stimulus-coupled cellular events. Such measurements are complicated. These events occur very rapidly and are thus difficult to monitor kinetically. In addition, only a small fraction of the cellular Ca^{2+} is in the rapidly exchanging pool, while a sizable pro-

Table 3-4. Volumetric Susceptibility of Tissue Measured by the Isomagnetic Method[a]

Tissue	Magnetizing agent	Susceptibility (SI $\times$ 10^{-6})
Yeast (*S. cerevisiae*)	0.05 M ErCl$_3$	20 ± 2
Human colonic carcinoma, fixed	0.01 M ErCl$_3$	13 ± 2.5
Rat liver, fixed	0.05 M ErCl$_3$	13.5
Bovine cartilage	0.05 M ErCl$_3$	50–125
Bovine subchondral bone	0.05 M ErCl$_3$	560–580

[a] From Russell *et al.* (1987).

portion of the exchangeable Ca^{2+} associated with cells is on the external cell surface. The lanthanum method was introduced by Van Breeman and McNaughton (1970) in an attempt to resolve these difficulties.

Four assumed properties of La^{3+} led to the development of this method. The first was the ability of La^{3+} to displace superficially bound Ca^{2+}. Thus, in flux experiments using $^{45}Ca^{2+}$, La^{3+} should be able to remove the externally sequestered $^{45}Ca^{2+}$ and permit measurement of the $^{45}Ca^{2+}$ which has been genuinely internalized by the cell. The next two important assumptions of the lanthanum method are the ability of La^{3+} to block completely both the influx and efflux of Ca^{2+}. Thus, by adding La^{3+}, it should be possible to stop Ca^{2+} exchange at will, thereby facilitating its accurate measurement. Lastly, it was assumed that La^{3+} has the important property of not entering living cells. It would thus not displace Ca^{2+} from intracellular sites and complicate the interpretation.

Much attention has been devoted to this method. It has received wide use, but considerable criticism, most of which has attacked the assumption that La^{3+} completely inhibits Ca^{2+} efflux. The first and second assumptions, that La^{3+} displaces surface Ca^{2+} and blocks Ca^{2+} uptake, have been well substantiated experimentally (Section 6.3). The last assumption, that La^{3+} does not penetrate living cells, is also well supported experimentally, although this is challenged from time to time (e.g., Hodgson *et al.*, 1972, discussed in Section 6.2).

The original application of the lanthanum method was to the measurement of the uptake of Ca^{2+} by smooth muscle. In these studies a LaCl$_3$ concentration of 2 mM was used. Realization that Ca^{2+} efflux was incompletely inhibited under these conditions (e.g., Burton and Godfraind, 1974; Karaki and Weiss, 1979) led to the La^{3+} concentration being increased firstly to 10 mM and then to 80.8 mM; it was also found to be necessary to lower the temperature to 0.5°C (Karaki and Weiss, 1979).

Under these conditions, satisfactory measurements were obtained with smooth muscle, even though Ca^{2+} efflux was still not completely eliminated. Similar results were subsequently obtained for renal arteries and veins (Karaki et al., 1980; Hester and Weiss, 1981). The lanthanum method has also been applied to the study of Ca^{2+} fluxes in nonexcitable tissues such as pancreatic islets (Hellman, 1978) and erythrocytes (Swislocki et al., 1983). In a more general way, La^{3+} has been employed to discriminate between intracellular and extracellular sources of "trigger" Ca^{2+}, and as a monitor of the integrity of cellular membranes.

3.10 Other Methods

3.10.1 X-Ray Diffraction

Lanthanides have been employed as X-ray heavy atoms in crystallographic studies of a number of proteins and tRNA. Sm^{3+}, Nd^{3+}, and Eu^{3+} produce especially large scattering and are thus particularly useful, especially in phase determination. The results of such investigations are discussed in Chapters 4 and 5. In several proteins, such as thermolysin (Section 4.3), the Ln^{3+} ions occupy Ca^{2+}-binding sites, while in enzymes such as lysozyme (Section 7.4), which do not require Ca^{2+}, the significance of the site is unclear. In tRNA (Section 5.3), Ln^{3+} ions are thought to occupy Mg^{2+} sites. Insertion of a Ln^{3+} ion into these molecules has generally been nondisruptive, supporting the view that Ln^{3+} ions can replace Ca^{2+}, and certain other metal ions, isomorphously. This conclusion helps justify the use of Ln^{3+} ions as replacement probes with which to study the interactions of metal ions with biological materials in a manner which retains biochemical fidelity.

Colman et al. (1972) have noted that the large scattering of thermal neutrons by ^{149}Sm and ^{157}Gd makes them the best-suited elements for neutron diffraction studies of macromolecules.

3.10.2 DNA Analysis

In Section 5.3 we shall discuss the way in which Tb^{3+}-DNA complexes luminesce when excited at appropriate wavelengths. Deoxyguanosyl residues appear to be the predominant energy donors. In addition, luminescence is much greater for single-stranded DNA. Both of these characteristics can be used analytically.

The large enhancement in Tb^{3+} luminescence that accompanies unwinding of the helix can reveal subtle conformational changes that are otherwise hard to detect. Another application of the phenomenon is in the measurement of DNA reannealing, where cot curves can be constructed by measuring changes in Tb^{3+} luminescence (Section 5.3; Fig. 5-5). Furthermore, luminescent staining with Tb^{3+} reveals bands of single-stranded DNA on agarose gels after electrophoresis. Paradoxically, with RNA it is the double-stranded species that Tb^{3+} detects on gels (Al-Hakeen and Sommer, 1987). In a more general way, Tb^{3+} can be used to measure the single-strandedness of any given preparation of DNA, to detect damage produced by carcinogens or changes resulting from the binding of drugs.

As deoxyguanosyl residues are the predominant chromophore for Tb^{3+} luminescence, it has been suggested that Tb^{3+} can be used to measure the G + C content of DNA (Cavatorta *et al.*, 1980). When used in conjunction with the dye diamidino-2-phenylindole, which is specific for A–T residues, the authors claimed accurate assessment of both the G + C and A + T content of calf thymus DNA.

Lanthanides have been used in the liquid chromatographic detection of polynucleotides and nucleic acids (Wenzel and Collette, 1988).

3.10.3 Miscellaneous Methods

When heated, Ln^{3+} can cleave the phosphodiester linkages of RNA and several nucleotides (Section 2.6). However, thioester and O-ester linkages are not cleaved. These properties have been exploited by Jackson and Voorheis (1985) to help identify the link between the protein and phospholipid components of a trypanosomal surface coat constituent as a phosphodiester bond.

The heavier lanthanides accumulate in the bones of experimental animals (Section 7.5). Thomasset *et al.* (1976) examined the possibility of labeling the bones of growing rats with $^{177}Lu^{3+}$ and then monitoring the subsequent release of $^{177}Lu^{3+}$ as an index of osteolysis. This method showed promise but does not appear to have been followed up.

Pita *et al.* (1978) used La^{3+} to displace Ca^{2+} from sequestering agents in nanoliter volumes of physiological fluids, thus permitting accurate measurement of their total Ca^{2+} contents. Good results were reported, despite using a method in which $LaCl_3$ was apparently dissolved in phosphate buffer. Cerium(IV) sulfate has been used to determine tannin (Kapel and Karunanithy, 1974), and the Russian literature, according to the review by Ellis (1977), contains references to the use of Ce^{3+} in measuring fluoride and glucose concentrations.

La^{3+} is used experimentally to dissociate calmodulin (Section 4.4.3) from the enzymes that it helps modulate, and from membranes (e.g., Gross *et al.*, 1987).

3.11 Summary

Much of the interest in the biochemistry of the lanthanides stems from their valuable experimental uses. Many, but not all, of these involve the ability of the lanthanides to serve as informative substituents for several physiologically important metal ions. By far the most prominent of these is Ca^{2+}, which is spectroscopically silent yet chemically similar to Ln^{3+} (Section 2.5; Table 2-11). The spectral properties of the lanthanides then permit structural and functional analysis of the metal–ligand interaction. NMR spectroscopy and luminescence spectroscopy have proved especially valuable in this context.

The introduction of a paramagnetic Ln^{3+} ion perturbs the magnetic environment around adjacent, resonating nuclei, thereby altering the NMR spectrum. Paramagnetic lanthanides influence NMR spectra in two ways: they may shift peaks in the upfield or downfield direction, or they may broaden peaks. Pr^{3+} and Eu^{3+} give appreciable shifts with negligible broadening, whereas Dy^{3+}, Ho^{3+}, and Yb^{3+} produce very large shifts with some degree of line broadening. Gd^{3+}, on the other hand, has a strong broadening effect but causes little or no shifting. Titrimetric analysis of these disturbances permits determination of quantitative binding parameters. In addition, if peak assignments are known, those nuclei close to the Ln^{3+}-binding site can be identified. Conformational analysis is also possible. The degree of shifting of a resonance is ideally a function of the distance of the resonating nucleus from the paramagnetic Ln^{3+} ion and their angular separation. With Gd^{3+}, the degree of broadening has no angular component but is dependent upon the sixth power of the distance of the Gd^{3+} ion from the affected nucleus. Quantitative analysis of the angles and distances thus obtained provides conformational information. Although the Ln^{3+}-NMR technique alone cannot provide the complete conformational description of a molecule, it remains a valuable adjunct to independent methods of conformational analysis. Examples are provided by nucleotides (Section 5.1) and lysozyme (Section 4.7). NMR spectroscopy has also been used to study the interactions of lanthanides with model membranes (Section 6.2) and the transport of Ln^{3+} ions and other substances into phospholipid vesicles (Section 6.2).

Much use has been made of luminescence spectroscopy in conjunction with Tb^{3+} or Eu^{3+} ions. These lanthanides may be excited directly with a pulsed laser or indirectly via suitable chromophores. Examination of the excitation spectrum is informative. With direct excitation, each peak in the excitation spectrum represents a discrete class of binding environment. With indirect excitation, the position of the peak often permits identification of the donor chromophore. Because the efficiency of energy transfer from the donor chromophore to the Ln^{3+} ion depends upon the distance between them, the Ln^{3+}–chromophore separation can be calculated. This principle still applies when the chromophore lies buried in, for example, an organelle. In such cases, a soluble chelate of the Ln^{3+} ion is introduced and indirectly excited at the wavelength of maximum absorption of the chromophore. By this process of "diffusion-enhanced fluorescent energy transfer," the depth to which the chromophore is buried in its parent structure can be determined. The luminescent emissions can also be quenched by absorptive moieties in the vicinity of the lanthanide. This phenomenon can also be analyzed quantitatively to obtain the distance between the Ln^{3+} ion and the quenching moiety. Such an approach has been especially useful in determining the distances between different metal-binding sites on the same protein. Comparison of the lifetime of the luminescent emission in H_2O and D_2O permits calculation of the number of water molecules coordinated to the Ln^{3+} ion. Additional information may be obtained if the emission is circularly polarized.

Spectrophotometry, magnetic circular dichroism spectroscopy, electron paramagnetic resonance spectroscopy, and Mössbauer spectroscopy have been relatively little used in studies of lanthanide biochemistry.

In addition to structural studies, lanthanides can also be used in functional studies. Thus, the alteration in kinetic parameters of a Ca^{2+}-requiring enzyme when substituted by a Ln^{3+} ion can provide information on the role of Ca^{2+} in the native reaction (Chapter 4). Aiding all these sorts of biochemical investigation is the gradual nature of the changes in most chemical properties that occur across the lanthanide series (Chapter 2). This permits nonparamagnetic Ln^{3+} ions to be used safely as controls in NMR studies employing paramagnetic members, for instance, and allows the spatial restriction upon Ln^{3+} sites to be investigated by exploiting the lanthanide contraction.

As well as providing information about isolated biochemicals *in vitro,* lanthanides have been extensively employed in studies of cellular metabolism (Chapter 6). These investigations rest upon the important discovery that Ln^{3+} ions can block Ca^{2+} channels on cell surfaces without themselves entering the cell. As the transmembrane transport of Ca^{2+} is a crucial event in many stimulus-coupled cellular responses, lanthanides have been widely applied to investigations of nerves, muscles, and hormone-responsive tissues. As part of this, the "lanthanum method" has been developed to aid the measurement of Ca^{2+} fluxes. The realization that Ln^{3+} ions are generally restricted to the extracellular space has also led to an extensive literature on their use as electron-dense markers in transmission electron microscopy, particularly in identifying "gap junctions" which are otherwise hard to see.

Lanthanides have also been employed as aids to the separation, preparation, and purification of molecules and cells, in immunoassays, in histocytochemistry, and in various types of enzyme assay. The medical uses of lanthanides in imaging are discussed in Chapter 9.

References

Al-Hakeen, M., and Sommer, S. S., 1987. Terbium identifies double-stranded RNA on gels by quenching the fluorescence of intercalated ethidium bromide, *Anal. Biochem.* 163: 433–439.

Angermüller, S., and Fahimi, H., 1984. A new cerium-based method for cytochemical localization of thiamine pyrophosphatase in the Golgi complex of rat hepatocytes. Comparison with the lead technique, *Histochemistry* 80:107–111.

Anteunis, A., and Vial, M., 1975. Cytochemical and ultrastructural studies concerning the cell coat glycoproteins in normal and transformed human blood lymphocytes. II. Comparison of lanthanum-retaining cell coat components in T and B lymphocytes transformed by various kinds of stimulating agents, *Exp. Cell Res.* 90:47–55.

Arnold, G., Liscum, L., and Holtzman, E., 1979. Ultrastructural localization of D-amino acid oxidase in microperoxisomes of rat nervous system, *J. Histochem. Cytochem.* 27: 735–745.

Bailey, M. P., Rooks, B. F., and Riley, C., 1984. Terbium chelate for use as a label in fluorescent immunoassays, *Analyst* 109:1449–1450.

Bailey, M. P., Rooks, B. F., and Riley, C., 1985. Terbium chelates for fluorescence immunoassay, *Analyst* 110:603–604.

Balschi, J., Cirillo, V., and Springer, C. S., 1982. Direct high resolution nuclear magnetic resonance studies of cation transport *in vivo*: Na^+ transport in yeast cells, *Biophys. J.* 38:323–326.

Barnett, M. L., and Szabo, G., 1973. Gap junctions in human gingival keratinized epithelium, *J. Peridontal Res.* 8:117–126.

Barry, C. D., North, A. C. T., Glasel, J. A., Williams, R. J. P., and Xavier, A. V., 1971. Quantitative determination of mononucleotide conformations in solution using lanthanide ion shift and broadening NMR probes, *Nature* 232:236–245.

Barry, C. D., Glasel, J. A., Williams, R. J. P., and Xavier, A. V., 1974. Quantitative determination of conformations of flexible molecules in solution using lanthanide ions as nuclear magnetic resonance probes: application to adenosine-5-monophosphate, *J. Mol. Biol.* 84:471–490.

Birnbaum, E. R., and Darnall, D. W., 1973. Carboxylic and amino acid complexes of neodymium(III) by difference absorption spectroscopy, *Bioinorg. Chem.* 3:15–22.

Birnbaum, E. R., Gomez, J. E., and Darnall, D. W., 1970. Rare earth metal ions as probes of electrostatic binding sites in proteins, *J. Am. Chem. Soc.* 92:5287–5288.

Birnbaum, E. R., Abbott, F., Gomez, J. E., and Darnall, D. W., 1977. The calcium ion binding site on bovine chymotrypsin A, *Arch. Biochem. Biophys.* 179:469–476.

Bishop, M. A., 1985. Vascular permeability to lanthanum in the rat incisor pulp. Comparison with endoneurial vessels in the inferior alveolar nerve, *Cell Tissue Res.* 239:131–136.

Bleaney, B., 1972. Nuclear magnetic resonance shifts in solution due to lanthanide ions, *J. Magn. Reson.* 8:91–100.

Blok, J., Orderwater, J. J. M., de Water, R., and Ginsel, L. A., 1982. A cytochemical method for the demonstration of 5′nucleotidase in mouse peritoneal macrophages, with cerium ions using a trapping agent, *Histochemistry* 75:437–443.

Blum, H., Leigh, J. S., Salerno, J. C., and Ohnishi, T., 1978. Orientation of bovine adrenal cortex cytochrome P-450 in submitochondrial particle multilayers, *Arch. Biochem. Biophys.* 187:153–157.

Bona, C., and Anteunis, A., 1973. Structure of the lymphocyte membrane. IV. Cell coat of lymphocytes obtained from various lymphoid organs in chickens and mice, *Ann. Pasteur* 1246:321–327.

Bouldin, T. W., and Krigman, M. R., 1975. Differential permeability of cerebral capillary and choroid plexus to lanthanum ion, *Brain Res.* 99:444–448.

Bowen, J. P., and Westcott, V. C., 1980. Magnetic techniques for separating non-magnetic material, U.S. Patent 4,187,170.

Boyd, K., Melnykovych, G., and Fiskin, A. M., 1972. Lanthanum as a stain for sialic acid residues in replicas of HeLa cell surfaces, *J. Cell Biol.* 55:25a.

Bradbury, J. H., Crompton, M. W., and Warren, B., 1974. Determination of the sequence of peptides by NMR spectroscopy, *Anal. Biochem.* 62:310–316.

Briggs, R. T., Drath, D. B., Karnovsky, M. L., and Karnovsky, M. J., 1975. Localization of NADH oxidase on the surface of human polymorphonuclear leukocytes by a new cytochemical method, *J. Cell Biol.* 67:566–586.

Brindle, K. M., Brown, F. F., Campbell, I. D., Grathwohl, C., and Kuchel, P. W., 1979. Application of spin-echo nuclear magnetic resonance to whole cell systems, *Biochem. J.* 180:37–44.

Brittain, H. G., and Richardson, F. S., 1977. Circularly polarized emission studies on Tb^{3+}: and Eu^{3+}:complexes with potentially terdentate amino acids in aqueous solution, *Bioinorg. Chem.* 7:233–243.

Brittain, H. G., Richardson, F. S., and Martin, R. B., 1976. Terbium(III) emission as a probe of calcium(II) binding sites in protein, *J. Am. Chem. Soc.* 98:8255–8260.

Buccigross, J. M., and Nelson, D. J., 1986. EPR studies show that lanthanides do not have the same order of binding to calmodulin, *Biochem. Biophys. Res. Commun.* 138:1243–1249.

Burton, J., and Godfraind, T., 1974. Sodium-Calcium sites in smooth muscle and their accessibility to lanthanum, *J. Physiol.* 241:287–298.

Bystrov, V. F., Dubrovina, N. I., Barsukov, L. I., and Bergelson, L. D., 1971. NMR differentiation of the internal and external phospholipid membrane surfaces using paramagnetic Mn^{2+} and Eu^{3+} ions, *Chem. Phys. Lipids* 6:343–350.

Calvet, F., Siegel, B. M., and Stern, K. G., 1948. Electron optical observations on chromosome structure in resting cells, *Nature* 162:305–306.

Case, G. D., Ohnishi, T., and Leigh, J. S., 1976. Intramitochondrial positions of ubiquinone and iron-sulphur centers determined by dipolar interactions with paramagnetic ions, *Biochem. J.* 160:785–795.

Cassone, A., and Garaci, E., 1974. Lanthanum staining of the intermediate region of the cell wall in *Escherichia coli, Experienta* 30:1230–1232.

Castel, M., Sahar, A., and Erlij, D., 1974. The movement of lanthanum across diffusion barriers in the choroid plexus of the cat, *Brain Res.* 67:178–184.

Cavatorta, P., Barcellona, M. L., Avitabile, M., Von Berger, J., and Masotti, L., 1980.

Terbium as a new probe to detect G + C content of DNA, *Biochem. Exp. Biol.* 16: 365–369.

Chargaff, E., Vischer, E., Doninger, R., Green, C., and Misani, F., 1949. The composition of the desoxypentose nucleic acids of thymus and spleen, *J. Biol. Chem.* 177:405–416.

Colman, P. M., Weaver, L. H., and Matthews, B. W., 1972. Rare earths as isomorphous calcium replacements for protein crystallography, *Biochem. Biophys. Res. Commun.* 46:1999–2005.

Darnall, D. W., Abbott, F., Gomez, J. E., and Birnbaum, E. R., 1976. Fluorescent energy transfer measurements between the calcium binding site and the specificity pocket of bovine trypsin using lanthanide probes, *Biochemistry* 15:5017–5023.

Davidson, J. N., and Waymouth, C., 1944. Tissue nucleic acids: ribonucleic acids and nucleotides in embryonic and adult tissues, *Biochem. J.* 38:39–50.

Davis, S. A., and Richardson, F. S., 1980. Circularly polarized luminescence induced by terbium–nucleoside interactions in aqueous solution, *J. Inorg. Nucl. Chem.* 42:1973–1975.

DeJersey, J., Morley, P. J., and Martin, R. B., 1981. Lanthanide probes in biological systems: characterization of luminescent excitation spectra of terbium complexes with proteins, *Biophys. Chem.* 13:233–243.

Dimicoli, J. L., and Bieth, J., 1977. Location of the calcium ion binding site in porcine pancreatic elastase using a lanthanide ion probe, *Biochemistry* 16:5532–5537.

Dobson, C. M., Williams, R. J. P., and Xavier, A. V., 1973. Separation of contact and pseudocontact contributions of shifts induced by lanthanide (III) ions in nuclear magnetic resonance spectra, *J. Chem. Soc., Dalton Trans.* 1973:2662–2664.

Doganges, P. T., and Schubert, M., 1964. The use of lanthanum to study degradation of a proteinpolysaccharide from cartilage, *J. Biol. Chem.* 239:1498–1503.

Doggenweiler, C. F., and Frank, S., 1965. Staining properties of lanthanum on cell membranes, *Proc. Natl. Acad. Sci. USA* 53:425–430.

Donato, H., and Martin, R. B., 1974. Conformations of carp muscle calcium binding parvalbumin, *Biochemistry* 13:4575–4579.

Dorovini-Zis, K., Sato, M., Goping, G., Rapoport, S., and Brightman, M., 1983. Ionic lanthanum passage across cerebral endothelium exposed to hyperosmotic arabinose, *Acta Neuropathol.* 60:49–60.

Dower, S. K., Dwek, R. A., McLaughlin, A. C., Mole, L. E., Press, E. M., and Sunderland, C. A., 1975. The binding of lanthanides to non-immune rabbit immunoglobulin G and its fragments, *Biochem. J.* 149:73–82.

Elias, P. M., and Friend, D. S, 1975. The permeability barrier in mammalian epidermis, *J. Cell Biol.* 65:180–191.

Ellis, K. J., 1977. The lanthanide elements in biochemistry, biology and medicine, *Inorg. Perspect. Biol. Med.* 1:101–135.

Eskola, J. V., Nevalainen, T. J., and Lovgren, T. N-E., 1983. Time-resolved fluoroimmunoassay of human pancreatic phospholipase A_2, *Clin. Chem.* 29:1777–1780.

Evans, C. H., 1981. Interactions of tervalent lanthanide ions with bacterial collagenase (clostridiopeptidase A), *Biochem. J.* 195:677–684.

Evans, C. H., 1985. The lanthanide enhanced affinity chromatography of clostridial collagenase, *Biochem. J.* 225:553–556.

Evans, C. H., and Tew, W. P., 1981. Isolation of biological materials by use of erbium(III)-induced magnetic susceptibilities, *Science* 213:653–654.

Evans, C. H., Bowen, E. R., Bowen, J., Tew, W. P., and Westcott, V. C., 1980. Synovial fluid analysis by ferrography, *J. Biochem. Biophys. Meth.* 2:11–18.

Evans, C. H., Mears, D. C., and Stanitski, C. L., 1982. Ferrographic analysis of wear in human joints: evaluation by comparison with arthroscopic examination of symptomatic knees, *J. Bone Joint Surg.* 64B:572–578.

Evans, C. H., Russell, A. P., and Westcott, V. C., 1986. Demonstration of the principle of paramagnetic chromatography for resolving mixtures of particles, *J. Chromatogr.* 351: 409–415.

Evans, C. H., Russell, A. P., and Westcott, V. C., 1989. Approaches to paramagnetic separations in biology and medicine, *Partic. Sci. Technol.* (in press).

Förster, T., 1959. Transfer mechanism of electronic excitation, *Disc. Faraday Soc.* 27:7–17.

Friedman, A. M., Sullivan, J. C., Ruby, S. L., Lindenbaum, A., Russell, J. J., Zabransky, B. S., and Raywolu, G. U., 1976. Studies on tumor metabolism. I: By use of Mössbauer spectroscopy and autoradiography of ^{153}Sm, *J. Nucl. Med. Biol.* 3:37–40.

Fujimoto, T., Inomata, K., and Ogawa, K., 1982. A cerium method for the ultrastructural localization of monoamine oxidase activity, *Histochem. J.* 14:87–98.

Furie, B. C., and Furie, B., 1975. Interaction of lanthanide ions with bovine factor X and their use in the affinity chromatography of the venom coagulant protein of *Vipera russelli*, *J. Biol. Chem.* 250:601–608.

Furie, B., Eastlake, A., Schechter, A. N., and Anfinsen, C. B., 1973. The interaction of the lanthanide ions with staphylococcal nuclease, *J. Biol. Chem.* 248:5821–5825.

Gallyas, F., and Wolff, J. R., 1986. Metal catalyzed oxidation renders silver intensification selective. Application for the histochemistry of diaminobenzidine and neurofibrillary changes, *J. Histochem. Cytochem.* 34:1667–1672.

Garant, P. R., 1972. The demonstration of complex gap junctions between the cells of the enamel organ with lanthanum nitrate, *J. Ultrastruct. Res.* 40:333–348.

Geraldes, C. F. G. C., and Williams, R. J. P., 1978. Nucleotide torsional flexibility in solution and the use of the lanthanides as nuclear-magnetic-resonance conformational probes, *Eur. J. Biochem.* 85:463–470.

Glasel, J. A., 1973. Lanthanide ions as nuclear magnetic resonance chemical shift probes in biological systems, in *Current Research Topics in Bioinorganic Chemistry* (S. J. Lippard, ed.), Vol. 18, Wiley, New York, pp. 383–413.

Göller-Walrand, C., DeMoitie-Neyt, N., and Beyers, Y., 1980. Parametrisation of the MCD-spectra of rare earths: example of Pr^{3+} in PVA-matrix, in *The Rare Earths in Modern Science and Technology* (G. J. McCarthy, H. B. Silber, and J. J. Rhyne, eds.), Vol. 3, Plenum Press, New York, pp. 109–114.

Gross, M. K., Toscano, D. G., and Toscano, W. A., 1987. Calmodulin-mediated adenylate cyclase from mammalian sperm, *J. Biol. Chem.* 262:8672–8676.

Gupta, R. K., and Gupta, P., 1982. Direct observations of resolved resonances from intra- and extra-cellular sodium-23 ions in N.M.R. studies of intact cells and tissues using dysprosium(III) tripolyphosphate as paramagnetic shift reagent, *J. Magn. Reson.* 47: 344–350.

Halbhuber, K. J., and Zimmerman, N., 1986. Light microscopical localization of acid and alkaline phosphatase activity by lanthanum–lead–(La-Pb)-methods, *Acta Histochem.* 80:35–40.

Hashimoto, K., 1971. Intercellular spaces of the human epidermis as demonstrated with lanthanum, *J. Invest. Derm.* 57:17–31.

Heller, A., and Wasserman, E., 1965. Intermolecular energy transfer from excited organic compounds to rare-earth ions in dilute solutions, *J. Chem. Phys.* 42:949–955.

Hellman, B., 1978. Calcium and pancreatic β-cell function. 3. Validity of the La^{3+}-wash technique for discriminating between superficial and intracellular $^{45}Ca^{2+}$, *Biochim. Biophys. Acta* 540:534–542.

Hemmila, I., Dakubu, S., Mukkala, V. M., Siitari, H, and Lovgren, T., 1984. Europium as a label in time-resolved immunofluorometric assays, *Anal. Biochem.* 137:335–343.

Hester, R. K., and Weiss, G. B., 1981. Comparison of degree of dependence of canine renal arteries and veins on high and low affinity calcium for responses to norepinephrine and potassium, *J. Pharmacol. Exp. Ther.* 216:239–246.

Heuring, M., and Bergmann, M., 1985. Differential permeability of pineal capillaries to lanthanum ion in the rat (*Rattus Norvegicus*), gerbil (*Meriones unguiculatus*) and golden hamster (*Mesocricetus auratus*), *Cell Tissue Res.* 241:149–154.

Hinckley, C. C., 1969. Paramagnetic shifts in solutions of cholesterol and the dipyridine adduct of trisdipivatomethanatoeuropium(III). A shift reagent, *J. Am. Chem. Soc.* 91: 5160–5162.

Hodgson, B. J., Kidwas, A. M., and Daniel, E. E., 1972. Uptake of lanthanum by smooth muscle, *Can. J. Physiol. Pharmacol.* 50:730–733.

Holmquist, B., and Horrocks, W. DeW., 1975. Magnetic circular dichroic spectra of lanthanide probes, *Fed. Proc.* 34:594.

Holmquist, B., and Vallee, B. L., 1978. Magnetic circular dichroism, in *Methods in Enzymology* (C. H. W. Hirs and S. N. Timasheff, eds.), Vol. XLIV G, Academic Press, New York, pp. 149–179.

Homer, R. B., and Mortimer, B. D., 1978. Europium II as a replacement for calcium II in conconavalin A. A precipitation assay and magnetic circular dichroism study, *FEBS Lett.* 87:69–72.

Horrocks, W. DeW., and Collier, W. E., 1981. Lanthanide ion luminescence probes of distance between intrinsic protein fluorophores and bound metal ions: quantitation of energy transfer between tryptophan and terbium(III) or europium(III) in the calcium-binding protein parvalbumin, *J. Am. Chem. Soc.* 103:2856–2862.

Horrocks, W. DeW., and Sipe, J. P., 1971. Lanthanide shift reagents. A survey, *J. Am. Chem. Soc.* 93:6800–6804.

Horrocks, W. DeW., and Snyder, A. P., 1981. Measurement of distance between fluorescent amino acid residues and metal ion binding sites. Quantitation of energy transfer between tryptophan and terbium(III) or europium(III) in thermolysin, *Biochem. Biophys. Res. Commun.* 100:111–117.

Horrocks, W. DeW., and Sudnick, D. R., 1981. Lanthanide ion luminescence probes of the structure of biological macromolecules, *Acc. Chem. Res.* 14:384–392.

Hulstaert, C. E., Kalicharan, D., and Hardonk, M. J., 1983. Cytochemical demonstration of phosphatases in the rat liver by a cerium-based method in combination with osmium tetroxide and potassium ferrocyanide postfixation, *Histochemistry* 78:71–79.

Jackson, D. G., and Voorheis, H. P., 1985. Release of the variable surface coat glycoprotein from *Trypanosoma brucei* requires the cleavage of a phosphate ester, *J. Biol. Chem.* 260:5179–5183.

Jung, K., 1986. An optimized micromethod for determining the the catalytic activity of serum ribonuclease, *J. Clin. Chem. Clin. Biochem.* 24:243–250.

Kapel, M., and Karunanithy, R., 1974. The determination of tannins with cerium(IV) sulphate, *Analyst* 99:661–665.

Karaki, H., and Weiss, G. B., 1979. Alterations in high and low affinity binding of ^{45}Ca in rabbit aortic smooth muscle by norepinephrine and potassium after exposure to lanthanum and low temperature, *J. Pharmacol. Exp. Ther.* 211:86–92.

Karaki, H., Hester, R. K., and Weiss, G. B., 1980. Cellular basis of nitropresside-induced relaxation of graded responses to norepinephrine and potassium in canine renal arteries, *Arch. Int. Pharmacodyn. Ther.* 245:198–210.

Karp, M. T., Suominen, A. I., Hemmila, I., and Mantsala, P. I., 1983. Time-resolved europium fluorescence in enzyme activity measurements: a sensitive protease assay, *J. Appl. Biochem.* 5:399–403.

Khan, T. A., and Overton J., 1970. Lanthanum staining of developing chick cartilage and reaggregating cartilage cells, *J. Cell Biol.* 44:433–438.

König, B., DeCamargo, A. M., and Ferraz de Carvalho, C. A., 1981. Morphological aspects and observations about the permeability of the rat mesentery to the lanthanium nitrate. Some comparisons with the morphology of the human mesentery, *Anat. Anz.* 149: 365–374.

LaMar, G. M., and Faller, J. W., 1973. Strategies for the study of structure using lanthanide reagents, *J. Am. Chem. Soc.* 95:3817–3818.

Lane, N. J., and Treherne, J. E., 1972. Studies on perineural functional complexes and the site of uptake of microperoxidase and lanthanum in the cockroach central nervous system, *Tissue Cell* 4:427–436.

Leif, R. C., Thomas, R. A., Yopp, T. A, Watson, B. D., Guarino, V. R., Hindman, D. H. K., Lefkove, N., and Vallarino, M., 1977. Development of instrumentation and fluorochromes for automated multiparameter analysis of cells, *Clin. Chem.* 23:1492–1498.

Lesseps, R. J., 1967. The removal by phospholipase C of a layer of lanthanum-staining material external to the cell membrane in embryonic chick cells, *J. Cell Biol.* 34:173–183.

Luk, C. K., and Richardson, F. S., 1975. Circularly polarized luminescence and energy transfer studies on carboxylic acid complexes of europium(III) and terbium(III) in solution, *J. Am. Chem. Soc.* 97:6666–6675.

Martin, R. B., 1983. Structural chemistry of calcium: lanthanides as probes, in *Calcium in Biology* (T. G. Spiro, ed.), Wiley, New York, pp. 237–270.

Martin, R. B., and Richardson, F. S., 1979. Lanthanides as probes for calcium in biological systems, *Quart. Rev. Biophys.* 12:181–209.

Mason, R. M., and Mayes, R. W., 1973. Extraction of cartilage protein-polysaccharides with inorganic salt solutions, *Biochem. J.* 131:535–540.

Matsuo, S., Nishikawa, S., Ichikawa, H., Nishimoto, T., Wakisaka, S., Kitano, E., Yamamoto, K., Nakata, T., and Akai, M., 1984. Ultrastructural study of developing rat molar tooth germ *in vitro*. 2. The influence of cultivation and culture medium on the penetration of lanthanium into the ameloblast layer, *J. Osaka Univ. Dent. Sch.* 24: 31–39.

Meurman, O. H., Hemmila, I. A., Lovgren, T. N-E., and Halonen, P. E., 1982. Time-resolved fluoroimmunoassay: a new test for rubella antibodies, *J. Clin. Microbiol.* 16: 920–925.

Milhorat, T. H., Davis, D. A., and Hammock, M. K., 1975. Experimental intracerebral movement of electron microscopic tracers of various molecular sizes, *J. Neurosurg.* 42:315–329.

Moore, G. R., Ratcliffe, R. G., and Williams, R. J. P., 1983. NMR and the biochemist, in *Essays in Biochemistry* (P. N. Campbell and R. D. Marshall, eds.), Vol. 19, Academic Press, New York, pp. 142–195.

Munakata, H, and Yosizawa, Z., 1980. Extraction of hyaluronic acid from rabbit skin with lanthanum chloride, *Tohoku J. Exp. Med.* 132:337–340.

Nag, S., Robertson, D. M., and Dinsdale, H. B., 1982. Intracerebral arteriolar permeability to lanthanum, *Am. J. Pathol.* 107:336–341.

Neaves, W. B., 1972. The passage of extracellular tracers through the follicular epithelium of lizard ovaries, *J. Exp. Zool.* 179:339–364.

Neuberg, C., and Grauer, A., 1953. Precipitability of small quantities of cell constituents and metabolites as insoluble radioactive compounds, *Biochim. Biophys. Acta* 12:265–272.

Oi, T., and Utsumi, N., 1980. Ultrastructure of hypertrophic chondrocytes of rat mandibular condyles using lanthanum-containing fixatives, *Arch. Oral. Biol.* 25:77–81.

Opie, L. E., and Lavin, G. J., 1946. Localization of ribonucleic acid in the cytoplasm of liver cells, *J. Exp. Med.* 84:107–112.

Payton, B. W., Bennett, M. U. L., and Pappas, G. D. 1969. Permeability and structure of junctional membranes at an electronic synapse, *Science* 166:1641–1643.

Pettersson, K., Siitari, H., Hemmila, I., Soini, E., Lovgren, T., Hanninen, V., Tanner, P., and Stenmark, V.-H., 1983. Time-resolved fluoroimmunoassay of human choriogonadotropin, *Clin. Chem.* 24:60–64.

Pike, M. M., and Springer, C. S., 1982. Aqueous shift reagents for high resolution cationic nuclear magnetic resonance, *J. Magn. Reson.* 46:348–353.

Pita, J. C., Blanco, L. N., and Howell, D. S., 1978. Determination of total and unbound calcium in nanoliter samples of cartilage fluid and serum, *Anal. Biochem.* 90:126–135.

Rae, J. L., and Stacey, T., 1979. Lanthanum and procion yellow as extracellular markers in the crystalline lens of the rat, *Exp. Eye Res.* 28:1–21.

Rechardt, L., and Hervonen, H., 1985. Cytochemical demonstration of adenylate cyclase activity with cerium, *Histochemistry* 82:501–505.

Reuben, J., 1971. Electron spin relaxation in aqueous solutions of gadolinium(III). Aquo, cacodylate and bovine serum albumin complexes, *J. Phys. Chem.* 75:3164–3172.

Reuben, J., 1973. Paramagnetic lanthanide shift reagents in NMR spectroscopy: principles, methodology and application, *Prog. NMR Spectrosc.* 9:1–16.

Reuben, J., 1979. Bioinorganic chemistry: lanthanides as probes in systems of biological interest, in *Handbook of Chemistry and Physics of Rare Earths* (K. A. Gschneidner and L. Eyring, eds.), Vol. 4, North-Holland, Amsterdam, pp. 515–552.

Reuben, J., and Elgavish, G. A., 1979. Shift reagents and NMR of paramagnetic lanthanide complexes, in *Handbook of Chemistry and Physics of the Rare Earths* (K. A. Gschneidner and L. Eyring, eds.), Vol. 4, North-Holland, Amsterdam, pp. 483–514.

Revel, J. P., and Karnovsky, M. J., 1967. Hexagonal array of subunits in intercellular junctions of the mouse heart and liver, *J. Cell Biol.* 33:C7–C12.

Ricci, R. W., and Kilichowski, K. B., 1974. Fluorescence quenching of the indole ring system by lanthanide ions, *J. Phys. Chem.* 78:1953–1962.

Richardson, F. S., 1982. Terbium(III) and europium(III) ions as luminescent probes and stains for biomolecular systems, *Chem. Rev.* 82:541–552.

Robinson, J. M., 1985. Improved localization of intracellular sites of phosphatases using cerium and cell permeabilization, *J. Histochem. Cytochem.* 33:749–754.

Robinson, J. M., and Karnovsky, M. J., 1983a. Ultrastructural localization of 5'-nucleotidase in guinea pig neutrophils based upon the use of cerium as capturing agent, *J. Histochem. Cytochem.* 31:1190–1196.

Robinson, J. M., and Karnovsky, M. J., 1983b. Ultrastructural localization of several phosphatases with cerium, *J. Histochem. Cytochem.* 31:1197–1208.

Robinson, J. M., Briggs, R. T., and Karnovsky, M. J., 1978. Localization of D-amino acid oxidase on the surface of human polymorphonuclear leukocytes, *J. Cell Biol.* 77:59–71.

Roughley, P. J., Hatt, M., and Mason, R. M., 1978. Physiological properties of cartilage proteoglycans extracted by lanthanum chloride, *Biochim. Biophys. Acta* 539:445–458.

Russell, A. P., Westcott, V. C., Demaria, A., and Johns, M., 1983. The concentration and separation of bacteria and cells by ferrography, *Wear* 90:159–165.

Russell, A. P., Evans, C. H., and Westcott, V. C., 1987. Measurement of the susceptibility of paramagnetically labeled cells with paramagnetic solutions, *Anal. Biochem.* 164: 181–189.

Russell, L. D., 1978. The blood–testis barrier and its formation relative to spermatocyte maturation in the adult rat: a lanthanum tracer study, *Anat. Rec.* 190:99–111.

Sara, T., Hyde, J. S., and Swartz, H. M., 1976. Ion-exchange in melanin: an electron spin resonance study with lanthanide probes, *Science* 192:1132–1134.

Scaff, W. L., Dyer, D. L., and Mori, K., 1969. Fluorescent europium chelate stain, *J. Bacteriol.* 98:246–248.

Shaklai, M., and Tavassoli, M., 1982. Lanthanum as an electron microscopic stain, *J. Histochem. Cytochem.* 30:1325–1330.

Shea, S. M., 1971. Lanthanum staining of the surface coat of cells, *J. Cell Biol.* 51:611–620.

Siitari, H., Hemmila, I., Soini, E., Lovgren, T., and Koistinen, V., 1983. Detection of hepatitis B surface antigen using time-resolved fluoroimmunoassay, *Nature* 301:258–260.

Soini, E. J., Pelliniemi, L. J., Hemmilä, I. A., Mukkala, V.-M., Kankare, J. J., and Fröjdman, K., 1988. Lanthanide chelates as new fluorochrome labels for cytochemistry, *J. Histochem. Cytochem.* 36:1449–1551.

Spartalian, K., and Oosterhuis, W. T., 1973. Mössbauer effect studies of transferrin, *J. Chem. Phys.* 59:617–622.

Sperling, R., Furie, B. C., Blumenstein, M., Keyt, B., and Furie, B., 1978. Metal binding properties of γ-carboxyglutamic acid. Implications for the vitamin K-dependent blood coagulant proteins, *J. Biol. Chem.* 253:3898–3906.

Squier, C., and Edie, J., 1983. Localization of lanthanum tracer in oral epithelium using transmission electron microscopy and the electron microprobe, *Histochem. J.* 15: 1123–1130.

Stephens, E. M., and Grisham, C. M., 1979. Lithium-7 nuclear magnetic resonance, water proton nuclear magnetic resonance and gadolinium electron paramagnetic resonance studies of the sarcoplasmic reticulum calcium ion transport adenosine triphosphatase, *Biochemistry* 18:4876–4885.

Stryer, L., Thomas, T. D., and Meares, C. F., 1982. Diffusion-enhanced fluorescent energy transfer, *Annu. Rev. Biophys. Bioeng.* 11:203–222.

Sudnick, D. R., and Horrocks, D. DeW., 1979. Lanthanide ion probes of structure in biology. Environmentally sensitive fine structure in laser-induced terbium(III) luminescence, *Biochim. Biophys. Acta* 578:135–144.

Swislocki, N. I., Kramer, J. J., O'Connell, M. A., and Cunningham, E. B., 1983. Covalent modification of membrane components in the regulation of erythrocyte shape, *Ann. N.Y. Acad. Sci.:* 662–678.

Syvänen, A. C., Tchen, P., Ranki, M., and Söderlund, H., 1986. Time-resolved fluorometry: a sensitive method to quantify DNA-hybrids, *Nucl. Acids Res.* 14:1017–1028.

Tavassoli, M., Aoki, M., and Shaklai, M., 1980. A novel stromal cell type in the rat marrow recognizable by its preferential uptake of lanthanum, *Exp. Hemat.* 8:568–577.

Taylor, A. R. D., and Hall, J. L., 1979. An ultrastructural comparison of lanthanum and silicotungstic acid/chromic acid as plasma membrane stains of isolated protoplasts, *Plant Sci. Lett.* 14:139–144.

Thomasset, M., Cuisinier-Gleizes, P., and Mathieu, H., 1976. Bone resorption measurement with unusual bone markers: critical evaluation of the method in phosphorus-deficient and calcium-deficient growing rats, *Calcif. Tissue Res.* 21:1–15.

Tisher, C. C., and Yarger, W. E., 1975. Lanthanum permeability of tight junctions along the collecting duct of the rat, *Kidney Int.* 7:35–43.

Valentini, M. A., and Wright, J. C., 1983. Site selective rare earth spectroscopy in proteins, *Chem. Phys. Lett.* 100:133–137.

Van Breeman, C., and McNaughton, E., 1970. The separation of cell membrane calcium transport from extracellular calcium exchange in vascular smooth muscle, *Biochem. Biophys. Res. Commun.* 39:567–574.

Veenhuis, M., Van Dijken, J. P., and Harder, W., 1980. A new method for the cytochemical demonstration of phosphatase activities in yeasts based on the use of cerous ions, *Microbiol. Lett.* 9:285–289.

Wenzel, T. J., 1987. *NMR Shift Reagents,* CRC Press, Boca Raton, FL.

Wenzel, T. J., and Collette, L.M., 1988. Lanthanide ions as luminescent chromophores for the liquid chromatographic detection of polynucleotides and nucleic acids, *J. Chromatog.* 436:299–307.

Wilmott, N. J., Miller, J. N., and Tyson, J. F., 1984. Potential use of a terbium-transferrin complex as a label in an immunoassay for gentamycin, *Analyst* 109:343–345.

Zacks, S. I., and Saito, A., 1970. Direct connections between the T system and the subneural apparatus in mouse neuromuscular junctions demonstrated by lanthanum, *J. Histochem. Cytochem.* 18:302–304.

Zimmerman, N., and Halbhuber, K-J., 1985. Light microscopical localization of enzymes by means of cerium-based methods. I. Detection of acid phosphatase by a new cerium–lead technique (Ce–Pb-method), *Acta Histochem.* 76:97–104.

The Interaction of Lanthanides with Amino Acids and Proteins

4.1 Introduction

Because Ca^{2+} is such a recalcitrant ion, Ln^{3+} ions have found extensive use as informative substituents with which to investigate various structural and functional aspects of the Ca^{2+}-binding sites on proteins. The chemical principles upon which this strategy rests are discussed in Chapter 2. In some cases, lanthanides have served to report on the binding sites of other metal ions, such as Fe^{3+} in transferrin (Section 4.8) and ferritin (Section 4.8), Mn^{2+} in pyruvate kinase (Section 4.24) and glutamine synthetase (Section 4.25), or Mg^{2+} in pyruvate kinase (Section 4.24) and alkaline phosphatase (Section 4.25). Lanthanides have also provided useful information about proteins, such as lysozyme (Section 4.7), which are not thought to interact specifically with Ca^{2+}, or any other metal ion, under physiological conditions. In addition, chemical modification of a protein can be used to introduce specific, strong binding sites for Ln^{3+} ions (e.g., Marinetti *et al.*, 1976; Leung and Meares, 1977; Bradbury *et al.*, 1978; Walter *et al.*, 1981).

A variety of experimental techniques have been brought to bear upon the structural properties of Ln^{3+}-substituted proteins. These include NMR spectroscopy, luminescence spectroscopy, and, less frequently, EPR spectroscopy, spectrophotometry, X-ray diffraction, Mössbauer spectroscopy, and magnetic circular dichroism (Chapter 3). None of these would provide as much useful information with the native, Ca^{2+}-containing protein. As described in Chapter 5, this experimental approach can be applied to substances other than proteins.

Functional perturbations produced by the substitution of Ln^{3+} ions for Ca^{2+} are also worth studying, as they often help determine the role of Ca^{2+} in the native structure. Such an approach has been especially valuable for Ca^{2+}-requiring enzymes. Martin and Richardson (1979) have suggested that the functional properties of the protein upon substitution by a lanthanide should be retained if the role of Ca^{2+} is purely structural, yet be lost if Ca^{2+} has an important functional role. Central to the success of many such investigations is the assumption that Ln^{3+} ions replace Ca^{2+} isomorphously. This assumption is supported by X-ray diffraction studies of thermolysin (Section 4.3), parvalbumin (Section 4.4.1), and troponin C (Section 4.4.2), showing that Ln^{3+} ions can indeed occupy certain Ca^{2+}-binding sites in a nondisruptive manner. Minor differences do exist between the Ca^{2+}- and Ln^{3+}-containing forms, but, overall, a high degree of isomorphism is conserved. Further evidence that Ln^{3+} ions specifically replace Ca^{2+} at precise locations has come from binding studies which demonstrate the competitive nature of the substitution. Often the conformational changes in the protein produced by metal ion binding are the same for Ca^{2+} and Ln^{3+} ions. Functional assays also support the claim of specificity in the attachment of Ln^{3+} ions to proteins. For example, enzymic activity is sometimes retained following the substitution of Ca^{2+} by Ln^{3+} ions. In addition, in cases where the substitution inhibits enzymic activity, kinetic analyses often show that inhibition by lanthanides is competitive with respect to Ca^{2+}.

4.2 Amino Acids and Small Peptides

As expected from the discussion in Section 2.3, lanthanides coordinate with the unprotonated carboxyl groups of amino acids. Weak interaction with the hydroxyl oxygen of serine (Prados *et al.*, 1974), and presumably also threonine and tyrosine, can occur too. Whether or not lanthanides are also capable of coordinating with unprotonated amino groups remains an unresolved controversy. The problem is complicated, as the pH values at which coordination to the α-amino nitrogen atom becomes a reasonable proposition are also those at which Ln^{3+} hydrolysis occurs.

Sinha (1966) states that both the amino and carboxyl components of amino acids contribute to complexing Ln^{3+} ions and uses this as an explanation of why amino acids form stronger complexes than monocarboxylic acids with Ln^{3+}. In support of this claim are the conclusions of Katzin and Gulyas (1968) and Katzin (1969) resulting from their spectroscopic studies of the interaction of Pr^{3+} and Eu^{3+} with several amino acids at pH 7. In keeping with coordination to unprotonated amino groups, their pK values help to determine the stability of Pr^{3+}–amino acid complexes in the sequence Asp (pK = 8.8) $\geqslant$ Ser (pK = 9.15) > Val (pK = 9.62) $\approx$ Leu (pK = 9.6) $\approx$ Ala (pK = 9.69) > Pro (pK = 10.6) (Katzin and Gulyas, 1968). At pH 7, histidine may also coordinate to Ln^{3+} through its imidazole nitrogen (Sherry *et al.*, 1973). Evidence that the N-terminal amino groups of peptides also coordinate Ln^{3+} ions comes from NMR studies of the interaction of Gd^{3+} with various small glycopeptides at pH 6–7 (Dill *et al.*, 1983), Section 5.4. However, according to Prados *et al.* (1974), the high pK_a values for the deprotonation of α-amino nitrogens ensure that hydrolysis of Ln^{3+} ions (Section 2.3) occurs before uncharged N donors are generated. No amino nitrogen involvement was reported for the coordination of Ln^{3+} ions to γ-carboxyglutamic acid (Gla), [Glu-4]oxytocin, or several other small peptides (Tanner and Choppin, 1968; Sperling *et al.*, 1978; Walter *et al.*, 1981).

To minimize coordination to the amino nitrogen, some investigators have employed pH values as low as is permitted by the pK_a of the carboxyl group (e.g., Tanner and Choppin, 1968). This tactic can, however, reduce the affinity of the Ln^{3+} ion for the zwitterion through the repulsive influence of the protonated amino group. Other researchers have covalently blocked the amino group by acetylation (Dill *et al.*, 1983; Shelling *et al.*, 1984). Nevertheless, interaction between the acetylated N-terminus of a peptide and Ln^{3+} ions has been detected (Dill *et al.*, 1983; Section 5.4). Some binding values are given in Table 4-1.

Stoichiometries of 1:1, 1:2, and 1:3 have been reported for Ln^{3+}–amino acid complexes (Sinha, 1966). Sherry and Pascnol (1977) suggested that alanine participates in monodentate coordination with lanthanides from Pr^{3+} to Tb^{3+}, and in bidentate coordination from Dy^{3+} to Yb^{3+}. Coordination to bicarboxylic amino acids is bidentate, involving both carboxyl groups (Moeller *et al.*, 1965). Sperling *et al.* (1978) reported that Tb^{3+} forms a ternary complex with Gla having the Tb^{3+}:Gla stoichiometry of 1:2. By paramagnetic relaxation enhancement experiments, Gd^{3+}—O distances of approximately 3.2 Å and 3.6 Å for each oxygen in each carboxyl group were determined. It is likely that a Gla residue acts as a bidentate ligand, displacing two molecules of water from the aqueous Ln^{3+} ion. From their circularly polarized emission (CPE) studies on Tb^{3+} and Eu^{3+} complexes of Asp, Ser, Thr, and His, Brittain and Richardson (1977) speculated that at pH values above 7, the amino acid:Ln^{3+} stoichiometry may change from 1:1 to 2:1.

Angiotensin II is an octapeptide with the sequence Asp-Arg-Val-Tyr-Ile-His-Pro-Phe. Its association with Ca^{2+} and Tb^{3+} has been monitored by measuring the changes in its Tyr fluorescence and Tb^{3+} luminescence. From titration experiments, K_d values for Ca^{2+} of 0.1 mM (Lenkinski *et al.*, 1978) and 15.5 mM (Canada, 1981) and for Tb^{3+} of 0.13 ± 0.2 mM (Lenkinski *et al.*, 1978) and 2.4 mM (Canada, 1981) have been calculated. Conversion of Asp-1 to an Asn residue eliminated the changes in fluo-

Table 4-1. Binding Data for Certain Complexes between Lanthanides and Amino Acids

Amino acid	Binding parameter	Conditions	Lanthanide^{3+}							Reference
			La^{3+}	Ce^{3+}	Pr^{3+}	Nd^{3+}	Pm^{3+}	Eu^{3+}	Tb^{3+}	
Glycine	Log stability constant	pH 3.64, 25°C, 2 M NaClO$_4$	—	3.4	—	—	4.7	5.0	—	Tanner and Choppin, 1968
Glycine	Log stability constant	0.1 M KCl, 30°C	3.2	3.4	3.6	3.7	—	—	—	Moeller et al., 1965
Aspartate	Log K_1, [Ln-Asp]$^+$	0.1 M KCl, 25°C	5.0	5.2	5.4	5.5	—	—	—	Moeller et al., 1965
	Log K_2, [Ln(Asp)$_2$]$^-$	0.1 M KCl, 25°C	4.2	4.6	4.8	4.9	—	—	—	Sinha, 1966
γ-Carboxyglutamate	K_d	pH 6.5	—	—	—	—	—	—	~50 μM	Sperling et al., 1978
Alanine	Stability constant	pH 3	All Ln^{3+} approx. 0.7 ± 0.1 M							Sherry and Pascnol, 1977
	Stability constant	pH 4	—	—	—	6.5	—	—	—	Sherry et al., 1973
Histidine	Stability constant	pH 4	—	—	—	1.8	—	—	—	Sherry et al., 1973
	Stability constant	pH 7	—	—	—	123.0	—	—	—	Sherry et al., 1973
Threonine	Stability constant	pH 4	—	—	—	7.6	—	—	—	Sherry et al., 1973
Serine	Stability constant	pH 4	—	—	—	12.6	—	—	—	Sherry et al., 1973

rescence produced by Ca^{2+} or Tb^{3+} (Lenkinski $et\,al.$, 1978). Most of the ^{13}C-NMR peaks of angiotensin II are shifted by Yb^{3+} (Lenkinski and Stephens, 1983). Depending on the pH, Yb^{3+} bound to the side-chain carboxyl group of Asp-1 ($K_d = 11.4\,mM$), the C-terminal carboxyl group of Phe-8 ($K_d = 2.8\,mM$), or to both of them in a bidentate complex. Similar results were found for Eu^{3+}, which had a K_d of 6.5 mM (Lenkinski and Stephens, 1982). La^{3+} produced few changes in the ^{13}C-NMR spectrum, while Gd^{3+} permitted the calculation of the distances between Gd^{3+} and various nuclei in the peptide. In this context, Bradbury $et\,al.$ (1974) have suggested that paramagnetic lanthanides could be used to sequence peptides from their C-termini.

The pentapeptide Arg-Lys-Asp-Val-Tyr is a biologically active fragment of the thymic hormone thymopoietin. Tb^{3+} binds to this fragment, undergoing luminescence enhancement upon irradiation at 270 nm due to energy transfer from the C-terminal Tyr residue. Proton NMR spectroscopy provided K_d values of $9.5 \times 10^{-3}M$ for Pr^{3+}, $1.2 \times 10^{-2}M$ for Yb^{3+}, and $7.1 \times 10^{-3}M$ for Eu^{3+}. Esterification of the carboxyl side chain of the Asp residue or the C-terminal carboxyl group of Tyr reduced the affinity of Pr^{3+} for the fragment, producing K_d values of $1.3 \times 10^{-1}M$ and $>0.5\,M$, respectively. These findings and the structural NMR data suggested that the predominant form of complex was a bidentate one in which Ln^{3+} bind to both carboxyl groups in the molecule (Vaughn et $al.$, 1981, 1982). A lanthanide-binding site was created in oxytocin by the introduction of a Glu residue at the 4-position. Information about the conformation of the resulting peptide, Cys-Tyr-Ile-Glu-Asn-Cys-Pro-Leu-Gly-NH_2, was then obtained by ^1H-NMR spectroscopy with Yb^{3+}. Lanthanides were shown to coordinate to the carboxyl group of Glu and the carbonyl side chain of Asn (Walter $et\,al.$, 1981; Lenkinski and Stephens, 1983).

In other studies, small peptides containing certain amino acids found at the Ca^{2+}-binding sites of proteins have been synthesized and studied. In this way, Shelling $et\,al.$ (1984) examined N-acetyl-Asp and N-acetyl-Asp-Gly-aspartylamide, which are associated with the EF Ca^{2+}-binding site of parvalbumin (Section 4.4.1). N-Acetyl-Asp formed 1:1 complexes with Yb^{3+}, Tm^{3+}, Er^{3+}, Ho^{3+}, and Dy^{3+}, with K_d values in the range 0.15–0.30 mM. For a 1:1 stoichiometry of Ln^{3+}:N-acetyl-Asp-Gly-aspartylamide, K_d values of 0.15–0.21 mM were calculated. With a 1:2 stoichiometry, K_d values were 7.45–7.48 mM. In a similar fashion, acetyl-Asp-Val-Asp-Ala has been shown to bind Pr^{3+} via both carboxyls of each Asp, in a bidentate fashion (Asso $et\,al.$, 1985). La^{3+} interactions with synthetic peptide analogs of the Ca^{2+}-binding site III of troponin C (Section 4.4.2) have also been studied (Marsden $et\,al.$, 1988).

Lenkinski *et al.* (1980) studied the binding of Ln^{3+} to bleomycin, a glycoprotein antibiotic which interacts with DNA. Potentiometry showed that attachment of one Ca^{2+} or Ln^{3+} ion to bleomycin displaced a proton from its diaminopropionamide group. Luminescence enhancement of Tb^{3+} was observed upon binding to bleomycin. Both potentiometry and luminescence measurements provided a K_a of $2.5 \times 10^4 M^{-1}$. Competition experiments with other Ln^{3+} ions revealed a marked increase in K_a with decreasing ionic size in going from La^{3+} ($K_a = 2.1 \times 10^2 M^{-1}$) via Pr^{3+} ($K_a = 1.6 \times 10^3 M^{-1}$) to Yb^{3+} ($K_a = 1.6 \times 10^6 M^{-1}$). Proton NMR spectroscopy revealed that Pr^{3+} and bleomycin underwent rapid exchange, whereas Yb^{3+} exchange was slow by NMR standards, with an estimated lifetime of 4 s for the Yb^{3+}-bleomycin complex. Based upon this value, lifetimes of 8×10^{-4} s for La^{3+} and 0.1 s for Tb^{3+} were estimated.

4.3 Thermolysin

Thermolysin (EC 3.4.24.4) is a heat-stable, extracellular, neutral endopeptidase produced by the thermophilic bacterium *Bacillus thermoproteolyticus*. In common with a number of other bacterial neutral proteinases, such as clostridiopeptidase, thermolysin is a zinc metalloenzyme which binds Ca^{2+} ions and which is susceptible to inhibition by chelators such as *o*-phenanthroline and EDTA. It comprises a single polypeptide chain of molecular weight 34,600, containing 316 amino acid residues. Each molecule of thermolysin contains one Zn^{2+} ion, which is at the active site of the enzyme and which plays a crucial role in scission of the substrate peptide bond. Four Ca^{2+} ions per molecule add thermal stability to the enzyme and prevent autolysis (Roche and Voordouw, 1978).

Thermolysin is particularly suited to the present discussion, as the three-dimensional structure of the native and Ln^{3+}-substituted enzyme has been determined by X-ray crystallography (Colman *et al.*, 1972a,b; Matthews *et al.*, 1972; Matthews and Weaver, 1974). Two of the Ca^{2+} ions are bound at a double site, designated S(1) and S(2), where both are coordinated by the same three carboxyl groups (Glu-177, Asp-185, and Glu-190) which act as bridges between the two ions. The inner Ca^{2+} ion at site S(1) is shielded from the solvent and is additionally coordinated by the carboxyl group of Asp-138, the backbone carbonyl oxygen of Glu-187, and an internal water molecule. The outer Ca^{2+} ion at site S(2) is also coordinated by the backbone carbonyl oxygen of Asn-183 and two water molecules. The Ca^{2+} ions at sites S(3) and S(4) are bound singly. At site S(3), Ca^{2+} is coordinated in an approximately octahedral fashion by the carboxyl groups of Asp-57 and Asp-59, the backbone carbonyl oxygen of Glu-61, and three water molecules. In a similar fashion, Ca^{2+} at site S(4) is coordinated by the backbone carbonyl oxygens of Tyr-193, Thr-194, and Ile-197, the hydroxyl group of Thr-194, the carboxyls of Asp-200, and two water molecules.

Although Nd^{3+}, Pr^{3+}, and Dy^{3+} ions are unable to restore the catalytic activity of the zinc-free apoenzyme (Holmquist and Vallee, 1978), all of 10 Ln^{3+} ions tested by Matthews and Weaver (1974) displaced Ca^{2+} ions from binding sites S(1), S(2), S(3), and S(4), although it was possible that the occupancy of S(4) by a Lu^{3+} ion was only partial. Replacement of the Ca^{2+} ion at site S(1) by a Ln^{3+} ion occurs with displacement of the Ca^{2+} ion at S(2). However, no Ln^{3+} ion occupies the S(2) site under these conditions. At site S(1), a shift in coordinates of about 0.33 Å occurs on Ln^{3+} binding. This is probably due to an increase in coordination number. If so, the extra ligands might be provided by acidic coordinating groups from the adjacent site S(2), which would hence be unavailable for coordination to a second Ln^{3+} ion. This, and the greater ionic repulsion existing between two adjacent Ln^{3+} ions than between two Ca^{2+} ions, may explain the inability of Ln^{3+} ions to occupy both sites S(1) and S(2). A similar circumstance may occur in lysozyme (Section 4.7). As Roche and Voordouw (1978) have commented, it would be interesting to know whether bivalent lanthanides, such as Eu^{2+}, can occupy both sites S(1) and S(2). The distances of Ca^{2+} and Eu^{3+} from the ligands at all four binding sites on thermolysin are shown in Table 4-2. Binding of Nd^{3+} to thermolysin could be monitored by magnetic circular dichroism (Holmquist and Vallee, 1978). Eu^{3+}-substituted thermolysin had normal proteolytic activity but was not as stable as the native enzyme (Matthews and Weaver, 1974).

The affinity of Eu^{3+} ions for site S(1) greatly exceeds that of Ca^{2+} ions for the double site S(1), S(2). Indeed, Eu^{3+} sequestered at this site cannot be removed with EDTA (Dahlquist et al., 1976). However, Eu^{3+} binds less strongly than Ca^{2+} at sites S(3) and S(4). At site S(3), a Eu^{3+} ion coordinates the same ligands as the Ca^{2+} ion it replaces, with only minor differences in ligand distances (Table 4-2). As a general trend, there is a progressive shift of about 0.35 Å at site S(3) in going from La^{3+} to Lu^{3+}, with the larger Ln^{3+} ions occupying positions successively farther from that of Ca^{2+}. A similar state of affairs obtains at site S(4), with a difference of about 0.5 Å between the position of Ca^{2+} and that of Ln^{3+}. Eu^{3+} increases its coordination number at site S(4) to 8 by coordinating an additional water molecule. Overall, the smaller Ln^{3+} ions replace Ca^{2+} in a more isomorphous manner than do the larger members of the series (Colman et al., 1972b; Matthews and Weaver, 1974).

Table 4-2. Calcium and Europium Distances in Thermolysin[a]

	Distance (Å)		
Ligand	Ca 1	Eu 1	Ca 2
Sites S(1) and S(2)			
Asp 185	3.6	3.3	2.4
	2.7	2.4	
Glu 177	2.1	2.1	
	2.6	2.4	2.1
Glu 190	1.9	2.1	
	2.6	2.6	2.1
Glu 187	2.0	2.1	
Asn 183			2.1
Asp 138	3.0	3.3	
H_2O 1	2.9	2.7	
H_2O 2			2.9
H_2O 3			2.6

	Ca 3	Eu 3
Site S(3)		
Asp 57	3.5	3.6
	2.6	3.0
Asp 59	2.0	2.4
	3.4	3.3
Glu 61	2.6	3.2
H_2O	2.3	2.0
H_2O	2.1	1.8
H_2O	2.1	1.7

	Ca 4	Eu 4
Site S(4)		
Tyr 193	2.5	2.4
Thr 194	1.8	1.9
	2.2	2.7
Ile 197	3.0	2.9
Asp 200	2.5	2.8
H_2O	2.5	2.2
H_2O	2.2	1.8
H_2O		2.8

From Matthews *et al.* (1974), with permission.

These experiments rigorously confirmed for the first time that the substitution of Ca^{2+} ions by Ln^{3+} ions on a protein molecule can occur with good isomorphism, thereby justifying, in principle, the use of Ln^{3+} ions as specific probes of the structure of Ca^{2+}-binding sites of this type. Horrocks and Snyder (1981) confirmed interion distances in thermolysin by measuring the transfer of energy from tryptophan to a Tb^{3+} or Eu^{3+} substituent at the S(1) site. In one series of experiments, a Tb^{3+} ion was selectively introduced at site S(1). This is possible because, as noted above, Ln^{3+} ions displace both Ca^{2+} ions from sites S(1) and S(2), yet occupy only the S(1) site with high affinity, having much lower affinity for sites S(3) and S(4). With a Tb^{3+} ion in the S(1) position, luminescence enhancement was observed with an excitation wavelength of 295 nm, suggesting tryptophan as the donor group (Section 3.3). This designation gains support from the elimination of luminescence upon treatment of the enzyme with N-bromosuccinimide (Horrocks et al., 1975), a reagent which primarily modifies tryptophan residues. From the luminescence data, a Trp–Tb^{3+} separation of 9.9 Å was calculated (Horrocks and Snyder, 1981). The donor in question may well be Trp-186, which is only 9 Å from S(1). This conclusion agrees with that obtained from the oxygen-quenching studies of Prendergast et al. (1983). In a second series of experiments, Eu^{3+} was used as the substituent at S(1). Transfer of energy from Trp-186 to Eu^{3+} did not enhance Eu^{3+} luminescence because, unlike the case with Tb^{3+}, this occurs by a charge transfer mechanism (Section 3.3). However, absorption of energy by Eu^{3+} could be monitored as quenching of protein fluorescence, which decreased by 14% after the addition of one equivalent of Eu^{3+} per mole of thermolysin. From these data, a Trp–Eu^{3+} separation of 10.5 Å was calculated. When the Zn^{2+} ion at the active site was replaced by a Co^{2+} ion, Tb^{3+} luminescence was quenched by nearly 90%. Quantitative studies of this phenomenon have produced a calculated Co^{2+}–Tb^{3+} distance of 13.7 Å (Horrocks et al., 1975), a figure which is identical to that determined for Zn^{2+}–Ca^{2+}[S(1)] by the more laborious X-ray diffraction studies.

In a subsequent study (Snyder et al., 1981), a luminescent Ln^{3+} probe (Eu^{3+} or Tb^{3+}) was inserted into site S(1) and an acceptor Ln^{3+} (Pr^{3+}, Nd^{3+}, Ho^{3+}, or Er^{3+}) was inserted into sites S(3) and S(4). Exchange of Ln^{3+} ions between sites S(1) and S(3) or S(4) was not observed. Virtually all the energy transfer under these conditions occurred between sites S(1) and S(4). Measurements of Eu^{3+} and Tb^{3+} emission lifetimes in H_2O and D_2O, following excitation with a pulsed laser, determined that 1.2 ± 0.5 water molecules are bound to the Ln^{3+} ion at S(1), a value consistent with the X-ray crystallographic data (Table 4-2). Similar experiments with Eu^{3+} at S(3) and S(4) suggest 3.1 and 4 coordinated water molecules,

respectively, again in quite good agreement with the X-ray crystallographic data. The S(1)–S(4) distance of 10.9 Å–11.8 Å calculated by Snyder *et al.* (1981) is in good agreement with the 10.9 Å determined crystallographically. However, for an unknown reason, $Tb^{3+} \rightarrow Nd^{3+}$, but
not $Eu^{3+} \rightarrow Nd^{3+}$, energy transfer was more efficient than theory would
predict, such that the S(1)–S(4) separation was slightly underestimated
as 8.6–8.8 Å.

4.4 The "E-F Hand" Proteins

The second Ca^{2+}-binding protein whose interaction with Ln^{3+} ions
has been rigorously studied by X-ray crystallography is parvalbumin. For
reasons discussed below, one of its two Ca^{2+}-binding sites is termed the
"E-F hand." Similar structures exist in other intracellular proteins with
high affinity for Ca^{2+} ions ($K_d \approx 1 \, \mu M$) and which undergo marked conformational changes upon binding Ca^{2+} ions. These proteins include calmodulin, troponin C, various intestinal Ca^{2+}-binding proteins, and probably S-100 and several others. They are presently the subject of intense
research, due to their crucial role in the regulation of stimulus-coupled
cellular responses.

4.4.1 Parvalbumin

The sarcoplasm of fast skeletal muscles in most vertebrates contains
a class of water-soluble proteins called parvalbumins. In fish, their concentration can be as high as 0.5–1 mM. Their molecular weights range
from 9000 to 13,000, and their isoelectric points from pH 4 to pH 5. About
9% of the amino acid residues are phenylalanine.

Carp parvalbumin exists in three forms, of which the B-form is the
best characterized. It has 108 amino acid residues. There is no tyrosine
or tryptophan, only one each of histidine, arginine, and cysteine, and 10
phenylalanine residues, which are not stacked. It binds two Ca^{2+} ions,
producing a net molecular weight of 11,489. Parvalbumins share a considerable degree of sequence homology with troponin C (Section 4.4.2).
The three-dimensional structure of carp muscle parvalbumin B (isotype
3, pI 4.2) has been refined to a resolution of 1.85 Å (Moews and Kretsinger,
1975). The molecule is nearly spherical. Its crystal structure consists of
six α-helical regions, labeled A–F. One Ca^{2+} ion is bound in the loop that
joins helices C and D, and a second Ca^{2+} ion is sequestered in the loop
that joins helices E and F. Ca^{2+} coordinates with six protein ligands in

the CD site and with seven protein ligands, plus a water molecule, in the EF site. As helix E, loop EF, and helix F resemble the extended forefinger, clenched middle finger, and extended thumb of a rigid hand, this domain is called the "E-F hand" (Kretsinger, 1976). This name is now used to designate all proteins with this type of Ca^{2+} binding domain. The affinity of both the C-D and E-F hands for Ca^{2+} is very high, with a K_d of the order of $10^{-8}-10^{-9} M$. Both loops contain 10 to 12 amino acid residues, with Ca^{2+} ligands being provided by approximately every other residue. Parvalbumin thus provides a wonderful model with which to study the interactions of lanthanides with proteins. It is small, has two very well defined Ca^{2+}-binding sites, contains only one type of aromatic amino acid, and its crystal structure is known in fine detail.

Using crystals grown at a molar Tb^{3+}:parvalbumin ratio of 0.5, Moews and Kretsinger (1975) found replacement of Ca^{2+} by Tb^{3+} only at the EF site. However, later work revealed that at a molar ratio of 1:1, Tb^{3+} also occupies the CD site (Sowadski et al., 1978). The CD Tb^{3+} ion sits at a position only 0.4–0.7 Å from the position that the Ca^{2+} ion would normally occupy. The EF Tb^{3+} ion lies between 0.2 and 0.6 Å from the Ca^{2+} position. Given an error of 0.15 Å in determining the Ca^{2+} locations and a slightly larger error in determining the site of Tb^{3+} attachment, the replacement of Ca^{2+} ions by Tb^{3+} ions at these two positions could well be isomorphous.

Upon binding to parvalbumin, Tb^{3+} ions produced changes in the circular dichroism spectrum similar to those produced by Ca^{2+} ions. Irradiation at 259 nm excited Tb^{3+} luminescence as a result of energy transfer from the aromatic side chain of Phe-57. Although the carbonyl group of Phe-57 serves as a ligand for metal binding at the CD site, the side chain is within 5 Å of the EF site at which luminescence occurs. The emission at 545 nm was circularly polarized, as shown in Fig. 4-1, and showed a fine structure which was not discernible from the emission spectrum alone (Donato and Martin, 1974). The titration curve of enhanced Tb^{3+} luminescence was unusual in that luminescence rose almost linearly, with a maximum at about 1.5 ions of Tb^{3+} per molecule of parvalbumin, and then decreased to 55% of this value at a Tb^{3+}:parvalbumin stoichiometry of 3:1 (Donato and Martin, 1974; Nelson et al., 1977). To explain their findings, Martin and co-workers postulated a third, quenching Tb^{3+}-binding site near the EF site. Evidence for a third Ln^{3+}-binding site on parvalbumin has also been provided by Rhee et al. (1981). They confirmed that luminescence of both Eu^{3+} and Tb^{3+} at pH 6.5 was quenched at Ln^{3+}:parvalbumin stoichiometries above 2. However, a pH 3.8, no quenching occurred. The authors thus postulated that the third, quenching Ln^{3+} ion bound to parvalbumin at pH 6.5, but not pH 3.8. By titration,

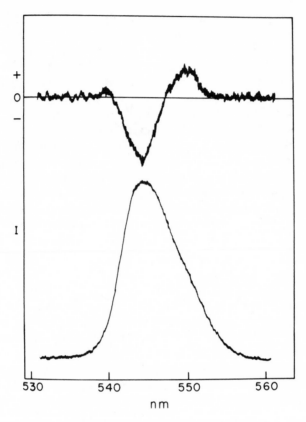

Figure 4-1. Emission spectrum (lower trace) and circularly polarized emission spectrum (upper trace) of Tb^{3+}-parvalbumin. From Donato and Martin (1974), with permission.

they arrived at a pK_a of about 4.5 for this binding site, implicating carboxyl groups as the ligands responsible. Comparison of the emission in H_2O and D_2O suggested that three water molecules were bound to the Eu^{3+} ion at this site, the protein itself presumably supplying an additional four or five ligands. At the CD and EF sites, Eu^{3+} coordinated 1.2 ± 0.5 water molecules. As luminescence at both the CD and EF sites was quenched by the apparent binding of the third Ln^{3+} ion, Rhee *et al.* (1981) suggested that the third site was close to both of the established sites. From an analysis of the primary structure of carp parvalbumin, these authors speculated on the involvement of Asp-53, Asp-59, Asp-62, and Asp-61 as ligands. As the first three of these are coordinated to metal ions at the CD site, there may be a "double" Ln^{3+}-binding site of the type described for the S(1) and S(2) sites of thermolysin (Section 4.3), involving carboxylate bridges. Further evidence for the existence of three binding sites was provided by examination of the $^7F_0 \rightarrow {}^5D_0$ excitation spectrum of Eu^{3+} (Rhee et al., 1981). Three peaks corresponding to Tb^{3+} at the CD site (579.2 nm), EF site (579.6 nm), and the presumptive third site (578.4 nm) were resolved.

In contrast to this, several groups have independently suggested that there is no third Ln^{3+}-binding site on parvalbumin and explain quenching by alternative mechanisms. Sowadski et al. (1978) examined the matter crystallographically, without finding evidence of a third site. Instead, they proposed a mechanism based upon quenching of the luminescence of the EF Tb^{3+} by the CD Tb^{3+}, although the experiments of Nelson et al. (1977) had apparently excluded this possibility. Atomic absorption measurements have shown that the most highly luminescent parvalbumin contains two Tb^{3+} ions (Miller et al., 1980). Martin (1983) has argued that the crystallographic studies would not detect a weak Tb^{3+}-binding site because of the presence of the high concentration of competing sulfate ions used in preparing the crystals. More recently, Henzl et al. (1985) have repeated the luminescence studies with pike parvalbumin, concluding that quenching results from collisional deactivation of the tightly bound ions by excess Tb^{3+} in free solution and not from binding of Tb^{3+} at a third site. They suggested that the changes in apparent binding behavior with pH which were reported by Rhee et al. (1981) instead reflected a spectral transition involving Eu^{3+} bound at both the CD and EF sites. Furthermore, when the concentration of parvalbumin was reduced from $175 \mu M$ to $10 \mu M$, the Tb^{3+} luminescence titration curve did not show a quenching component. The NMR data of Cavé et al. (1979) indeed demonstrated the existence of a third site of metal-ion binding, with a K_d for Mn^{2+} of 0.6 mM. However, although also displaying a weak affinity for Ca^{2+} ($K_d \approx$ 4–5 mM), this site apparently did not accept Gd^{3+} or La^{3+}. Proton NMR investigations have also failed to detect a third binding site for Yb^{3+}:parvalbumin ratios as high as 4.5:1 (Lee and Sykes, 1981).

The problem thus remains unsettled, with the balance of the evidence weighing against a third Ln^{3+}-binding site. The mysterious quenching phenomenon at high Ln^{3+}:parvalbumin ratios has also been reported for troponin C, a synthetic Ca^{2+}-binding peptide based upon one of its Ca^{2+}-binding sites, calbindin (Section 4.4.4) and even EDTA at high Tb^{3+} concentration (Henzl et al., 1985). However, even this is disputed by Miller et al. (1975), who did not observe quenching with troponin C.

Unlike carp parvalbumin, cod parvalbumin III contains a single tryptophan residue which replaces Phe-102 in the primary sequence of the former. Calculations based upon the efficiency of transfer of energy from

Trp-102 to Tb^{3+} or Eu^{3+} give donor–acceptor separations of 11.8 and 10.2 Å, respectively (Horrocks and Collier, 1981). These values are in good agreement with a distance of 10.6 Å estimated from X-ray structural data. It is worth noting that these studies were important in identifying the energy transfer mechanism as one involving a Förster dipole–dipole mechanism (Section 3.3). Quenching studies with Tb^{3+} or Eu^{3+} as the energy donor and Pr^{3+}, Nd^{3+}, Ho^{3+}, or Er^{3+} as the acceptor have provided estimates of the intersite distance on parvalbumin (Rhee *et al.,* 1981). With Eu^{3+} as donor, this distance averaged 12.1 Å, while with Tb^{3+} as donor it averaged 9.4 Å. These values are in quite good agreement with the X-ray crystallographic distance of 11.8 Å.

Through the efforts of Sykes and his colleagues at the University of Alberta, much information has been gathered on the solution structure and metal-ion binding properties of various parvalbumins by NMR spectroscopy. In its bi-Ca^{2+} state, the ^{1}H-NMR spectrum of parvalbumin shows chemical shifts in the −0.5 to 10 ppm range. However, when Yb^{3+} replaces one or both of the Ca^{2+} ions, specific peaks are shifted to new positions in the −19 to 32 ppm range (Lee and Sykes, 1980; Lee *et al.,* 1985) (Fig. 4-2). The ^{13}C-NMR spectrum is also altered. Diamagnetic Ln^{3+} ions, such as Lu^{3+} and La^{3+}, also produce subtle changes in the NMR resonances, indicative of conformational changes.

Dissociation constants shown in Table 4-3 indicate that Ln^{3+} ions bind more tightly than Ca^{2+} to each site. Whereas the affinity of the EF site for Ln^{3+} ions does not alter throughout the lanthanide series, the larger lanthanides have a greater affinity than the smaller members for the CD site. These differences mean that the heavy and light lanthanides displace Ca^{2+} from parvalbumin in a sequential manner (Lee and Sykes, 1981; Corson *et al.,* 1983b). Using stopped-flow methods, Corson *et al.* (1983b) demonstrated clear differences in the dissociation rate constants of the Ln^{3+} ions at the CD and EF sites (Table 4-4). It can be seen that Ln^{3+} ions dissociate much more quickly from the CD site and that throughout the lanthanide series, off rates decrease markedly at the EF site, yet slightly increase at the CD site. These results do not in themselves explain the changes in K_d values (Table 4-3). To do so, the authors suggest, by analogy with Ln^{3+}-murexide complex formation, that the on rates show a linear decrease at the EF site in going from La^{3+} to Lu^{3+}. In this way, the K_d would remain about the same for each lanthanide in the EF binding site. At the fast-release, CD site, a constant k_{off} across the lanthanide series would produce the observed increase in K_d in going from La^{3+} to Lu^{3+}. Release kinetics were biphasic for most lanthanides, with both a fast ($-1.2 \leq \log k_{fast} \leq -0.7$) and a slow ($-1.2 \geq \log k_{slow} \geq -2.9$) component. During these studies, Corson *et al.* (1983a) determined acti-

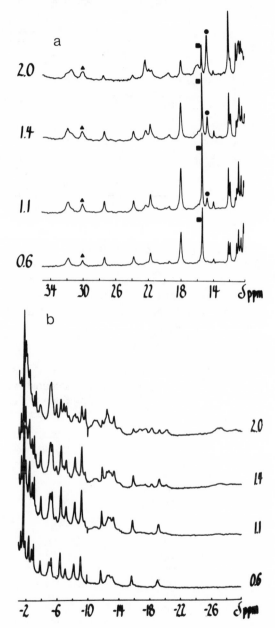

Figure 4-2. Proton NMR spectra of carp parvalbumin at increasing Yb³⁺ concentrations. (a) Downfield region of spectrum; (b) upfield region of spectrum. Yb³⁺:protein ratios are 0.6, 1.1, 1.4, and 2.0 as indicated. From Lee and Sykes (1981), with permission.

Table 4-3. Dissociation Constants of Various
Lanthanides for the Two Ca^{2+}-Binding
Sites of Parvalbumin

Ln^{3+}	K_d ($M \times 10^{-11}$)		Reference
	CD site	EF site	
La^{3+}	2.0	4.8	Williams et al., 1984
Ce^{3+}	3.2	4.8	Williams et al., 1984
Gd^{3+}	5.0	5.0	Williams et al., 1984
	0.5	0.5	Cavé et al., 1979
Tb^{3+}	600	2.0	Lee and Sykes, 1981[a]
Yb^{3+}	52	4.0	Williams et al., 1984
Lu^{3+}	36	4.8	Williams et al., 1984

[a] Calculated from data of Nelson et al. (1977).

vation energies of 2.6 kcal/mol for the release of Yb^{3+} and 1.6 kcal/mol for La^{3+}.

Because of its paramagnetic properties (Section 2.2; Table 2-2) and its selectivity for the EF site (Table 4-3), Yb^{3+} has been extensively used in NMR studies aimed at obtaining structural information about parvalbumin. Also aiding these studies is the fact that of all the paramagnetic

Table 4-4. Off Rate Constants for the
Dissociation of Lanthanides from the Two
Ca^{2+}-Binding Sites of Parvalbumin[a]

Ln^{3+}	k_{off} ($s^1 \times 10^{-3}$)	
	CD site	EF site
La^{3+}		61
Pr^{3+}		32
Nd^{3+}		26
Sm^{3+}		15.7
Eu^{3+}	94	16.4
Gd^{3+}	90	13.0
Tb^{3+}	97	7.0
Dy^{3+}	119	5.4
Ho^{3+}	124	5.0
Er^{3+}	175	2.5
Tm^{3+}	148	2.0
Yb^{3+}	146	1.3
Lu^{3+}	179	1.2

[a] From Corson et al. (1983b), with permission.

Ln^{3+} ions, Yb^{3+} has the largest ratio of pseudocontact to contact shifts (Section 3.2). Certain useful peak assignments have proved possible. For instance, the C_4-H and C_2-H resonances are unambiguously assigned in the ^{1}H-NMR spectrum of carp parvalbumin (Williams *et al.*, 1983). As X-ray crystallography shows this residue to be closer to the EF site than the CD site, it can be used to monitor changes at the EF site. Assignments have also been possible for the amino terminal protons and seven methyl groups. As a major contribution to the linewidths of the shifted peaks comes from susceptibility relaxation mechanisms, the Ln^{3+}–H distance can be obtained. The results show that, although the solution and crystal structures of parvalbumin are very similar, differences exist.

Corson *et al.* (1986) have recently produced a series of modified parvalbumins by carboxypeptidase treatment of carp parvalbumin. These derivatives, differing only in the length of their C-terminal F helix, were compared with respect to the Ln^{3+} off rates and binding constants at the EF site. Removal of the first two amino acids, Ala-108 and Lys-107, had very little effect on these parameters. However, further deletions seriously impaired Ln^{3+} binding and produced major alterations in the solution structure of the molecule. Furthermore, the CD site was remarkably sensitive to alterations in the F helix. NMR spectroscopy of native and Ln^{3+}-substituted parvalbumins has also been used to compare the solution structures of different types of parvalbumins from various species (Lee *et al.*, 1985; Williams *et al.*, 1986).

4.4.2 Troponin C

Troponin (TN) is a trimer found in skeletal muscle, where it helps to regulate muscle contraction by controlling the interaction of actin and myosin. It has three dissimilar subunits, TN-T (M.W. 37,000), TN-I (M.W. 23,000), and TN-C (M.W. 17,846). Rabbit skeletal muscle TN-C is a single polypeptide of 159 amino acid residues. It has no tryptophan residues and only two tyrosines, one of which is located in Ca^{2+}-binding site III. There are four internally homologous regions labeled I–IV from the N-terminus. Each region has a 10-residue Ca^{2+}-binding loop, with carboxyl side chains, flanked on both sides by α-helical segments. Sites I and II each bind Ca^{2+} specifically with a K_d of about 3 μM. Sites III and IV bind Ca^{2+} strongly, with a K_d of 0.05 μM, and Mg^{2+} weakly, with a K_d of 0.2 mM. The first two sites control the regulatory activity of troponin C. In bovine cardiac TN-C, amino acid substitutions in region I have destroyed its Ca^{2+}-binding capability, and the protein binds only three Ca^{2+} ions, two with high affinity and one with low affinity (Kretsinger, 1976).

The crystal structure of turkey TN-C at a resolution of 2.8 Å has recently been published (Herzberg and James, 1985). TN-C is shaped as a dumbbell, 75 Å long with about 67% of the amino acids in a helical conformation. It has two domains joined by a nine-turn helix. Although crystals were grown in the presence of excess Ca^{2+}, only sites III and IV were occupied in the final product. Eu^{3+} and Lu^{3+} were able to replace Ca^{2+} at sites III and IV, while Tm^{3+} replaced Ca^{2+} only at site III. Of these, replacement by Tm^{3+} was the most isomorphous and Eu^{3+} the least. At site III, the Ln^{3+} ion was bound within 1 Å of the position of Ca^{2+}, whereas at site IV it was displaced by 5.5 Å. In addition, another Ln^{3+}-binding site outside loop I was detected.

TN-C enhanced Tb^{3+} luminescence through energy transfer from Tyr-109 (Donato and Martin, 1974; Leavis et al., 1980), which is homologous to the Phe-57 residue in the CD Ca^{2+} site of parvalbumin (Section 4.4.1). Quenching of protein fluorescence at 310 nm also occurs. Each of the first two Tb^{3+} ions added to TN-C accepted energy from Tyr-109, whereas the third and fourth Tb^{3+} ions did not luminesce (Donato and Martin, 1974; Miller et al., 1975; Leavis et al., 1980). The Tb^{3+} emission was circularly polarized, the CPE spectrum revealing three components to the $^5D_4 \rightarrow {}^7F_5$ transition. As CPE from Tb^{3+}-substituted TN-C was extinguished upon combination with the other TN subunits, this may provide a sensitive monitor of TN subunit interactions. Whereas Ca^{2+} enhanced the endogenous Tyr fluorescence of TN-C, various Ln^{3+} ions did not.

Analysis of the stopped-flow kinetics of the fluorescent changes and stoichiometric considerations suggested that Tb^{3+} preferentially bound to the high-affinity, Ca^{2+}–Mg^{2+} sites III and IV. The affinity of Tb^{3+} for these sites was considerably higher than that of Ca^{2+}. However, Tb^{3+} produced similar changes to Ca^{2+} in the far-UV circular dichroism (CD) spectrum of TN-C, indicative of greater α-helicity (Leavis et al., 1980). The largest conformational changes followed addition of the first two ions. Comparison of the UV absorption difference spectra indicated subtle differences in the effects of Ca^{2+} and Tb^{3+}. In particular, two positive peaks at 274 and 280 nm in the Tb^{3+}-substituted protein were only 65–70% of those of the Ca^{2+} form. Such changes could affect the environment of Tyr-109, leading to the differential effects of Tb^{3+} and Ca^{2+} upon Tyr fluorescence.

Direct laser excitation of Tb^{3+} and Eu^{3+} permits study of Ln^{3+} ions at the "silent" sites I and II. Wang et al. (1981) determined K_a values at the high-affinity sites III and IV of $5.2 \times 10^8 M^{-1}$ for Tb^{3+} and $4.7 \times 10^9 M^{-1}$ for Eu^{3+}. At the low-affinity sites I and II, the K_a values were $9.7 \times 10^6 M^{-1}$ for Tb^{3+} and $5.3 \times 10^7 M^{-1}$ for Eu^{3+}. The excitation spectrum for TN-C:Eu^{3+} contained only two peaks, suggesting that the binding environment at each of the two individual sites within each class was identical. In support of this, they reported that a proteolytic fragment of TN-C which contained only the low-affinity sites I and II gave only one excitation peak. Measurements of the emission lifetimes in D_2O and H_2O provided hydration numbers for Eu^{3+} of 2.1 ± 0.2 at the high-affinity sites and 2.8 ± 0.5 at the low-affinity sites. According to Wang *et al.* (1981), the higher affinity of Eu^{3+} relative to Ca^{2+} can be accounted for by positive entropy changes due to the dehydration which results from binding. Aqueous Eu^{3+} ions are coordinated to eight or nine water molecules (Section 2.3). Upon binding to TN-C, six or seven of these molecules are displaced, whereas Ca^{2+} ions probably lose only five water molecules upon binding. Dehydration of Tb^{3+} ions is not as intense as that of Eu^{3+} ions, accounting for the slightly lower affinity of the former. Similarly, differences in the hydration of Eu^{3+} ions at sites I and II and those at sites III and IV might explain the different affinities of Eu^{3+} for these sites. Lanthanides interact with synthetic peptide analogs of site III (Marsden *et al.*, 1988).

TN-C from bovine cardiac muscle lacks the Ca^{2+} site I. However, its Tb^{3+} luminescence behavior is similar to that of lapine TN-C (Brittain *et al.*, 1976), as might be predicted from the foregoing discussion. Although the amino acid sequences of bovine and lapine TN-C differ by 35%, the arrangement whereby Tyr-109 interacts with sites III and IV is strongly conserved. It seems likely that the peptide carbonyl oxygen of Tyr-109 is part of the site III Ca^{2+}-binding loop of lapine TN-C, while its aromatic side chain overlies Ca^{2+} site IV. This configuration is analogous to that described for parvalbumin (Section 4.4.1), where the side chain of Phe-57 in the CD loop overlies the EF site. In the presence of $6M$ urea, both Ca^{2+} and Tb^{3+} preferentially bind to a single site, probably site III, on TN-C (Leavis *et al.*, 1980). As mentioned at the end of Section 4.4.1, there is disputed evidence of quenching of Tb^{3+} luminescence at high Tb^{3+}:TN-C stoichiometries.

4.4.3 Calmodulin

Calmodulin (CaM) is a ubiquitous Ca^{2+}-binding protein which mediates intracellular responses to Ca^{2+} fluxes. It does this by interacting with specific receptor proteins in a Ca^{2+}-dependent manner. Bovine brain CaM has a M.W. of 16,700 and 148 amino acid residues. Its complete amino acid sequence has been determined by Watterson *et al.* (1980). It is a highly acidic protein of pI 4.1, with an unusual amino acid, trimethyllysine, at position 115. There are two tyrosine residues at positions 99 and 138, but no cysteine, hydroxyproline, or tryptophan.

CaM has been crystallized and its three dimensional structure determined at 3.0 Å resolution (Babu *et al.*, 1985). It is a dumbbell-shaped molecule, with two globular lobes connected by a long, exposed α-helix. Each lobe binds two Ca^{2+} ions through the helix-loop-helix structure that is typical of "E-F" hand proteins. Ca^{2+}-binding domains I and II reside in the N-terminal globular region, and Ca^{2+}-binding domains III and IV in the C-terminal globular region. Both X-ray crystallography and luminescence studies with Tb^{3+} and Eu^{3+} (Wang, 1986) agree that sites I and II on the N-terminal domain are 1 nm from sites III and IV at the C-terminal domain. Further luminescence studies with Ln^{3+} ions have demonstrated that the inhibitory component of troponin I binds to CaM at a position such that its Cys-133 residue lies 2.7 nm from sites I and II, and 2.5 nm from sites III and IV (Wang, 1986).

Upon binding Ca^{2+}, the protein undergoes marked conformational changes, exposing a hydrophobic domain and producing an 8–10% increase in α-helical content with a corresponding decrease in random coil content (Klee, 1977). At the same time, the intrinsic tyrosine fluorescence of the protein is doubled. Selective nitration has shown that Tyr-138 is responsible for this (Wang and Aquaron, 1980). In this state, the protein is more compact and remains stable in 8 M urea. Most of the conformational changes occur during the binding of only one or two Ca^{2+} ions to CaM (Klee, 1977; Dedman *et al.*, 1977), although other studies have claimed that such changes occur over the whole binding range (Wolf *et al.*, 1977; Walsh *et al.*, 1978).

The presence of two specific tyrosine residues in mammalian CaM has been exploited for studies of Tb^{3+} luminescence. Tyr-99 is located in the Ca^{2+}-binding site III, and Tyr-138 in site IV. At a stoichiometry of two Tb^{3+} ions per molecule of CaM, there is little change in luminescence, suggesting that Tb^{3+} binds first to sites where there is no energy transfer, presumably sites I and II. Selective nitrosylation of Tyr-99 or Tyr-138 confirms this (Wang *et al.*, 1982). Addition of three and four ions of Tb^{3+} per molecule produces large increases in luminescence as sites III and IV are occupied (Wang *et al.*, 1982; Kilhoffer *et al.*, 1980), although others have claimed an appreciable increase in luminescence upon adding two and three Tb^{3+} ions (Wallace *et al.*, 1982). Binding of Tb^{3+} to CaM produces a considerable degree of conformation rearrangement, as evidenced by changes in its tyrosine fluorescence and circular dichroism at 222 nm (Wang *et al.*, 1982). The excitation spectrum of Eu^{3+}-saturated CaM shows only one peak, suggesting that, despite the different affinities of lanthanides for the high- and low-affinity sites, the microenvironment of the four sites is remarkably similar. However, Horrocks *et al.* (1983) have challenged this result, claiming that there are differences, depending on whether two or four Eu^{3+} ions are bound. Each Eu^{3+} ion coordinates 1.38 ± 0.07 water molecules (Wang *et al.*, 1982). From computer simulations, Wang and Aquaron (1980) have estimated that Tb^{3+} binds to the high-affinity sites with a K_d of $10^{-8}M$, a figure compatible with the estimate of $<20\,nM$ provided by Wallace *et al.* (1982). The K_d for Ca^{2+} is around $1\,\mu M$. Wang *et al.* (1984) subsequently used the preference of Ln^{3+} ions for sites III and IV to help confirm that Ca^{2+} shows greater affinity for sites I and II. As the authors point out, in view of the chemical similarities between Ca^{2+} and Ln^{3+} ions (Section 2.5; Table 2-11), it is surprising to find this dichotomy in binding site preference. This is especially so as the site specificity of Ca^{2+} and Tb^{3+} is the same in troponin C (Section 4.4.2). Martin (1984) has suggested that the greater affinity of Tb^{3+} for sites I and II in CaM results from their -4 charge. Sites III and IV have only a -3 charge. However, not all Ln^{3+} behave like Tb^{3+}. Results with spin-labelled CaM show that, while Eu^{3+} and Tb^{3+} bind in the opposite order to Ca^{2+}, Lu^{3+} and Er^{3+} have the same site preference as Ca^{2+}. Furthermore, Nd^{3+} and La^{3+} behave differently from both Ca^{2+} and Eu^{3+} and Tb^{3+} (Buccigross and Nelson, 1986).

Octopus CaM, unlike the mammalian form, has only one Tyr residue, located in a position homologous to the Tyr-138 residue of bovine brain CaM (Kilhoffer *et al.*, 1980). The tyrosine fluorescence of octopus CaM increased in a biphasic fashion upon binding Tb^{3+}. A plateau value was reached at a Tb^{3+}:CaM stoichiometry of about 3.5. As quenching was not observed, the reduction in Tyr fluorescence of mammalian CaM at high Tb^{3+} stoichiometries probably involved Tyr-99. Octopus CaM also enhanced Tb^{3+} luminescence upon excitation at $280\,nm$, luminescence being mostly associated with Tb^{3+}:CaM stoichiometries of $4:1$ and $5:1$. The authors suggest that this reflects binding of Tb^{3+} to site III, which lacks Tyr, before site IV, which contains the only Tyr in the molecule. As the enhancement of Tb^{3+} luminescence is small compared to that observed with mammalian CaM, domain IV on octopus CaM may have a much lower affinity for Tb^{3+} than domain III (Kilhoffer *et al.*, 1980). This might also explain why the dose–response curve for Tb^{3+} luminescence continues to rise beyond a Tb^{3+}:CaM molar ratio of 4.

There is evidence that Ln^{3+}-substituted CaM may be functionally active. In addition to inducing similar changes to Ca^{2+} in Tyr fluorescence, the CD spectrum, and the UV absorption spectrum of CaM, it also produces the characteristic change in electrophoretic mobility of CaM on polyacrylamide gels (Wallace *et al.*, 1982). In the presence of Tb^{3+}, CaM forms complexes with troponin I and stimulates phosphodiesterase, although high concentrations of Tb^{3+}, but not Ca^{2+}, are inhibitory (Tallant *et al.*, 1980). La^{3+} can also substitute for Ca^{2+} in the Ca^{2+}-dependent attachment of CaM to cell membranes (Lau and Gnegy, 1980). At suboptimal concentrations, Ca^{2+} and La^{3+} have an additive effect. However, nanomolar concentrations of La^{3+}, Ce^{3+}, or Tb^{3+} inhibit the CaM-regulated guanylate cyclase of *Paramecium* by dissociating CaM from the enzyme (Klumpp *et al.*, 1983). Regardless of any stimulatory effect that low concentrations of Ln^{3+} might have, high concentrations usually inhibit CaM-mediated processes. Indeed, treatment with millimolar concentrations of La^{3+} is one technique used to dissociate CaM from enzymes and membranes (e.g., Gross *et al.*, 1987).

4.4.4 Calbindin

Calbindin (formerly known as intestinal Ca^{2+}-binding protein) (1CaBP) is an "E-F hand" protein of M.W. 9000 with two high-affinity Ca^{2+}-binding sites. The C-terminal site has a classical "E-F hand" configuration. Here the Ca^{2+} ion is coordinated by three negatively charged oxygen ions and one main-chain carbonyl oxygen. At the N-terminal site, the E-F hand is distorted into a "pseudo E-F hand" by substitution of Pro-23 in the binding loop. Here the Ca^{2+} ion is coordinated by only one negatively charged oxygen from a side-chain carboxylate and by four main-chain carbonyl oxygens (Szebenyi *et al.*, 1981). There are several phenylalanine residues and a single tyrosine residue, estimated to be about 10 Å from the N-terminal and 20 Å from the C-terminal sites. K_d values of 1×10^{-7} to $5 \times 10^{-7} M$ have been obtained for each Ca^{2+}-binding site. Slightly greater affinity may exist for the EF site as, at a Ca^{2+}:calbindin molar ratio of 1:1, 65% of the Ca^{2+} binds to the EF site and 35% to the pseudo EF site (Vogel *et al.*, 1985). The two sites are interactive, suggesting possible cooperativity.

Equilibrium dialysis experiments suggest that porcine calbindin binds two equivalents of Tb^{3+} per mole, with K_d values of 0.29 μM and 3.51 μM (Chiba *et al.*, 1984). A typical Tb^{3+} emission spectrum was obtained upon excitation at around 260 nm, titration of which provided K_d values of 0.21 μM and 1.78 μM. According to O'Neil *et al.* (1984), Tb^{3+} has a $K_a > 10^7 M^{-1}$ at the high-affinity site and a K_a of about $10^5 M^{-1}$ at the low-affinity site. Titration profiles of Tb^{3+} luminescence and tyrosine fluorescence of the Tb^{3+}–calbindin complex in the presence of different concentrations of Ca^{2+} suggested that Tb^{3+} had a greater affinity than Ca^{2+} for one of the two sites, but a weaker affinity than Ca^{2+} for the other. The pH dependence of Tb^{3+} luminescence suggested that carboxyl groups provided the ligands for metal binding, as expected from the X-ray crystallographic data. Inspection of the excitation spectrum suggested phenylalanine as quantitatively the major primary chromophore (Chiba et al., 1984). Luminescence data suggest that Tb^{3+} binds first to the C-terminal EF site and secondly to the N-terminal, pseudo EF site. This finding agrees with that of Szebenyi et al. (1981), who, upon soaking crystals of bovine calbindin in 1 mM $NdCl_3$, achieved replacement of Ca^{2+} by Nd^{3+} only at the C-terminal site. O'Neil and co-workers (1984) have suggested that the higher affinity of Tb^{3+} ions for the C-terminal site results from its -3 charge, the N-terminal site having only a -1 charge, a similar explanation having been proposed to explain site selectivity in CaM (Martin, 1984).

Proton NMR spectroscopic investigations have supported the conclusions of these studies. Using $^{43}Ca^{2+}$, it was shown that La^{3+} displaced Ca^{2+} from the EF site (Fig. 4-3). At La^{3+}:calbindin ratios above 1, precipitation occurred (Vogel et al., 1985). NMR studies have provided K_d values of $(4.9 \pm 6.7) \times 10^{-4} M$ and $(2.3 \pm 1.2) \times 10^{-4} M$ for Lu^{3+} at each site (Shelling et al., 1985). At Lu^{3+}:calbindin ratios above 2, the protein began to aggregate. Binding of Yb^{3+} was also sequential, with K_d values of $<10^{-7} M$ and $(5.0 \pm 1.6) \times 10^{-5} M$. From competitive studies with Ca^{2+}, the former value was refined to $\sim 2 \times 10^{-9} M$. Because of the relatively low affinity of Yb^{3+} for the second site, it could only displace one Ca^{2+} from calbindin.

Thus, all studies agree that binding of Ln^{3+} to calbindin is sequential, with the C-terminal EF site having the greater affinity. Ca^{2+} has a greater affinity than Ln^{3+} ions for the N-terminal, pseudo EF site, while the reverse is true at the EF site. It is of interest that Ln^{3+} ions show a strong site selectivity that Ca^{2+} lacks. In Section 4.4.3, we saw that the site selectivity of Ca^{2+} and Ln^{3+} ions for CaM is different, while for TN-C (Section 4.4.2) it is the same. Suggestions of communication between the two binding sites on calbindin have been corroborated by NMR studies and laser-excited luminescence studies with lanthanides (Vogel et al., 1985). With ^{43}Ca present in the pseudo EF site, the NMR spectrum is sensitive to whether Ca^{2+} or La^{3+} exists in the other site. Furthermore, when La^{3+} occupies the EF site, Ca^{2+} can no longer bind to the pseudo EF site.

Ca^{2+} and Ln^{3+} ions appear to produce similar conformational shifts in calbindin. For example, they both produce a strong positive ellipticity band at 276 nm upon binding to calbindin, due to perturbations in tyrosine ellipticity. The endogenous tyrosine fluorescence of the protein is greatly enhanced on binding the first Tb^{3+} ion, but addition of the second Tb^{3+} ion quenches (O'Neil et al., 1984). Proton NMR spectra are affected in a similar fashion by the addition of Ln^{3+} or Ca^{2+} (Shelling et al., 1985).

Abnormalities in the Tb^{3+} luminescence titration curves at high pro-

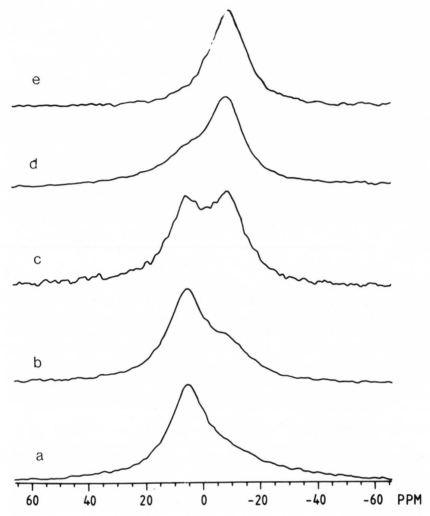

Figure 4-3. Calcium-43 NMR spectra of calbindin at increasing La^{3+} concentrations: (a) 0 La^{3+}; (b) 0.2 equiv. La^{3+}; (c) 0.5 equiv. La^{3+}; (d) 0.7 equiv. La^{3+}; (e) 1.0 equiv. La^{3+}. [ICaBP] = 2 mM. Spectral changes reflect displacement of Ca^{2+} from the EF site to the pseudo EF site. From Vogel *et al.* (1985), with permission.

tein concentrations suggest there may be a third site of attachment at which Tb^{3+} luminescence is quenched. In discussing their NMR data, Shelling *et al.* (1985) have also speculated about a putative third site. The X-ray crystallographic data (Szebenyi *et al.*, 1981) support this suggestion.

As already described, similar speculations have been discussed for parvalbumin and TN-C.

4.4.5 Oncomodulin

Oncomodulin is an "E-F hand" protein found in malignant and fetal tissue. Rat tumor oncomodulin has 108 amino acid residues, showing a high degree of sequence homology with parvalbumin. Both of its Ca^{2+}-binding sites have high affinity for Ln^{3+} (Henzel $et\ al.$, 1986), but unlike parvalbumin, the CD-site shows only a low preference for Ln^{3+} over Ca^{2+}, with a K_{Ca}/K_{Eu} ratio of 11 ± 1. At the EF site of oncomodulin, this ratio is 300 ± 80 (Henzel and Birnbaum, 1988).

4.5 The "γ-Carboxyglutamate" (Gla) Proteins

One group of Ca^{2+}-binding proteins contains an unusual amino acid, γ-carboxyglutamic acid (Gla) (Section 4.2), formed post-translationally in liver by a vitamin K-dependent process. Several of these "Gla-proteins" are included in the so-called vitamin K-dependent plasma clotting proteins. These are protein C, protein S, and the clotting factors II (prothrombin), VII, IX, and X. The role of each clotting factor in blood coagulation is indicated in Fig. 8-6. Factor XIII, which is not a Gla-protein, is discussed in Section 4.6. All 10–12 of the Gla residues are concentrated in the N-terminal region of each protein. Nine of them are identically placed in each, six Gla residues occurring in pairs.

All the vitamin K-dependent plasma proteins bind to membranes in a Ca^{2+}-dependent manner, Ca^{2+} ions inducing the conformational changes necessary to do this. There are multiple (5–15) Ca^{2+}-binding sites, two of which have relatively high affinity, with a K_d of $0.2\,mM$ (Bloom and Mann, 1978). Several weaker sites have K_d values of up to $1\,mM$. Lanthanides are potent anticoagulants (Section 8.8).

4.5.1 Prothrombin (Factor II)

Prothrombin is present in plasma at a concentration of about 150 μg/ml. It is a glycoprotein of M.W. 70,000, with 10 Gla residues among its 582 amino acids; these reside in the N-terminal region. Prothrombin binds approximately 10 Ca^{2+} ions cooperatively, with a Hill coefficient of about 1.5. Half-saturation occurs at $0.6\,mM$ Ca^{2+}. Thrombin cleaves

Table 4-5. Binding Data for Gd^{3+} with Bovine
Prothrombin and Its Derivatives[a]

	K_d (μM)	mol Gd^{3+}/mol protein
Prothrombin	0.75	1.88 ± 0.32
Prethrombin 1	1.1	0.95 ± 0.23
Prothrombin fragment 1	>0.16	1.85 ± 0.37
Prethrombin 2	<0.81	0.85 ± 0.2
Prothrombin fragment 2	1.6	
α-Thrombin	0.1	

[a] From Furie et al. (1976), with permission.

prothrombin into prethrombin 1 (M.W. 51,000) and prothrombin fragment 1 (M.W. 23,000), which contains residues 1–156. Daughter peptides from these products include prethrombin 2 (M.W. 41,000), prothrombin fragment 1.2 (M.W. 36,000), and prothrombin fragment 2 (M.W. 13,000). Fragment 1 contains the 10 Gla residues known to exist near the N-terminus of prothrombin. The conversion of prothrombin to thrombin is mediated by factor Xa, being greatly accelerated by Ca^{2+} and phospholipid.

Furie et al. (1976) examined the binding of $^{153}Gd^{3+}$ to each of these molecules, deriving the results shown in Table 4-5. The authors interpreted their data as indicating the presence of a large, heterogeneous population of weak binding sites on thrombin. Further information on the role of Gla residues in metal-ion binding was obtained by using a fragment obtained from thrombin by tryptic digestion. This fragment, comprising amino acids 12–44, contained 8 of the 10 Gla residues of thrombin (Furie et al., 1979). Scatchard analysis of the binding of $^{153}Gd^{3+}$ to this fragment provided evidence of a single high-affinity site of $K_d = 0.55\,\mu M$ and four to six lower-affinity sites of $K_d = 4$–$8\,\mu M$. The S-carboxymethyl derivative of this fragment retained only the lower-affinity sites. The authors suggested that the high-affinity site involved two Gla residues, while the low-affinity sites involved only one Gla residue. Paramagnetic Ln^{3+} ions, but not the diamagnetic member, La^{3+}, perturbed the ^{13}C-NMR spectrum of the peptide. Making use of the broadening of the resonances induced by Gd^{3+}, it proved possible to identify the ϵ carbon atoms of Arg-16 and Arg-25, as well as the γ, β, and α carbons of Gla residues, as lying close to the Ln^{3+}-binding site. From such data, a three-dimensional model of the conformation of the fragment and its binding site was proposed (Furie et al., 1979).

Bovine prothrombin residues 1–39 contain all the Gla residues of the protein. At low Eu^{3+} :peptide 1–39 ratios, about six water molecules were displaced from the Eu^{3+} ion. For prothrombin fragment 1, this value averaged 7.5 (Marsh et al., 1981). Eu^{3+} luminescence titration of human prothrombin indicated that four equivalents of Eu^{3+} bound per mole of protein (Rhee et al., 1982). Analysis of the excitation spectrum suggested that two different binding environments existed. The binding curve suggested positive cooperativity. Each of the sequestered Eu^{3+} ions retained 2.4 ± 0.5 coordinated water molecules at one of the classes of binding sites, and 2.6 ± 0.5 at the other class. From such data, it was deduced that at least three Gla residues were necessary to displace the required number of water molecules from hydrated, aqueous Eu^{3+}.

Rather different results were obtained from luminescence studies of the binding of Tb^{3+} to bovine prothrombin and fragment 1 (Sommerville et al., 1985). Approximately 10 to 12 Ln^{3+}-binding sites were detected on prothrombin, with about 6 of them present in fragment 1. Three classes of Tb^{3+}-binding site were identified. Three Tb^{3+} ions bound to the high-affinity sites in a noncooperative manner. Quenching experiments with Ho^{3+} suggested that the three high-affinity sites were clustered to within 6–12 Å of each other. The three medium-affinity sites were characterized by greater Tb^{3+} luminescence due to energy transfer from Trp residues. These too appeared to be clustered. There were four to six weak-affinity sites which, unlike the other sets of binding site, were not present in fragment 1. Although Tb^{3+} luminescence at these sites by energy transfer from Trp residues was strong, direct excitation produced only weak luminescence. Competition experiments confirmed that Tb^{3+} was binding to Ca^{2+} sites.

Some patients suffering from clotting disorders produce abnormal prothrombin molecules with altered Gla contents. Examination of such molecules has impressively confirmed the importance of Gla residues in Ca^{2+}- and La^{3+}-binding and in physiological activity (Table 4-6).

Ca^{2+} and Ln^{3+} ions show similar activity in reducing the intrinsic fluorescence of prothrombin and fragment 1 (Prendergast and Mann, 1977; Sommerville et al., 1985). Antibodies specific for the conformation of bovine fragment 1 in the presence of Ca^{2+} have permitted this matter to be addressed in some detail. The data of Marsh et al. (1981) show that Eu^{3+}-substituted fragment 1 has a lower affinity than the Ca^{2+}-containing peptide for the antibody. Thus, the conformational changes induced experimentally by Ln^{3+} ions are not identical to the ones induced physiologically by Ca^{2+}. This finding may be relevant to the functional activity of Ln^{3+}-substituted prothrombin, the consideration of which follows next.

The conversion of prothrombin to thrombin is mediated by activated

Table 4-6. Gla Content and Gd^{3+}-Binding Characteristics of Normal and Variant Human Prothrombin[a]

Form of prothrombin	Gla content	High-affinity sites		Low-affinity sites	
		N^b	K_d (nM)	N^b	K_d (nM)
Normal	10.4 ± 0.7	2.1	36	3.9	1017
"Abnormal"	0.6 ± 0.3	0		0.54	1025
Variant 1	7.2 ± 0.8	0.72	15	2.8	2750
Variant 2	4.7 ± 0.6	0.88	10	1.8	4050
Variant 3	6	0.95	47	2.2	3500

[a] From Borowski et al. (1985), with permission.
[b] N = Number of sites per protein molecule.

factor X (X_a) and requires activated factor V (V_a), Ca^{2+}, and phospholipid. A variety of lanthanides (Fig. 4-4) may substitute for Ca^{2+} in the activation of bovine prethrombin 1 or prothrombin to thrombin by factor X_a in the presence of factor V_a and phospholipid. However, the reaction rate is reduced to about 10% of normal. The larger lanthanides are less effective than the smaller ones as Ca^{2+} substitutes. Furthermore, the yield of thrombin is much lower in the presence of the larger Ln^{3+} ions (Furie et al., 1976). In contrast, Rhee et al. (1982) found that Eu^{3+} could support human prothrombin conversion as effectively as Ca^{2+}. Whether this reflects differences between bovine and human prothrombin is unknown. Electrophoresis confirmed that the same reaction products were seen whether Ca^{2+} or Gd^{3+} ions were used as cofactors in the conversion of thrombin. Although the initial rate of thrombin formation from prothrombin with $20\,\mu M$ Gd^{3+} was only 17% of that seen with $2\,mM$ Ca^{2+}, the final yield was the same if the reaction was allowed to run long enough. The dose–response curves for the generation of products from Gd^{3+}-thrombin are shown in Fig. 4-5. Suboptimal concentrations of Gd^{3+} ($<20\,\mu M$) or Ca^{2+} ($<0.7\,mM$) reduced both the rate of reaction and the final yield of thrombin. However, with thrombin formation from prethrombin 1, maximum rates of conversion (25% with $2\,mM$ Ca^{2+}) were seen with $10\,\mu M$ Gd^{3+}, higher concentrations being inhibitory.

The requirement for Ca^{2+} in prothrombin conversion has been ascribed to the necessary formation of a prothrombin–Ca^{2+}–phospholipid ternary structure, with Ca^{2+} acting as a bridge. Lanthanides can substitute for Ca^{2+} in promoting the binding of prothrombin to phospholipid (Rhee et al., 1982; Prendergast and Mann, 1977; Nelsestuen et al., 1976). By measuring the emission lifetimes of excited Eu^{3+} in the presence of D_2O and H_2O, Rhee et al. (1982) were able to determine the hydration numbers

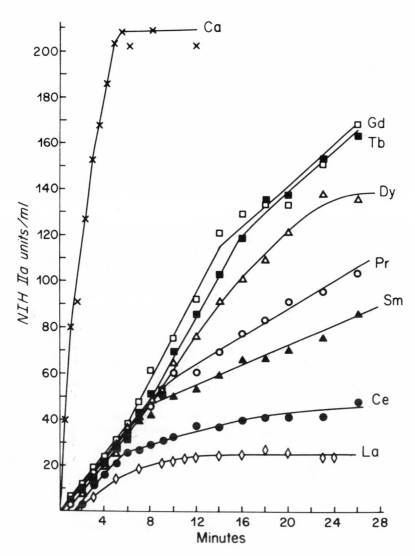

Figure 4-4. Activation of prothrombin by factor X_a in the presence of Ca^{2+} or lanthanide ions. No activity was observed in the absence of metal ions. From Furie *et al.* (1976), with permission.

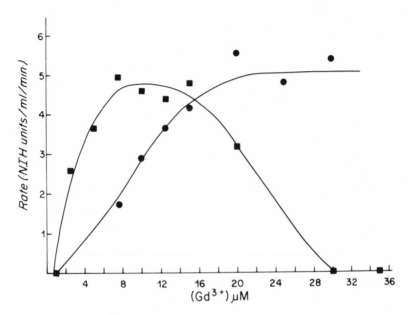

Figure 4-5. Rate of thrombin generation from prothrombin (●) and prethrombin 1 (■) at various Gd³⁺ concentrations. From Furie *et al.* (1976), with permission.

of Eu^{3+} under various circumstances. Binding of Eu^{3+} to prothrombin displaced approximately six water molecules from the hydration shell of the free, aqueous ion. Upon binding of Eu^{3+} to phospholipid, about two to three molecules were displaced, indicating weaker association with phospholipid than with prothrombin. When Eu^{3+} was mixed with both prothrombin and phospholipid, it maintained the same hydration number as when mixed with prothrombin alone. As the addition of phospholipid failed to displace water molecules from the inner coordination sphere of prothrombin-bound Eu^{3+}, phospholipid was probably not bound directly to Eu^{3+} in the ternary complex. Thus, the role of Eu^{3+}, and hence Ca^{2+}, is probably not that of a bridge (Rhee *et al.*, 1982). Instead, the authors support an alternative mechanism whereby Ca^{2+} and Ln^{3+} ions produce a conformation change, as indicated by the altered intrinsic fluorescence of prothrombin, exposing a domain which interacts with phospholipid.

4.5.2 Factor X (Stuart–Prower Factor)

Bovine factor X is a disulfide-linked dimeric glycoprotein with a molecular weight of about 56,000 which is present in the plasma at a concentration of 10–15 μg/ml.

Factor X can be activated by both the intrinsic and extrinsic coagulation pathways (Section 8.8; Fig. 8-6). In the former, activation requires factor IX, factor VIII, phospholipid, and Ca^{2+}. In the latter, activation is mediated by a tissue factor (factor III) in the presence of factor VII. Factor X may also be activated nonphysiologically by trypsin or by Russell's viper venom. The latter has attracted attention as a potential source of an alternative activator of possible therapeutic use in certain hemophilias. The venom contains a serine endopeptidase of M.W. 60,000 which activates factor X, the only known substrate for this enzyme. Also known as venom coagulant protein, it activates factor X by cleaving the same heavy chain Arg–Ile bond that is hydrolyzed during physiological activation. This reaction has an absolute requirement for Ca^{2+} and is inhibited by lanthanides (Furie and Furie, 1975).

Interaction between lanthanides and the substrate, factor X, appears to be at the crux of the inhibition. Although Gd^{3+} has little affinity for the venom coagulant protein, this Ln^{3+}, Sm^{3+}, and Yb^{3+} each bind readily to factor X. Two high-affinity Ln^{3+}-binding sites of K_d 2×10^{-7}–$4 \times 10^{-7} M$ and four to six low-affinity sites of K_d 1×10^{-5}–$6 \times 10^{-5} M$ have been identified. There are two Ca^{2+}-binding sites with $K_d \approx 3 \times 10^{-4} M$ and several weaker sites with $K_d > 1 mM$. Tb^{3+} luminescence studies suggest that at least one Ln^{3+}-binding site is close to a Tyr or Trp residue. Little change in the UV absorption spectrum of factor X was found on adding Tb^{3+}, indicating that minimal conformational change occurs. Luminescence titration studies suggest a K_d of 3–19 μM. This value is about the same as that determined by rate dialysis for the lower-affinity Ln^{3+}-binding sites. Presumably, the high-affinity Ln^{3+}-binding sites are insufficiently close to a suitable aromatic amino acid residue to be titratable by Tb^{3+} luminescence. Kinetic analyses of the inhibition of factor X activation by Nd^{3+} yield a K_i of 1–4 μM, a range of values intermediate between the K_d values for the high- and low-affinity sites.

Ca^{2+} is required for binding of the coagulant protein to factor X. Nd^{3+} inhibition of activation, while competitive with respect to Ca^{2+}, still permits binding of substrate and enzyme. However, as with staphylococcal nuclease (Section 4.14) and clostridiopeptidase A (Section 4.11.1), the enzyme–Ln^{3+}–substrate complex is abortive. The ability of Ln^{3+} ions to serve as cofactors for the binding of substrate and enzyme yet to inhibit digestion of the substrate has found use in certain affinity chromatographic methods (see Section 3.8.2).

If Ca^{2+} were merely acting as a "glue," linking enzyme to substrate, we might expect at least some enzymic activity to occur in the presence of lanthanides. That this is not so indicates that Ca^{2+} may play an additional role, possibly as a participant in catalysis itself. This suggestion is interesting in light of the effects of Ln^{3+} on trypsinogen activation.

Both factor X and trypsinogen are zymogens of serine proteinases which are proteolytically activated in a Ca^{2+}-dependent manner. Furthermore, their active sites and N-terminal amino acid sequences show marked homology. However, as we have seen, Ln^{3+} ions inhibit factor X activation but accelerate trypsinogen activation, where the mechanism is thought to be a simpler one of charge masking (Section 4.9).

Activated factor X (X_a) is able to activate factor VII and, in the presence of Ca^{2+}, phospholipid, and factor V, to convert prothrombin to thrombin (Section 4.5.1). It retains the two high-affinity Ln^{3+}-binding sites of factor X, although the lower-affinity sites may be missing (Rhee $et\,al.$, 1984). Examination of the excitation spectrum of the $^7F_0 \rightarrow {}^5D_0$ transition of Eu^{3+} reveals only one maximum, suggesting that the molecular environments of the two high-affinity binding sites are very similar. Addition of more than two equivalents of Ln^{3+} per mole of factor X causes precipitation. Each sequestered Eu^{3+} ion retains two to three molecules of coordinated water, suggesting that three to six ligands are provided by factor X. By using Tb^{3+} as an energy donor at one Ln^{3+} site, and Nd^{3+}, Ho^{3+}, or Er^{3+} as an energy acceptor at the other, an intersite distance of 10.7 Å was calculated. Based on these data, the authors suggested that Gla-19 and Gla-20 provided the ligands for one of the high-affinity sites, with Gla-25 and Gla-26 doing this at the other site.

4.5.3 Factor IX (Christmas Factor)

Factor IX consists of a single polypeptide chain of M.W. 55,400 which contains 26% carbohydrate. In the presence of Ca^{2+}, it is proteolytically activated to factor IX_a by factor XI_a. Factor IX can also be activated by tissue factor in the presence of Ca^{2+} and factor VII. The blood clotting cascade is shown in Fig. 8-6. Factor IX_a is a serine proteinase which proteolytically activates factor X in the presence of factor VIII, phospholipids, and Ca^{2+}.

Factor IX contains two high-affinity Ca^{2+}-binding sites, with K_d values of 0.1 ± 0.02 mM, and several low-affinity sites. Tb^{3+} and Sm^{3+} were able to displace $^{45}Ca^{2+}$ from factor IX; no K_d values were given for the Ln^{3+} ions, but they showed greater affinity than Ca^{2+}. Although Gd^{3+}, Sm^{3+}, and Tb^{3+} had no effect on factor IX_a coagulant activity or factor XI_a esterase activity, they strongly inhibited the activation of factor IX by factor XI_a. In the presence of 2 mM Ca^{2+}, 0.1 mM Gd^{3+} completely inhibited activation. With 0.18 mM Tb^{3+} or Sm^{3+}, the rate of activation was only 35% of normal and the final yield only 15% of normal (Amphlett $et\,al.$, 1978).

4.6 Factor XIII (Fibrin Stabilizing Factor)

Factor XIII is not a Gla protein. Present in the plasma at a concentration of 10–$20\,\mu g/ml$, it is activated by thrombin in the presence of Ca^{2+}. Activated factor XIII plays a role in cross-linking fibrin clots. It enhanced Tb^{3+} luminescence, with the excitation spectrum suggesting a Trp donor. Quenching of the intrinsic fluorescence of the protein by Tb^{3+} was greater than that calculated from energy transfer considerations. This appears to have been due to conformational changes induced by the lanthanide.

Luminescence lifetime comparisons revealed that at low Eu^{3+} : factor XIII molar rations, the tightest-bound Eu^{3+} interacted with at least six protein ligands. As the Eu:protein ratio increased, increasingly numbers of Eu^{3+} were bound with lower coordination to protein ligands. At lower concentrations of protein, less nonspecific binding occurred. As less self-association occurred under these conditions, it was concluded that the low-affinity sites were associated with the aggregated form of the protein. The maximum rate of activation of factor XIII by thrombin in the presence of Eu^{3+} or Tb^{3+} was only 25% and 40%, respectively, of that catalyzed by Ca^{2+}. Activation by Eu^{3+} was maximal with $10^{-4} M\ Eu^{3+}$. At $1\,mM$ Eu^{3+}, activation was only 5% of that found with Ca^{2+} (Sarasua *et al.*, 1982).

4.7 Lysozyme

Lysozyme (EC 3.2.1.17) hydrolyzes primarily the β-1,4-glucosidic linkages between *N*-acetylmuramic and *N*-acetylglucosamine. Although not requiring Ca^{2+} for enzymic activity, lysozyme presented several advantages to Williams and his colleagues (Dwek *et al.*, 1971) in their pioneering studies of the possible use of Ln^{3+} ions as NMR shift and relaxation probes of the solution structures and functions of macromolecules. Not only is lysozyme's molecular weight of 14,500 low for an enzyme, but its three-dimensional structure, including a Gd^{3+}-binding site, was in the process of being determined by X-ray diffraction. In addition, the mechanism of action of the enzyme was well understood.

Accumulated X-ray crystallographic and NMR spectroscopic data indicate that lysozyme has two binding sites for Gd^{3+} (Fig. 4-6). At the higher-affinity site, Gd^{3+} coordinates with the carboxyl groups of Asp-52 and Glu-35. The weaker site lies $3.6\,Å$ away, near Asp-101. Gd^{3+} cannot bind simultaneously to both sites, probably because of repulsion between two trivalent cations so close together. Similar exclusion between two adjacent binding sites occurs with thermolysin (Section 4.3). However,

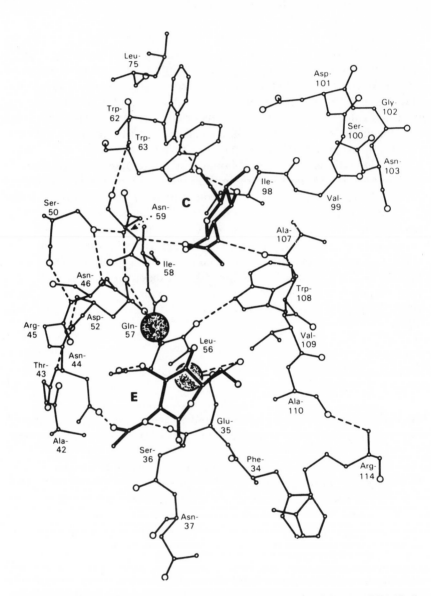

Figure 4-6. Active site cleft of lysozyme, showing the positions of the two Gd^{3+}-binding sites. Binding of Gd^{3+} competitively inhibits the binding of sugar monomers at E. From Perkins *et al.* (1981), with permission.

Table 4-7. Calculated Association
Constants for Ln^{3+} and
Lysozyme[a]

Ln^{3+}	K_a $(M^{-1} \times 10^3)$
La^{3+}	5.0 ± 1.4
Pr^{3+}	7.5 ± 3.7
Nd^{3+}	3.7 ± 2.1
Gd^{3+}	7.0 ± 2.6
Tb^{3+}	4.0 ± 1.7
Dy^{3+}	8.7 ± 4.0
Er^{3+}	6.6 ± 1.5
Tm^{3+}	4.1 ± 1.9

[a] From Ostroy et al. (1978), with permission.

when the strong binding site was weakened by chemical modification of Trp-108, the affinity of Gd^{3+} for the weak site was unaltered (Dill and Allerhand, 1977). Mutual exclusion between the two sites is consistent with the X-ray crystallographic data of Kurachi et al. (1975), indicating low occupancy of the second site. Solution studies have also indicated a Gd^{3+}:lysozyme ratio of 1:1 (Secemski and Lienhard, 1974).

As the higher-affinity Gd^{3+} site is also the substrate-binding site, it is not surprising that lanthanides inhibit the enzymic activity of lysozyme (Ostroy et al., 1978). Inhibition data were fitted to several models, the simplest one that described the experimental findings being a one-site model. Thus, the inhibition of lysozyme by Gd^{3+} could be ascribed to the binding of one Gd^{3+} ion at the active site. Association constants calculated from the inhibition data are given in Table 4-7. These values agree quite well with the K_a value of $2.1 \times 10^3 M^{-1}$ provided by UV difference spectroscopy (Secemski and Lienhard, 1974). However, proton relaxation experiments (Jones et al., 1974) suggested a K_d value for Gd^{3+} of 12.75 ± 0.71 mM at 25°C. These studies suggested that three molecules of water were coordinated with the Gd^{3+} bound to lysozyme.

A series of Ln^{3+} ions were employed to determine, by NMR spectroscopy, the structure of lysozyme in solution (Campbell et al., 1973, 1975). The authors found good agreement between the solution and crystal structure of the enzyme, although one of the assumptions of the analysis, that the magnetic susceptibility tensor of the paramagnetic Ln^{3+} ion is axially symmetric, has been challenged (Agresti et al., 1977). The binding of N-acetylglucosamine inhibitors to lysozyme has been studied with the aid of a variety of different Ln^{3+} ions, using both ^{1}H- and ^{13}C-NMR spectroscopy (Perkins et al., 1979, 1981). The inhibitors were found to bind at two independent sites, one of which appeared, from competition studies, to be close to the Ln^{3+}-binding site.

4.8 Proteins of Iron Transport and Storage

Serum contains an iron-transporting protein called transferrin. Human transferrin has a molecular weight of 80,000 and binds two equivalents of Fe^{3+} per mole. The two metal-binding sites are similar, but not identical. It will also bind several Ln^{3+} ions, with an order of efficiency $Er^{3+} \geqslant Ho^{3+} \geqslant Tb^{3+} > Er^{3+} > Nd^{3+} \approx Pr^{3+}$ (Luk, 1971). Two ions per mole of Tb^{3+}, Eu^{3+}, Er^{3+}, and Ho^{3+} are bound, but transferrin will only bind one ion per mole of the larger Ln^{3+} ions, such as Nd^{3+} and Pr^{3+}. The excitation spectrum of Eu^{3+} luminescence shows that the molecular environments around the two Eu^{3+}-binding sites on transferrin differ (O'Hara and Bersohn, 1982). This difference was explained by the possible greater negative environment of one site. UV difference spectra suggest that Tyr residues are influenced by Ln^{3+} binding (Luk, 1971), a conclusion supported by Tb^{3+} luminescence studies. Two tyrosine residues are involved in binding Tb^{3+}, Ho^{3+}, Er^{3+}, and one of the Eu^{3+} ions, but only one tyrosine residue is involved in binding the other Eu^{3+} ion, Nd^{3+}, and Pr^{3+} (Luk, 1971; Pecoraro et al., 1981). Binding of Tb^{3+} to transferrin slightly quenches its intrinsic tryptophan fluorescence. Comparison of the luminescence in D_2O and H_2O suggests that 1.4 water molecules are bound to each Tb^{3+} ion.

Attempts were made to measure the distance between the two metal-binding sites on transferrin, by inserting a Tb^{3+} ion at one site and a Fe^{3+} or Cu^{2+} ion as an energy acceptor at the other. Such studies have provided values of 25 ± 2 Å (Meares and Ledbetter, 1977) and 35.5 Å (O'Hara et al., 1981). Diffusion-enhanced energy transfer (Section 3.3) experiments have suggested that the Fe^{3+}-binding sites are probably buried 10–15 Å below the surface of the protein. The dissociation constants of Tb^{3+} for the two binding sites of transferrin are $4 \times 10^{-7} M$ and approximately $4 \times 10^{-5} M$ (Meares and Ledbetter, 1977). The luminescence of Tb^{3+} bound to transferrin is highly circularly polarized (Gafni and Steinberg, 1974). In contrast to other workers, Gafni and Steinberg (1974) concluded that the two Tb^{3+}-binding sites on transferrin are identical. Mössbauer spectroscopy has been applied to $^{151}Eu^{3+}$-substituted transferrin (Spartalian and Oosterhuis, 1973).

Conalbumin is a glycoprotein from egg white, with very similar properties to transferrin. The emission spectra of Tb^{3+} luminescence from Tb^{3+}-transferrin and Tb^{3+}-conalbumin are identical (Gafni and Steinberg,

1974). Furthermore, their circularly polarized emission spectra are extremely similar. Another ironbinding protein of this type is lactoferrin, found in external secretions and neutrophils. It, too, binds Tb^{3+} with the production of a difference spectrum qualitatively equivalent to that of Tb^{3+}-transferrin but with a smaller absorption difference (Teuwissen et al., 1972). As for transferrin, the spectral data suggested the involvement of tyrosyl residues at the two metal-binding sites.

Ferritin is the iron storage protein of mammals. It comprises 24 subunits, each with a molecular weight of 20,000, forming a cavity within which up to 4500 Fe^{3+} ions can be stored. Apoferritin binds Fe^{2+} ions and oxidizes them to Fe^{3+} before storage. Tb^{3+}, unlike Ca^{2+}, inhibits the uptake of Fe^{2+} by apoferritin (Macara et al., 1973; Stefanini et al., 1983). Binding of Tb^{3+} to apoferritin does not alter the protein's intrinsic fluorescence or its optical absorption spectrum, although Tb^{3+} luminesence is enhanced upon irradiation at 285 nm (Stefanini et al., 1983). Titration of Tb^{3+} luminescence presented evidence of one Tb^{3+}-binding site of $K_a = 5.2 \times 10^3 M^{-1}$ and two sites of $K_a = 1.5 \times 10^5 M^{-1}$. This agrees with X-ray crystallographic data showing the presence of three Tb^{3+}-sites on apoferritin (Banyard et al., 1978). At least some of these sites appear to be accessible to Fe^{2+}, but not to Ca^{2+}, which failed to reduce Tb^{3+} luminescence. Tb^{3+} may also share binding sites with Fe^{3+} and vanadate, but probably not Zn^{2+} (Chasteen and Theil, 1982; Treffry and Harrison, 1984). The pK for Tb^{3+} binding indicated that carboxyl groups were important donor ligands.

4.9 Trypsin and Related Enzymes

Among the first detailed studies of the effects of lanthanides on protein function were those of Darnall and his colleagues concerning trypsinogen activation (Darnall and Birnbaum, 1970, 1973; Darnall et al., 1973; Gomez et al., 1974). Trypsinogen, the inactive, zymogen precursor of the neutral, serine proteinase trypsin, is secreted into the mammalian digestive tract by the pancreas. The active proteinase, trypsin, is formed by the specific, peptidolytic removal of an acidic hexapeptide (Val-Asp_4-Lys) from the N-terminus of trypsinogen (Desnuelle and Fabre, 1955). Under physiological conditions, trypsinogen activation is initiated by enteropeptidase (q.v.). Trypsin, thus formed, autocatalytically activates further trypsinogen in a reaction which is accelerated by Ca^{2+} (McDonald and Kunitz, 1941). Calcium ions also protect trypsin from autodigestion (Buck et al., 1962). Trypsinogen has two Ca^{2+}-binding sites. The stronger one is retained in trypsin, while the weaker one resides in the acidic hexa-

Table 4-8. Association
Constants of
Lanthanide–Trypsin
Complexes[a]

Ln^{3+}	K_a (M^{-1})
La^{3+}	180
Pr^{3+}	372
Nd^{3+}	540
Sm^{3+}	668
Gd^{3+}	360
Tb^{3+}	900
Ho^{3+}	750
Er^{3+}	962
Tm^{3+}	1630
Yb^{3+}	1860
Lu^{3+}	2110
Y^{3+}	823

[a] From Epstein *et al.* (1973).

peptide extension peptide and is thus lost upon activation. The latter site is thought to involve the two Asp residues which lie closest to the Lys-6–Ile-7 cleavage site (Radhakrishnan *et al.*, 1969). Bode and Schwager (1975) identified a surface loop containing amino acid residues 70–80 as a possible Ca^{2+}-binding site on bovine trypsin. According to this suggestion, donor groups are the carboxyls of Glu-70 and Glu-80 and the carbonyl oxygens of Asn-72 and Val-75 with two water molecules being coordinated.

Unlike Ca^{2+}, Tb^{3+} and Yb^{3+} quench endogenous trypsin fluorescence (Epstein *et al.*, 1974, 1977). Tb^{3+} luminescence is enhanced by a factor of 10^5 upon binding to trypsin. This involves energy transfer from a Trp residue (Epstein *et al.*, 1974), possibly Trp-141, which lies about 8.5 Å from the putative Ca^{2+}-binding site in the 70–80 loop (DeJersey *et al.*, 1980). Luminescence titration experiments with Tb^{3+}, and its competitive displacement by other Ln^{3+} ions or Ca^{2+}, permitted the calculation of the association constants shown in Table 4-8 (Epstein *et al.*, 1973). Efficiency of binding was inversely related to the size of the Ln^{3+} ion. By a similar method, DeJersey *et al.* (1980) determined the much higher K_a values for Tb^{3+} of $(4.8 \pm 0.3) \times 10^4 M^{-1}$ with porcine trypsin and $(10 \pm 2) \times 10^4 M^{-1}$ with bovine trypsin. The corresponding values with Ca^{2+} were $(22 \pm 4) \times 10^4 M^{-1}$ for porcine trypsin and $(11 \pm 3) \times 10^4 M^{-1}$ for bovine trypsin. Addition of benzamidine or butylamine, which are active site inhibitors of trypsin, did not alter the intensity of Tb^{3+} luminescence.

Luminescence of Tb^{3+} was also enhanced upon binding to α-chymotrypsin but not to the bacterial enzyme α-lytic protease (DeJersey et al., 1980). The excitation spectrum of Tb^{3+}-chymotrypsin also suggested energy transfer via a Trp residue. K_a values of $(0.90 \pm 0.03) \times 10^4 M^{-1}$ for Tb^{3+} and $(0.92 \pm 0.06) \times 10^4 M^{-1}$ for Ca^{2+} were calculated. The weaker binding of Tb^{3+} and Ca^{2+} to chymotrypsin than to trypsin may result from the substitution of the Glu-80 of trypsin by Ile in chymotrypsin.

Abbott and colleagues (1975a) have measured the distance from the Ca^{2+}-binding site to the active site in commercially available bovine trypsin. To do this, Gd^{3+} was substituted at the former location and a specific active site inhibitor, p-toluamidine, at the latter. Measurement of the effect of Gd^{3+} on the NMR relaxation time of the ortho- and methyl protons of the inhibitor gave distances of 8.8 ± 0.5 Å and 10.0 ± 0.5 Å, respectively. This suggested to the authors that Asp-194 and Ser-190 were likely ligands for the metal-binding site on trypsin. In support of this assignment, Abbott et al. (1975b) cited the proximity of Trp-141, which lies 6–9 Å from the proposed site and which, according to DeJersey et al. (1980), donates energy for Tb^{3+} luminescence. However, the X-ray crystallographic data of Bode and Schwager (1975) disagree with this conclusion. It is possible that the solution structure and crystal structure of trypsin differ. The subsequent identification of a second Ln^{3+}-binding site on trypsin (Epstein et al., 1977) may also explain this discrepancy. Spectroscopic evidence indicates a primary binding site for Gd^{3+} with $K_a = 800 M^{-1}$ and a secondary site with $K_a = 100 M^{-1}$. Proton relaxation data show that aquo-Gd^{3+} loses six water molecules upon binding to trypsin (Epstein et al., 1977).

As commercial trypsin contains a mixture of β-trypsin, its proteolyzed derivative α-trypsin, and further, inert degradation products, subsequent studies were conducted with purified α- and β-trypsin (Abbott et al., 1975). The data suggested a single high-affinity binding site for Gd^{3+} with a K_a value of $3.3 \times 10^3 M^{-1}$ for α-trypsin and $4.1 \times 10^3 M^{-1}$ for β-trypsin at pH 6. The corresponding value for the binding of Ca^{2+} to β-trypsin was $260 M^{-1}$. The Gd^{3+}–inhibitor distance on α- and β-trypsin was the same and equal to that reported by Abbott et al. (1975a) for unfractionated trypsin.

Lanthanides accelerate the autocatalytic conversion of trypsinogen to trypsin more efficiently than Ca^{2+}. Not only is trypsinogen activation quicker, but Ln^{3+} ions are active at 100-fold lower concentrations (Darnall and Birnbaum, 1970; Darnall et al., 1973; Gomez et al., 1974). In the original studies, the smallest Ln^{3+} ions appeared to promote trypsinogen activation most efficiently. However, Gomez et al. (1974) noted that the higher concentrations of lanthanides inhibited trypsin. Because the smaller

Ln^{3+} ions are the best inhibitors (Gomez *et al.*, 1974; Evans, 1981), the activating effect of the smaller Ln^{3+} ions was underestimated. At a lanthanide concentration of 0.1 mM, which does not inhibit trypsin, there was no simple relationship between the radius of the Ln^{3+} ion and its ability to activate trypsinogen. Gomez *et al.* (1974) speculated that this complex pattern could reflect not only the size of each Ln^{3+} ion, but also the extent of its dehydration upon binding to trypsinogen.

Both Ca^{2+} and Ln^{3+} ions accelerate trypsinogen activation by lowering K_m with regard to trypsinogen, without changing V_{max} (Gomez *et al.*, 1974). Their mode of action is thought to involve masking the negative charges on the β-carboxyl groups of the Asp residues in the N-terminal extension peptides of trypsinogen. In support of this is the observation that chemical blocking of these groups renders trypsinogen activation independent of Ca^{2+} (Radhakrishnan *et al.*, 1969). Difference absorption spectroscopic studies of Nd^{3+} upon binding to trypsin and trypsinogen indicate the involvement of carboxyl groups (Birnbaum *et al.*, 1970). These changes are reduced by Ca^{2+}, consistent with the suggestion that Nd^{3+} occupies the Ca^{2+}-binding sites on these proteins. By virtue of their higher charge-to-volume ratios, Ln^{3+} ions mask better the Asp residues on the propeptide and consequently lower the K_m to a greater degree. Thus, whereas 50 mM Ca^{2+} reduces the K_m threefold, 0.5 mM Nd^{3+} does so 14-fold (Gomez *et al.*, 1974). The possible influence of Ln^{3+} ion hydrolysis, which should have been extensive at the assay pH of 8 (Section 2.3), was not discussed by the authors. Binding of Ln^{3+} ions to the Ca^{2+} site on trypsinogen, which is conserved in trypsin, is not thought to influence activation. Although Ca^{2+} is not required by trypsin, a serine proteinase, high concentrations of lanthanides inhibit the enzyme. Both Gomez *et al.* (1974) and Evans (1981) have shown that inhibition increases as the radius of the Ln^{3+} ion decreases. This trend is reflected in the relative affinities of different Ln^{3+} ions for trypsin (Table 4-8) (Epstein *et al.*, 1973).

In view of their strong enhancement of trypsinogen activation by trypsin, it is surprising that lanthanides strongly inhibit activation by enteropeptidase (EC 3.4.21.9) (Rinderknecht and Friedman, 1976). Inhibition was noncompetitive, with 50% inhibition occurring at approximately $3 \mu M$ Ln^{3+}. As lanthanides had little or no effect on the hydrolysis of synthetic substrates by enteropeptidase, it is the interaction of Ln^{3+} with trypsinogen which prevents activation by enteropeptidase.

4.10 Elastase

Like trypsin and chymotrypsin, elastase (EC 3.4.21.11) is a neutral serine proteinase which, following studies with lanthanides, was unexpectedly found to bind Ca^{2+}. Neither Ca^{2+} nor Ln^{3+} ions affect the activity of elastase. These metals also fail to alter the rate of autolysis of the enzyme (Dimicoli and Bieth, 1977). Gd^{3+} and Ca^{2+} bind to identical sites on elastase, with K_d values of $4.5 \times 10^{-5} M$ and $2 \times 10^{-5} M$, respectively, at pH 5 (Dimicoli and Bieth, 1977). These values are not affected by the presence of the active site inhibitor trifluoroacetylalanine, nor are inhibitor or substrate binding affected by Ca^{2+} or Gd^{3+}. This independence may be partly explained by the large Gd^{3+}–inhibitor distance of 20 Å estimated by ^{19}F-NMR spectroscopy. By analogy with trypsin, Glu-70 and Glu-80 may coordinate to Ca^{2+} or Gd^{3+} in elastase. These ligands have been shown to be adjacent to a uranyl ion in the crystal structure of the enzyme. From their data, Dimicoli and Bieth (1977) ruled out Asp-194 as a ligand, a possibility suggested for trypsin by Abbott et al. (1974, 1975) (Section 4.9).

Luminescence studies with Tb^{3+} have also confirmed the binding of lanthanides to elastase. K_a values of $3 \times 10^5 M^{-1}$ (DeJersey and Martin, 1980) and $2.9 \times 10^5 M^{-1}$ (Duportail et al., 1980) were obtained for Tb^{3+}, and $2.1 \times 10^5 M^{-1}$ for Ca^{2+} (DeJersey and Martin, 1980). A tryptophan residue, possibly Trp-141, was identified as the donor amino acid. The luminescent emission from Tb^{3+} was strongly circularly polarized. These studies revealed the presence of a previously unsuspected second binding site on elastase for the inhibitory peptide acetyltrialanine (DeJersey and Martin, 1980). Several small competitive inhibitors increased the luminescence intensity of Tb^{3+}-elastase. α_1-Proteinase inhibitor and α_2-macroglobulin also altered Tb^{3+} luminescence (Duportail et al., 1980). The affinity of Tb^{3+} for elastase was 10-fold lower in the presence of α_2-macroglobulin. Of interest is the finding that quantitative changes in the luminescence of Tb^{3+} were more sensitive to conformational change than was the circularly polarized emission spectrum (Duportail et al., 1980). Chemical modification of one Arg residue in elastase decreased its affinity for Tb^{3+} by a factor of 10 and altered the circularly polarized emission spectrum (Davril et al., 1984). However, in this case, the maximum luminescence intensity was not affected.

4.11 Collagenases

4.11.1 Clostridiopeptidase A

Clostridiopeptidase A (EC 3.4.24.3) is a neutral, collagenolytic metalloproteinase produced by C. histolyticum. Recognizing the sequence Gly-Pro-y, it endopeptidolytically cleaves collagen and gelatin (denatured collagen) into short peptides. As the helical domain of collagen contains a large number of proline residues and the repeating triplet Gly-x-y, it possesses many cleavage sites for the enzyme. Each enzyme molecule contains one equivalent of Zn^{2+}, which is thought to reside at the active site, while there is an absolute requirement for Ca^{2+}. There is also evidence for a third regulatory site to which transition metals, but not Ca^{2+}, may bind (Evans and Mason, 1986). It has proved remarkably difficult to arrive at a consensus on the molecular weight and subunit structure of clostridial collagenase. Estimates vary from a tetramer of unit M.W. 25,000 to monomeric molecules of M.W. from 68,000 to 150,000.

Clostridiopeptidase A is inhibited by lanthanides (Evans, 1981). This holds whether collagen, gelatin, or a synthetic pentapeptide is used as substrate. As Ln^{3+} ions interact with collagen and gelatin (Section 4.12), the synthetic pentapeptide phenylazobenzyloxycarbonyl-L-Pro-L-Leu-Gly-L-Pro-D-Arg (Pz-peptide) has been used for kinetic analysis of this inhibition. Of the four lanthanides tested (Lu^{3+}, Er^{3+}, Sm^{3+}, and La^{3+}), Sm^{3+} proved the best inhibitor, with a K_i of $12\,\mu M$. Inhibition was competitive with Ca^{2+}, which displayed a K_d of $0.27\,mM$. Whereas Er^{3+} and Lu^{3+} were also good inhibitors, La^{3+} inhibited clostridiopeptidase A only weakly. This finding is interesting in light of the relative sizes of these Ln^{3+} ions in comparison to Ca^{2+} (Table 2-12; Section 2.5) and suggests that the size of the Ca^{2+}-binding site(s) to which Ln^{3+} ions attach in inhibiting the enzyme shows strict spatial specificity, excluding ions which are larger than Ca^{2+}. A similar dimensional constraint has been described for other systems (Section 2.5).

The kinetics of the inhibition of clostridial collagenase by Sm^{3+} were uncompetitive with regard to the substrate, a circumstance normally ascribed to the formation of an abortive enzyme–substrate–inhibitor ternary complex. This has interesting parallels with the proposed mechanism of inhibition, by lanthanides, of staphylococcal nuclease (Section 4.14), activation of factor X (Section 4.5.2), and vertebrate collagenase (Section 4.11.2). Given that Sm^{3+} appears to compete with Ca^{2+} for its binding site(s) on clostridiopeptidase A, these data further support the conclusion of Seifter and Harper (1959) that Ca^{2+} is involved in the binding of this enzyme to its substrate. Ca^{2+} also stabilizes clostridial collagenase against thermodenaturation, a function which Sm^{3+} was able to fulfill partially (Evans, 1981).

Some of the conclusions from the above kinetic analysis have been confirmed directly. Ca^{2+}-free clostridiopeptidase A failed to bind to a collagen affinity column (Evans, 1985). Sm^{3+} was able to replace Ca^{2+} in promoting the attachment of the enzyme to the column. Er^{3+} and Lu^{3+} could also do this, but La^{3+}, as predicted from the kinetic data, was only weakly active. This property of Sm^{3+} permitted its use in the purification of clostridiopeptidase A by "lanthanide enhanced affinity chromatography" (Section 3.8.2). Neither the way in which Ca^{2+} promotes the binding of clostridial collagenase to collagen nor the reason why binding in the presence of Sm^{3+} proves abortive is known. Studies using UV absorption spectroscopy and fluorescence spectroscopy have failed to provide evidence of a major conformational change in the enzyme upon binding Ca^{2+} or Sm^{3+} (Mason and Evans, unpublished).

4.11.2 Mammalian Collagenases

Although zinc-containing metalloproteinases with an absolute requirement for Ca^{2+}, mammalian collagenases (EC 3.4.24.7) show several differences from microbial forms of the enzyme. The most striking is the nature of the digestion products. Whereas clostridial collagenase produces many small peptides by cleavage at multiple sites, vertebrate collagenases cleave just one bond, Gly–Leu or Gly–Ile, depending on the type of collagen. This bond is located at a site one-quarter of the distance from the C-terminus of the molecule. Estimates of the molecular weight of the active enzyme range from 33,000 to 60,000. There is no evidence of subunits.

Lanthanides inhibit partially purified preparations of the collagenase secreted by cultures of human, rheumatoid synovium (Evans and Ridella, 1985). Ca^{2+} relieves the inhibition in a dose-dependent manner, suggesting that Ln^{3+} ions act by competitive attachment to the Ca^{2+}-binding site(s) on the enzyme. Among Lu^{3+}, Sm^{3+}, Er^{3+}, and La^{3+}, the strongest inhibitor is La^{3+}, with 50% inhibition occurring at about $100\,\mu M$ $LaCl_3$. Inhibition of the synovial enzyme by La^{3+} is uncompetitive, again suggesting the formation of an abortive collagen–Ln^{3+}–collagenase complex. Direct binding assays confirmed that collagenase bound more completely to collagen in the presence of La^{3+} than in the presence of Ca^{2+}. But unlike clostridiopeptidase A, synovial collagenase also bound appreciably to collagen in the absence of Ca^{2+} or Ln^{3+}, confirming the Ca^{2+} independence of this step, as reported for human skin collagenase by Seltzer *et al.* (1976).

4.12 Collagen

Neutral salt solutions of type I collagen polymerize upon warming to form opaque collagen gels. Gelling curves, showing the increase in optical density of collagen solutions as they polymerize, are sigmoidal.

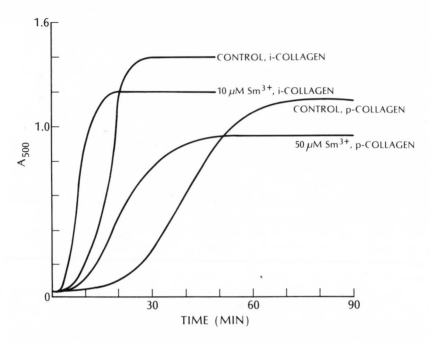

Figure 4-7. Effect of Sm^{3+} on the gelling of intact (i-) and pepsinized (p-) collagen molecules. From Drouven and Evans (1986), with permission.

Such curves contain a "lag" or "nucleation" phase, during which the collagen molecules form linear filaments, without increase in optical density. During the subsequent "growth" phase, the turbidity increases quickly as the filaments aggregate laterally into fibrils. Finally, a plateau is reached (A_{max}). Covalent cross-links quickly form which insolubilize the fibrils and protect them from thermodenaturation. The fibrils are highly ordered and show a characteristic banding pattern under transmission electron microscopy.

If the majority of the terminal, nonhelical domains of collagen are removed with pepsin, the resulting, attenuated collagen molecules gel less quickly to form fibrils with a poor banding pattern. Lanthanides, and to a lesser extent Ca^{2+}, restore the quantitative aspects of gelling which are lost after pepsinization (Evans and Drouven, 1983). However, these cations do not improve banding. In addition, Ln^{3+} ions, but not Ca^{2+}, accelerate the gelling of unpepsinized, native collagen (Drouven and Evans, 1986) (Fig. 4-7). The larger Ln^{3+} ions are better able to promote the growth of pepsinized collagen fibrils; La^{3+} and Sm^{3+} accelerate the growth phase to a greater degree than Er^{3+} does, while Lu^{3+} is inhibitory. A similar trend is seen with intact collagen, although none of the lanthanides are inhibitory in this case. The shorter lag phase and quicker growth phase in the presence of Sm^{3+} reflect lower activation energies for these processes. The effects of Sm^{3+} on the growth phase appear to be independent of its shortening of the lag phase.

Collagen gels formed in the presence of Sm^{3+} have lower optical density. For pepsinized collagen, this appears to reflect the formation of thinner fibrils. However, Sm^{3+} does not affect the width of fibrils formed from intact collagen. Instead, the differences in A_{max} are probably due to subtle alterations in the organization of the fibrils, as Sm^{3+} promotes the formation of fibrils with the native banding pattern. In addition, Sm^{3+} protects nascent fibrils formed from pepsinized or intact collagen against thermodenaturation. In this regard, it is interesting that a new tanning agent based on lanthanides has been announced. Leather tanned with lanthanides is more resistant to heat (Yici, 1986). Lanthanides have no direct effect on the rate of formation of covalent cross-links between the nonhelical domains of collagen molecules within nascent fibrils.

The mechanisms through which lanthanides accelerate collagen gelling are unknown, but their elucidation promises to shed light on the physiological mode of polymerization. Quantitative considerations suggest that as few as five Sm^{3+}-binding sites per collagen molecule may be involved. Tb^{3+} luminescence data hint that these are more likely to be in the helical domain of collagen than in the nonhelical, terminal regions (Drouven and Evans, 1986). Once again, the "bridging" and "charge masking" properties of Ln^{3+} ions are worthy of consideration as modes of action. Lanthanides promote the formation of calcified deposits along collagen fibers *in vivo* (Section 8.6) and associate with the collagenous matrix of connective tissues (e.g., Bonucci *et al.*, 1988).

4.13 α-Amylase

α-Amylase [α(1 → 4)-glucan 4-glucanohydrolase; EC 3.2.1.1] degrades α 1 → 4 linkages in starch, yielding a mixture of glucose and maltose. The enzyme purified from *Bacillus subtilis* is a dimer with a total M.W. of 48,000, which binds four to five ions of Ca^{2+}. Although α-amylase has an absolute requirement for Ca^{2+} (Imanishi, 1966), the role of Ca^{2+} in catalysis is unknown. Smolka *et al.* (1971) reported that Nd^{3+} inhibited α-amylase, with 50% inhibition occurring at a concentration of Nd^{3+} between 10^{-3} and $10^{-4} M$. As Nd^{3+} was able partially to reactivate the Ca^{2+}-free apoenzyme, it seemed likely that Nd^{3+} replaced Ca^{2+} in the holoenzyme, forming a substituted species with lower enzymic activity.

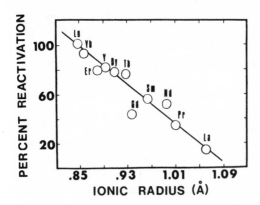

Figure 4-8. Reactivation of Ca^{2+}-free α-amylase by lanthanides. Ionic radii are indicative only (Section 2.2; Table 2-1). From Smolka *et al.* (1971), with permission.

The ability of Ln^{3+} to reactivate Ca^{2+}-free α-amylase was inversely proportional to ionic radius (Fig. 4-8), with Lu^{3+}, the smallest Ln^{3+} ion, restoring full enzymic activity. The circular dichroism spectra of native α-amylase, its Ca^{2+}-free apoenzyme, and its Nd^{3+}-substituted forms were identical, suggesting that this enzyme undergoes no major conformational change in the presence or absence of Ca^{2+} or Ln^{3+} ions.

Levitzki and Reuben (1973) and Steer and Levitzki (1973) were unable to confirm the enzyme kinetics of Smolka *et al.* (1971). According to the data of Darnall and Birnbaum (1973), the discrepancy turns on the choice of buffer. Whereas the original observations were made with maleate buffer pH 6.9, Levitzki and co-workers attempted to repeat them with HEPES buffer pH 6.9. As maleate chelates Ln^{3+} ions, whereas HEPES does not, the latter should have been the better choice of buffer (Section 2.8). Paradoxically, for the purposes of demonstrating Ln^{3+} reactivation of α-amylase, this was not so. Although concentrations of Lu^{3+} below 0.1 mM restored full enzymic activity to apo-α-amylase, high concentrations inhibited the enzyme. By forming complexes with lanthanides, maleate prevents the appearance of inhibitory concentrations of free Ln^{3+} ions, even when the total lanthanide concentration in the assay is as high as 1 mM. HEPES fails to act as a lanthanide "sink," thus restricting the activation of α-amylase by Ln^{3+} ions to a narrow concentration range (Fig. 4-9).

Choosri (1981) has subsequently investigated these matters in more detail. In agreement with earlier results for Nd^{3+} (Smolka *et al.*, 1971), Tb^{3+} inhibited the α-amylase of *B. subtilis*. Although confirming that lanthanides partially reactivated apo-amylase, Choosri reported that the larger Ln^{3+} ions were better activators than the smaller ones, a finding in complete contrast to that reported by Smolka *et al.* (1971) (Fig. 4-8).

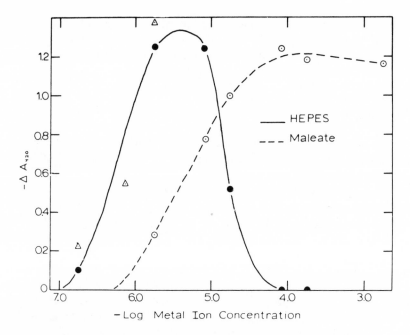

Figure 4-9. Reactivation of Ca^{2+}-free α-amylase by Lu^{3+} in HEPES or maleate buffer. From Darnall and Birnbaum (1973), with permission.

Unlike the latter authors, Choosri used HEPES buffer. Luminescence titration with Tb^{3+} or Eu^{3+} suggested three to four Ln^{3+} binding sites per enzyme dimer (Choosri, 1981). Direct laser excitation of Eu^{3+} revealed two different classes of Eu^{3+}-binding site, with a third site appearing under certain conditions. It was suggested that the third site could be the Zn^{2+}-binding site of the enzyme. A fourth site at which Eu^{3+} is very loosely bound was also indicated. Eu^{3+} at the first two sites contained 2.3 ± 0.5 and 1.9 ± 0.5 coordinated water molecules. At the third site, Eu^{3+} coordinated 2.2 ± 0.5 water molecules. As interlanthanide energy transfer was undetectable, the three Ln^{3+}-binding sites are probably separated by at least 15 Å.

Choosri (1981) also studied the α-amylase of *Aspergillus oryzae*. Here, Eu^{3+} was also a weak inhibitor of the enzyme, and all the Ln^{3+} ions tested were able partially to reactivate the apoenzyme. However, unlike the bacterial enzyme, the α-amylase of *A. oryzae* was reactivated almost equally by each lanthanide. Luminescence titration experiments revealed that four Ln^{3+} ions were bound to the enzyme before precipitation occurred. The excitation spectrum showed Trp to be the major donor. In agreement with this, Eu^{3+} quenched the intrinsic Trp fluorescence of the protein. Two classes of Eu^{3+}-binding site were determined. At one of these, 7.1 ± 0.5 water molecules were coordinated to Eu^{3+}; at the other, 1.9 ± 0.5 water molecules were coordinated.

4.14 Staphylococcal Nuclease

Staphylococcus aureus produces an extracellular nuclease (EC 3.1.4.7) which hydrolyzes specific phosphodiester bands in RNA and DNA. The enzyme has a single polypeptide chain of M.W. 16,900 and an absolute requirement for Ca^{2+} (Arnone *et al.*, 1971), without which the enzyme fails to bind to substrate or to substrate analogues such as deoxythymidine 3′,5′-diphosphate. Although able to restore the ability of the Ca^{2+}-free apoenzyme to bind to its substrate, Ln^{3+} ions are potent inhibitors of the nucleolytic reaction, with K_i values in the millimolar range. The mechanism of the inhibition appears to involve the competitive displacement of Ca^{2+} from the enzyme (Furie *et al.*, 1973), with the formation of an abortive enzyme–Ln^{3+}–substrate complex. Such a mechanism has also been proposed for lanthanide inhibition of factor X activation (Section 4.5.2) and clostridiopeptidase A (Section 4.11.1). The nuclease–Ln^{3+}–substrate complex is more resistant to trypsin than is the native ternary complex. In the absence of substrate, Nd^{3+}, but not Ca^{2+}, confers an intermediate degree of trypsin resistance. This property may be related to the conformational shift that Nd^{3+} or Gd^{3+}, but not Ca^{2+}, induces in staphylococcal nuclease. Inspection of the changes in the UV absorption spectra of the enzyme in the presence or absence of Nd^{3+} suggested perturbations in tyrosyl residues. These are thought to be Tyr-115 and Tyr-85, which lie in the active site, 13.3 and 12.4 Å, respectively, from the Ca^{2+} site. These two adjacent Tyr residues have been implicated in substrate binding (Cuatrecasas *et al.*, 1967). As the K_i values are slightly lower than the K_m values calculated in the absence of substrate, it is possible that the presence of substrate increases the affinity of the nuclease for lanthanides.

By substituting Gd^{3+} for Ca^{2+}, and inserting the nucleotide inhibitor 3′,5′-dTDP at the active site, Furie *et al.* (1974) were able to use NMR paramagnetic relaxation rate enhancement measurements to obtain structural information about the nuclease–Gd^{3+}–3′,5′-dTDP complex. The data obtained agreed well with X-ray crystallographic findings.

4.15 Ca^{2+}/Mg^{2+}-ATPase

Cells need to regulate very closely their cytosolic Ca^{2+} concentrations, usually maintaining them at submicromolar levels. This is achieved despite the presence of an extracellular milieu where the Ca^{2+} concentration is often in the millimolar range. Such regulation is aided by specialized Ca^{2+} "pumps" which harness the energy released by hydrolysis of the terminal phosphodiester bond of ATP to transport Ca^{2+} across cellular membranes. These enzymes hence show ATPase activity; as they require both Ca^{2+} and Mg^{2+}, they are referred to as Ca^{2+}/Mg^{2+}-ATPases. Such enzymes are associated with cell membranes and may transport Ca^{2+} across the limiting membranes of intracellular organelles, such as mitochondria and the endoplasmic reticulum, or across the plasma membrane, thus ejecting Ca^{2+} from the cell altogether. There also exist Mg^{2+}-ATPases, which do not require Ca^{2+}, and Na^+/K^+-ATPases, which transport monovalent cations (q.v.). Many of the physiological effects of lanthanides reflect their influence on cellular calcium homeostasis (Section 6.3).

The best-studied example is the Ca^{2+}/Mg^{2+}-ATPase (EC 3.6.1.3) of the sarcoplasmic reticulum (SR), an intracellular, membranous organelle of skeletal muscle cells, corresponding to the endoplasmic reticulum of nonmuscle cells. During muscle contraction, the SR releases Ca^{2+} which triggers the interaction between the thick and thin myofilaments. During muscle relaxation, Ca^{2+} is transported, against a Ca^{2+} gradient, across the SR. It is this second step which is the energy-dependent process catalyzed by Ca^{2+}/Mg^{2+}-ATPase. Two moles of Ca^{2+} are transported per mole of ATP hydrolyzed.

Four distinct steps within the overall ATPase, Ca^{2+}-transporting reaction have been identified, as shown in the following scheme:

$$
\begin{array}{ccc}
 & \text{ADP} & \\
 & \nearrow & \\
E_1 + \text{ATP} \overset{}{\underset{\text{I}}{\rightleftharpoons}} & & E_1\!-\!P \\
\uparrow & & \Big\Updownarrow \ \text{II} \\
\text{IV} & & \\
E_2 + P_i \underset{Mg^{2+}}{\overset{\text{III}}{\rightleftharpoons}} & & E_2\!-\!P
\end{array}
$$

The first reaction is the ATPase step, which produces ADP and a phosphorylated form of the enzyme ($E_1\!-\!P$). This reaction can be reversed by adding ADP. During the second reaction, the $E_1\!-\!P$ intermediate undergoes a conformational shift to form an altered intermediate designated $E_2\!-\!P$, which cannot be dephosphorylated by ADP. Phosphate is cleaved from E_2 during the Mg^{2+}-requiring third reaction, and the original form of the enzyme (E_1) is subsequently regenerated. Only the last reaction is irreversible.

It has been known since the work of Chevallier and Butow (1971) that lanthanides inhibit the overall reaction. They reported that Ln^{3+} ions compete with Ca^{2+} for the Ca^{2+}-binding ("transporter") sites on the enzyme, forming a Ln^{3+}-substituted derivative which undergoes reaction I to form the phosphorylated enzyme intermediate but which fails to dephosphorylate. Using purified enzyme preparations, Yamada and Tonomura (1972) confirmed this conclusion, providing $K_{50\%}$ values of $2.5\,\mu M$ for both Ce^{3+} and La^{3+}. However, Itoh and Kawakita (1984) have recently challenged these conclusions. Although confirming that Gd^{3+} and Tb^{3+} were able to compete with Ca^{2+} for the transporter site, inhibition of ATPase activity occurred at Ln^{3+} concentrations well below those needed to displace Ca^{2+}; 50% inhibition occurred at about $20\,\mu M\,Ln^{3+}$. Inhibition was reversible with Gd^{3+} concentrations below $30\,\mu M$ but irreversible at higher concentrations. Unlike the first two reports, that of Itoh and Kawakita (1984) indicated that Gd^{3+} also inhibited E_1—P formation, with a $K_{50\%}$ of about $10\,\mu M$, as well as inhibiting dephosphorylation. They further suggested that E_1—P formed in the presence of Gd^{3+} did not undergo reaction II to form E_2—P. This inhibition was not reversed by EDTA, indicating that the Gd^{3+} had somehow been rendered inaccessible by the conformational shift. In the absence of Ca^{2+}, step I will not occur; Gd^{3+} cannot substitute for it.

Itoh and Kawakita (1984) concluded that there are three classes of Ln^{3+}-binding site on SR ATPase. Binding to the highest-affinity sites is responsible for inhibition of ATPase activity. Binding to the highest-affinity site also inhibited both E_1—P formation and the conversion of E_1—P to E_2—P, but not the ADP-dependent decomposition. The Ln^{3+}-binding site of intermediate affinity is also the Ca^{2+}-binding site, while the weakest Ln^{3+}-binding site sequesters Mg^{2+} under physiological conditions. Itoh and Kawakita (1984) also noted enhanced Tb^{3+} luminescence upon binding to the enzyme. As this was diminished by Mn^{2+} or Gd^{3+}, but not Ca^{2+}, they concluded that the Mg^{2+} site was the one at which Tb^{3+} luminesced. The results of Abramson and Shamoo (1980) further complicate matters by suggesting that at low concentrations (2–$10\,\mu M$), La^{3+} stimulates ATPase activity by about 50%, without affecting Ca^{2+} transport, while $20\,\mu M\,La^{3+}$ inhibits both. According to Dos Remedios (1977), the ability of lanthanides to inhibit SR Ca^{2+}/Mg^{2+}-ATPase increases with increasing ionic radius.

Tb^{3+}-binding studies by Highsmith and Head (1983) do not support Itoh and Kawakita's conclusions. Using SR vesicles, Highsmith and Head reported 50% inhibition of SR ATPase by Tb^{3+} in the 0.1–$0.5\,mM$ concentration range. Ca^{2+}, but not Mg^{2+} (2–$20\,mM$), was able to reverse the inhibition. The insensitivity to Mg^{2+} rules out Tb^{3+}-ATP as the inhibitory agent. Finding an apparent K_a for Ca^{2+} of $6 \times 10^6\,M^{-1}$, Highsmith and

Head (1983) concluded, unlike Itoh and Kawakita (1984), that Tb^{3+} inhibited ATPase by binding to the high-affinity Ca^{2+}-binding sites. The K_a for Tb^{3+} was $10^9 M^{-1}$. This binding occurred without cooperativity or enhancement of Tb^{3+} luminescence. A second group of weaker Tb^{3+}-binding sites with a K_a in the presence of 5 mM Mg^{2+} of 1×10^4–$5 \times 10^4 M^{-1}$ enhanced Tb^{3+} luminescence 950-fold. In agreement with the results of Itoh and Kawakita (1984), Mg^{2+} ($K_a = 94 M^{-1}$) competed with Tb^{3+} for the sites at which luminescence was enhanced. The nature of Tb^{3+} association to the Mg^{2+} sites depended upon the form of the ATPase and the ionic environment. When present in vesicles, Tb^{3+} binding showed negative cooperativity. When the enzyme was solubilized in a nonionic detergent in the absence of Ca^{2+}, only one class of binding site was seen. In the presence of Ca^{2+}, Tb^{3+} binding to the Mg^{2+} sites displayed positive cooperativity with a Hill coefficient of 2.1.

Further studies on the interaction of Nd^{3+} with the ATPase present in SR vesicles indicated that Nd^{3+} inhibited ATPase activity by binding to the high-affinity Ca^{2+} sites (Highsmith and Murphy, 1984). K_a for the binding of Nd^{3+} to these sites was at least $2 \times 10^9 M^{-1}$. When the active site was labeled with a fluorescent probe, no energy transfer to Nd^{3+} could be detected, indicating a separation of at least 21 Å between the active site and high-affinity Ca^{2+} sites. Direct laser excitation of Tb^{3+} luminescence has confirmed a K_a value of 3×10^7–$5 \times 10^8 M^{-1}$ for the binding of Tb^{3+} to SR ATPase and suggested that two molecules of water are coordinated to Tb^{3+} at the high-affinity Ca^{2+} sites (Scott, 1984).

In view of the ability of lanthanides to trap SR ATPase in the E_1 form, they have been used in structural studies comparing the E_1 and E_2 forms of the enzyme (Dux et al., 1985). The E_2 form is stabilized by vanadate in the absence of Ca^{2+} or Ln^{3+} ions. Trypsin cleaves SR ATPase at two sites. Vanadate protects the enzyme from the second cleavage. Ca^{2+} and Ln^{3+} antagonize this effect, permitting the second cleavage reaction (Dux et al., 1985). A variety of lanthanides were shown to promote the crystallization of the enzyme. For Pr^{3+}, Gd^{3+}, and La^{3+}, crystallization was optimal at equimolar concentrations of Ln^{3+} and enzyme. A 10-fold higher concentration of Ca^{2+} was required. The crystal forms obtained with Ca^{2+} and Ln^{3+} ions were similar.

The Ca^{2+} pump of erythrocytes is a similar enzyme, which transports Ca^{2+} from the cytosol to the extracellular compartment. Lanthanides suppress its ATPase and Ca^{2+}-transporting activities, with 50% inhibition of transport occurring at around 0.1 mM Ln^{3+} (Schatzmann and Tschabold, 1971; Luterbacher and Schatzmann, 1983). Lanthanides produce a fourfold increase in the concentration of the phosphorylated enzyme intermediate. The $K_{50\%}$ for this increase is 23 μM La^{3+}; inhibition of the

E_2—P phosphatase reaction has a $K_{50\%}$ of $6\,\mu M$ La^{3+}. As with SR ATPase, E_1—P formed in the presence of La^{3+} can be reversibly decomposed in response to ADP, suggesting that La^{3+} interferes with step II, the conversion of E_1—P to E_2—P (Luterbacher and Schatzmann, 1983). This agrees with the conclusions of Itoh and Kawakita (1984) for SR ATPase. Both groups also agree that La^{3+} does not inhibit step IV. As La^{3+} inhibits Ca^{2+} transport from the inside of "inside-out" vesicles and from the outside of intact cells, Luterbacher and Schatzmann (1983) suggest that La^{3+} binds at or near the Ca^{2+} transport site, a domain which faces alternatively inside and outside.

La^{3+} has also been shown to inhibit the Mg^{2+}-ATPase and Na^+/K^+-ATPase of rat heart sarcolemma (Takeo et al., 1979), and erythrocyte ghosts (Weiner and Lee, 1972). However, these two enzymes are much less sensitive to lanthanides; it is possible to suppress completely Ca^{2+} efflux from erythrocytes at La^{3+} concentrations where the total ATPase is only reduced by 50% (Quist and Roufogalis, 1975). Inhibition of Na^+/K^+-ATPase from heart microsomes is uncompetitive (Nayler and Harris, 1976).

4.16 Actin and Myosin

Over 80% of the contractile apparatus of skeletal muscle is accounted for by two filamentous proteins, myosin and actin. In conjunction with several other proteins, including troponin (Section 4.4.2), actin and myosin interact to bring about muscle contraction. The effects of lanthanides on muscle physiology are discussed in Sections 6.3 and 6.4.

Actin can occur as a globular monomer (G-actin) or a fibrous polymer (F-actin). The monomer is a single polypeptide of M.W. 42,000, is rich in Pro and Cys, and contains the unusual amino acid 3-methylhistidine.

Studies of the interaction of lanthanides with actin were first reported by Dos Remedios and Barden (1977), who described a triphasic response. In these studies, solutions of G-actin were polymerized in the presence of ATP by the addition of KCl to a final concentration of $0.1\,M$. Addition to this system of up to four equivalents of Gd^{3+} per mole of actin had no effect on changes in the viscosity of the solution which accompany the formation of F-actin. At ratios of four to eight equivalents of Gd^{3+} per mole of actin, the solution attained excessive viscosity, apparently due to aggregation of the F-actin filaments. When Gd^{3+} was present in excess of eight equivalents per mole of actin, polymerization was inhibited, and the solution attained the viscosity of G-actin (Fig. 4-10). Phase 1 involved the association of Gd^{3+} with the ATP present in the solution. The onset

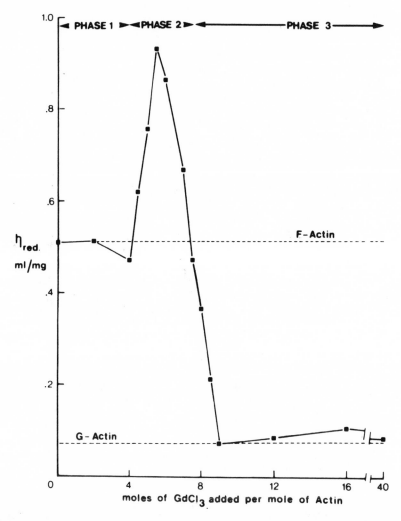

Figure 4-10. Effect of Gd^{3+} on the viscosity reduction of actin in response to KCl. From Dos Remedios and Barden (1977), with permission.

of phase 2 coincided with the saturation of the ATP and the appearance of Gd^{3+} able to attach to actin. The Gd^{3+}-actin of phase 3 was apparently undenatured, and its ability to promote the ATPase activity of myosin was unaltered. However, Gershman *et al.* (1979) reported that, while G-actin had no effect on the ATPase activity of myosin subfragment 1, the Gd^{3+}-actin of phase 3 was slightly inhibitory. These results stimulated consid-

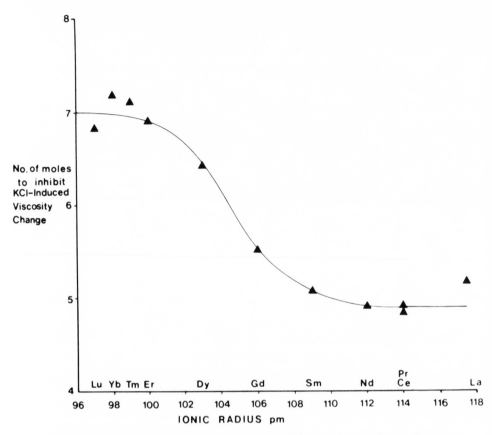

Figure 4-11. Abilities of different lanthanides to inhibit the KCl-induced viscosity change of G-actin. Ionic radii are indicative only (Section 2.2; Table 2-1). From Barden and Dos Remedios (1978), with permission.

erable further research into the nature of the aggregates formed in the presence of Gd^{3+} and the details of the binding of Gd^{3+} to actin. In addition, the ability of Ln^{3+} ions to inhibit the formation of F-actin has served as a useful adjunct to the preparation of actin for X-ray crystallographic studies (Aebi *et al.*, 1980).

Inhibitory activity varied across the lanthanide series (Fig. 4-11; Barden and Dos Remedios, 1978). Electron microscopic examination of the aggregates formed by Gd^{3+} in the absence of KCl showed the presence of multilayered sheets distinct from the filaments of F-actin (Dos Remedios and Dickens, 1978). In the presence of 0.1 *M* KCl, the aggregates were quite different, appearing as tubes. These tubes share a number of properties with F-actin, including the ability to bind to myosin or its subfragment 1, slow rate of exchange of bound nucleotide and metal ions, and greater resistance to digestion by chymotrypsin (Barden and Dos Remedios, 1980). The ability of Gd^{3+}-actin tubes to activate the ATPase of myosin was about twice as great as the activating ability of F-actin. Addition of myosin caused the conversion of the actin tubes to F-actin filaments.

Only the lanthanides from Ce^{3+} to Ho^{3+} were able to promote the formation of actin tubes (Dos Remedios *et al.*, 1980). La^{3+} and the members from Er^{3+} to Lu^{3+} produced only amorphous precipitates. This may be connected with the stoichiometries of their binding to actin. Lanthanides Ce^{3+} to Eu^{3+} bound to actin with a 5:1 molar ratio; for Gd^{3+} to Ho^{3+} this ratio was 6:1, while for Er^{3+} to Lu^{3+} and La^{3+}, a 7:1 ratio was determined. Dos Remedios *et al.* (1980) suggested that binding of seven equivalents of Ln^{3+} inhibited tube formation by conferring too great a positive charge on the actin monomers. Barden and Dos Remedios (1979) showed that Gd^{3+} would replace Ca^{2+} at both its high- and low-affinity sites on actin, with a final stoichiometry of about six to seven equivalents of Gd^{3+} per mole of actin. Ferri and Grazi (1981) employed Tb^{3+} luminescence to demonstrate, in agreement with Dos Remedios *et al.* (1980), a binding stoichiometry of six equivalents of Tb^{3+} per mole of actin. Association constants of $0.8 \times 10^6 M^{-1}$ for Tb^{3+} and $0.2 \times 10^3 M^{-1}$ for Ca^{2+} were measured. The figure for Ca^{2+} being indicative of the low-affinity Ca^{2+} sites, the authors investigated whether Tb^{3+} could replace Ca^{2+} at its high-affinity site. In contrast to Barden and Dos Remedios (1979), they did not detect any displacement of $^{45}Ca^{2+}$ from this site by Tb^{3+}.

Luminescence enhancement data for Tb^{3+} disagree on the nature of the environment of the high-affinity Ca^{2+} site that it is usually assumed to occupy. According to Curmi *et al.* (1982), this site does not have adjacent aromatic amino acids, so that the binding of Tb^{3+} here is luminescently "silent." However, Burtnick (1982) stated that luminescence enhancement of Tb^{3+} did occur at this site, without presenting independent evidence of this. According to Ferri and Grazi (1981), polymerization of actin in the presence of $0.1 M$ KCl does not affect the number of Tb^{3+}-binding sites, although the increased ionic strength lowers the apparent K_a. However, polymers induced by $2 mM$ Mg^{2+} have drastically lowered K_a for Tb^{3+}, and the number of binding sites is reduced to four. Tb^{3+} luminescence at the high-affinity site claimed by Burtnick (1982) probably occurs via energy transfer from a Trp residue. Polymerization of actin increased the luminescent emission. Tb^{3+}-actin was equal to G-actin in its ability to inhibit DNase I. Although Tb^{3+} luminescence was enhanced upon binding to DNase I, the binding of DNase I to Tb^{3+}-actin decreased Tb^{3+} luminescence (Burtnick, 1982).

Other studies of the conformational changes induced in actin by lanthanides have employed CD spectroscopy and UV difference spectroscopy. Tb^{3+} was found to reduce both the α-helicity and content of β-sheet structure in G-actin and F-actin (McCubbin $et\,al.$, 1981). When the A_1 subfragment-1 of myosin was combined with G-actin, Tb^{3+} induced a large conformational change in the complex. Tb^{3+} disrupted the complex between F-actin and the A_1 subfragment of myosin. Changes in the UV absorption spectrum upon adding Tb^{3+} were different from those induced by polymerization (Curmi $et\,al.$, 1982). The authors attributed the changes to mobilization of the adenosine moiety of ATP bound to the actin moiety, an explanation supported by NMR spectroscopy (Barden $et\,al.$, 1980). NMR data suggest that the ATP- and lanthanide-binding sites on actin are separated by at least 16 Å. The mobilization of adenosine apparently exposes a previously buried Trp residue.

As we have seen, there is evidence, albeit disputed, that Gd^{3+}-actin retains the ability to stimulate myosin ATPase activity. Tb^{3+} and La^{3+} both inhibit the endogenous ATPase activity of the subfragment 1 of myosin (Oikawa $et\,al.$, 1980). The inhibitory effects of Ln^{3+} on the Ca^{2+}/Mg^{2+}-ATPase, Na^{+}/K^{+}-ATPase, and Mg^{2+}-ATPase activities of SR and erythrocyte membranes were discussed in Section 4.15. According to Oikawa $et\,al.$ (1980), La^{3+} and Tb^{3+} also inhibited the Mg^{2+}-ATPase activity of the actin–myosin fragment 1 complex. In the presence of a partially inhibitory concentration of TN-1 (Section 4.4.2), both Tb^{3+} and La^{3+} were able to inhibit the actin-activated ATPase activity of myosin fragment 1. However, in each case, TN-C was able to restore full activity. With such a complicated mixture of effectors, it was impossible to identify the level at which the lanthanides were acting.

Lanthanides were not able to replace Ca^{2+} in the actin-activated Ca^{2+}-ATPase of scallop myosin. Further investigation revealed that Ln^{3+} ions were not able to replace Ca^{2+} at either of the two Ca^{2+}-binding sites on scallop myosin. High concentrations of lanthanides inhibited ATPase activity, presumably by binding to other sites on the molecule (Chantler, 1983).

Myosin light-chain kinase is discussed in Section 4.24.

4.17 Hemocyanin

Hemocyanin is the molecular oxygen carrier found in the blood of many invertebrates. Ca^{2+} promotes the aggregation of hemocyanin and

Table 4-9. Characteristic Parameters for the Binding of O_2 to
Panulirus Hemocyanin in the Absence or Presence of Various
Ln^{3+} Ions and Ca^{2+}

Ion	Concentration (mM)	$p_{1/2}$ (mm Hg)	$n_{1/2}$	K_T (mm Hg)	ΔF_1 (kcal/mol)
None		16	1.2	40	0.5–0.6
Tb^{3+}	4	28	1.8	80	0.9–1.0
Pr^{3+}	1	28	1.8	80	0.9–1.0
Gd^{3+}	3.8	28	1.8	80	0.9–1.0
Eu^{3+}	1	40	2.2	100	1.1
La^{3+}	1	45	2.2	100	1.1
Er^{3+}	0.8	45	2.2	100	1.1
Ca^{2+}	10	67	2.2	200	1.5

[a] From Kuiper *et al.* (1981), with permission.

acts as an allosteric effector, increasing its affinity for O_2. A number of lanthanides, including Pr^{3+}, Er^{3+}, Tb^{3+}, Gd^{3+}, and Eu^{3+}, can substitute for Ca^{2+} as allosteric effectors (Kuiper *et al.*, 1981; Table 4-9).

Hemocyanin from *Panulirus interruptus* enhances Tb^{3+} luminescence by energy transfer from tryptophan. Titration experiments provide a K_d value of 0.2 mM for Tb^{3+} binding. Addition of Ca^{2+} or nonluminescent Ln^{3+} ions displaces Tb^{3+} from hemocyanin with a concomitant decrease in luminescence. Such experiments permit the determination of the apparent K_d values shown in Table 4-10. The value of 16.67 mM calculated for Ca^{2+} differs considerably from that of 67 μM determined with a Ca^{2+}-sensitive electrode. This discrepancy may reflect the complex nature of the quenching of Tb^{3+} luminescence by Ca^{2+}. It is biphasic, with a residual 25% luminescence which resists the action of saturating amounts of Ca^{2+}. Quenching by Eu^{3+} is also biphasic. Eu^{3+} is also far more potent, at low concentrations, than the other Ln^{3+} ions, with about 50% quenching at 5 μM Eu^{3+}. As the quantitative effects of Eu^{3+} as an allosteric effector are in the normal lanthanide range (Table 4-10), and as Eu^{3+} has absorption bands which overlap the Tb^{3+} emission, Kuiper *et al.* (1981) suggest that low concentrations of Eu^{3+} quench Tb^{3+} luminescence by intercepting and absorbing its emitted radiation. Such a suggestion leads to the general conclusion that at least some Ca^{2+}-binding sites are clustered.

The usefulness of Tb^{3+} luminescence in studying hemocyanin has been confirmed by Nelson *et al.* (1981), using material isolated from the horseshoe crab, *Limulus polyphemus*. Energy transfer from a Trp residue to Tb^{3+} was confirmed with the reagent *N*-bromosuccinimide, which spe-

Table 4-10. Dissociation
Constants for the Binding of
Ln^{3+} and Ca^{2+} to
P. interruptus Hemocyanin

Ion	K_d (mM)[a]
Tb^{3+}	0.20
Gd^{3+}	0.67
La^{3+}	0.10
Pr^{3+}	0.07
Er^{3+}	0.02
Ca^{2+}	16.67
	0.07[b]

[a] Data from Kuiper *et al.* (1981). K_d
values were determined from the
decrease in Tb^{3+} luminescence.
[b] Determined with a Ca^{2+}-selective
electrode.

cifically blocks tryptophan side chains. These authors reported that, at pH 7, Ca^{2+} increases Tb^{3+} luminescence, apparently by altering the conformation of hemocyanin. They failed to address the matter of why adding Ca^{2+} did not have the expected effect of displacing Tb^{3+}, thereby lowering luminescence as Kuiper *et al.* (1981) had found. Nelson *et al.* (1981) did find quenching by Ca^{2+} at pH 8.9, but few conclusions can be drawn from experiments with lanthanides conducted at such a high pH (Section 2.3). Molluskan hemocyanins are normally present physiologically as aggregates. Ca^{2+} promotes this aggregation *in vitro*, in a reaction which can also be mediated by Tb^{3+}.

The Tb^{3+} luminescence experiments described above were conducted with deoxygenated hemocyanin. As oxygen quenches Tb^{3+} luminescence by 65–70%, the method is not useful in investigations of oxyhemocyanin (Kuiper *et al.*, 1979). However, direct laser stimulation of Eu^{3+} luminescence provides information about both oxygenated and deoxygenated forms of hemocyanin. The hemocyanin of *Busycon canaliculatum* contains two types of high-affinity Eu^{3+}-binding sites, with K_a values of $10^5 M^{-1}$ and $10^4 M^{-1}$. At Eu^{3+}:hemocyanin ratios greater than 4.5:1, the protein aggregates (Hwang *et al.*, 1984). There exists a derivatized form of hemocyanin, known as "spectral probe hemocyanin," which gives EPR signals which change upon oxygenation or addition of Ca^{2+}. Eu^{3+} produces the same changes as Ca^{2+} and affects O_2 binding, as monitored by the EPR signal, in the same manner. Such experiments confirm that Eu^{3+} exhibits a heterotrophic effector role parallel to that of Ca^{2+}.

Upon binding to oxyhemocyanin, Eu^{3+} ions lose two to three molecules of bound water, to coordinate an average of 6.5 molecules of water per bound Eu^{3+} ion. A further one to two molecules of water are lost upon deoxygenation, with only an average of 5.1 molecules of water coordinated with each Eu^{3+} ion. Only one of the two Eu^{3+} sites is competitive with respect to Ca^{2+}. At the noncompetitive site, each Eu^{3+} ion coordinates an average of 5.4 water molecules in oxyhemocyanin, and 5.0 water molecules in deoxyhemocyanin. Quenching studies have been used to estimate the distance between the Cu site and the competitive Eu^{3+} site as 32 Å.

The hemocyanin of *Megathura crenulata* forms ionic channels in lipid bilayers. Tb^{3+} promotes the formation of new channels by binding to the phosphatidylcholine bilayer and masking its negative charge (Menestrina, 1983).

4.18 Phospholipase A₂

Phospholipase A_2 (EC 3.1.1.4) is a Ca^{2+}-requiring lipase which can be found both extracellularly and intracellularly. The extracellular species have a molecular weight of about 14,000, are water-soluble, and are very stable. The pancreatic form of phospholipase A_2 is secreted as a zymogen which is converted to the active form through limited proteolysis by trypsin. Pancreatic phospholipase A_2 shows low activity toward monomeric substrates but high activity toward micellar aggregates. X-ray diffraction studies have shown that the lipid-binding domain of the bovine enzyme forms a hydrophobic edge surrounding the cavity of the active site. The Ca^{2+} ion binds in the active site of porcine phospholipase A_2 to the carboxyl side chain of the Asp-49, the carbonyl oxygens of Tyr-28, Gly-30, and Gly-32, and two water molecules. The K_d for Ca^{2+} is about 12.5 mM at pH 6, and 1 mM at pH 10; at neutral pH, the affinity for Ca^{2+} is enhanced by micellar substrate analogues. Ca^{2+} functions to bind lipid molecules to the active center.

Lanthanides are poor functional substitutes for the Ca^{2+} ion of phospholipase A_2. Eu^{3+}- or Gd^{3+}-substituted bovine or porcine phospholipase A_2 show only 4–5% of their maximum activity, although this is still 20 to 30 times that of the zymogen. The K_{act} for Gd^{3+} is 25 μM, compared to 550 μM for Ca^{2+} (Hershberg *et al.*, 1976). The K_d values for the binding of these ions to the enzyme are much higher. For Gd^{3+}, K_d is 0.5 mM for the porcine enzyme, and 0.18 mM for the proenzyme (Hershberg *et al.*,

1976). Eu^{3+} binds to the bovine and the porcine enzyme with K_d values of 0.22 mM and 0.16 mM, respectively. Ca^{2+} competitively displaces Eu^{3+} from phospholipase A_2, yielding apparent K_d values for Ca^{2+} of 4.7 mM for the bovine enzyme and 1.4 mM for the porcine enzyme, by Eu^{3+} excitation spectroscopy. Competitive reduction of the Gd^{3+}-proton relaxation rate has yielded K_d values of 0.08 mM for Tb^{3+} and 0.07 mM for Eu^{3+}. The discrepancies between the K_{act} values for the enzymic reactions and the K_d values of the metal–enzyme interaction suggest that the presence of substrate alters the metal-binding site. Other evidence supports this conclusion. Although Gd^{3+} has only weak affinity for monomeric and micellar n-alkylphosphorylcholines (K_d = 2–8 mM), micelles increase significantly the PRR (proton relaxation rate) enhancement upon binding to enzyme–Gd^{3+}–monomer complexes. The affinity of the micelles for the enzyme is much higher in the presence of Gd^{3+}. It is possible that phospholipase A_2 is sensitive to the changes in curvature that lanthanides produce in phospholipid vesicles (Section 6.2.1).

Eu^{3+} excitation spectroscopy of both the bovine and the porcine enzyme yields one peak, representing a single binding site on the enzyme. A small, second peak is thought to result from aggregation of the enzyme. This method shows that Eu^{3+} is bound to 4.6 molecules of water in porcine phospholipase A_2, 4.1 in isophospholipase A_2, and 5.2 in bovine phospholipase A_2. On adding a monomeric substrate analogue, one molecule of H_2O is excluded in each case. However, the addition of the micellar form of the analogue produces severe dehydration of Eu^{3+}, with only one H_2O molecule remaining as a ligand. Dehydration is less pronounced with the bovine enzyme. Van Scharrenburg et al. (1985) speculate that the decreased activity of the Eu^{3+}-substituted form may reflect its lower rate of ligand exchange.

Phospholipase D also requires Ca^{2+}, which participates in the formation of an enzyme–substrate complex. La^{3+} can displace Ca^{2+} from its binding site in producing a catalytically inactive complex. Ca^{2+} not only binds the substrate to the enzyme but neutralizes extra charge on substrate micelles. Ca^{2+} forms a recognition site in the active center of phospholipase D (Rakhimov et al., 1982).

4.19 Acetylcholinesterase

Ca^{2+} ions accelerate the activity of acetylcholinesterase (EC 3.1.1.7) by an unknown mechanism. Activation is reversed by EDTA. Tb^{3+} (1 μM) also activates the enzyme by a V_{max} effect. However, at a concentration of 100 μM, Tb^{3+} acts as an uncompetitive inhibitor, and at

$500\,\mu M$, as a competitive inhibitor. Scatchard plots based on Tb^{3+} luminescence give high-affinity sites with $K_d \approx 7.6\,\mu M$ and low-affinity sites with $K_d \approx 49.6\,\mu M$ (Marquis, 1984). Tomlinson et al. (1982), using *Electrophorus electricus,* confirmed that La^{3+} activates acetylcholinesterase at low concentration and inhibits at high concentration. However, these authors also found that the activation effect only held for freshly mixed solutions of La^{3+} and enzyme. After preincubating such mixtures, the effect of La^{3+} was inhibitory. However, as these authors used a pH of 8 for these experiments, the delayed effect could result from the formation of precipitates of $La(OH)_3$ (Section 2.3; Table 2-10).

Despite inhibiting acetylcholinesterase, La^{3+} partially relieves the inhibition of this enzyme by decamethonium (Marquis and Webb, 1976).

4.20 Concanavalin A

A lectin isolated from jack bean, concanavalin A binds specifically to polysaccharides containing α-D-mannopyranoside or β-D-glucopyranoside moieties. Its binding activity requires the presence of a transition metal at a site designated S_1 and the presence of Ca^{2+} at a separate site, S_2. Following the observation that lanthanides could restore partial saccharide-binding activity to Ca^{2+}-free concanavalin A (Sherry et al., 1975), it was assumed that Ln^{3+} ions could occupy the S_2 site on this protein. However, proton resonance relaxation studies demonstrated that Gd^{3+} binds instead to a third site, designated S_3, with a K_d of about $15\,\mu M$. Several additional weak sites were also identified (Barber et al., 1975). Crystallographic (Becker et al., 1975) and Tb^{3+} luminescence (Richardson and Behnke, 1978; Avigliano et al., 1984) data have confirmed the existence of a separate Ln^{3+}-binding site.

The luminescence experiments suggested that Tb^{3+} occupied a position close to a Trp residue, possibly Trp-88. Endogenous protein fluorescence was reduced by 10–20% in the presence of Tb^{3+}. Scatchard analysis revealed a single high-affinity Tb^{3+}-binding site of $K_a = (3.2 \pm 0.5) \times 10^4\,M^{-1}$ and at least four weaker sites of $K_a \approx (5 \pm 1) \times 10^3\,M^{-1}$. The presence of millimolar concentrations of Mn^{2+} or Ca^{2+} had no effect on the binding of Tb^{3+} to its high-affinity site but reduced the apparent number of low-affinity sites without affecting the K_a values (Richardson and Behnke, 1978). Attempts to detect quenching of Tb^{3+} luminescence by Co^{2+} at site S_1 failed, suggesting that sites S_1 and S_3 were separated by at least 38 Å, a distance compatible with the crystallographic evidence (Richardson and Behnke, 1978; Barber et al., 1975). Although other lanthanides could compete away Tb^{3+} and thus lower its luminescence, Ca^{2+}

could not, a finding consistent with the existence of separate binding sites for Ca^{2+} and Ln^{3+} ions. Transition metals had a limited capacity to displace Tb^{3+} from concanavalin A, indicating that Ln^{3+} ions may have a weak affinity for site S_1. Studies on the magnetic circular dichroism spectrum of Pr^{3+}-concanavalin A confirmed the conclusions derived from Tb^{3+} luminescence. Although lanthanides can only partially reactivate Ca^{2+}-free concanavalin A, they do not affect the polysaccharide-binding properties of native concanavalin A (Avigliano *et al.*, 1984).

Of great interest is the observation that although Ln^{3+} ions cannot occupy the S_2 site on concanavalin A which accepts Ca^{2+}, Eu^{2+} can do so (Homer and Mortimer, 1978). A proposed explanation is that the extra positive charge of the trivalent lanthanide ions creates excessive electrostatic repulsion from the divalent ion at S_1. However, the results of Richardson and Behnke (1978) suggest that Ln^{3+} ions fail to occupy site S_2 even in the absence of a transition metal. Because of its ability to bind at site S_2, Eu^{2+} ions were much better than Ln^{3+} ions at activating Ca^{2+}-free concanavalin A. Time-dependent changes in the MCD spectrum of Eu^{2+}-concanavalin A suggested that the binding of Eu^{2+} induces a slow conformational change in the protein (Homer and Mortimer, 1978).

4.21 Serum Albumin

Bovine serum albumin (BSA) shares with lysozyme the distinction of being the first purified protein to have its interaction with the lanthanides studied in a serious manner. Birnbaum *et al.* (1970) selected this protein to examine the possibility that changes in the absorption spectrum of a Ln^{3+} ion upon binding to a protein could provide information on the environment of the binding site. Although this was accepted as a theoretical possibility, it was known that the changes would be slight, perhaps too small to be of experimental value (see discussions in Sections 2.2 and 3.4.1). The absorption spectrum of Nd^{3+} (Fig. 2-1) in the presence or absence of BSA did not look different upon visual inspection. However, the difference absorption spectrum (Fig. 4-12) revealed perturbations, especially at 520 nm and 580 nm. By comparing the difference absorption spectra of Nd^{3+} in combination with a number of model ligands, the authors concluded that Nd^{3+} was bound to BSA through only carboxyl ligands. Reuben (1971) used proton relaxation methods to confirm the presence of four binding sites for Gd^{3+} on BSA. The apparent K_d was 0.13 mM. As a 120-fold excess of Ca^{2+} did not influence the enhancement of the relaxation rate produced by Gd^{3+}, it appeared that Ca^{2+} did not strongly compete for Gd^{3+} at these sites. Later NMR studies with $^{139}La^{3+}$

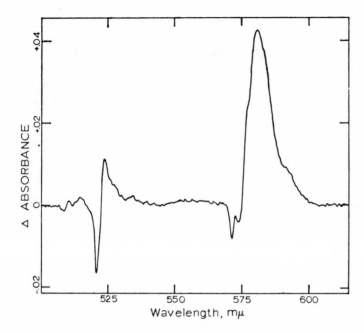

Figure 4-12. Perturbations in the absorption spectrum of Nd^{3+} induced by bovine serum albumin. From Birnbaum *et al.* (1970), with permission.

(Reuben and Luz, 1976) could only detect weak interaction between BSA and this probe, with a K_d of 460 mM.

Lanthanides induce conformational changes in bovine and human serum albumin (Shahid *et al.*, 1982). These are similar to the expansion of the protein's structure that occurs upon lowering the pH. The effectiveness in producing this change followed the sequence $Lu^{3+} > Tb^{3+} > Nd^{3+} > La^{3+}$. A variety of other metal ions, including Al^{3+}, Ca^{2+}, and several transition metal ions, were unable to do this. The authors explained the conformational changes produced by H^+ and the Ln^{3+} ions in terms of their ability to bind to carboxyl groups and thus neutralize negative charges in the protein. However, it is unclear why other metal ions should not be able to do this too.

Strong attachment sites for lanthanides have been introduced into human serum albumin by the covalent attachment of bifunctional chelating agents based upon EDTA (Leung and Meares, 1977). Luminescence from sequestered Eu^{3+} or Tb^{3+} permitted the detection of as little as $10^{-5} M$ or $10^{-7} M$, respectively, of bound lanthanide.

4.22 Immunoglobulin

Dower *et al.* (1975) employed proton relaxation enhancement and EPR techniques to investigate the binding of Gd^{3+} to rabbit nonimmune immunoglobulin G (IgG). At pH 5.5, there were six binding sites for Gd^{3+} on IgG. Two of these were high-affinity sites with K_d values of about 5 μM, located in the F_c domain. Two weaker sites with K_d values of about 140 μM were present in the F_{ab} domain. The $(F_{ab}')_2$ fragment contained four sites of K_d about 100 μM. The binding characteristics of intact IgG molecules were the sum of the isolated F_{ab} and F_c fragments, suggesting little interaction between the high-affinity and low-affinity Gd^{3+} sites. Ca^{2+} ions did not compete with Gd^{3+} for these sites. The addition of Gd^{3+} to F_{ab} or F_c fragments increased the intrinsic Trp fluorescence of these proteins by 10%. Tb^{3+} luminescence was enhanced approximately 100-fold upon binding to these proteins.

These sorts of experiments were repeated with IgG from an animal which had been immunized against a pneumococcal polysaccharide (Willan *et al.*, 1977). Water solvent proton relaxation analysis confirmed the presence of two high-affinity Gd^{3+} sites on this IgG, each with a K_d value of 3.75 μM. These appeared to be present in the F_c domain of the molecule. Isolated F_{ab} fragments had two binding sites for Gd^{3+}, with a K_d value of 250 $\pm$ 150 μM. Binding of an octasaccharide constituent of the antigen to the IgG did not affect Gd^{3+} binding. However, addition of a 28-unit antigenic saccharide did so. The effects of this molecule might have reflected alterations in the tumbling time of the IgG (Willan *et al.*, 1977).

Enhanced protein fluorescence was found on the binding of certain lanthanides to the variable region fragment of the IgA myeloma protein MOPC 315 (Dwek *et al.*, 1976). From such changes, K_d values for Gd^{3+} of 700 μM at pH 4.8 and 10 μM at pH 7.0 were determined. Each fragment had one binding site for Gd^{3+}. Tb^{3+} had similar binding characteristics to Gd^{3+}, but La^{3+}, although having a K_d of about 500 μM at pH 5.1, had a relatively high K_d of 100 μM at pH 6.8. Eu^{3+} produced no change in the intrinsic fluorescence of the protein but competitively inhibited the enhancement due to Gd^{3+}. This permitted calculation of a K_d for Eu^{3+} of 50–80 μM at pH 5.9. Ca^{2+} did not compete for the Ln^{3+}-binding site on this protein. Although hapten was sequestered at a separate site, Ln^{3+} ions and hapten mutually affected each other's binding. EPR studies with spin-labeled hapten demonstrated that a distance of 12 Å separated the Ln^{3+}-binding site from the spin label. Time-dependent changes in the fluorescence of the protein upon binding hapten and Ln^{3+} ions suggested that the fragment could adopt more than one conformation with different affinities for Ln^{3+} and hapten. In this sense, lanthanides could be used to monitor changes in conformation that arise upon binding hapten. High-resolution NMR studies indicated that Ln^{3+} binding produced only small, localized conformational changes (Dwek $et\,al.$, 1976).

Burton $et\,al.$ (1977a) employed a detailed proton relaxation enhancement analysis to determine the molecular motion parameters and the hydration number of the Gd^{3+} ion. When bound to IgG or its F_c or pF_c (C-terminal quarter of the heavy chain dimer) fragment, 3.8–5.8 water molecules were coordinated to Gd^{3+}. With fragment F_{ab}, the hydration number was 1. As Horrocks (1982) has pointed out, these values are surprising, as Gd^{3+} binds much more tightly to the first group of proteins than to the F_{ab} fragment. It appears that the hydration numbers generated for IgG–Gd^{3+} complexes were susceptible to large uncertainties. To address this limitation, Burton $et\,al.$ (1977b) developed a method based upon comparisons between solvent ^1H and ^2H relaxation measurements.

4.23 Enolase

Enolase (2-phospho-D-glycerate hydrolase; EC 4.2.1.11) is a dimeric enzyme catalyzing the reversible conversion of 2-phosphoglycerate to phosphoenolpyruvate. It has an absolute requirement for a divalent cation, which is satisfied physiologically by Mg^{2+}. There appear to be two metal-binding sites on the enzyme. When a suitable metal ion binds to the site known as the "conformational" site, the tertiary structure of enolase alters, permitting the binding of substrate. This in turn permits the binding of an additional metal ion at the "catalytic" site. It is the metal ion at the conformational site which determines whether or not enzymic activity will occur. The metal ion at the catalytic site determines the rate of reaction (Brewer, 1985). Tb^{3+} binds to the conformational site of enolase but fails to support enzymic activity (Brewer $et\,al.$, 1981). Addition of Tb^{3+} to the apoenzyme increased the intrinsic fluorescence of the protein in the same manner as Mg^{2+} and, like Mg^{2+}, enhanced its thermostability. Two equivalents of Tb^{3+} were bound per mole of dimer. No enhancement of Tb^{3+} luminescence occurred upon binding to enolase. Measurements of the luminescent emission lifetimes of Tb^{3+} following direct laser excitation suggested the coordination of three molecules of water to each sequestered Tb^{3+} ion. Calorimetric measurements indicated that Tb^{3+} facilitates the binding of substrate to enolase, even though no reaction occurs. The inhibitory effects of Tb^{3+} are reversible, confirming that binding of Tb^{3+} does not denature the enzyme.

In addition to the "conformational" and "catalytic" sites for metal ions, enolase has an "inhibitory" site with a high affinity for transition metals. Quantitative analysis of the quenching of Tb^{3+} luminescence by Co^{2+} indicated a Tb–Co distance of $17 \pm 4\,\text{Å}$ (Brewer et al., 1981). With Co^{2+} in one of the two conformational sites and Tb^{3+} in the other, no quenching occurred, suggesting a large separation between these sites.

4.24 Kinases

Phosphoglycerate kinase (EC 2.7.2.3) catalyzes the reversible phosphorylation of 3-phosphoglycerate by ATP. Lanthanides inhibit the forward reaction by forming $Ln\text{-}ATP^-$ complexes which compete with Mg-ATP as substrates for the reaction. The K_i values for La-, Eu-, Pr-, and Yb-ATP$^-$ were all about 0.04 ± 0.01 mM, compared with a K_m value of 0.073 mM for Mg-ATP (Tanswell et al., 1974). Proton NMR spectroscopy provided K_d values of 0.15 ± 0.07 mM for Mg-ATP$^-$ and 0.25 ± 0.13 mM for Mg-ADP$^-$, but La-ATP$^-$ bound too tightly to permit the determination of its K_d value. The alterations in the resonances produced by La-ATP were similar to those produced by Mg-ATP, suggesting a common binding site. NMR spectroscopy with Gd-, Pr-, and Eu-ATP has been used to identify certain nuclei in the vicinity of the active site of the enzyme and to make distance assignments. No conformational change in the Ln-ATP$^-$ complex upon binding to the enzyme was detected. Ln-pyrophosphate enabled the phosphate-binding site of the enzyme to be examined by NMR spectroscopy (Tanswell et al., 1976).

Ln-ATP$^-$ complexes also inhibit hexokinase, the enzyme which catalyzes the reversible phosphorylation of glucose (Viola et al., 1980; Morrison and Cleland, 1980). Inhibition is competitive with respect to Mg-ATP; the K_i values are shown in Table 4-11. When free Mg^{2+} was added, enzymic activity reappeared due to the competitive displacement of Ln^{3+} ions from ATP. Morrison and Cleland (1980, 1983) used this phenomenon to calculate K_d values for Ln-ATP$^-$ complexes. These are difficult to measure by standard techniques due to their low values (Section 5.1; Table 5-3). Although all Ln-ATP$^-$ complexes are inhibitors of hexokinase, only those complexes containing Sm^{3+} to Lu^{3+} are of the "slow-binding" type (Morrison, 1982).

Pyruvate kinase (EC 2.7.1.40) reversibly catalyzes the phosphorylation of ADP by phosphoenolpyruvate. It, too, is strongly inhibited by lanthanides. This enzyme requires Mn^{2+} or Mg^{2+} for activity. Water proton relaxation measurements have identified four binding sites for Gd^{3+} on pyruvate kinase, each with a K_d of $26 \pm 10\,\mu M$ (Valentine and Cottam, 1973). Competition experiments showed that Zn^{2+}, Mn^{2+}, Mg^{2+}, Ca^{2+}, Nd^{3+}, Tm^{3+}, and La^{3+} also bound to this site. The K_d value for the

Table 4-11. Inhibitor
Constants for the Inhibition of
Yeast Hexokinase by Ln-
ATP$^-$ Complexes[a]

Ln^{3+}	K_i (μM)
La^{3+}	174 ± 8
Ce^{3+}	149 ± 8
Nd^{3+}	105 ± 4
Eu^{3+}	36 ± 4
Gd^{3+}	32 ± 1
Tb^{3+}	10 ± 1
Er^{3+}	1.16 ± 0.33
Yb^{3+}	0.84 ± 0.22
Lu^{3+}	0.84 ± 0.36

[a] From Viola *et al.* (1980), with permission.

interaction of Gd-ATP$^-$ with pyruvate kinase was estimated as 13 ± 4 μM. It was suggested that Gd^{3+} acted as a metal bridge in the formation of the ternary complex. No change in the water proton relaxation rate enhancement was detected upon forming a substrate–Gd^{3+}–enzyme ternary complex. However, broadening of the ^{1}H-NMR peaks attributable to the vinyl protons of phosphoenolpyruvate confirms that this ternary complex does indeed form (Cottam *et al.*, 1974), with a vinyl proton–Gd^{3+} distance of 7.9 Å. Cottam and Ward (1975) confirmed the existence of this complex by ^{31}P-NMR spectroscopy, providing a Gd^{3+}–P distance of 5.2 Å. The authors suggested that Gd^{3+} may inhibit enzymic activity by blocking a conformational change induced by the binding of substrate.

The interactions between several protein kinases and the lanthanides have been studied. Myosin light chain kinase is regulated by Ca^{2+}-calmodulin. At a concentration of 100 μM, La^{3+} or Tb^{3+} were able to support over 60% of the enzymic activity produced by the same concentration of Ca^{2+} (Mazzei *et al.*, 1983). Low concentrations of Ln^{3+} potentiated the stimulatory effects of suboptimal concentrations of Ca^{2+}, but high concentrations of Ln^{3+} were inhibitory.

Protein kinase C requires both phospholipid and Ca^{2+}. According to Mazzei *et al.* (1983), La^{3+} and Tb^{3+} (100 μM) are only weakly able to replace Ca^{2+} in supporting the protein kinase C activity of bovine and porcine spleen, heart, and brain. However, La^{3+} (10–300 μM) appears to support a high activity of protein kinase C partially purified from murine T-lymphocytes (Smith *et al.*, 1986). La–ATP could not serve as a phosphate donor for protein kinase C. Mazzei *et al.* (1983) further reported that La^{3+} or Tb^{3+} at concentrations up to 200 μM greatly potentiated the activity of protein kinase C in the presence of suboptimal concentrations of Ca^{2+}. Higher concentrations of Ln^{3+} were inhibitory.

La^{3+} and Tb^{3+} (100 μM) slightly inhibited cGMP- and cAMP-dependent protein kinases (Mazzei *et al.*, 1983).

4.25 Other Proteins

Crystallographic studies on human muscle glyceraldehyde phosphate dehydrogenase (EC 1.2.1.12) have identified a binding site for Sm^{3+} at the active center (Watson *et al.*, 1972). Addition of Gd^{3+} to spin-labeled enzyme decreased the height of the EPR spectal lines produced by the label (Dwek *et al.*, 1975). La^{3+} had a much smaller effect, although enough to suggest that a conformational change occurs. Based upon their effects on the EPR spectra, K_d values of 0.73 ± 0.04 mM and 1.4 ± 0.2 mM were calculated for Gd^{3+} and La^{3+}, respectively; the K_d value for the binding of Gd^{3+} to the human enzyme was 0.5 ± 0.08 mM. The enzyme was not inhibited by La^{3+} (Dwek *et al.*, 1975).

X-ray crystallographic analysis of ferredoxin suggested two main sites and two minor sites for Sm^{3+}, with two main and six minor sites for Pr^{3+} (Sieker *et al.*, 1972). The iron in this protein gives rise to EPR signals whose relaxation and linewidths are modified in the presence of Dy-EDTA (Blum *et al.*, 1981). Based upon the r^{-6} dependence of the magnitude of the perturbing effects of Dy^{3+}, the Fe–S cluster of ferredoxin from *Clostridium pasteurianum* appeared to lie 10.6 Å below the surface of the protein, to which Dy-EDTA is presumably restricted. A similar approach has been applied to various cytochromes, which are also iron-containing proteins. Blum *et al.* (1980) estimated that a distance of 5 Å separated Dy^{3+} on the surface of cytochrome c from the heme iron within it. For cytochrome oxidase $(a + a_3)$, this distance was 6.5 Å. When cytochrome c and cytochrome oxidase combine, Dy^{3+} can only approach to 12.6 Å from the heme iron of cytochrome c. By an analogous procedure, Dockter (1982) employed luminescent chelates of Tb^{3+} to measure, by quenching, the iron–Tb^{3+} distances in various cytochromes. Values of 10, 16, and 16 Å were reported for cytochrome c, cytochrome c peroxidase, and cytochrome c oxidase, respectively.

Both the acetylcholine receptor (Rübsamen *et al.*, 1976) and insulin receptor (Williams and Turtle, 1984) bind Ln^{3+}. Luminesent enhancement via energy transfer from Tyr or Trp residues suggests that there are 10 Tb^{3+}-binding sites per acetylcholine receptor, each with a K_d of 18 ±

$0.5\,\mu M$. This value was independent of ionic strength. In the presence of $8\,\text{m}M\ Ca^{2+}$, two classes of Tb^{3+}-binding site appeared, only one of which interacted with Ca^{2+}. The latter represented 60% of the total Tb^{3+} sites and had a K_d for Tb^{3+} of $1.5 \times 10^{-4}M$, the K_d for Ca^{2+} being $1.1 \pm 0.1\,\text{m}M$. Acetylcholine agonists, but not antagonists, displaced Tb^{3+} from the Ca^{2+}-sensitive sites. Neutron activation analysis (Section 2.7.7) confirmed the existence of 10 Tb^{3+}-binding sites, 6 of which also accept Ca^{2+} (Rübsamen $et\,al.$, 1976). The insulin receptor shows similar behavior (Williams and Turtle, 1984). Tb^{3+} luminescence after indirect excitation via Trp chromophores was half-maximal at $0.15\,\text{m}M\ Tb^{3+}$. About 60% of the Tb^{3+} could be displaced by Ca^{2+} or Mg^{2+}. Insulin and other agonists displaced Tb^{3+} from its Ca^{2+}-sensitive binding sites on the receptor, with those agonists showing the greatest biological activity being the most effective displacers of Tb^{3+} and Ca^{2+}. In addition, Tb^{3+} and Ca^{2+} increased the affinity of insulin for its receptor, without altering the number of receptors. Tb^{3+} was 50-fold more active than Ca^{2+} in this respect.

Insulin is stored in the β-cells of the pancreas as a hexamer containing two Zn^{2+} ions and three Ca^{2+} ions. Emission lifetime measurements following direct laser excitation of the Eu^{3+}-substituted hexamer suggest that 2.0 ± 0.5 water molecules are bound to each Eu^{3+} ion (Evelhoch, 1981). Quenching studies using Co^{2+}- and Eu^{3+}-substituted insulin hexamer provided a Co^{2+}–Eu^{3+} distance of $9.6 \pm 0.5\,\text{Å}$. These studies indicated that there are three symmetrically oriented Eu^{3+}-binding sites near the center of the hexamer. The effects of lanthanides upon insulin and glucose homeostasis are discussed in Chapters 6 and 8, respectively.

Lanthanides bind to yeast inorganic pyrophosphatase, quenching the intrinsic protein fluorescence (Sperow and Butler, 1972). Quenching by Nd^{3+}, Sm^{3+}, and Gd^{3+} was attributed to denaturation of the enzyme, but Eu^{3+} appeared to be more specific. Its binding to pyrophosphatase was too strong to permit accurate determination of K_d, but an estimate of $0.23\,\mu M$ was obtained; the K_d for Ca^{2+} is $800\,\mu M$. Two metal-ion binding sites accepting Eu^{3+}, Ca^{2+}, or Mg^{2+} exist on the enzyme. Under physiological conditions, Mg^{2+} is the activating metal ion, while the Eu^{3+}-substituted enzyme had no enzymic activity.

Several lanthanides have been shown to inhibit the serine proteinases plasminogen activator and plasmin (Dano and Reich, 1979). This is surprising in view of the independence from major metal-ion effects of serine proteinases. Trypsin and elastase, for instance, are only weakly inhibited by lanthanides (Sections 4.9 and 4.10). However, the use of a pH of 8.1 in the plasmin and plasminogen activator assays renders difficult the interpretation of the lanthanide inhibition (Section 2.3). Plasmin was inhibited 50% by La^{3+}, Eu^{3+}, Pr^{3+}, Yb^{3+}, or Y^{3+} at concentrations of 68, 220,

110, and 37 μM, respectively. For plasminogen activator, the corresponding concentrations were 70, 140, 23, 31, and 31 μM.

Lanthanides are unable to replace Mg^{2+} in the C3 convertase reaction during the activation of complement by the alternative pathway (Fishelson and Muller-Eberhard, 1983). However, the product of this reaction, C3b,Bb, is thermally stabilized by Ln^{3+} ions and rendered partially resistant to inactivation by factors H and I.

Catechol-O-methyltransferase (EC 2.1.1.7) is a Mg^{2+}-requiring enzyme catalyzing the O-methylation of catechol compounds such as epinephrine. It is inhibited by lanthanides with 50% inhibition in the presence of 1 mM Mg^{2+} occurring at $3 \times 10^{-6} M$ La^{3+}, $1.2 \times 10^{-6} M$ Nd^{3+}, and $8.5 \times 10^{-7} M$ Eu^{3+}. Ca^{2+} is also inhibitory, with 50% effectiveness at a concentration of $4.5 \times 10^{-4} M$. Inhibition by lanthanides was of the "mixed" type with respect to Mg^{2+}, the artificial substrate dihydroxybenzoic acid, and the methyl donor S-adenosylmethionine (Quiram and Weinshilboum, 1976). Lanthanides also inhibit NAD-specific isocitrate dehydrogenase (EC 1.1.1.41), with K_i values in the 1.81–3.10 μM range. Ca^{2+} did not reverse the inhibition, but inhibition by La^{3+} was competitive with respect to Mg^{2+}-isocitrate (Aogaichi et al., 1980).

The alkaline phosphatase (EC 3.1.3.1) of E. coli contains four Zn^{2+} ions and requires the binding of one Mg^{2+} to attain enzymic activity. Tb^{3+} is able to occupy the Mg^{2+} site, where it is almost as effective in promoting catalysis. However, neither Mg^{2+} nor Tb^{3+} produces phosphatase activity in the Zn^{2+}-free enzyme. Luminescence enhancement titrations provide a K_d of $0.16 \pm 0.03 \mu M$ for Tb^{3+}. Substrate and Zn^{2+} are both effective quenchers of Tb^{3+} luminescence, via a mechanism involving conformational changes in the enzyme. The major drawback in these studies is the use of 0.1 M Tris–HCl buffer pH 8 (Section 2.3). As the authors describe, under these conditions Tb^{3+} was nondialyzable, even though there was no visible precipitate (Sherry et al., 1978).

Lactalbumin can exist in one of several different conformations depending upon the presence or absence of Ca^{2+}, Zn^{2+}, and other metal ions. The apoprotein has three binding sites for Tb^{3+}, which luminesce through energy transfer from Trp residues. The first Tb^{3+} ion appears to displace Ca^{2+} from its binding site on lactalbumin. Binding of the second Tb^{3+} ion reduces the affinity of Tb^{3+} at the first site. At high concentrations of Tb^{3+}, a third Tb^{3+} ion binds to lactalbumin. This binding is initially weak, but the conformation change it induces in the protein leads to an increase in its strength (Kronman et al., 1981; Bratcher and Kronman, 1984; Kronman and Bratcher, 1984).

The glutamine synthetase (EC 6.3.1.2) of E. coli is a dodecamer which requires Mn^{2+} or Mg^{2+} for enzymic activity. A variety of lanthanides can replace the divalent metal in the adenylated form of the enzyme, producing up to about 20% of the activity of the native enzyme (Wedler and D'Aurora, 1974). Much lower stimulation of the unadenylated enzyme occurred. EPR and NMR spectroscopy showed that Nd^{3+} did not displace Mn^{2+} from the enzyme. Difference absorption spectra revealed small differences in the absorption of Nd^{3+} upon binding to the enzyme. However, Tb^{3+} luminescence was not enhanced (Wedler and D'Aurora, 1974).

Other proteins whose interaction with lanthanides has been reported are listed in Table 4-12.

4.26 Summary

The major ligand for Ln^{3+} ions on proteins is the carboxyl group. Additional coordination may occur through carbonyl and hydroxyl oxygens, with a variable number of water molecules completing the coordination sphere. Although there is some evidence that lanthanides can interact with unprotonated amino nitrogens of amino acids and the imidazole nitrogen of histidine, these do not appear to contribute to lanthanide binding in proteins.

In many cases, lanthanides occupy Ca^{2+}-binding sites on proteins. X-ray crystallographic analysis of Ln^{3+}-substituted thermolysin, parvalbumin, troponin C, and calmodulin has confirmed that the replacement of Ca^{2+} by Ln^{3+} ions can occur isomorphously. Although the match is not perfect in every case, any perturbations are usually minor. Because of their higher charge-to-volume ratio, Ln^{3+} ions usually have higher affinities than Ca^{2+} itself for the Ca^{2+}-binding sites. However, there are exceptions to this generalization, such as the "pseudo E-F hand" site of intestinal Ca^{2+}-binding protein. An extreme example is provided by concanavalin A, where Ln^{3+} ions do not attach to the Ca^{2+} site. Other metal ions, such as Mg^{2+}, Mn^{2+}, and Fe^{3+}, may be displaced from certain proteins by lanthanides. In addition, lanthanides sometimes bind to unique sites on proteins which are not known to accept other metal ions. Examples include lysozyme and the S_3 site of concanavalin A. The affinities of Ln^{3+} ions for different proteins vary enormously, with K_d values below $1 \mu M$ reported for the "E-F hand" proteins. In general, the affinity increases with decreasing hydration of the sequestered Ln^{3+} ion and with increasing cationic charge at the binding site. The smaller Ln^{3+} ions often bind more strongly than the larger ones due to their greater charge-to-volume ratio. However, spatial considerations sometimes predominate over this trend. An interesting example is provided by clostridiopeptidase A, which appears to select Ln^{3+} ions closest in size to Ca^{2+}.

Table 4-12. Other Ln^{3+}–Protein Interactions

Protein	Observation	Reference
C-Reactive protein	Enhanced Tb^{3+} luminescence	Short and Osmand, 1983
α-Globulin	Binding of ^{140}La, ^{90}Y[a]	Rosoff and Spencer, 1975
Leaf peroxidase	Inhibition by La^{3+}	Castillo et al., 1984
Tooth phosphoprotein	Binding of La^{3+}	Cookson et al., 1980
Aequorin	Various Ln^{3+}, but not Ce^{4+}, promote light emission	Izutsu et al., 1972
Glycophorin A	Binding of Gd^{3+}[b]	Daman and Dill, 1983
Lipase	Ce^{3+} does not affect activity[a]	Renaud et al., 1980
Pancreatic trypsin inhibitor	Tyr nitration introduces strong Ln^{3+}-binding site	Marinetti et al., 1976
Lactate dehydrogenase Isocitrate dehydrogenase Glucose-6-phosphate dehydrogenase	Inhibition by various Ln^{3+}	Holten et al., 1966
Glutamate dehydrogenase	Reverse reaction strongly inhibited Forward reaction, little inhibition	Holten et al., 1966
Interferon	Stabilization by various Ln^{3+}	Sedmak and Grossberg, 1981
RNA polymerase of isolated nuclei	60% inhibition by 1 mM La^{3+}[a]	Novello and Stirpe, 1969
Subtilisin	Tb^{3+}, Nd^{3+} do not alter CD spectrum or activity but render enzyme thermolabile	Genov et al., 1985
Bone glycoprotein	Binding of Y^{3+}	Peacocke and Williams, 1966
Ribonuclease A	Strong Ln^{3+}-binding site introduced by reaction with phthalyl isocyanate	Bradbury et al., 1978
Succinic dehydrogenase	Activated by La^{3+}, Nd^{3+} (N.B. phosphate buffer used)	Horecker et al., 1939
Brain metalloproteinase	Activated by La^{3+}	Zimmerman and Schlaepfer, 1982
Adenylate cyclase	Inhibited by Ln^{3+}[c]	Nathanson et al., 1976 Best et al., 1980
Guanylate cyclase	Inhibited by Ln^{3+}[c,d]	El-Fakahany et al., 1984 Klumpp et al., 1983
Isoleucyl-tRNA synthetase	Ln^{3+} replace Mg^{2+} on tRNA[e]	Kayne and Cohn, 1972
RNA polymerase	Analysis of inhibitor sites by diffusion-enhanced energy transfer	Meares and Rice, 1981
Bacteriorhodopsin	Proton-pumping inhibited	Drachev et al., 1984

Table 4-12. (*Continued*)

Protein	Observation	Reference
Deoxyribonuclease, glucose-6-phosphate dehydrogenase, lactate dehydrogenase, pyrophosphatase, alcohol dehydrogenase, catalase, polyphenol oxidase	Inhibited by Y^{3+}, Ln^{3+}	Clayton, 1959
Hexokinase, invertase	Uninhibited by Y^{3+}, Ln^{3+}	Clayton, 1959

[a] See Chapter 8.
[b] See Section 5.4.
[c] See Chapter 6.
[d] See Section 4.4.3.
[e] See Section 5.3.

Lanthanides facilitate the structural and functional analysis of many proteins. By far the most useful techniques in this regard are luminescence spectroscopy (Section 3.3) and NMR spectroscopy (Section 3.2). With small proteins, such as lysozyme and parvalbumin, where peak assignments are more easily made, NMR spectroscopy has provided detailed structural information. Where this is not possible, spectroscopic techniques nevertheless permit selective monitoring of conformational changes and help to identify amino acid residues in the vicinity of the binding site. If the Ln^{3+}-binding site is close to the active site of an enzyme, or some other important domain of a protein, very valuable analyses are possible. Spectroscopy also provides estimates of the hydration numbers of sequestered Ln^{3+} ions and, where more than one metal-binding site exists, a measure of the interionic distance. Lanthanides often alter the properties of proteins in ways which provide functional information. For instance, Ln^{3+} ions are better cofactors than Ca^{2+} for the activation of trypsinogen by trypsin. This observation is compatible with a "charge-masking" role for metal ions in this reaction. Other proteins, such as calmodulin, retain more or less complete biochemical activity following substitution by a lanthanide. This is so because both Ca^{2+} and Ln^{3+} ions provoke an equivalent conformational change, which permits calmodulin to interact with those proteins whose functions it modulates. However, most Ca^{2+}-requiring proteins function less well when substituted by lanthanides. Nevertheless, the nature of the inhibition by lanthanides is often informative, as it may identify the role that Ca^{2+} plays in the native protein.

Examples include staphylococcal nuclease and clostridiopeptidase A, where such studies helped to confirm a role for Ca^{2+} in substrate binding.

References

Abbott, F., Dasnall, D. W., and Birnbaum, E. R., 1975a. The location of the lanthanide ion binding site on bovine trypsin. *Biochem. Biophys. Res. Commun.* 65:241–247.

Abbott, F., Gomez, J. E., Birnbaum, E. R., and Darnall, D. W., 1975b. The location of the calcium ion binding site in bovine alpha-trypsin and beta-trypsin using lanthanide ion probes, *Biochemistry* 14:4935–4943.

Abramson, J. J., and Shamoo, A. E., 1980. The effect of divalent and trivalent cation binding on the transport of calcium- and magnesium-dependent adenosine triphosphatase, *Ann. N.Y. Acad. Sci.* 358:322–323.

Aebi, U., Smith, P. R., Isenberg, G., and Pollard, C. G., 1980. Structure of crystalline actin sheets, *Nature* 288:296–298.

Agresti, D. G., Lenkinski, R. E., and Glickson, J. D., 1977. Lanthanide induced N.M.R. perturbations of HEW lysozyme: evidence for non-axial symmetry, *Biochem. Biophys. Res. Commun.* 76:711–719.

Amphlett, G. W., Byrne, R., and Castellino, F. J., 1978. The binding of metal ions to bovine factor IX, *J. Biol. Chem.* 253:6774–6779.

Aogaichi, T., Evans, J., Gabriel, J., and Plaut, G. W., 1980. The effects of calcium and lanthanide ions on the activity of bovine heart nicotinamide adenine dinucleotide-specific isocitrate dehydrogenase, *Arch. Biochem. Biophys.* 204:350–356.

Arnone, A., Bier, C. J., Cotton, F. A., Day, V. W., Hazen, E. E., Richardson, D. C., Richardson, J. S., and Yonath, A., 1971. A high resolution structure of an inhibitor complex of the extracellular nuclease of *Staphylococcus aureus*. I. Experimental procedures and chain tracing, *J. Biol. Chem.* 246:2302–2316.

Asso, M., Granier, C., Van Rietschoten, J., and Benlian, D., 1985. Calcium and praseodymium complexes in solution. ^{1}H N.M.R. conformational study of the model tetrapeptide acetyl-aspartyl-valyl-aspartyl-alanine, *Int. J. Pept. Protein Res.* 26:10–20.

Avigliano, L., Aducci, P., Sirianni, P., and Finazzi-Agro, A., 1984. A fluorometric study of the lanthanides binding to conconavalin A, *Int. J. Biochem.* 16:1409–1413.

Banyard, S., Stammers, D. K., and Harrison, P. M., 1978. Electron density map of apoferritin at 2.8 Å resolution, *Nature* 271:282–284.

Babu, Y. S., Sack, J. S., Greenhough, T. J., Bugg, C. E., Means, A. R., and Cook, W. J., 1985. Three-dimensional structure of calmodulin, *Nature* 315:37–40.

Barber, B. H., Fuhr, B., and Carver, J. P., 1975. A magnetic resonance study of conconavalin A. Identification of a lanthanide binding site, *Biochemistry* 14:4075–4082.

Barden, J. A., and Dos Remedios, C. G., 1978. Evidence for the non-filamentous aggregation of actin induced by lanthanide ions, *Biochim. Biophys. Acta* 537:417–427.

Barden, J. A., and Dos Remedios, C. G., 1979. Binding stoichiometry of gadolinium to actin: its effect on the actin-bound divalent cation, *Biochem. Biophys. Res. Commun.* 86:529–535.

Barden, J. A., and Dos Remedios, C. G., 1980. Crystalline actin tubes. I. Is the conformation of the lanthanide-induced actin tube monomer more like F-actin than G-actin? *Biochim. Biophys. Acta* 629:163–173.

Barden, J. A., Cooke, R., Wright, P. E., and Dos Remedios, C. G., 1980. Proton nuclear magnetic resonance and electron paramagnetic resonance studies on skeletal muscle actin indicate that the metal and nucleotide binding sites are separate, *Biochemistry* 19: 5912–5916.

Becker, J. W., Reeke, G. N., Wang, J. L., Cunningham, B. A., and Edelman, G. M., 1975. The covalent and three-dimensional structure of conconavalin A. III. Structure of the monomer and its interactions with metal and saccharides, *J. Biol. Chem.* 250:1513–1524.

Best, L. C., Bone, E. A., Jones, P. B., and Russell, R. G., 1980. Lanthanum stimulates the accumulation of cyclic AMP and inhibits secretion and thromboxane B2 formation in human platelets, *Biochim. Biophys. Acta* 632:336–342.

Birnbaum, E. R., Gomez, J. E., and Darnall, D. W., 1970. Rare earth metal ions as probes of electrostatic binding sites in proteins, *J. Am. Chem. Soc.* 92:5287–5288.

Bloom, J. W., and Mann, K. B., 1978. Metal ion induced conformational transitions of prothrombin and thrombin fragment 1, *Biochemistry* 17:4430–4438.

Blum, H., Leigh, J. S., and Ohnishi, T., 1980. Effect of dysprosium on the spin-lattice relaxation time of cytochrome c and cytochrome a, *Biochim. Biophys. Acta* 626:31–40.

Blum, H., Cusanovich, M. A., Sweeney, W. V., and Ohnishi, T., 1981. Magnetic interactions between dysprosium complexes and two soluble iron-sulfur proteins, *J. Biol. Chem.* 256:2199–2206.

Bode, W., and Schwager, P., 1975. The refined crystal structure of bovine β-trypsin at 1.8 Å resolution. II. Crystallographic refinement, calcium binding site, benzamidine binding site and active site at pH 7.0, *J. Mol. Biol.* 98:693–717.

Bonucci, E., Silvestrini, G., and DiGrezia, R., 1988. The ultrastructure of the organic phase associated with the inorganic substance in calcified tissue, *Clin. Orthop. Rel. Res.* 233: 243–261.

Borowski, M., Furie, B. C., Goldsmith, G. H., and Furie, B., 1985. Metal and phospholipid binding properties of partially carboxylated human prothrombin variants, *J. Biol. Chem.* 260:9258–9264.

Bradbury, J. H., Brown, L. R., Crompton, M. W., and Warren, B., 1974. Determination of the sequence of peptides by PMR spectroscopy, *Anal. Biochem.* 62:310–316.

Bradbury, J. H., Howell, J. R., Johnson, R. N., and Warren, B., 1978. Introduction of a strong binding site for lanthanides at the N-terminus of peptides and ribonuclease A, *Eur. J. Biochem.* 84:503–511.

Bratcher, S. C., and Kronman, M. J., 1984. Metal ion binding to the N and A conformers of bovine α-lactalbumin, *J. Biol. Chem.* 259:10875–10886.

Bresnahan, S. J., Baugh, L. E., and Borowitz, J. L., 1980. Mechanisms of La^{3+}-induced adrenal catecholamine release, *Res. Commun. Chem. Pathol. Pharmacol.* 28:229–244.

Brewer, J. M., 1985. Specificity and mechanism of action of metal ions in yeast enolase, *FEBS Lett.* 182:8–14.

Brewer, J. M., Carreira, L. A., Irwin, R. M., and Elliot, J. I., 1981. Binding of terbium(III) to yeast enolase, *J. Inorg. Biochem.* 14:33–44.

Brittain, H. G., and Richardson, F. S., 1977. Circularly polarized emission studies on Tb^{3+} and Eu^{3+} complexes with potentially terdentate amino acids in aqueous solution, *Bioinorg. Chem.* 7:233–243.

Brittain, H. G., Richardson, F. S., Martin, R. B., Burtnik, L. D., and Kay, C. M., 1976. Circularly polarized emission of terbium(III) substituted bovine cardiac troponin-C, *Biochem. Biophys. Res. Commun.* 68:1013–1019.

Buccigross, J. M., and Nelson, D. J., 1986. EPR studies show that all lanthanides do not have the same order of binding to calmodulin, *Biochem. Biophys. Res. Commun.* 138: 1243–1249.

Buck, F. F., Vithayathil, A., Bier, M., and Nord, F. F., 1962. On the mechanism of enzyme action. LXXIII. Studies on trypsins from beef, sheep and pig pancreas, *Arch. Biochem. Biophys.* 97:417–424.

Burtnick, L. D., 1982. Tb^{3+} as a luminescent probe of actin structure: effects of polymerization, KI and the binding of deoxyribonuclease I, *Arch. Biochem. Biophys.* 216:81–87.

Burton, D. R., Forsen, S., Karlstrom, G., Dwek, R. A., McLaughlin, A. C., and Wain-Hobson, S., 1977a. The determination of molecular-motion parameters from proton-relaxation-enhancement measurements in a number of Gd(III)–antibody–fragment complexes. A comparative study, *Eur. J. Biochem.* 75:445–453.

Burton, D. R., Dwek, R. A., Forsen, S., and Karlstrom, G., 1977b. A novel approach to water proton relaxation in paramagnetic ion–macromolecule complexes, *Biochemistry* 16:250–258.

Campbell, I. D., Dobson, C. M., Williams, R. J. P., and Xavier, A. V., 1973. Determination of the structure of proteins in solution. Lysozyme, *Ann. N.Y. Acad. Sci.* 222:163–174.

Campbell, I. D., Dobson, C. M., and Williams, R. J. P., 1975. Studies of exchangeable hydrogens in lysozyme by means of Fourier transform proton magnetic resonance, *Proc. Roy. Soc. B* 189:484–502.

Canada, R. G., 1981. Terbium fluorescence studies of the metal-angiotensin II complex, *Biochem. Biophys. Res. Commun.* 99:913–919.

Castillo, F. J., Penel, C., and Greppin, H., 1984. Peroxidase release induced by ozone in Sedum album leaves. Involvement of Ca^{2+}, *Plant Physiol.* 74:846–851.

Cavé, A., Daures, M. F., Parello, J., Saint-Yves, A., and Sempere, R., 1979. NMR studies of primary and secondary sites of parvalbumins using the two paramagnetic probes Gd(III) and Mn(II), *Biochimie* 61:755–765.

Chantler, P. D., 1983. Lanthanides do not function as calcium analogues in scallop myosin, *J. Biol. Chem.* 258:4702–4705.

Chasteen, N. D., and Theil, E. C., 1982. Iron binding by horse spleen apoferritin. A vanadyl(IV) EPR spin probe study, *J. Biol. Chem.* 257:7672–7677.

Chevallier, J., and Butow, R. A., 1971. Calcium binding to the sarcoplasmic reticulum of rabbit skeletal muscle, *Biochemistry* 10:2733–2737.

Chiba, K., Ohyashiki, T., and Mohri, T., 1984. Stoichiometry and location of terbium and calcium bindings to porcine intestinal calcium-binding protein, *J. Biochem.* 95:1767–1774.

Choosri, T., 1981. The study of α-amylases using lanthanide(III) ions as substitutional probes, Ph.D. thesis, Pennsylvania State University.

Clayton, R. A., 1959. *In vitro* inhibition of selected enzymes by rare earth chlorides, *Arch. Biochem. Biophys.* 85:559–560.

Colman, P. M., Weaver, L. H., and Matthews, B. W., 1972a. Rare earths as isomorphous calcium replacements for protein crystallography, *Biochem. Biophys. Res. Commun.* 46:1999–2005.

Colman, P. M., Jansonius, J. N., and Matthews, B. W., 1972b. The structure of thermolysin: an electron density map at 2.3 Å resolution, *J. Mol. Biol.* 70:701–724.

Colman, P. M., Weaver, L. H., and Matthews, B. W., 1974. Binding of lanthanide ions to thermolysin, *Biochemistry* 13:1719–1725.

Cookson, D. J., Levine, B. A., Williams, R. J. P., Jontell, M., Linde, A., and deBarnard, B., 1980. Cation binding by the rat incisor dentine phosphoprotein, in *Calcium-Binding Proteins: Structure and Function,* (F. L. Siegel, E. Carafoli, R. H. Kretsinger, D. H. MacLennan, and R. H. Wasserman, eds), Elsevier/North-Holland, Amsterdam, pp. 483–484.

Corson, D. C., Lee, L., McQuaid, G. A., and Sykes, B. D., 1983a. An optical stopped-flow and ^{1}H and ^{113}Cd nuclear magnetic resonance study of the kinetics and stoichiometry of the interaction of the lanthanide Yb^{3+} with carp parvalbumin, *Can. J. Biochem. Cell. Biol.* 61:860–870.

Corson, D. C., Williams, T. C., and Sykes, B. D., 1983b. Calcium binding proteins: optical stopped-flow and proton nuclear magnetic resonance studies of the binding of the lanthanide series of metal ions to parvalbumin, *Biochemistry* 22:5882–5889.

Corson, D. C., Williams, T. C., Kay, L. E., and Sykes, B. D., 1986. ^{1}H NMR spectroscopic studies of calcium-binding proteins. I. Stepwise proteolysis of the C-terminal α-helix of a helix–loop–helix metal-binding domain, *Biochemistry* 25:1817–1826.

Cottam, G. L., and Ward, R. L., 1975. Fourier transform phosphorus magnetic resonance study of the interaction of P-enolpyruvate with the muscle pyruvate kinase–gadolinium complex, *Biochem. Biophys. Res. Commun.* 64:797–802.

Cottam, G. L., Valentine, K. M., Thompson, B. C., and Sherry, A. D., 1974. Magnetic resonance studies of the formation of the ternary phosphoenolpyruvate–gadolinium–muscle pyruvate kinase complex, *Biochemistry* 13:3532–3537.

Cuatrecasas, P., Fuchs, S., and Anfinsen, C. B., 1967. Catalytic properties and specificity of the extracellular nuclease of *Staphylococcus aureus, J. Biol. Chem.* 242:1541–1547.

Curmi, P. M., Barden, J. A., and Dos Remedios, C. G., 1982. Conformational studies of G-actin containing bound lanthanide, *Eur. J. Biochem.* 122:239–244.

Dahlquist, F. W., Long, J. W., and Bigbee, W. L., 1976. Role of calcium ions in the thermal stability of thermolysin, *Biochemistry* 15:1103–1111.

Daman, M. E., and Dill, K., 1983. ^{13}C-n.m.r.-spectral study of the binding of Gd^{3+} to glycophorin, *Carbohydr. Res.* 111:205–214.

Dano, K., and Reich, E., 1979. Plasminogen activator from cells transformed by an oncogenic virus. Inhibitors of the activation reaction, *Biochim. Biophys. Acta* 566:138–151.

Darnall, D. W., and Birnbaum, E. R., 1970. Rare earth metal ions as probes of calcium ion binding sites in proteins, *J. Biol. Chem.* 245:6484–6488.

Darnall, D. W., and Birnbaum, E. R., 1973. Lanthanide ions activate α-amylase, *Biochemistry* 12:3489–3491.

Darnall, D. W., Birnbaum, E. R., Sherry, A. D., and Gomez, J. E., 1973. Lanthanide ions as calcium ion substitutes in trypsin and trypsinogen, *Proceedings of the 10th Rare Earth Research Conference,* Vol. 1, pp. 117–126.

Davril, M., Jung, M. L., Duportail, G., Lohez, M., Han, K. K., and Bieth, J. G., 1984. Arginine modification in elastase. Effect on catalytic activity and conformation of the calcium-binding site, *J. Biol. Chem.* 259:3851–3857.

Dedman, J. R., Potter, J. D., Jackson, R. L., Johnson, J. D., and Means, A. R., 1977. Physicochemical properties of rat testis Ca^{2+}-dependent regulator protein of cyclic nucleotide phosphodiesterase. Relationship of Ca^{2+}-binding, conformational changes and phosphodiesterase activity, *J. Biol. Chem.* 252:8415–8422.

DeJersey, J., and Martin, R. B., 1980. Lanthanide probes in biological systems: the calcium binding site of pancreatic elastase as studied by terbium luminescence, *Biochemistry* 19:1127–1132.

DeJersey, J., Lahue, R. S., and Martin, R. B., 1980. Terbium luminescence as a probe of the calcium binding site of trypsin and alpha-chymotrypsin, *Arch. Biochem. Biophys.* 205:536–542.

Desmuelle, P., and Fabre, C., 1955. Sur la séquence N-terminale du trypsinogène et son ablation pendant l'activation de ce zymogène, *Biochim. Biophys. Acta* 18:49–57.

Dill, K., and Allerhand, A., 1977. Effect of chemical modifications at tryptophan-108 on binding of lanthanide ions to hen egg-white lysozyme. Application of natural-abundance carbon-13 nuclear magnetic resonance spectroscopy, *Biochemistry* 16:5711–5716.

Dill, K., Daman, M. E., Batstone-Cunningham, R. L., Lacombe, J. M., and Pavia, A. A., 1983. ^{13}C-n.m.r.-spectral study of the mode of binding of Gd^{3+} to various glycopeptides, *Carbohydr. Res.* 123:123–135.

Dimicoli, J. L., and Bieth, J., 1977. Location of the calcium ion binding site in porcine pancreatic elastase using a lanthanide ion probe, *Biochemistry* 16:5532–5537.

Dockter, M. E., 1982. Diffusion-enhanced energy transfer characterization of heme locations in yeast cytochromes, *Fed. Proc.* 41:749.

Donato, H., and Martin, R. B., 1974. Conformations of carp muscle calcium binding parvalbumin, *Biochemistry* 13:4575–4579.

Dos Remedios, C. G., 1977. Ionic radius selectivity of skeletal muscle membranes, *Nature* 170:750–751.

Dos Remedios, C. G., and Barden, J. A., 1977. Effects of Gd(III) on G-actin: inhibition of polymerization of G-actin and activation of myosin ATPase activity by Gd-G-actin, *Biochem. Biophys. Res. Commun.* 77:1339–1346.

Dos Remedios, C. G., and Dickens, M. J., 1978. Actin microcrystals and tubes formed in the presence of gadolinium ions, *Nature* 276:731–733.

Dos Remedios, C. G., Barden, J. A., and Valois, A. A., 1980. Crystalline actin tubes. II. The effect of various lanthanide ions on actin tube formation, *Biochim. Biophys. Acta* 624:174–186.

Dower, S. K., Dwek, R. A., McLaughlin, A. C., Mole, L. E., Press, E. M., and Sunderland, C. A., 1975. The binding of lanthanides to non-immune rabbit immunoglobulin G and its fragments, *Biochem. J.* 149:73–82.

Drachev, A. L., Drachev, L. A., Kaulen, A. D., and Khitrina, L. V., 1984. The action of lanthanum ions and formaldehyde on the proton-pumping function of bacteriorhodopsin, *Eur. J. Biochem.* 138:349–356.

Drouven, B. J., and Evans, C. H., 1986. Collagen fibrillogenesis in the presence of lanthanides, *J. Biol. Chem.* 261:11792–11797.

Duportail, G., Lefevre, J. F., Lestienne, P., Dimicoli, J. L., and Bieth, J. G., 1980. Binding of terbium to porcine pancreatic elastase. Ligand-induced changes in the stability, the maximum luminescence intensity, and the circularly polarized luminescence spectrum of the complex, *Biochemistry* 19:1377–1382.

Dux, L., Taylor, K. A., Ting-Beall, H. P., and Martonosi, A., 1985. Crystallization of the Ca^{2+}-ATPase of sarcoplasmic reticulum by calcium and lanthanide ions, *J. Biol. Chem.* 260:11730–11743.

Dwek, R. A., Richards, R. E., Morallee, K. G., Neiboer, E., Williams, R. J. P., and Xavier, A. V., 1971. The lanthanide cations as probes in biological systems. Proton relaxation enhancement studies for model systems and lysozyme, *Eur. J. Biochem.* 21:204–209.

Dwek, R. A., Levy, H. R., Radda, G. K., and Seeley, P. J., 1975. Spin label and lanthanide binding sites on glyceraldehyde-3-phosphate dehydrogenase, *Biochim. Biophys. Acta* 377:26–33.

Dwek, R. A., Grivol, D., Jones, R., McLaughlin, A. C., Wain-Hobson, S., White, A. I., and White, C., 1976. Interactions of the lanthanide- and hapten-binding sites in the F_v fragment from the myeloma protein MOPC 315, *Biochem. J.* 155:37–53.

El-Fakahany, E. E., Pfenning, M., and Richelson, E., 1984. Kinetic effects of terbium on muscarine acetylcholine receptors of murine neuroblastoma cells, *J. Neurochem.* 42: 863–869.

Epstein, M., Levitzki, A., and Reuben, J., 1973. Studies of the calcium binding site of trypsin using rare earth ions, *Proceedings of the 10th Rare Earth Research Conference*, Vol. 1, Plenum Press, New York, pp. 124–126.

Epstein, M., Levitzki, A., and Reuben, J., 1974. Binding of lanthanides and of divalent metal ions to porcine trypsin, *Biochemistry* 13:1777–1782.

Epstein, M., Reuben, J., and Levitski, A., 1977. Calcium binding site of trypsin as probed by lanthanides, *Biochemistry* 16:2449–2457.

Evans, C. H., 1981. Interactions of tervalent lanthanide ions with bacterial collagenase (clostridiopeptidase A), *Biochem. J.* 195:677–684.

Evans, C. H., 1985. The lanthanide-enhanced affinity chromatography of clostridial collagenase, *Biochem. J.* 225:553–556.

Evans, C. H., and Drouven, B. J., 1983. The enhancement of the rate of collagen polymerization by calcium and lanthanide ions, *Biochem. J.* 213:751–758.

Evans, C. H., and Mason, G. C., 1986. Studies on the stimulation of the bacterial collagenolytic enzyme clostridiopeptidase A by cobalt(II) ions, *Int. J. Biochem.* 18:89–92.

Evans, C. H., and Ridella, J. D., 1985. Inhibition, by lanthanides, of neutral proteinases secreted by human, rheumatoid synovium, *Eur. J. Biochem.* 151:29–32.

Evelhoch, J. C., 1981. Spectoscopic studies of zinc and calcium binding proteins, Ph.D. thesis, University of California, Riverside.

Ferri, A., and Grazi, E., 1981. Different polymeric forms of actin detected by the fluorescent probe terbium ion, *Biochemistry* 20:6362–6366.

Fishelson, Z., and Muller-Eberhard, H. J., 1983. The C3/C5 convertase of the alternative pathway of complement: stabilization and restriction of control by lanthanide ions, *Mol. Immunol.* 20:309–315.

Furie, B., Eastlake, A., Schechter, A. N., and Anfinsen, C. B., 1973. The interaction of the lanthanide ions with staphylococcal nuclease, *J. Biol. Chem.* 248:5821–5825.

Furie, B., Griffen, J. H., Feldman, R., Sokoloski, E. A., and Schechter, A. N., 1974. The active site of staphylococcal nuclease: paramagnetic relaxation of bound nucleotide inhibitor nuclei by lanthanide ions, *Proc. Natl. Acad. Sci. USA* 71:2833–2837.

Furie, B. C., and Furie, B., 1975. Interaction of lanthanide ions with bovine factor X and their use in the affinity chromatography of the venom coagulant protein of *Vipera russelli*, *J. Biol. Chem.* 250:601–608.

Furie, B. C., Mann, K. G., and Furie, B., 1976. Substitution of lanthanide ions for calcium ions in the activation of bovine prothrombin by activated factor X, *J. Biol. Chem.* 254:3235–3241.

Furie, B. C., Blumenstein, H., and Furie, B., 1979. Metal binding sites of a γ-carboxyglutamic acid-rich fragment of bovine prothrombin, *J. Biol. Chem.* 254:12521–12530.

Gafni, A., and Steinberg, I. Z., 1974. Optical activity of terbium ions bound to transferrin and conalbumin studied by circular polarization of luminescence, *Biochemistry* 13:800–803.

Genov, N., Shopova, M., and Boteva, R., 1985. Studies on the lanthanide complexes of subtilisins, *Rev. Port. Quim.* 27:266–267.

Gershman, L. C., Selden, L. A., and Estes, J. E., 1979. On the interaction of muscle actin with gadolinium, *Biochem. Biophys. Res. Commun.* 91:1280–1287.

Gomez, J. E., Birnbaum, E. R., and Darnall, D. W., 1974. The metal ion acceleration of the conversion of trypsinogen to trypsin. Lanthanide ions as calcium ion substitutes, *Biochemistry* 13:3745–3750.

Gross, M. K., Toscano, D. G., and Toscano, W. A., 1987. Calmodulin-mediated adenylate cyclase from mammalian sperm, *J. Biol. Chem.* 262:8672–8676.

Henzl, M. T., and Birnbaum, E. R., 1988. Oncomodulin and Parvalbumin. A comparison of their interactions with europium ion, *J. Biol. Chem.* 263:10674–10680.

Henzl, M. T., Hapak, R. C. and Birnbaum, E. R., 1986. Lanthanide binding properties of rat calmodulin, *Biochim. Biophys. Acta* 872:16–23.

Henzl, M. T., McCubbin, W. D., Kay, C. M., and Birnbaum, E. R., 1985. Luminescence studies of lanthanide ion binding to parvalbumin, *J. Biol. Chem.* 260:8447–8455.

Hershberg, R. D., Reed, G. H., Slotboom, A. J., and DeHaas, G. H., 1976. Phospholipase A₂ complexes with gadolinium(III) and interaction of the enzyme–metal ion complex with monomeric and micellar alkylphosphorylcholines. Water proton nuclear magnetic relaxation studies, *Biochemistry* 15:2268–2274.

Herzberg, O., and James, M. N. G., 1985. Structure of the calcium regulatory muscle protein troponin-C at 2.8 Å resolution, *Nature* 313:653–659.

Highsmith, S. R., and Head, M. R., 1983. Terbium(3 +) binding to calcium and magnesium binding sites on sarcoplasmic reticulum ATPase, *J. Biol. Chem.* 258:6858–6862.

Highsmith, S., and Murphy, A. J., 1984. Nd^{3+} and Co^{2+} binding to sarcoplasmic reticulum Ca ATPase. An estimation of the distance from the ATP binding site to the high-affinity calcium binding sites, *J. Biol. Chem.* 259:14651–14656.

Holmquist, B., and Vallee, B. L., 1978. Magnetic circular dichroism, in *Methods in Enzymology* (C. H. W. Hirs and S. N Timasheff, eds.), Vol. XLIV G, Academic Press, New York, pp. 149–179.

Holten, V. Z., Kyker, G. C., and Pulliam, M., 1966. Effects of lanthanum chlorides on selected enzymes, *Proc. Soc. Exp. Biol. Med.* 123:913–919.

Homer, R. B., and Mortimer, B. D., 1978. Europium II as a replacement for calcium II in conconavalin A. A precipitation assay and magnetic circular dichroism study, *FEBS Lett.* 87:69–72.

Horecker, B. L., Stotz, E., and Hogness, T. R., 1939. The promoting effect of aluminum, chromium and the rare earths in the succinic dehydrogenase–cytochrome system, *J. Biol. Chem.* 128:251–256.

Horrocks, W. DeW., 1982. Lanthanide ion probes of biomolecular structure, in *Advances in Inorganic Biochemistry* (G. L. Eichhorn and L. G. Marzilli, eds.), Vol. 4, Elsevier, New York, pp. 201–261.

Horrocks, W. DeW., and Collier, W. E., 1981. Lanthanide ion luminescence probes. Measurement of distance between intrinsic protein fluorophores and bound metal ions: quantitation of energy transfer between tryptophan and terbium(III) or europium(III) in the calcium-binding protein parvalbumin, *J. Am. Chem. Soc.* 103:2856–2862.

Horrocks, W. D., and Snyder, A. P., 1981. Measurement of distance between fluorescent amino acid residues and metal ion binding sites. Quantitation of energy transfer between tryptophan and terbium(III) or europium(III) in thermolysin, *Biochem. Biophys. Res. Commun.* 100:111–117.

Horrocks, W. DeW., Holmquist, B., and Vallee, B. L., 1975. Energy transfer between terbium(III) and cobalt(II) in thermolysin: a new class of metal–metal distance probes, *Proc. Natl. Acad. Sci. USA* 72:4764–4768.

Horrocks, W. DeW., Mulqueen, P., Rhee, M. J., Breen, P. J., and Hild, E. K., 1983. Europium(III) laser luminescence excitation spectroscopy of calcium-modulated proteins: parvalbumin and calmodulin, *Inorg. Chim. Acta* 79:24–25.

Hwang, Y. T., Andrews, L. J., and Solomon, E. I., 1984. Resonant fluorescence study of the Eu^{3+}-substituted Ca^{2+} site in *Busycon* hemocyanin: structural coupling between the heterotropic allosteric effector and the coupled binuclear copper active site, *J. Am. Chem. Soc.* 106:3832–3838.

Imanishi, A., 1966. Calcium binding by bacterial α-amylase, *J. Biochem.* 60:381–390.

Itoh, N., and Kawakita, M., 1984. Characterization of Gd^{3+} and Tb^{3+} binding sites on Ca^{2+}-Mg^{2+}-adenosine triphosphatase of sarcoplasmic reticulum, *J. Biochem.* 95:661–669.

Izutsu, K. T., Felton, S. P., Siegel, I. A., Yoda, W. T., and Chen, A. C. N., 1972. Aequorin: its ionic specificity, *Biochem. Biophys. Res. Commun.* 49:1034–1039.

Jones, R., Dwek, R. A., and Forsen, S., 1974. The mechanism of water-proton relaxation in enzyme–paramagnetic-ion complexes. I. The Gd(III) lysozyme complex, *Eur. J. Biochem.* 47:271–283.

Katzin, L. I., 1969. Absorption and circular dichroic spectral studies of europium(III) complexes with sugar acids and amino acids, with remarks on hypersensitivity, *Inorg. Chem.* 8:1649–1654.

Katzin, L. I., and Gulyas, E., 1968. Absorption and circular dichroism spectral studies of chelate complexes of praseodymium(III) with α-amino acids, *Inorg. Chem.* 7:2442–2446.

Kayne, M. S., and Cohn, M., 1972. Cation requirements of isoleucyl-RNA synthetase from *Escherichia coli*, *Biochem. Biophys. Res. Commun.* 46:1285–1291.

Kilhoffer, M. C., Gerard, D., and Demaille, J. G., 1980. Terbium binding to octopus calmodulin provides the complete sequence of ion binding, *FEBS Lett.* 120:99–103.

Klee, C. B., 1977. Conformational transition accompanying the binding of Ca^{2+} to the protein activator of 3',5'-cyclic adenosine monophosphate phosphodiesterase, *Biochemistry* 16: 1017–1024.

Klumpp, S., Kleerfeld, G., and Schultz, J. E., 1983. Calcium/calmodulin-regulated guanylate cyclase of the excitable ciliary membrane from paramecium. Dissociation of calmodulin by La^{3+}: calmodulin specificity and properties of the reconstituted guanylate cyclase, *J. Biol. Chem.* 258:12455–12459.

Kretsinger, R. H., 1976. Calcium binding proteins, *Annu. Rev. Biochem.* 45:239–266.

Kronman, M. J., and Bratcher, S. C., 1984. Conformational changes induced by zinc and terbium binding to native bovine alpha-lactalbumin and calcium-free alpha-lactalbumin, *J. Biol. Chem.* 259:10887–10895.

Kronman, M. J., Sinha, S. K., and Brew, K., 1981. Characteristics of the binding of Ca^{2+} and other divalent metal ions to bovine α-lactalbumin, *J. Biol. Chem.* 256:8582–8587.

Kuiper, H., Finazzi-Agro, A., Antonini, E., and Brunori, M., 1979. The replacement of calcium by terbium as an allosteric effector of hemocyanins, *FEBS Lett.* 99:317–320.

Kuiper, H. A., Zolla, L., Finazzi-Agro, A., and Brunori, M., 1981. Interaction of lanthanide ions with *Panulirus interruptus* hemocyanin: evidence for vicinity of some of the cation binding sites, *J. Mol. Biol.* 149:805–812.

Kurachi, K., Sieker, L. C., and Jensen, L. H., 1975. Metal ion binding in triclinic lysozyme, *J. Biol. Chem.* 250:7663–7667.

Lau, Y. S., and Gnegy, M. E., 1980. Effects of lanthanum and trifluoperazine on ^{125}I calmodulin binding to rat striated particulates, *J. Pharmacol. Exp. Ther.* 215:28–34.

Leavis, P. C., Nagy, B., Lehrer, S. S., Bialkowska, H., and Gergely, J., 1980. Terbium binding to troponin C: binding stoichiometry and structural changes induced in the protein, *Arch. Biochem. Biophys.* 200:17–21.

Lee, L., and Sykes, B. D., 1980. The use of lanthanide NMR shift probes in the determination of the structure of calcium binding proteins in solution; application to the EF calcium binding site of carp parvalbumin, *Develop. Biochem.* 14:323–326.

Lee, L., and Sykes, B. D., 1981. Proton nuclear magnetic resonance determination of the sequential ytterbium replacement of calcium in carp parvalbumin, *Biochemistry* 20: 1156–1162.

Lee, L., Corson, D. C., and Sykes, B. D., 1985. Structural studies of calcium binding proteins using nuclear magnetic resonance, *Biophys. J.* 47:139–142.

Lenkinski, R. E., and Stephens, R. C., 1982. The lanthanides as structural probes in peptides, in *The Rare Earths in Modern Science and Technology* (G. J. McCarthy, H. B. Silber, and J. J. Rhyne, eds.), Vol. 3, Plenum Press, New York, pp. 45–51.

Lenkinski, R. E., and Stephens, R. L., 1983. The nature of the Ln^{3+}-angiotensin II complex. A ^{13}C nmr study of the binding of Yb^{3+} to angiotensin II, *J. Inorg. Biochem.* 18:175–180.

Lenkinski, R. E., Glickson, J. D., and Walter, R., 1978. A fluorescence study of the binding of calcium and terbium ions to angiotensin, *Bioinorg. Chem.* 8:363–368.

Lenkinski, R. E., Peerce, B. E., Pillai, R. P., and Glickson, J. D., 1980. Calcium(II) and the trivalent lanthanide ion complexes of the bleomycin antibiotics. Potentiometric, fluorescence and ^{1}H NMR studies, *J. Am. Chem. Soc.* 102:7088–7093.

Leung, C. S. H., and Meares, C. F., 1977. Attachment of fluorescent metal chelates to macromolecules using "bifunctional" chelating agents, *Biochem. Biophys. Res. Commun.* 75:149–155.

Levitzki, A., and Reuben, J., 1973. Abortive complexes of α-amylases with lanthanides, *Biochemistry* 12:41–44.

Luk, C. K., 1971. Study of the nature of the metal-binding sites and estimate of the distance between the metal-binding sites in transferrin using trivalent lanthanide ions as fluorescent probes, *Biochemistry* 10:2838–2843.

Luterbacher, S., and Schatzmann, H. J., 1983. The site of action of lanthanum in the reaction cycle of the human red cell membrane calcium-pump ATPase, *Experientia* 39:311–312.

Macara, I. G., Hoy, T. G., and Harrison, P. M., 1973. The formation of ferritin from apoferritin. Inhibition and metal-ion binding sites, *Biochem. J.* 135:785–789.

McCubbin, W. D., Oikawa, K., and Kay, C. M., 1981. The effect of terbium on the structure of actin and myosin subfragment 1 as measured by circular dichroism, *FEBS Lett.* 127: 245–249.

McDonald, M. R., and Kunitz, M., 1941. The effect of calcium and other ions on the autocatalytic formation of trypsin from trypsinogen, *J. Gen. Physiol.* 25:53–73.

Marinetti, T. D., Snyder, G. H., and Sykes, B. D., 1976. Nuclear magnetic resonance determination of intramolecular distances in bovine pancreatic trypsin inhibitor using nitrotyrosine chelation of lanthanides, *Biochemistry* 15:4600–4608.

Marquis, J. K., 1984. Terbium binding to rat brain acetylcholinesterase. A fluorescent probe of anionic sites, *Comp. Biochem. Physiol. C* 78:335–338.

Marquis, J. K., and Webb, G. D., 1976. The effects of calcium and lanthanum on the interaction of decamethonium with soluble acetylcholinesterase from *Electrophorus electricus, J. Neurochem.* 27:329–331.

Marsden, B. J., Hodges, R. S., and Sykes, B. D., 1988. ^{1}H-NMR Studies of synthetic peptide analogues of calcium-binding site III of rabbit skeletal troponin C: effect on the lanthanum affinity of the interchange of aspartic acid and asparagine residues at the metal ion coordinating positions, *Biochemistry* 27:4198–4206.

Marsh, H. C., Sarasua, M. M., Madar, D. A., Hiskey, R. G., and Koehler, K. A., 1981. Europium(III) binding to bovine prothrombin residues 1–39 and to bovine prothrombin fragment 1, *J. Biol. Chem.* 256:7863–7870.

Martin, R. B., 1983. Structural chemistry of calcium: lanthanides as probes, in *Calcium in Biology* (T. G. Spiro, ed.), Wiley, New York, pp. 237–270.

Martin, R. B., 1984. Bioinorganic chemistry of calcium, in *Metal Ions in Biological Systems,* (H. Sigel, ed.), Vol. 17, Marcel Dekker, Basel, pp. 1–50.

Martin, R. B., and Richardson, F. S., 1979. Lanthanides as probes for calcium in biological systems, *Quart. Rev. Biophys.* 12:181–209.

Matthews, B. W., and Weaver, L. H., 1974. Binding of lanthanide ions to thermolysin, *Biochemistry* 13:1719–1725.

Matthews, B. W., Colman, P. M., Jansonius, J. N., Titani, K., Walsh, K. A., and Neurath, H., 1972. Structure of thermolysin, *Nature New Biol.* 238:41–43.

Matthews, B. W., Weaver, L. H., and Kester, W. R., 1974. The conformation of thermolylsin, *J. Biol. Chem.* 249:8030–8044.

Mazzei, G. J., Qi, D. F., Schatzman, R. C., Raynor, R. L., Turner, R. S., and Kuo, J. F., 1983. Comparative abilities of lanthanide ions, La^{3+} and Tb^{3+} to substitute for calcium in regulating phospholipid-sensitive Ca^{2+}-dependent protein kinase and myosin light chain kinase, *Life Sci.* 33:119–129.

Meares, C. F., and Ledbetter, J. E., 1977. Energy transfer between terbium and iron bound to transferrin: reinvestigation of the distance between metal-binding sites, *Biochemistry* 16:5178–5180.

Meares, C. F., and Rice, L. S., 1981. Diffusion-enhanced energy transfer shows accessibility of ribonucleic acid polymerase inhibitor binding sites, *Biochemistry* 20:610–617.

Menestrina, G., 1983. Effects of terbium on the hemocyanin pore formation rate in phosphatidylcholine planar bilayers, *Biochim. Biophys. Acta* 735:297–301.

Miller, T. L., Nelson, D. J., Brittain, H. G., Richardson, F. S., Martin, R. B., and Kay, C., 1975. Calcium binding sites of rabbit troponin and carp parvalbumin, *FEBS Lett.* 58: 262–264.

Miller, T. L., Cook, R. M., Nelson, D. J., and Theoharides, A. D., 1980. Terbium luminescence from the calcium binding sites of parvalbumin, *J. Mol. Biol.* 141:223–226.

Moeller, T., Martin, D. F., Thompson, L. C., Ferrus, R., Feistel, G. F., and Randall, W. J., 1965. The coordination chemistry of yttrium and the rare earth metal ions, *Chem. Rev.* 65:1–50.

Moews, P. C., and Kretsinger, R. H., 1975. Terbium replacement of calcium in carp muscle calcium-binding parvalbumin: an X-ray crystallographic study, *J. Mol. Biol.* 91:229–232.

Morrison, J. F., and Cleland, W. W., 1980. A kinetic method for determining dissociation constants for metal complexes of adenosine 5'-triphosphate and adenosine 5'-diphosphate, *Biochemistry* 19:3128–3131.

Morrison, J. F., 1982. The slow-binding and and slow tight-binding inhibition of enzyme-catalyzed reactions, *Trends Biochem. Sci.* 7:102–105.

Morrison, J. F., and Cleland, W. W., 1983. Lanthanide–adenosine 5'-triphosphate complexes: determination of their dissociation constants and mechanism of action as inhibitors of yeast hexokinase, *Biochemistry* 22:5507–5513.

Nathanson, J. A., Freedman, R., and Hoffer, B. J., 1976. Lanthanum inhibits brain adenylate cyclase and blocks noradrenergic depression of Purkinje cell discharge independent of calcium, *Nature* 261:330–332.

Nayler, W. G., and Harris, J. P., 1976. Inhibition of lanthanum of the Na^+-K^+ activated, ouabain-sensitive adenosinetriphosphatase enzyme, *J. Mol. Cell Cardiol.* 8:811–822.

Nelsestuen, G. L., Broderius, M., and Martin, G., 1976. Role of γ-carboxyglutamic acid. Cation specificity of prothrombin and factor X–phospholipid binding, *J. Biol. Chem.* 251:6886–6893.

Nelson, B. E., Gan, S. J., and Strothkamp, K. G., 1981. Terbium ion binding to *Limulus polyphemus* hemocyanin, *Biochem. Biophys. Res. Commun.* 100:1305–1313.

Nelson, D. J., Miller, T. L., and Martin, R. B., 1977. Noncooperative Ca(II) removal and terbium(III) substitution in carp muscle calcium binding parvalbumin, *Bioinorg. Chem.* 7:325–334.

Novello, F., and Stirpe, F., 1969. The effects of copper and other ions on the ribonucleic acid polymerase activity of rat liver nuclei, *Biochem. J.* 111:115–119.

O'Hara, P. B., and Bersohn, R., 1982. Resolution of the two metal binding sites of human serum transferrin by low-temperature excitation of bound europium(III), *Biochemistry* 21:5269–5272.

O'Hara, P., Yeh, S. M., Meares, C. F., and Bersohn, R., 1981. Distance between metal-binding sites in transferrin: energy transfer from bound terbium(III) to iron(III) or manganese(III), *Biochemistry* 20:4704–4708.

Oikawa, K., McCubbin, W. D., and Kay, C. M., 1980. The effects of terbium and lanthanum on the biological activity of some representative muscle protein systems, *FEBS Lett.* 118:137–140.

O'Neil, J. D., Dorrington, K. J., and Hofmann, T., 1984. Luminescence and circular-dichroism analysis of terbium binding by pig intestinal calcium-binding protein (relative mass = 9000), *Can. J. Biochem. Cell Biol.* 62:434–442.

Ostroy, F., Gams, R. A., Glickson, J. D., and Lenkinski, R. E., 1978. Inhibition of lysozyme by polyvalent metal ions, *Biochim. Biophys. Acta* 527:56–62.

Peacocke, A. R., and Williams, P. A., 1966. Binding of calcium, yttrium and thorium to a glycoprotein from bovine cortical bone, *Nature* 211:1040–1041.

Pecoraro, V. L., Harris, W. R., Corrano, C. J., and Raymond, K. N., 1981. Siderophilin metal coordination. Difference ultraviolet spectroscopy of di-, tri-, and tetravalent metal ions with ethylenebis[(o-hydroxyphenyl)glycine], *Biochemistry* 20:7033–7042.

Perkins, S. J., Johnson, L. N., Machin, P. A., and Phillips, D. C., 1979. Crystal structures of hen egg-white lysozyme complexes with gadolinium(III) and gadolinium(III)-N-acetyl-D-glucosamine, *Biochem. J.* 181:21–36.

Perkins, S. J., Johnson, L. N., Phillips, D. C., and Dwek, R. A., 1981. The simultaneous binding of lanthanide and N-acetylglucosamine inhibitors to hen egg-white lysozyme in solution by ^{1}H and ^{13}C nuclear magnetic resonance, *Biochem. J.* 193:573–588.

Prados, R., Stadtherr, L. G., Donato, H., and Martin, R. B., 1974. Lanthanide complexes of amino acids, *J. Inorg. Nucl. Chem.* 36:689–693.

Prendergast, F. G., and Mann, K. G., 1977. Differentiation of metal ion-induced transitions of prothrombin fragment 1, *J. Biol. Chem.* 252:840–850.

Prendergast, F. G., Lu, J., and Callahan, P. J., 1983. Oxygen quenching of sensitized terbium luminescence in complexes of terbium with small organic ligands and proteins, *J. Biol. Chem.* 258:4075–4078.

Quiram, D. R., and Weinshilboum, R. M., 1976. Inhibition of rat liver catechol-O-methyltransferase by lanthanum, neodymium and europium, *Biochem. Pharmacol.* 25:1727–1732.

Quist, E. E., and Roufogalis, B. D., 1975. Determination of the stoichiometry of the calcium pump in human erythrocytes using lanthanum as a selective inhibitor, *FEBS Lett.* 50:135–139.

Radhakrishnan, T. M., Walsh, K. A., and Neurath, H., 1969. The promotion of activation of bovine trypsinogen by specific modification of aspartyl residues, *Biochemistry* 8:4020–4027.

Rakhimov, M. M., Kalendareva, T. I., Rashidova, S. Sh., and Mad'yarov, Sh. R., 1982. Role of calcium ions in the catalytic activity of phospholipase D, *Biokhimiya* 47:1649–1662; *Chem. Abstr.* 98:13588w (1983).

Renaud, G., Soler-Argilaga, C., and Infante, R., 1980. Effect of cerium on liver lipids metabolism and plasma lipoproteins synthesis in the rat, *Biochem. Biophys. Res. Commun.* 95:220–227.

Reuben, J., 1971. Gadolinium(III) as a paramagnetic probe for proton relaxation studies of biological macromolecules. Binding to bovine serum albumin, *Biochemistry* 10:2834–2838.

Reuben, J., and Luz, Z., 1976. Longitudinal relaxation in spin 7/2 systems. Frequency dependence of lanthanum-139 relaxation times in protein solutions as a method of studying macromolecular dynamics, *J. Phys. Chem.* 80:1357–1369.

Rhee, M-J., Sudnick, D. R., Arkle, V. K., and Horrocks, W. DeW., 1981. Lanthanide ion luminescence probes. Characterization of metal ion binding sites and intermetal energy transfer distance measurements in calcium-binding proteins. I. Parvalbumin, *Biochemistry* 20:3328–3334.

Rhee, M. J., Horrocks, W. D., and Kosow, D. P., 1982. Laser-induced europium(III) luminescence as a probe of the metal ion mediated association of human prothrombin with phospholipid, *Biochemistry* 21:4524–4528.

Rhee, M. J., Horrocks, W. DeW., and Kosow, D. P., 1984. Laser-induced lanthanide luminescence as a probe of metal ion-binding sites of human factor X, *J. Biol. Chem.* 259:7404–7408.

Richardson, C. E., and Behnke, W. D., 1978. Physical studies of lanthanide binding to conconavalin A, *Biochim. Biophys. Acta* 534:267–274.

Rinderknecht, H., and Friedman, R. M., 1976. The effect of lanthanide ions on enteropeptidase-catalyzed activation of trypsinogen, *Biochim. Biophys. Acta* 452:497–502.

Roche, R. S., and Voordouw, G., 1978. The structural and functional roles of metal ions in thermolysin, *CRC Crit. Rev. Biochem.* 5:1–23.

Rosoff, B., and Spencer, H., 1975. Studies on electrophoretic binding of radioactive rare earths, *Health Phys.* 28:611–612.

Rübsamen, H., Hess, G. P., Eldefrawi, A. T., and Eldefrawi, M. E., 1976. Interaction between calcium and ligand-binding sites of the purified acetylcholine receptor studied by use of a fluorescent lanthanide, *Biochem. Biophys. Res. Commun.* 68:56–62.

Sarasua, M. M., Koehler, K. A., Skrzynia, C., and McDonagh, J. M., 1982. Human factor XIII–metal ion interactions. A luminescence and nuclear magnetic resonance study, *J. Biol. Chem.* 257:14102–14109.

Schatzmann, H. J., and Tschabold, M., 1971. The lanthanides Ho^{3+} and Pr^{3+} as inhibitors of calcium transport in human red cells, *Experientia* 27:59–61.

Scott, T. L., 1984. Luminescence studies of Tb^{3+} bound to the high affinity sites of the Ca^{2+}-ATPase of sarcoplasmic reticulum, *J. Biol. Chem.* 259:4035–4037.

Secemski, I. I., and Lienhard, G. E., 1974. The effect of gadolinium ion on the binding of inhibitors and substrates to lysozyme, *J. Biol. Chem.* 249:2932–2938.

Sedmak, J. J., and Grossberg, S. E., 1981. Interferon stabilization and enhancement by rare earth salts, *J. Gen. Virol.* 52:195–198.

Seifter, S., and Harper, E., 1959. Collagenases, in *Methods in Enzymology* (S. P. Colowick and N. O. Kaplan, eds.), Vol. IXX, Academic Press, New York, pp. 613–635.

Seltzer, J. L., Welgus, H. G., Jeffrey, J. J., and Eisen, A. Z., 1976. The function of Ca^{2+} in the action of mammalian collagenases, *Arch. Biochem. Biophys.* 173:355–361.

Shahid, F., Gomez, J. E., Birnbaum, E. R., and Darnall, D. W., 1982. The lanthanide-induced N → F transition and acid expansion of serum albumin, *J. Biol. Chem.* 257: 5618–5622.

Shelling, J. G., Bjornson, M. E., Hodges, R. S., Taneja, A. K., and Sykes, B. D., 1984. Contact and dipolar contributions to lanthanide induced NMR shifts of amino acid and peptide models for calcium binding sites in proteins, *J. Magn. Reson.* 57:99–114.

Shelling, J. G., Hofmann, T., and Sykes, B. D., 1985. Proton nuclear magnetic resonance studies of the interaction of the lanthanide ions ytterbium and lutetium with apo- and calcium-saturated porcine intestinal calcium binding protein, *Biochemistry* 24:2332–2338.

Sherry, A. D., and Pascnol, E., 1977. Proton and carbon lanthanide-induced shifts in aqueous alanine. Evidence for structural changes along the lanthanide series, *J. Am. Chem. Soc.* 99:5871–5876.

Sherry, A. D., Yoshida, C., Birnbaum, E. R., and Darnall, D. W., 1973. Nuclear magnetic resonance study of the interaction of neodymium(III) with amino acids and carboxylic acids. An aqueous shift reagent, *J. Am. Chem. Soc.* 95:3011–3014.

Sherry, A. D., Newman, A. D., and Gutz, C. G., 1975. The activation of concanavalin A by lanthanide ions, *Biochemistry* 14:2191–2196.

Sherry, A. D., Au-Young, S., and Cottam, G. L., 1978. Fluorescence properties of terbium-alkaline phosphatase, *Arch. Biochem. Biophys.* 189:277–282.

Short, M. T., and Osmand, A. P., 1983. Luminescence energy transfer studies of C-reactive protein. Binding of terbium(III) ions in C-reactive protein, *Immunol. Commun.* 12: 291–300.

Sieker, L. C., Adman, E., and Jensen, L. H., 1972. Structure of the Fe–S complex in a bacterial ferrodoxin, *Nature* 235:40–42.

Sinha, S. P., 1966. *Complexes of the Rare Earths,* Pergamon Press, New York.

Smith, B. M., Gindhart, T., and Colburn, N. H., 1986. Possible involvement of a lanthanide-sensitive protein kinase C substrate in lanthanide promotion of neoplastic transformation, *Carcinogenesis* 7:1949–1956.

Smolka, G. E., Birnbaum, E. R., and Darnall, D. W., 1971. Rare earth metal ions as substitutes for the calcium ion in *Bacillus subtilis* α-amylase, *Biochemistry* 10:4556–4561.

Snyder, A. P., Sudnick, D. R., Arkle, V. K., and Horrocks, W. DeW., 1981. Lanthanide ion luminescence probes. Characterization of metal ion binding sites and intermetal energy transfer distance measurements in calcium-binding proteins. 2. Thermolysin, *Biochemistry* 20:3334–3339.

Sommerville, L. E., Thomas, D. D., and Nelsentuen, G. L., 1985. Tb^{3+} binding to bovine prothrombin and bovine prothrombin fragment 1, *J. Biol. Chem.* 260:10444–10452.

Sowadski, J., Cornick, G., and Kretsinger, R. H., 1978. Terbium replacement of calcium in parvalbumin, *J. Mol. Biol.* 124:123–132.

Spartalian, K., and Oosterhuis, W. T., 1973. Moessbauer effect studies of transferrin, *J. Chem. Phys.* 59:617–622.

Sperling, R., Furie, B. C., Blumenstein, M., Keyt, B., and Furie, B., 1978. Metal binding properties of γ-carboxyglutamic acid. Implications for the vitamin-K dependent blood coagulation proteins, *J. Biol. Chem.* 253:3898–3906.

Sperow, J. W., and Butler, L. G., 1972. Yeast inorganic pyrophosphatase V. Binding of Eu^{3+}, *Bioinorg. Chem.* 2:87–91.

Steer, M. L., and Levitzki, H., 1973. Metal specificity of mammalian α-amylase as revealed by enzyme activity and structural probes, *FEBS Lett.* 31:89–93.

Stefanini, S., Chiancone, E., Antonini, E., and Finazzi-Agro, A., 1983. Binding of terbium to apoferritin: a fluorescence study, *Arch. Biochem. Biophys.* 222:430–434.

Stephens, E. M., and Grisham, C. M., 1979. Lithium-7 nuclear magnetic resonance, water proton nuclear magnetic resonance and gadolinium electron paramagnetic resonance studies of the sarcoplasmic reticulum calcium ion transport adenosine triphosphatase, *Biochemistry* 18:4876–4885.

Szebenyi, D. M. E., Oberdorf, S. K., and Moffat, K., 1981. Structure of vitamin-D dependent calcium-binding protein from bovine intestine, *Nature* 294:327–332.

Takeo, S., Duke, P., Tamm, G. M., Singal, P. K., and Dhalla, N. S., 1979. Effects of lanthanum on the heart sarcolemmal ATPase and calcium binding activities, *Can. J. Physiol. Pharmacol.* 57:496–503.

Tallant, E. A., Wallace, R. W., Dockter, M. E., and Cheung, W. Y., 1980. Activation of calmodulin by terbium (Tb^{3+}) and its use as a fluorescence probe, *Ann. N.Y. Acad. Sci.* 356:436.

Tanner, S. P., and Choppin, G. R., 1968. Lanthanide and actinide complexes of glycine. Determination of stability constants and thermodynamic parameters by a solvent extraction method, *Inorg. Chem.* 7:2046–2051.

Tanswell, P., Westhead, E. W., and Williams, R. J. P., 1974. Inhibition of yeast phosphoglycerate kinase by lanthanide–ATP complexes, *FEBS Lett.* 48:60–63.

Teuwissen, B., Masson, P. L., Osinski, P., and Heremans, J. F., 1972. Metal-combining properties of human lactoferrin. The possible involvement of tyrosyl residues in the binding sites. Spectrophotometric titration, *Eur. J. Biochem.* 31:239–245.

Tomlinson, G., Mutus, B., McLeman, I., and Mooibroek, M. J., 1982. Activation and inactivation of purified acetylcholinesterase from *Electrophorus electricus* by lanthanum(III), *Biochim. Biophys. Acta* 703:142–148.

Treffry, A., and Harrison, P. M., 1984. Spectroscopic studies on the binding of iron, terbium and zinc by apoferritin, *J. Inorg. Biochem.* 21:9–20.

Valentine, K. M., and Cottam, G. L., 1973. Gadolinium as a probe of the alkaline earth and ATP-metal binding sites in pyruvate kinase, *Arch. Biochem. Biophys.* 158:346–354.

Van Scharrenburg, G. J. M., Slotboom, A. J., de Haas, G. H., Mulqueen, P., Breen, P. J., and Horrocks, W. DeW., 1985. Catalytic Ca^{2+}-binding site of pancreatic phospholipase A_2: laser-induced Eu^{3+} luminescence study, *Biochemistry* 24:334–339.

Vaughn, J. B., Stephens, R. L., Lenkinski, R. E., Krishna, M. R., Heavner, G. A., and Goldstein, G., 1981. Proton NMR investigation of Ln^{3+} complexes of thymopoietin 32-36, *Biochim. Biophys. Acta* 671:50–60.

Vaughn, J. B., Stephens, R. L., Lenkinski, R. E., Heavner, G. A., Goldstein, G., and Krishna, N. R., 1982. Nuclear magnetic resonance analysis of Gd^{3+}-induced perturbations in thymopoietin 32-36: a study of amide and aromatic proton resonances, *Arch. Biochem. Biophys.* 217:468–472.

Viola, R. E., Morrison, J. F., and Cleland, W. W., 1980. Interaction of metal(III)–adenosine 5′-triphosphate complexes with yeast hexokinase, *Biochemistry* 19:3131–3137.

Vogel, H. J., Drakenbert, T., Forsen, S., O'Neil, J. D. J., and Hofmann, T., 1985. Structural differences in the two calcium binding sites of the porcine intestinal calcium binding protein: a multinuclear NMR study, *Biochemistry* 24:3870–3876.

Wallace, R. W., Tallant, E. A., Dockter, M. E., and Cheung, W. Y., 1982. Calcium binding domains of calmodulin, *J. Biol. Chem.* 257:1845–1854.

Walsh, M., Stevens, F. C., Oikawa, K., and Kay, C. M., 1978. Circular dichroism studies of native and chemically modified Ca^{2+}-dependent protein modulator, *Can. J. Biochem.* 57:267–273.

Walter, R., Smith, C. W., Sarathy, K. P., Pillai, R. P., Krishna, N. R., Lenkinski, R. E., Glickson, J. D., and Hruby, V. J., 1981. ¹H N.M.R. study of the conformation of [Glu 4] oxytocin and its lanthanide complexes in aqueous solution, *Int. J. Pept. Protein Res.* 17:56–64.

Wang, C-L. A., 1986. Distance measurements between metal-binding sites of calmodulin and from these sites to cys-133 to troponin I in the binary complex, *J. Biol. Chem.* 261:11106–11109.

Wang, C. L. A., and Aquaron, R. R., 1980. Binding of calcium and terbium to native and nitrated calmodulin, in *Calcium-Binding Proteins: Structure and Function* (F. L. Siegel, E. Carafoli, R. H. Kretsinger, D. H. MacLennan, and R. H. Wasserman, eds.), Elsevier/North-Holland, New York.

Wang, C. L. A., Leavis, P. C., Horrocks, W. DeW., and Gergely, J., 1981. Binding of lanthanide ions to troponin C, *Biochemistry* 20:2439–2444.

Wang, C. L., Aquaron, R. R., Leavis, P. C., and Gergely, J., 1982. Metal-binding properties of calmodulin, *Eur. J. Biochem.* 124:7–12.

Wang, C. L., Leavis, P. C., and Gergely, J., 1984. Kinetic studies show that Ca^{2+} and Tb^{3+} have different binding preferences toward the four Ca^{2+}-binding sites of calmodulin, *Biochemistry* 23:6410–6415.

Watson, H. C., Duee, E., and Mercer, W. D., 1972. Low resolution structure of glyceraldehyde 3-phosphate dehydrogenase, *Nature New Biol.* 240:130–133.

Watterson, D. M., Sharief, F., and Vanaman, T. C., 1980. The complete amino acid sequence of Ca^{2+}-dependent modulator protein (calmodulin) of bovine brain, *J. Biol. Chem.* 255: 462–475.

Wedler, F. C., and D'Aurora, V., 1974. Spectroscopic probes of *Escherichia coli* glutamine synthetase. Rare earth ions by difference absorption, *Biochim. Biophys. Acta* 371: 432–441.

Weiner, M. L., and Lee, K. S., 1972. Active calcium uptake by inside-out and right side-out vesicles of red blood cell membranes, *J. Gen. Physiol.* 59:462–475.

Willan, K. J., Wallace, K. H., Jaton, J. C., and Dwek, R. A., 1977. The use of gadolinium as a probe in the Fc region of a homogeneous anti-(type-III pneumococcal polysaccharide) antibody, *Biochem. J.* 161:205–211.

Williams, P. F., and Turtle, J. R., 1984. Terbium, a fluorescent probe for insulin receptor binding. Evidence for a conformational change in the receptor protein due to insulin binding, *Diabetes* 33:1106–1111.

Williams, T. C., Corson, D. C., and Sykes, B. D., 1983. Calcium binding sites on proteins: conclusions from studies of the interaction of the lanthanides with carp parvalbumin, *Develop. Biochem.* 25:57–58.

Williams, T. C., Corson, D. C., and Sykes, B. D., 1984. Calcium-binding proteins: calcium(II)–lanthanide(III) exchange in carp parvalbumin, *J. Am. Chem. Soc.* 106: 5698–5702.

Williams, T. C., Corson, D. C., Oikawa, K., McCubbin, W. D., Kay, C. M., and Sykes, B. D., 1986. ^{1}H NMR spectroscopic studies of calcium-binding proteins. 3. Solution conformations of rat apo-α-parvalbumin and metal-bound rat α-parvalbumin, *Biochemistry* 25:1835–1846.

Wolf, D. J., Poirier, P. G., Brostrom, C. O., and Brostrom, M. A., 1977. Divalent cation binding properties of bovine brain Ca^{2+}-dependent regulator protein, *J. Biol. Chem.* 252:4108–4116.

Yamada, S., and Tonomura, Y., 1972. Reaction mechanism of Ca^{2+}-dependent ATPase of sarcoplasmic reticulum from skeletal muscle. VII. Recognition and release of Ca^{2+} ions, *J. Biochem.* 72:417–425.

Yeh, S. M., and Meares, C. F., 1980. Characterization of transferrin metal-binding sites by diffusion-enhanced energy transfer, *Biochemistry* 19:5057–5062.

Yici, G., 1986. Work cited in *China Rare Earth Information*, No. 3, p. 4.

Zimmerman, U. J. P., and Schlaepfer, W. W., 1982. Characterization of a brain calcium-activated protease that degrades neurofilament proteins, *Biochemistry* 21:3977–3983.

Interactions of Lanthanides with Other Molecules of Biochemical Interest

5.1 Mononucleosides and Mononucleotides

Two main motives have encouraged studies of the interactions of the lanthanides with mononucleosides and mononucleotides. The first follows from the realization that the metabolically active forms of many nucleotides are ones in which a metal ion is coordinated. Best studied is the ATP-Mg complex which forms the physiologically active substrate for ATP-requiring reactions. As Mg^{2+}, like Ca^{2+}, is spectroscopically uninformative, it, too, has been replaced by Ln^{3+} ions for experimental studies. The second motive has been to use paramagnetic lanthanides in NMR spectroscopic investigations aimed at determining the solution structures of these molecules.

Lanthanides bind poorly to purine and pyrimidine monoribonucleosides. For instance, Barry *et al.* (1974a), using ^{1}H-NMR spectroscopy, could detect no significant shifting of adenosine protons, although various Ln^{3+} ions strongly shifted certain protons on AMP. However, a weak interaction is suggested by the ability of high concentrations of Eu^{3+} to alter the circular dichroism (CD) spectrum of adenosine (Bayley and Debenham, 1974; Table 5-1). Certain nucleosides also enhance Tb^{3+} luminescence to a small degree (Formoso, 1973; Davis and Richardson, 1980), although Ringer *et al.* (1980) could not confirm this for guanosine or deoxyguanosine. Circularly polarized luminescence (CPL) emissions were recorded for 1:2 and 1:5, but not 1:1, Tb^{3+}-nucleoside complexes containing cytidine, inosine, or uridine. These emissions required a pH greater than 6 and were maximal at pH 7. No circularly polarized emissions were found with the Tb^{3+} complexes of adenosine or deoxynucleosides. The excitation spectra indicated energy transfer from the pyrimidine and purine rings, while structural comparisons suggested terdentate chelation of Tb^{3+} via a base carbonyl group and two ribosyl hydroxyl groups. The

Table 5-1. The Effect of Eu^{3+} on the Intensity of the 260-nm CD Band of Adenosine Derivatives and Related Compounds[a]

Compound	Area of CD band (arbitrary units)		A_2/A_1 (%)
	A_1 $[Eu^{3+}] = 0$	A_2 $[Eu^{3+}] = 10\ mM$	
AMP	65	24	37
ADP	69	26	38
ADP-Rib	55	25	45
3':5'-cAMP	103	98	95
Adenosine	63	48	76
Deoxyadenosine	26	18	69
dAMP	22	20[b]	91[b]
NMN	28[b]	12[b], 14[c]	43[b], 50[c]
	A_1 $[Eu^{3+}] = 0$	A_2 $[Eu^{3+}] = 1\ mM$	A_2/A_1 (%)
AMP	65	49	75
IMP	43	28	65
XMP	55	6, 4[b], 6[c]	11, 7[b], 11[c]

[a] From Bayley and Debenham (1974), with permission.
[b] Denotes a positive band; all others negative.
[c] Multiple entries indicate crossing over in spectra.

absence of the 2'-hydroxyl group in deoxynucleosides accounts for their much lower affinity for Tb^{3+}. Similarly, adenosine, which lacks a base carbonyl group, did not enhance CPL emission from Tb^{3+} (Davis and Richardson, 1980). However, according to Martin (1983), these data are artifactual and cannot be taken as evidence of terdentate chelation of Tb^{3+} by nucleosides.

The presence of one or more phosphate groups on purine and pyrimidine nucleotides gives them a higher affinity than their corresponding nucleosides for lanthanides. As expected, the relative order of affinity for Ln^{3+} ions is nucleotide triphosphate (NTP) $\geqslant$ nucleotide diphosphate (NDP) > nucleotide monophosphate (NMP) (Tables 5-2 and 5-3). As the mononucleotides exist as NMP^- at pH 2 and NMP^{2-} at neutral pH, their affinity for Ln^{3+} ions is greatly affected by pH (Table 5-2). However, the nature of the Ln^{3+} ion and the purine or pyrimidine base have relatively little effect on the K_a values. The affinity of Ln^{3+} ions for ATP at neutral pH values is so high that the accurate determination of binding constants has been difficult. Using a method based upon the inhibition of hexokinase (Section 4.24) by Ln^{3+}-ATP complexes in the presence of various con-

Table 5-2. Association Constants for Lanthanide Complexes of
Mono- and Dinucleotides

Nucleotide	pH	Ln^{3+}	K_a (M^{-1})	Reference
5'-AMP	2	Ho	6 ± 2	Barry et al. (1971)
5'-AMP	2	Eu	10 ± 2	Barry et al. (1971)
5'-AMP	1.5–1.8	Eu	10	Barry et al. (1972)
5'-A,G,CMP	2	Eu,Pr	15 ± 5	Dobson et al. (1978)
5'-AMP	6.7	Eu	4.2×10^5	Galea et al. (1978)
5'-A,G,C,UMP	5.6	Tb	$\sim 10^3$	Formoso (1973)
5'-GMP	1.5–1.8	Eu	14	Barry et al. (1972)
5'-TMP	1.5–1.8	Eu	17	Barry et al. (1972)
5'-TMP	2	Eu	17 ± 3	Barry et al. (1971)
3':5'-cAMP	4.9	Ho	13.0 ± 1.7	Lavallee and Zeltmann (1974)
2':3'-cAMP	2.3	Eu	7.9 ± 1.2	Geraldes and Williams (1978)
2':3'-cCMP	2.3	Eu	8.6 ± 1.01	Geraldes and Williams (1978)
5'-ADP	7	Eu	$(58 \pm 3 \times 10^4$	Ellis and Morrison (1974)
5'-ADP	8	Eu	$(7.7 \pm 0.3) \times 10^4$	Ellis and Morrison (1974)
5'-ADP	6.7	Eu	7.1×10^6	Galea et al. (1978)

Table 5-3. Dissociation Constants of Ln-ATP
Complexes at pH 8[a]

Ln of Ln-ATP complex	Dissociation constant (μM)
La	0.33 ± 0.06
Ce	0.35 ± 0.06
Pr	0.31 ± 0.07
Nd	0.29 ± 0.05
Sm	0.22 ± 0.01
Eu	0.16 ± 0.04
Gd	0.087 ± 0.005
Tb	0.094 ± 0.024
Dy	0.049 ± 0.009
Ho	0.099 ± 0.010
Er	0.086 ± 0.019
Tm	0.038 ± 0.010
Yb	0.024 ± 0.006
Lu	0.044 ± 0.011

[a] From Morrison and Cleland (1983), with permission.

centrations of Mg^{2+}, Morrison and Cleland (1983) have provided the values shown in Table 5-3.

The various mononucleotides show large differences in their abilities to enhance Tb^{3+} luminescence. All investigators agree that 5'-XMP, 5'-GMP, and 5'-dGMP produce enhancements of approximately 10^2-10^3, while other NMP have little or no activity. However, there is some disagreement over the quantitative details. Formoso (1973) measured an order of effectiveness of GMP $\gg$ UMP $>$ CMP $>$ AMP, while Topal and Fresco (1980) determined the sequence dGMP $>$ GMP $\gg$ UMP $>$ CMP $>$ AMP $\approx$ dTMP (Table 5-4). However, Ringer *et al.* (1978) recorded the ranking GMP $>$ XMP $>$ dGMP, with CMP, dCMP, AMP, dAMP, UMP, dTMP, and IMP being inactive. This sequence contradicts Gross *et al.* (1982), who found XMP to be much more effective than GMP, with UMP being inactive. Disagreements over the relative effectiveness of NMP other than XMP, GMP, and dGMP in enhancing Tb^{3+} luminescence probably reflect the extremely low enhancements produced by these compounds. Restriction of strong enhancing ability to XMP, GMP, and dGMP suggests that electronegative substituents at the C-2 and C-6 positions in the purine ring are important in this phenomenon.

From Table 5-4 it can also be seen that the order of effectiveness in enhancing Tb^{3+} luminescence is NTP $>$ NDP $>$ NMP. Ribonucleotides are better chromophores than deoxyribonucleotides, although there is disagreement over the relative effectiveness of GMP and dGMP (Ringer *et al.*, 1978; Topal and Fresco, 1980). The 2'-OH group thus seems important in the interaction of Ln^{3+} ions with NMP, just as it was for nucleosides. The position of the phosphate group is also crucial, as luminescence enhancement by 2'-GMP, 3'-GMP, and 3'-dGMP is much less than that produced by their 5' isomers. However, Tb^{3+} luminescence enhancement by 2'-CMP exceeds that of 5'-CMP. Various chemical modifications to the guanine ring also lower the degree of Tb^{3+} luminescence enhancement by 5'-GMP (Ringer *et al.*, 1978). Neither 2':3'- nor 3':5'-cyclic NMP enhances Tb^{3+} luminescence (Ringer *et al.*, 1978; Topal and Fresco, 1980; Table 5-4), despite possessing the ability to bind Ln^{3+} ions (Table 5-2). Polynucleotides are better enhancers of Tb^{3+} luminescence than the corresponding NMP (Table 5-4; Section 5.3). Insignificant luminescent enhancement occurs with Sm^{3+}, Lu^{3+}, Eu^{3+}, and Dy^{3+} complexes of NMP (Topal and Fresco, 1980).

Tb^{3+} is a useful adjunct to studies of nucleotides, as it not only luminesces but, being paramagnetic, can also be used in conjunction with NMR spectroscopy (Section 3.2). Indeed, Tb^{3+} is 10 times more effective than Eu^{3+} in inducing shift perturbations (Barry *et al.*, 1974a). The NMR spectroscopic results for the interaction between Tb^{3+} $(0-36\,\mu M)$ and

Table 5-4. Luminescence of Complexes between Tb^{3+} and Nucleotides, Polynucleotides, and DNA[a,b]

Monomer (10⁻⁴ M/L)	Excitation λ_{max} (nm)	Luminescent emission, λ_{545}	
		$[Tb^{3+}] = 1.5 \times 10^{-4}\,M$	$[Tb^{3+}] = 5 \times 10^{-4}\,M$
Nucleotides			
dAMP	—	0	0
AMP	284	1	5
dCMP	—	0	0
CMP	285	4	7
dTMP	265	2	2
UMP	265	14	15
IMP	—	0	0
G	—	0	0
3:5-cGMP	—	0	0
dGMP	287	23	80
GMP	287	13	35
GDP	287	73	85
GTP	286	53	116

Polymer (10⁻⁴ M residues/L)	Excitation λ_{max} (nm)	Luminescent emission, λ_{545} $[Tb^{3+}] = 1.5 \times 10^{-4}\,M$
Polynucleotides		
Poly(dA)	276	18
Poly(A)	276	24
Poly(dC)	283	53
Poly(C)	283	210
Poly(dT)	280	113
Oligo(I)	275	18
Poly(dI)	274	13
Poly(dG)	287	103
Poly(G)	287	224 (391)[c]
DNA[d]		
Nat. T-7 DNA	—	0
Den. T-7 DNA	287	380
Nat. T-5 DNA	—	0
Den. T-5 DNA	287	350
Nat. CT DNA	—	0
Den. CT DNA	287	365

[a] From Topal and Fresco (1980), with permission.
[b] In 0.01 M cacodylate buffer, pH 7.
[c] Value in parentheses reflects luminescence in presence of 1 M urea.
[d] Nat. = native; Den. = denatured.

5'-UMP (2 mM) at pH 7 were consistent with binding of Tb^{3+} to the phosphate moiety, producing prominent shifts of both the 5' H atoms on the ribose ring (Gross et al., 1982). Proton H-6 on the uracil ring was the next most prominently shifted, which is in agreement with a largely anti conformation. Although the H-6 proton peak was shifted downfield, that of H-5 was shifted upfield, a result confirmed by other workers for 5'-UMP (Inagaki et al., 1978) and 3'-UMP (Yokoyama et al., 1981) at pH 1.7–2.0. Data for 5'-dGMP and 5'-XMP were also quite similar. Shift changes for UMP were greater than for dGMP, suggesting that UMP may have the higher affinity for Tb^{3+}. Whereas the H-8 peak was shifted slightly downfield in 5'-XMP, it moved upfield in dGMP, suggesting a conformational difference between the respective Tb^{3+} complexes. The peak most extensively broadened by Tb^{3+} was that of the H-5 proton of UMP, dGMP, and XMP, again consistent with binding to the phosphate moiety.

Attempts to use lanthanides to obtain information on the solution structures of nucleotides were pioneered by Williams and his colleagues (Barry et al., 1971). Quantitative analysis of the shifting and broadening of specific resonances in response to various Ln^{3+} ions permitted the determination of intramolecular distances and angles. These were then compared to various possible configurations, and that providing the best fit with the experimental data was identified (Section 3.2). Satisfactory fits between the experimental data and possible conformations were obtained with the Ln^{3+} ion lying in the O—P—O plane and with the effective symmetry axis passing through the Ln^{3+} ion and bisecting the O—P—O plane. The lanthanide coordinated with the two oxygen atoms of the phosphate group (Fig. 5-1). Such a conformation was similar to the anti conformations found by crystallographic means (Barry et al. 1974a). However, it became apparent that an equilibrium mixture of a number of different conformers probably exists in solution. Geraldes and Williams (1978) compared the solution conformation of 5'-AMP obtained by Ln^{3+}-NMR methods with those determined by independent NMR techniques. Although the lanthanide data were satisfied by postulating a single family of very similar conformations, the independent evidence strongly suggested that an equilibrium mixture of different 5'-AMP conformers existed in solution. The authors concluded that the Ln^{3+}-NMR data alone were insufficient for a complete conformational description of the molecule and that these data should be combined with the results of independent conformational analysis before ascribing final structures. The Ln^{3+}-NMR method actually complements other NMR techniques quite nicely. Whereas the latter reveal details of local structure, the lanthanide method provides conformational data on a larger scale, potentially determining the three-dimensional arrangement of nuclei within about 10 Å of the Ln^{3+} ion.

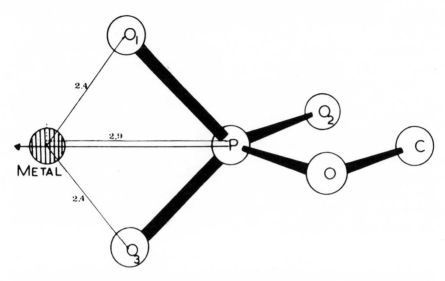

Figure 5-1. Diagram showing the Ln^{3+}-binding position with respect to the phosphate group of AMP and TMP. Interatomic distances are shown in Å. From Barry *et al.* (1971), with permission.

Numerous studies have confirmed the value of combining both Ln^{3+}-NMR and independent approaches to the conformational analysis of a number of different nucleotides (Geraldes and Williams, 1978; Dobson *et al.*, 1978; Inagaki *et al.*, 1978; Geraldes, 1979; Yokoyama *et al.*, 1981), cyclic NMP (Barry *et al.*, 1974b; Lavallee and Zeltmann, 1974; Robins *et al.*, 1977; Fazakerley *et al.*, 1977a,b; Inagaki *et al.*, 1978; Geraldes and Williams, 1978), and ADP and dADP (Fazakerley and Reid, 1979).

One assumption in such studies is that the lanthanide does not alter the conformation of the nucleotide in question. NMR data (e.g., Dobson *et al.*, 1978; Geraldes and Williams, 1978; Inagaki *et al.*, 1978) are consistent with this assumption. However, NMR evidence also suggests that the structure of Ln-NMP complexes may differ between the early and late members of the lanthanide series. Under conditions of large lanthanide excess at pH 5.7, Eu^{3+} strongly altered the CD spectra of AMP, IMP, and XMP (Bayley and Debenham, 1974). Smaller changes were seen for dAMP, while the spectrum of 3':5'-cAMP was virtually unaffected (Table 5-1).

Ln^{3+} ions form 1:1 or 1:2 complexes with NTP, depending upon the pH and concentration. Phosphorus-31 NMR data showed that the Ln^{3+}–phosphorus distances in Ln-ATP complexes involving Pr^{3+}, Nd^{3+},

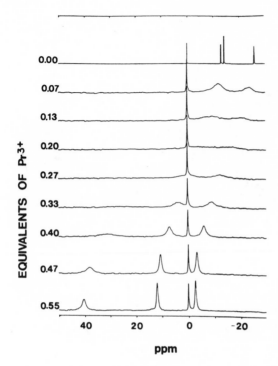

Figure 5-2. Effects of Pr^{3+} on the ^{31}P-NMR spectrum of ATP. pH = 6, [ATP] = 20 mM, 20% D_2O. From Eads *et al*. (1984), with permission.

Eu^{3+}, or Yb^{3+} were 5.1, 3.5, and 3.3 Å for the α, β, and γ phosphoryl groups, respectively (Tanswell *et al.*, 1975). Spin-echo data independently supported a Nd^{3+}–P distance of 3 Å (Shimizu *et al.*, 1979). As there was no interaction of the Ln^{3+} ion with the adenine ring, the binding site appeared to involve the β and γ phosphoryl groups. It was suggested that the two oxygen atoms of the γ-phosphate, and possibly the β-γ bridge oxygen, were in the inner coordination sphere (Tanswell *et al.*, 1975). The ^1H resonances in ATP had the opposite sign to those in AMP, while the signs of the ^{31}P dipolar shifts for a given Ln^{3+} ion were opposite to those in the ^1H resonances.

Titration of Eu^{3+} luminescence suggested that, at low concentrations of Eu^{3+}, a 1:2 Eu-ATP complex formed at pH 6 and at pH 8 (Eads *et al.*, 1984). At stoichiometries above 0.5 Eu^{3+}/ATP, a new complex was formed which was soluble at pH 8, but insoluble at pH 6. The authors suggested that this new form could be an oligomer held together by Eu^{3+} bridges. Titration of ATP with Pr^{3+} (Fig. 5-2) by ^{31}P-NMR spectroscopy revealed progressive broadening and then shifting of the peaks, which can be interpreted in favor of a 1:2 Pr^{3+}-ATP complex. In agreement with Tanswell *et al.* (1975), a Pr^{3+}–phosphorus distance of 3.5 Å was calculated, suggesting direct coordination by a β-phosphorus oxygen. The reciprocal excited state lifetimes of Eu^{3+} were used to monitor the relative populations of the 1:2 and 1:1 Ln-ATP complexes and to calculate an equilibrium constant of 300 $\pm$ 50 μM relating the interconversion of the two. Lifetimes of 240 $\pm$ 15 μs and 440 $\pm$ 10 μs for the 1:1 and 1:2 complexes, respectively, were calculated. EPR studies have independently confirmed the existence of 1:2 Ln-ATP complexes with Ce^{3+}, Nd^{3+}, Er^{3+}, and Yb^{3+} but also suggested the possible formation of 1:3 complexes (Shimizu *et al.*, 1979; Shimizu *et al.*, 1983). Relaxation enhancement data using Gd^{3+} provided a hydration number of 1.6 $\pm$ 0.5 for Eu^{3+} in the Eu-ATP complex. However, H_2O/D_2O excited state luminescence lifetimes gave values of 2.4 $\pm$ 0.5 at pH 6 (Eads *et al.*, 1984). The discrepancy probably reflects the ability of Eu^{3+} luminescence to report the total number of coordinated water molecules, while NMR with Gd^{3+} only detects those molecules which exchange rapidly. In agreement with the higher figure, Gutman and Levy (1983) determined a hydration number of 2.6 for Eu^{3+} complexed to ATP at pH 8.

Ln-ATP complexes inhibit the enzymes pyruvate kinase, creatine kinase, hexokinase, and phosphoglycerate kinase. They cannot serve as substrates for protein kinase C (Section 4.24).

5.2 Other Nucleotides

Barry *et al.* (1972) have studied the Eu^{3+} complex of adenylyl-$(3' \rightarrow 5')$-adenosine (ApA) at pH 1.5–1.8, to gain conformational information. Assuming the Ln^{3+} ion to bind to the two oxygen atoms of the phosphate group as shown for NMP (Section 5.1; Fig. 5-1), a rigid *anti* conformation was deduced. Similar studies with ApC, GpC, CpA, and CpG also showed *anti* conformations, with a K_a for Eu^{3+} of 4–6 M^{-1}. Eu^{3+} bound more weakly to these dinucleotides than to AMP, GMP, or TMP, whose K_a values were 10–17 M^{-1} (Table 5-2). Eu^{3+} (10 mM) had no effect on the CD spectrum of ApA (60 μM) at pH 5.7 (Bayley and Debenham, 1974). Ln^{3+}-EDTA shift and relaxation probes have also been applied to adenylyl-$(3' \rightarrow 5')$-adenosine 2'-phosphate (ApA2 p) at pH 7.5, where binding of the Ln^{3+} ion was restricted to the terminal phosphate group (Geraldes and Williams, 1978).

Increasing concentrations of Eu^{3+} displaced the CD spectrum of diadenosine-5'-pyrophosphate (AppA) to progressively longer wavelengths (Bayley and Debenham, 1974). At an AppA concentration of 64 μM, a 50% change in the spectrum occurred with 0.1–0.5 mM Eu^{3+} or

1–2 mM La^{3+}. Ca^{2+} had no effect on the CD spectrum of AppA at these concentrations. The change in the CD spectrum produced by Ln^{3+} ions decreased as the pH was lowered from pH 5.6. At pH 2, there was no effect at all, but the spectral change reappeared at pH 1. Eu^{3+} did not appear to affect the stacking of the bases and seemed to have a similar affinity for the stacked and unstacked forms of AppA. From the titration curve, Bayley and Debenham (1974) concluded that Eu^{3+} was likely to form 1:1 complexes with AppA. The simplest interpretation offered for the changes in the CD spectrum was that coordination of Eu^{3+} to the pyrophosphate group altered the relationship between the two adenine chromophores (Bayley and Debenham, 1974). Binding of Tb^{3+} to guanosyl-(3′ → 5′)-guanine can be inferred from the observed enhancement of luminescence (Burns, 1985).

Eu^{3+} or La^{3+} displaced the CD spectrum of NAD toward lower wavelengths, with 50% of the maximum shift occurring at a lanthanide concentration of 0.1–1.0 mM (Bayley and Debenham, 1974). As with AppA, the magnitude of the spectral change decreased with pH. The authors concluded that lanthanides bind to the phosphate moieties of NAD, a conclusion shared by Lee and Raszka (1975) for a number of diphosphopyridine coenzymes, including NAD. At pH 4.4, lanthanides bind to the pyrophosphate moiety of nicotinamide mononucleotide in a 1:1 stoichiometry. The data from NMR spectroscopy suggest that Ln^{3+} ions do not perturb the conformation of the nucleotide (Birdsall et al., 1975). Ln^{3+} ions also bind to the four oxygen donors of the pyrophosphate group in coenzyme A, forming 1:1 and 1:2 Ln^{3+}–ligand complexes at pH 6.4 (Fazakerley et al., 1977b).

5.3 Polynucleotides and Nucleic Acids

The affinities of lanthanides for polynucleotides exceed their affinities for individual nucleotides, with tRNA providing stronger ligands than DNA (cf. Tables 5-6 and 5-9). Most studies of these interactions have made use of Tb^{3+} luminescence, which has provided the unexpected, but very important, finding that Tb^{3+} luminescence can be used to discriminate between double-stranded (ds) and single-stranded (ss) nucleic acids (Section 3.10.2). These points are well illustrated in Table 5-4 (Topal and Fresco, 1980). This table also reveals that the enhancement of Tb^{3+} luminescence produced by polynucleotides greatly exceeds enhancement by the individual mononucleotides, with polyribonucleotides being more effective than the corresponding polydeoxyribonucleotides. The differences between the various polynucleotides in their abilities to enhance

Table 5-5. Comparison of Tb^{3+} Binding and Tb^{3+} Luminescence Enhancement by DNA and Polynucleotides[a]

Polynucleotide	v[b]	N[c]	F[d]	F/Tb^{3+}[e]
Poly(C)	0.059 ± 0.007	7	1.1	20
Poly(A)	0.134 ± 0.024	4	0.7	5
Poly(U)	0.166 ± 0.011	2	3.5	21
Poly(G)	0.182 ± 0.040	15	100.0	550
Poly(A,U,G)	0.202 ± 0.057	3	75.9	375
Poly(X)	0.273 ± 0.030	5	1065	3900
Poly(I)	0.275 ± 0.032	5	0.2	0.7
Poly(G):poly(C)	0.260 ± 0.040	3		
Micrococcus luteus DNA	0.278 ± 0.014	4		
E. coli DNA	0.310 ± 0.028	3		
Calf thymus DNA	0.313 ± 0.025	5		
Avian erythrocyte DNA	0.341 ± 0.016	2		

[a] From Gross and Simpkins (1981), with permission.
[b] Molar $^{160}Tb^{3+}$ bound/molar polynucleotide residue, presented as means ± SD.
[c] Number of determinations.
[d] Normalized fluorescence enhancement with respect to that of poly(G), which is assigned 100.0 arbitrarily; values represent means of at least two experiments.
[e] Fluorescence intensity/bound terbium ion.

Tb^{3+} luminescence do not result from differences in affinity (Table 5-5; Gross and Simpkins, 1981). In addition, ds molecules bind $^{160}Tb^{3+}$ with greater affinity than ss molecules, despite the large difference in luminescence enhancement. Urea increases the luminescence of Tb^{3+}-poly(G) by preventing the formation of four-stranded helices. This is consistent with the general observation that the greater the degree of secondary structure in a nucleic acid or polynucleotide, the lower is its ability to enhance Tb^{3+} luminescence. Binding data for Tb^{3+}-nucleic acid and Tb^{3+}-polynucleotide complexes are given in Table 5-6.

The observation that Tb^{3+} luminescence was specific for ss polynucleotides was made independently by Topal and Fresco (1980) and Ringer et al. (1980). The former authors employed a polynucleotide containing 75% C and 25% G. This existed 50% in a G–C paired ds form, and 50% in a ss poly(C) form. Titration of this polymer with Tb^{3+} produced a maximum luminescence which was 50% that of pure poly(C) (Fig. 5-3). No luminescence was observed until a Tb^{3+}/nucleotide molar ratio of 0.17 had been reached (inset Fig. 5-3). As the affinity of the ds forms for Tb^{3+} exceeded that of the ss forms, it was suggested that the lag reflected preferential binding of Tb^{3+}, at low concentration, to the ds G–C regions, which did not lead to luminescence. Ringer et al. (1980) arrived at the same conclusion by showing a large increase in Tb^{3+} luminescence upon

Table 5-6. Binding Parameters for Tb^{3+} Complexes of DNA
and Polynucleotides

Polynucleotide or DNA	K_a (M^{-1})	n (residues/Tb^{3+})	Reference
Poly(C)	8.2×10^5	3.7	Haertlé et al. (1981)
Poly(C)	8.2×10^5	3.7	Topal and Fresco (1980)
Oligo(dG)$_6$	2.8×10^4	1.3	Gross and Simpkins (1981)
Oligo(dG)$_8$	2.9×10^4	1.5	Gross and Simpkins (1981)
Poly(dG)[a]	3.0×10^4	3.3	Gross and Simpkins (1981)
Poly(dA-dT)	7.0×10^4	5.0	Haertlé et al. (1981)
Poly(A,G)	5.0×10^6	2.6	Haertlé et al. (1981)
E. coli DNA	4.1×10^4	3.7	Gross and Simpkins (1981)
Plasmid, supercoiled DNA	1.2×10^5	5.5	Gross and Simpkins (1981)
Calf thymus DNA[a]	NG[b]	3	Yonuschot et al. (1978)
Calf thymus DNA (27 mM NaCl)	10^7	NG	Draper (1985)
Calf thymus DNA (133 mM NaCl)	1.6×10^5	NG	Draper (1985)
Calf thymus DNA (200 mM NaCl)	1.2×10^4	NG	Draper (1985)
Calf thymus DNA (400 mM NaCl)	3.5×10^3	NG	Draper (1985)

[a] Determined with $^{160}Tb^{3+}$. All others determined by Tb^{3+} luminescence enhancement.
[b] NG: Not given.

thermally denaturing DNA. Tb^{3+} luminescence can be used experimentally to monitor rates of DNA reannealing (Fig. 5-4), to detect subtle changes in conformation which lead to local melting of the helix, and to detect ss DNA bands on gels after electrophoresis (Section 3.10.2).

Although Topal and Fresco (1980) claimed that ds DNA was completely inactive in enhancing Tb^{3+} luminescence, others (e.g., Yonuschot and Mushrush, 1975; Houssier et al., 1983; Stokke and Steen, 1985; Ringer et al., 1978; Gross et al., 1981; Haertlé et al., 1981) have reported low to moderate luminescence for intact DNA or ds polynucleotides. It is possible that "breathing" of ds DNA permits some energy transfer to Tb^{3+}. However, Ringer et al. (1985) have recently shown that luminescence of Tb^{3+} from dG residues in four different polynucleotides only occurred upon thermally denaturing the helices (Fig. 5-5). Neither the Z nor the B forms of these helices supported dG-mediated Tb^{3+} luminescence. Topal and Fresco (1980) suggested that Tb^{3+} luminescence enhancement by preparations of ds DNA may be due to the presence of damaged DNA or contamination. Supporting evidence comes from the observation that

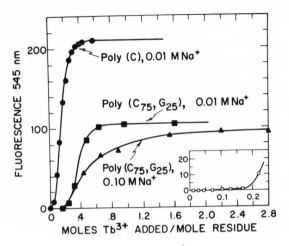

Figure 5-3. Titration of poly(C) and poly(C_{75}, G_{25}) with Tb^{3+} luminescence. Polynucleotide concentration in each case is 10^{-4} M. 0.01 M cacodylate buffer, pH 7.8. From Topal and Fresco (1980), with permission.

treatment of commercial DNA preparations with S_1 nuclease greatly reduces their ability to enhance Tb^{3+} luminescence (Gross and Simpkins, 1981). Supercoiled DNA, which contains regions of ss DNA, enhances Tb^{3+} luminescence more effectively than nonsupercoiled DNA does. Circular DNA is more effective than linear DNA, although titration curves for circular DNA do not show saturation of Tb^{3+} luminescence at high Tb^{3+} concentrations (Gross and Simpkins, 1981). The explanation of these differences between ds and ss DNA may lie in the nature of the ligands for Tb^{3+} in each case. In ds structures, Tb^{3+} ions bind only to the phosphate groups without enhancement of luminescence. However, in ss structures, Tb^{3+} ions are also able to coordinate with the bases from which they accept the energy to increase their luminescence.

Draper (1985) titrated calf thymus DNA with Eu^{3+}, observing that the apparent K_a varied with salt concentration (Table 5-6). Titration was possible up to 1 Eu^{3+} per 3 DNA phosphates, at which point the DNA aggregated. As there was only one excitation peak, all the Eu^{3+} ions were probably attaching to the same type of ligand, presumably the phosphate moieties of DNA. Lifetime measurements of the luminescent emisson suggested that seven water molecules were coordinated to each Eu^{3+} ion. As Eu^{3+} probably has a coordination number of 8 or 9 (Section 2.3), this means that only one or two ligands were provided by the DNA. Such numbers agree with calculations based upon the red shift of the $^7F_0 \rightarrow$

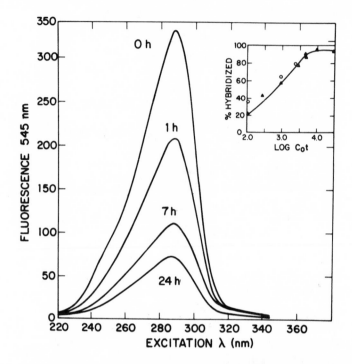

Figure 5-4. Tb^{3+} luminescence during the reassociation of chick red blood cell DNA (10 μg/ml). 50 mM Tris buffer pH 7.4, 0.3 M NaCl. DNA was denatured at 100°C and allowed to anneal at 65°C. Aliquots were removed at the indicated times, cooled, and their Tb^{3+} luminescence measured. The inset shows the percent hybridization calculated by luminescence data (O) or by S$_1$ nuclease digestion (▲). From Topal and Fresco (1980), with permission.

$^5D_0$ transition which suggested that 1.3 ± 0.5 negative charges coordinated to each Eu^{3+} (Draper, 1985).

From the data in Table 5-4, each of the nucleotide bases in DNA and RNA would seem able to contribute, to a greater or lesser degree, to the enhancement of Tb^{3+} luminescence. However, Ringer et al. (1978) claimed that nearly all the luminescence enhancement in Tb^{3+}-RNA and Tb^{3+}-DNA resulted from energy transfer from G and dG residues, respectively. The nature of the neighboring bases did not greatly influence the ability of dG residues in ss DNA to enhance Tb^{3+} luminescence (Ringer et al., 1985).

Yonuschot and Mushrush (1975) and Draper (1985) both found that calf thymus DNA precipitated when titrated with Tb^{3+} beyond a Tb^{3+}/DNA phosphate ratio of 0.3. Precipitation of DNA by Ln^{3+} had been observed in the early decades of this century and had been used by Chargaff et al.

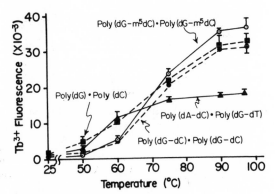

Figure 5-5. Enhancement of Tb^{3+} luminescence by synthetic DNA duplexes undergoing thermal denaturation. From Ringer *et al.* (1985), with permission.

(1949) in the purification of DNA (Section 3.8.1). Luminescence studies on the precipitated DNA (Yonuschot and Mushrush, 1975) showed that the reaction of Tb^{3+} with DNA was linear up to a Tb^{3+}/DNA phosphate stoichiometry of 1:1, after which no further binding occurred. When intact chromatin was employed instead of DNA, the luminescence reached a plateau at a Tb^{3+}/DNA phosphate ratio of 0.48, suggesting that 52% of the DNA phosphate was rendered unavailable to Tb^{3+} by the presence of chromatin proteins. Possible complications from interactions between Tb^{3+} and chromatin proteins were not discussed. Also, in view of the later discovery that Tb^{3+} luminescence preferentially detects ss DNA, we may wonder to what degree these results can be explained by an alternative mechanism on the basis that DNA in chromatin is protected from "nicking" or prevented from undergoing subtle conformational unwinding. Nevertheless, the figure of 52% agrees well with other estimates of the protein coverage of DNA phosphate obtained by alternative methods. However, as in a later paper Yonuschot *et al.* (1978) revised their Tb^{3+}:phosphate stoichiometry from 1:1 to 1:3, the meaning of these data is uncertain. Gross *et al.* (1981) found that histone H1 was more effective than either H2B or H4 in quenching Tb^{3+} luminescence of thymus DNA. Spermidine strongly inhibited Tb^{3+} luminescence, with 50% inhibition occurring with 0.3 mM spermidine. Concentrations of spermidine which strongly suppressed Tb^{3+} luminescence were those which also stimulate transcriptional activity in other systems. Proteins, such as lysozyme, ribonuclease A, and bovine serum albumin, which do not normally associate with DNA, had no effect on luminescence. Histones suppressed the luminescence of Tb^{3+} bound to *E. coli* DNA, *Micrococcus luteus* DNA, and calf thymus DNA with an approximately equal efficiency.

Evidence that Ln^{3+} ions could coordinate with ligands other than phosphate on polynucleotides was first provided by Gross and Simpkins (1981) and Gross *et al.* (1981; 1982), using Tb^{3+} luminescence and NMR spectroscopy. Lowering the pH was shown to inhibit Tb^{3+} luminescence by protonating N-7 of poly(X) and poly(G). This finding is consistent with the coordination by Tb^{3+} of N-7 and the carbonyl group of C-6 in X and G. Cu^{2+} is known to do this and is a stronger competitor for Tb^{3+}-binding sites on DNA. As discussed in Section 5.1, NMR spectroscopy had shown the interaction of Tb^{3+} with the phosphate groups of 5'-UMP and 5'-XMP (Gross *et al.*, 1982). However, when Tb^{3+}–polynucleotide interactions were studied, the spins of protons on the uracil or xanthine bases were also strongly shifted. For poly(U), the H-5 proton of uracil was more strongly affected than H-6, suggesting that Tb^{3+} coordinated with the electron donor sites of the C-4 oxygen atom of the uracil ring. For poly(X), there was an analogous enhancement of H-8, which the authors suggested resulted from the coordination of Tb^{3+} to N-7 of xanthine, despite the poor affinity that Ln^{3+} ions generally have for N donor atoms (Section 2.3). Ribose ring protons were more strongly influenced in the polynucleotides than in the monomer solutions, because of the attachment of Tb^{3+} to the adjacent phosphodiester linkages. Nuclear relaxation analysis showed increases in linewidth which generally preceded and paralleled the induced chemical shifts. Again, the ribose ring protons of poly(U) and poly(X) were more sensitive to these effects than were the ribosyl protons of their respective monomers (Gross *et al.*, 1982). The ability of Tb^{3+} to coordinate with purine or pyrimidine bases in ss polynucleotides, but not in the ds polynucleotides, seems to explain satisfactorily their different abilities to enhance luminescence. However, on the basis of CD and electro-optical techniques, Gersanovski *et al.* (1985) concluded that Tb^{3+} gained access to N-7 of guanine and other nucleotide bases, even in ds DNA. If so, the reason for the difference in Tb^{3+} luminescence between ss and ds DNA becomes again obscure.

There is evidence that binding of Tb^{3+} alters the conformation of DNA. Using CD and electronic dichroism and birefringence techniques, Gersanovski *et al.* (1985) observed that as the Tb^{3+}/DNA phosphate ratio increased, DNA underwent a transition from the B to the ψ form. Binding of Tb^{3+} to the DNA presumably altered the geometry of the sugar-phosphate backbone and influenced the stacking of the bases. However, Gross and Simpkins (1981) did not observe significant hyperchromicity at Tb^{3+}/DNA phosphate ratios of ≤ 0.5. At the high Tb^{3+}/DNA phosphate ratio of 10:1, Tb^{3+} caused poly(dG-dC):poly(dG-dC) to adopt the Z conformation (Haertlé *et al.*, 1981). A number of different metal ions can effect this switch, which is not specific for the lanthanides. Gross *et al.* (1981) suggested that Tb^{3+} may shift the helix–coil transition in favor of

Table 5-7. Influence of Tb^{3+} on the
Melting Temperature of
Poly(dA-dC):Poly(dG-dT)[a]

Tb^{3+}:DNA phosphate	T_M (°C)
0	82.7 ± 0.5
1:3	73.1 ± 1.2
1:1	53.0 ± 0.6
2:1	49.1 ± 0.8

[a] From Ringer et al. (1985), with permission.

the coil. Tb^{3+} destabilizes the DNA helix (Ringer et al., 1980; Gross and Simpkins, 1981; Ringer et al., 1985), progressively lowering the melting temperature (Table 5-7), sharpening the melting transition, and retarding reannealing on cooling. However, under conditions of high salt, the melting temperature is only weakly affected (Topal and Fresco, 1980).

Other metal ions compete with Tb^{3+} for binding to the DNA. For poly(dG-dC), which is ds, 50% inhibition of the luminescence enhancement at $100 \mu M$ Tb^{3+} required $5 mM$ Ca^{2+} or Mg^{2+}. For poly(dG), which is ss, these values increased to $46 mM$ for Ca^{2+} and $82.5 mM$ for Mg^{2+}. Both Ca^{2+} and Mg^{2+} interact almost exclusively with the phosphate groups of DNA. However Cu^{2+}, which also binds to the nucleoside bases, is a much more effective displacer of Tb^{3+}. Only $20 \mu M$ Cu^{2+} was necessary to inhibit Tb^{3+} luminescence by 50%, with both ds and ss polynucleotides (Gross et al., 1981). In view of the supposed differences in the manner in which Tb^{3+} binds to ds and ss polynucleotides, it is difficult to understand why Cu^{2+} should displace Tb^{3+} with equal efficiency from both forms. One explanation, which agrees with the suggestion of Topal and Fresco (1980), would be that Tb^{3+} binds to ss contaminants or locally denatured domains within supposedly pure preparations of ds polynucleotides.

Various substances which bind to polynucleotides and DNA alter Tb^{3+} luminescence. Histones and spermidine have already been discussed. Burns (1985) calculated that at a Tb^{3+}/poly(G) ratio of <0.4, most of the energy absorbed by nucleotides was not transferred to Tb^{3+} but dissipated nonradiatively. Addition of Tl^+ was found to increase Tb^{3+} luminescence. The author suggested that the mechanism involved increasing the intersystem transfer of energy from excited singlet to excited triplet states (Section 3.3). Tl^+ also enhanced luminescence of Tb^{3+} with poly(X), poly(G_1,U_2), and poly(G,U), but not with GMP, poly(C), or guanosylguanine. Luminescence of Tb^{3+} attached to denatured DNA was quenched by Tl^+.

Platinum anticancer drugs produce intrastrand DNA cross-links, the formation of which requires local denaturation of the helix. Five Pt coordination complexes were tested to determine whether Tb^{3+} luminescence could detect this denaturation (Houssier *et al.*, 1983). Diethylenetriamine-Pt had the weakest activity, enhancing Tb^{3+} luminescence by only a factor of two. *cis*-Dichlorodiammine-Pt, but not *trans*-dichlorodiammine-Pt (Arquilla *et al.*, 1983), displayed the largest effect, enhancing Tb^{3+} luminescence eightfold. For comparison, luminescence of Tb^{3+} attached to thermally denatured DNA was 15-fold greater than that of Tb^{3+} attached to native DNA. In contrast to this, Thompson *et al.* (1982) have reported that *cis*- and *trans*-dichlorodiammine-Pt quenched Tb^{3+} luminescence, possibly by displacing Tb^{3+} from the DNA or by cross-linking the DNA to prevent access of Tb^{3+} to the bases. Quenching of Tb^{3+}-DNA luminescence has also been observed for adriamycin (Simpkins *et al.*, 1984; Simpkins and Pearlman, 1984) and actinomycin D (Simpkins and Pearlman, 1984), but not ethidium bromide (Simpkins *et al.*, 1984). However, later work showed that the mechanism of quenching by adriamycin or actinomycin D did not involve displacement of Tb^{3+} from the DNA. Instead, there appeared to be energy transfer from Tb^{3+} to the drug, resulting in much faster decay of luminescence (Stokke and Steen, 1985). Because chemical modifications to the guanine ring reduce its ability to promote Tb^{3+} luminescence, it has been suggested that luminescence studies could be used to detect certain types of DNA damage (Ringer *et al.*, 1978; Gross and Simpkins, 1981).

The interaction of lanthanides with native RNAs has been studied in much greater detail than their interaction with DNA. Such studies date back to the observations of Kayne and Cohn (1972) that a certain Ln^{3+} ions could replace Mg^{2+} as a cofactor for the isoleucyl-tRNA synthetase of *E. coli*, although the rate of reaction was not as high as for the native species. The metal ion was required to stabilize the tRNA rather than to act upon the enzyme itself. X-ray crystallographic studies of tRNA^Phe from yeast have shown four or five different sites occupied by Gd^{3+}, Lu^{3+}, or Sm^{3+} (Holbrook *et al.*, 1977; Jack *et al.*, 1977; Suddarth *et al.*, 1974; Robertus *et al.*, 1974; Stout *et al.*, 1978). The bases around the five sites identified by Jack *et al.* (1977) are listed in Table 5-8. In all but the fourth site, Sm^{3+} is directly coordinated to oxygen atoms of phosphate groups. At the fourth binding site, Sm^{3+} lies close to O-6 and N-7 of a guanine base. The Sm^{3+} sites 1 and 2 hold together phosphate oxygens from different parts of the molecule and thus, like Mg^{2+}, stabilize the secondary structure. Although these are strong binding sites for Mg^{2+} under physiological conditions, Ln^{3+} ions are able to displace Mg^{2+} from these positions. However, the binding sites of Os, Pt, and Au are quite

Table 5-8. Samarium Nearest Neighbors in
Yeast tRNA[Phe] [a]

Site no.	Neighbors	Distance (Å)
Sm 1[b]	U8-OR	2.2
	A9-OR	2.3
Sm 2[b]	G20-OR	2.3
	A21-OL	2.6
Sm 3	U7-OR	2.5
	A14-OL	2.6
Sm 4[c]	G45-N7	3.3
	G45-O6	2.4
	A44-N7	4.4
	A23-OL	4.3
	A44-OL	4.8
Sm 5	A14-OR	2.4
	G57-OR	2.4 (symmetry-related)

[a] From Jack *et al.* (1977), with permission.
[b] Strong magnesium binding sites.
[c] Weak magnesium binding sites.

different from the lanthanide sites (Stout *et al.*, 1978). NMR spectroscopic studies of the shifts produced by adding Eu^{3+} to yeast tRNA[Phe] confirm the presence of four or five tight Ln^{3+}-binding sites. Binding was independent and sequential rather than cooperative, as is the case for Mg^{2+} (Jones and Kearns, 1974). In agreement with the crystallographic data, the NMR data suggested that the strongest Eu^{3+} and Gd^{3+} sites were near to residue U-8.

It has been possible to follow the association of yeast tRNA[Phe] with Eu^{3+} by monitoring the fluorescence of the Y base in the anticodon loop. Although addition of up to three Eu^{3+} ions per tRNA[Phe] molecule did not affect fluorescence, addition of three to eight Eu^{3+} per molecule halved Y fluorescence. The quenching curves were interpreted by postulating three strong Eu^{3+}-binding sites ($K_a \approx 10^8 M^{-1}$), two intermediate sites ($K_a \approx 2 \times 10^6 M^{-1}$), one of which quenches Y fluorescence, and about nine weak sites of $K_a \approx 4 \times 10^5 M^{-1}$ (Kearns and Bolton, 1978). Draper (1985) also interpreted the sigmoidal quenching curves in terms of the interaction of Eu^{3+} with three classes of binding site, the intermediate one affecting Y fluorescence. Affinity constants for the three classes were calculated to be 10^8, 2×10^7, and $5 \times 10^6 M^{-1}$. Thus, NMR, crystallographic, and fluorescence data all indicate that there is a Ln^{3+}-binding site near the Y base in the anticodon loop of yeast tRNA[Phe]. Both this tRNA and unfractionated yeast tRNA enhance Tb^{3+} luminescence when excited at 290 nm (Haertlé *et al.*, 1980).

Studies of the interaction of lanthanides with *E. coli* tRNA have been greatly facilitated by the independent discovery by Kayne and Cohn (1974) and by Wolfson and Kearns (1975) that enhanced Tb^{3+} and Eu^{3+} luminescence occurs through energy transfer from a single 4-thiouracil residue at position 8 on the anticodon loop. About half the species of *E. coli* tRNA possess this base, while yeast tRNAs lack it. Thus, the major peak in the excitation spectrum of enhanced Eu^{3+} or Tb^{3+} luminescence with *E. coli* tRNA occurs at about 340–345 nm, where thiouracil absorbs maximally. A smaller peak at 300 nm suggested that the common bases of *E. coli* tRNA can also transfer energy to Tb^{3+} and Eu^{3+}, as the detailed studies of Morley *et al.* (1981) indeed confirmed. Quenching of phosphorescence from A and G residues in tRNA by Tb^{3+} also occurs (Kayne and Cohn, 1974).

Titration experiments using *E. coli* tRNA showed that luminescence reached a maximum at a Eu^{3+}/RNA ratio of 3:1, and then declined. This phenomenon was explained by the existence of more than one class of binding site on the tRNA. Confirmation came with the demonstration of two types of attachment site, one showing fast decay of luminescence, and the other a slow decay rate. During titration with Eu^{3+}, there was a five-fold increase in the short-lived component until a Eu^{3+}/tRNA ratio of 3:1 was reached. This was followed by a decrease in the short-lived component and a concomitant increase in the long-lived component which leveled off at a 1:1 ratio of the two components (Wolfson and Kearns, 1975).

Photolysis, which cross-linked thiouridine at position 8 with cytosine at position 13, reduced luminescence. The shape of the luminescence titration curve did not alter after photolysis, indicating that the K_a of Eu^{3+} was not altered, only the efficiency of energy transfer. Fluorescence of thiouracil in tRNA was enhanced by the addition of the first Eu^{3+} per molecule, followed by a decrease, with complete extinction occurring after addition of six ions of Eu^{3+} per tRNA molecule. Quenching of thiouracil fluorescence paralleled the enhancement of Eu^{3+} and Tb^{3+} luminescence. The addition of up to three ions per tRNA of Sm^{3+}, Yb^{3+}, and Gd^{3+}, which do not luminesce, also enhanced thiouracil fluorescence. Mg^{2+} did so too, but at a much higher concentration (Kayne and Cohn, 1974).

Wolfson and Kearns (1975) suggested that *E. coli* tRNA possesses three strong binding sites for Eu^{3+}, each with a K_a of about $10^8 M^{-1}$. However, Kayne and Cohn (1974) proposed four independent binding sites, each with a K_d of about $(6.5 \pm 2.2) \times 10^{-6} M$. It was suggested

Table 5-9. Affinity Constants for Tb^{3+} and $tRNA^{fMet}$ at Different Salt Concentrations[a,b]

[NaCl] (M)	$K_a (M^{-1})$[c]		
	K_1	K_2	K_3
0.2	1.5×10^8	2.5×10^7	2.5×10^7
0.6	6×10^6	8×10^5	3×10^4
1.2	7.5×10^5	7.2×10^4	—
2.0	1.22×10^4	—	—

[a] Data taken from Draper (1985).
[b] pH = 6.
[c] These values have been calculated making three assumptions: (1) the salt dependence of the different affinity sites is $d(\log K)/d(\log Na^+)$; (2) the intermediate thiouracil sensitized size affinity is extrapolated from the data at 1.2 and 2 M NaCl; (3) 25 independent weak sites are assumed. Fifteen sites with 50% greater affinity fit the data nearly as well.

that these were sites which also accepted Mg^{2+} with a K_a of 2×10^5 M^{-1}. The strongest and weakest of the three Eu^{3+} sites should be near the thiouracil residue and thus contribute to the luminescence data discussed above. It was proposed that the weaker of these two sites has the shorter lifetime. Although the binding of Mg^{2+} to tRNA shows cooperativity, that of Eu^{3+} does not. Wolfson and Kearns (1975) also examined E. coli $tRNA^{Glu}$ which has no 4-thiouracil but has a modified 2-thiouracil in the anticodon loop. Maximum luminescence was obtained at a Eu^{3+}/tRNA ratio of 4–5. Unfractionated E. coli tRNA was almost identical to $tRNA^{fMet}$, with peak luminescence at three Eu^{3+} per RNA molecule. The binding behavior of denatured E. coli $tRNA^{fMet}$ was quite different, with peak luminescence occurring at a Eu^{3+}/$tRNA^{fMet}$ ratio of 20. Competition experiments suggested that yeast $tRNA^{Phe}$ and E. coli $tRNA^{fMet}$ had about the same affinity for Eu^{3+}.

Draper (1985) used luminescence titration to determine the binding constants for Tb^{3+}-E. coli $tRNA^{fMet}$ at different salt concentrations (Table 5-9). The values reported for 0.2 M NaCl agree quite well with the K_a value of $\sim 10^8 M^{-1}$ provided by Wolfson and Kearns (1975), whose solutions contained 0.1 M NaCl and 0.05 M sodium cacodylate at pH 7. These authors are also in agreement about the existence of three different classes of binding site. The two strongest classes of binding site were both sensitized by 4-thiourea. The third class of sites was not and, for theoretical reasons, was assigned a number of 25. The affinity (of the weakest class of sites) for Eu^{3+} was similar to that of the Eu^{3+}-binding sites on DNA. The highest-affinity site had two to five nonphosphate ligands in its inner coordination sphere and was strongly dehydrated. Mg^{2+} bound to the same sites with a 10-fold weaker affinity. As Draper (1985) discussed, his conclusions concerning the high-affinity Eu^{3+} site contained some unexpected details. Despite the known higher affinity of lanthanides for phosphate than for nucleic acid bases, the highest-affinity Eu^{3+} site apparently involved no inner-sphere phosphate. Secondly, none of the metal-ion sites identified in crystallized $tRNA^{Phe}$ by X-ray diffraction provided a minimum of two nonphosphate inner-sphere ligands. According to the crystallographic data, Sm^{3+} was liganded to phosphate. The author considered that the lanthanide binding properties of tRNA in solution may differ from those in crystals.

Eu^{3+} and Tb^{3+} luminescence have been used to monitor possible conformational changes between aminoacylated and deacylated tRNA and to follow their thermodenaturation (Pavlick and Formoso, 1978). Rordorf and Kearns (1976) showed that at Eu^{3+} : tRNA ratios of less than 6, Eu^{3+} stabilized tRNA against thermal denaturation. However, at a ratio of 8–10, there was a destabilizing effect. At 50°C and pH 7, Eu^{3+} catalyzed the hydrolysis of tRNA (Section 2.6). The authors concluded that Eu^{3+} bound to tRNA in a sequential manner with the first ion binding near to the two dihydroxyuridine residues. The first two and the fifth bound Eu^{3+} were most active in promoting hydrolysis. Cleavage of the 5'-phosphate linkages of poly(A), poly(C), poly(U), and poly(I) by a number of Ln^{3+} ions has been noted. DNA is not hydrolyzed (Izatt et al., 1971).

Tb^{3+} luminescence was also enhanced upon binding to ribosomes. Examination of the excitation spectra suggested that rRNAs, rather than ribosomal proteins, were the energy donors (Barela et al., 1975). In a similar fashion, enhanced Tb^{3+} luminescence upon binding to two icosahedral RNA viruses suggested energy transfer from RNA rather than from viral proteins (Morley et al., 1981).

5.4 Carbohydrates

Lanthanides have a very low affinity for simple uncharged saccharides. However, their interaction can be detected by NMR spectroscopy and circularly polarized Tb^{3+} luminescence spectroscopy. Among the first such studies were those of Angyal (1972). He determined stability constants of 10.4, 8.7, and 11.5 M^{-1} for the La^{3+} complexes of allopyranose, α-D-allofuranose, and D-glycero-D-guloheptose, respectively. According to Angyal (1972), before a simple, uncharged monosaccharide can bind Ln^{3+} ions, the three —OH ligands on the sugar have to exist in an axial–equatorial–axial configuration in at least one chain conformational anomer. Thus, the α-pyranose ring form, which has this configuration, binds Ln^{3+} ions, unlike the β form, which does not. This rule was confirmed by the results from studies of Tb^{3+} CPL discussed by Davis and Richardson (1980). NMR spectroscopy showed Yb^{3+} to be superior to La^{3+}, Pr^{3+}, and Eu^{3+} in shifting the ^{13}C resonances of sorbitol. The shift data suggested a terdentate complex, with Yb^{3+} resting 2.3 Å from O-2, O-3, and O-4 of the sugar ring (Kieboom $et\,al.$, 1977). Consistent with this interpretation, Gd^{3+} was shown to broaden the ^{13}C resonances of C-2, C-3, and C-4 or C-5 (Beattie and Kelso, 1981).

Introduction of a carboxyl group greatly increases the affinity of a sugar for lanthanides. NMR studies on the binding of Ln^{3+} ions to α-D-glucopyranuronate and α-D-galactopyranuronate confirm bidentate chelation by the ring O-5 atom and carboxyl oxygen (Izumi, 1980). However, Anthonsen $et\,al.$ (1972) suggested that these two ligands were supplemented by the hydroxyl atoms on C-4 in Eu^{3+}-α-D-galactouronate. Experiments with methyl galactouronate supported this view (Anthonsen $et\,al.$, 1973). For poly(α-1,4-L-guluronic acid), it was proposed that Eu^{3+} bound in a hexadentate fashion in cavities between adjacent sugar units (Anthonsen $et\,al.$, 1972). Ribose 5'-phosphate binds a Ln^{3+} ion to its phosphate moiety (Barry $et\,al.$, 1974a).

As part of a larger investigation of the binding of lanthanides to glycoproteins, Dill $et\,al.$ (1983a) studied the binding of Gd^{3+} to methyl N-acetyl-3-O-α-D-galactopyranosyl-L-serinate (Fig. 5-6a). As it had been shown that Gd^{3+}–amino acid interactions predominate over Gd^{3+}–saccharide interactions, the amino and carboxy termini of the serine residue in this compound were blocked by acetylation and esterification, respectively. The effect of Gd^{3+} on the ^{13}C-NMR spectrum of this substance is shown in Fig. 5-6b. Gd^{3+} broadened the signals of C-1' and C-2' in the sugar ring, the methyl group of the acetamido group of the Ser residue, and the methyl group of the ester group of serine. A weak interaction of O-6' was possibly indicated by the signal of C-6'. All other resonances were eventually broadened due to outer-sphere interactions with the metal ion. The authors suggested that there may be one or more weak binding sites in the vicinity of Ser C-2 and C-2' involving the carbonyl oxygen and the oxygen atoms on C-2 and C-1'; alternatively, there may be a second binding site involving the oxygen atom on C-6'.

When an N-acetyl group was introduced at position 2' on the sugar ring, to form O-α-D-GalpNAc-(1 → 3)-L-Ser(OMe)(NHAc), the effect of Gd^{3+} was a little different and suggested strong binding in the vicinity of Ser C-2 and C-1', involving the carbonyls of the ester and acetamido groups, with a second weak binding site near C-4'.

NMR data suggested that Gd^{3+} interacted strongly with the free carboxyl and amino termini of O-D-galactosylated glycylglycylthreonine

a

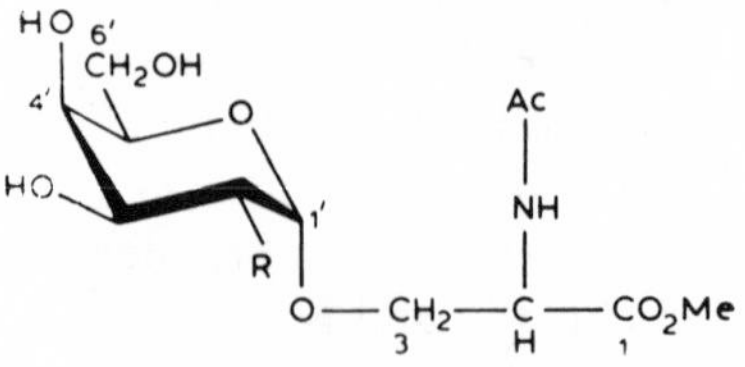

1 R = OH

2 R = NHAc

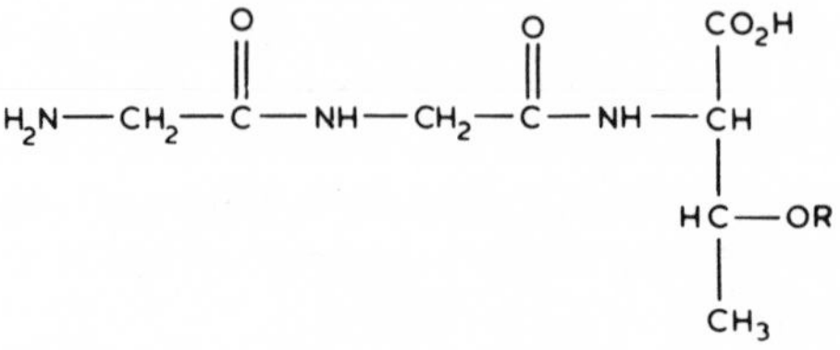

3 R = H

4 R = α-D-Gal*p*

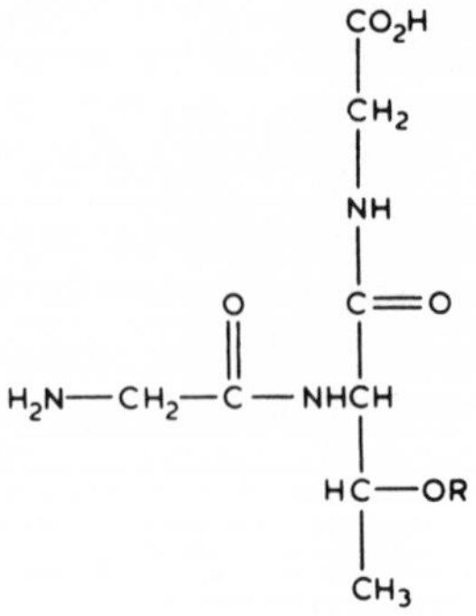

5 R = H

6 R = α-D-Gal*p*

Figure 5-6. (a) Structures of peptides and glycopeptides studied by Dill *et al.* (1983a): (1) methyl *N*-acetyl-3-*O*-α-D-galactopyranosyl-L-serinate; (3) glycylglycyl-L-threonine; (5) glycyl-L-threonylglycine. From Dill *et al.* (1983a), with permission. (b) Effect of Gd^{3+} on the ^{13}C-NMR spectrum of compound 1 shown in (a). From Dill *et al.* (1983a), with permission.

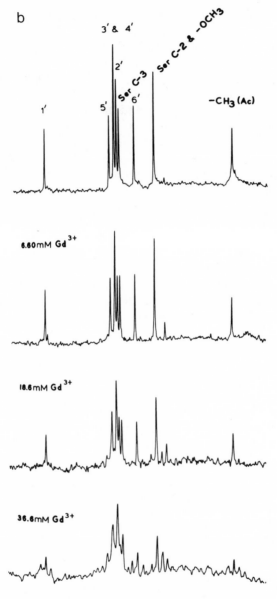

Figure 5-6. (*Continued*)

(Fig. 5-6a), as well as having a possible weak interaction with the galacto-syl group. This suggested either a separate binding site for Gd^{3+} on the galactose group near C-1' and C-2' that is not influenced by Gd^{3+} binding near the amino acid moieties, or that the Gd^{3+} bound to the amino acid groups influenced C-1' and C-2'. Addition of Gd^{3+} to compound 6 (Fig. 5-6a) gave a more complicated NMR pattern. Gd^{3+} interacted strongly with the C-terminal carboxyl group and the N-terminal amino group and weakly with C-2', C-1', and possibly C-5' of the galactosyl residue.

Thus, Gd^{3+} had a much greater affinity for the carboxyl and amino groups of the amino acid residues (Section 4.2) than for the sugar residues. Even when the amino acid function groups were blocked, the acetyl and ester blocking groups aided donors in the sugar residue in binding Gd^{3+}. To help distance the amino acid ligands from those of the sugar, Dill *et al.* (1983b) synthesized a hexapeptide, which contained three vicinyl α-D-galactopyranose residues. It was hoped that the presence of vicinyl sugar groups would increase sugar–Gd^{3+} interaction. However, the ^{13}C-NMR indicated that Gd^{3+} continued to interact with the blocked N-terminal and C-terminal amino acid residues.

The introduction of suitable acidic substituents into sugars greatly increases their affinity for lanthanides. Examples include glycoproteins which contain sialic acid and glycosaminoglycans. In the former category is glycophorin A, a 31,000-M.W. transmembrane component of erythro-cytes which binds Ca^{2+}. It comprises 60% carbohydrate, by weight, about half of which is N-acetylneuraminic acid (NeuAc). These sugar residues are thought to be responsible for binding Ca^{2+}. Daman and Dill (1983) established that Gd^{3+} also bound to α-NeuAc, occupying a position above the plane of the molecule and interacting with the carboxylate anion and the glycerol-1-yl side chain. In further studies (Daman and Dill, 1982), they examined the interaction of Gd^{3+} with native glycophorin and with its isolated alkali labile oligosaccharides. The tetrasaccharide:

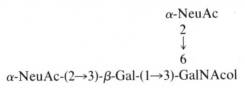

was shown to be important in Gd^{3+} binding.

NMR data suggested that glucuronate binds Yb^{3+} in a stoichiometry of at least two sugars per Yb^{3+} ion (Balt *et al.*, 1983), with the carboxyl groups forming the likely ligands. Izumi (1980), while agreeing that the carboxyl oxygens of β-D-glucuronate may be the sole ligands for Eu^{3+}, Pr^{3+}, and Nd^{3+}, assumed a 1:1 stoichiometry. Yb^{3+}-induced shifts of

β-D-glucose 6-sulfate were so small that it could not even be determined whether a complex was indeed formed (Balt *et al.*, 1983). These differences reveal that carboxyl oxygens are much stronger ligands for Ln^{3+} than the oxygen atoms on organic sulfates, as discussed in Section 2.3. The major anionic sites in cartilage are provided by glycosaminoglycans, the most common of which are chondroitin 4-sulfate and chondroitin 6-sulfate. Their ability to bind Ca^{2+} is important for a number of physiological processes, including calcification. Carbon-13 NMR spectral analysis of the interaction of Yb^{3+} with chondroitin 4-sulfate provided the structure shown in Fig. 5-7 (Balt *et al.*, 1983). The Yb^{3+}–oxygen minimal distances for chondroitin sulfate and chondroitin and the Ca^{2+}–oxygen minimal distances for chondroitin 4-sulfate are given in Table 5-10. Although the structure shown in Fig. 5-7 implies liganding by sulfate oxygen as well as carbonyl oxygen, the authors state that the sulfate group may be only accidentally directed to Yb^{3+}. Yb^{3+} coordinated to chondroitin (Table 5-10) in an identical way, with the carboxyl groups providing the most important ligands. Strong solutions of $LaCl_3$ have been used to extract proteoglycans from cartilage and hyaluronic acid from skin (Section 3.8.1).

5.5 Porphyrins

A number of porphyrin derivatives have been synthesized in which the Fe^{3+} ion has been replaced by a Ln^{3+} ion. The aim has been to insert this substituted prosthetic group back into heme proteins as internal NMR shift reagents. A series of Ln^{3+}-tetraarylporphine complexes has been synthesized and shown to produce the expected large shifts of ^1H resonances (Horrocks and Wong, 1976), thus providing structural information. Measurement of the dipolar shift ratios suggested that Eu^{3+} and Yb^{3+} lay 1.8 and 1.6 Å, respectively, out of the plane of the porphyrin ring. Complexes of Ln^{3+} ions and tetraarylporphine were soluble in organic solvents and could be used to shift NMR resonances in various organic molecules. Water-soluble lanthanide complexes with tetra-*p*-sulfonatophenylporphine were also synthesized (Horrocks and Hove, 1978).

Horrocks *et al.* (1975) were able to insert Yb^{3+}-mesoporphyrin IX into apomyoglobin, forming a 1:1 complex. Magnetic circular dichroism spectroscopy (Section 3.4.4) confirmed that Yb^{3+}-mesoporphyrin IX occupied the same position in the myoglobin as hemin normally would. In a similar fashion, Lu^{3+}- and Gd^{3+}-mesoporphyrin IX have been shown to insert into apomyoglobin (Srivastava, 1980). Hemin displaced the Ln^{3+}-substituted prosthetic group, forming metmyoglobin, again suggesting that

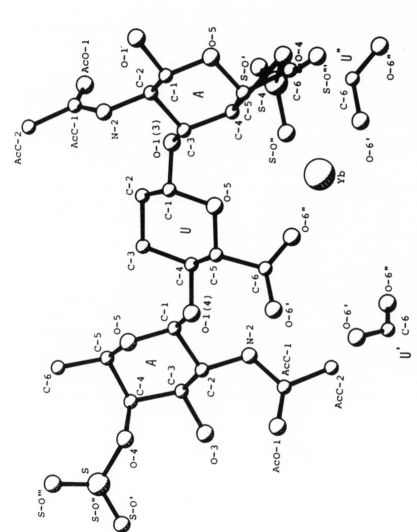

Figure 5-7. Structure of Yb^{3+}-chondroitin 4-sulfate in solution, from ^{13}C-NMR data. U, glucuronate residue; A, 2-amino-2-deoxygalactose-4-sulfate residue; U′, COO$^-$ group of a glucuronate residue of an antiparallel chain in the same unit cell; U″, COO$^-$ group of a glucuronate residue in an adjacent unit cell. From Balt *et al.* (1983), with permission.

Table 5-10. Yb^{3+}–Oxygen Minimal Distances (pm) for
Ytterbium Ch(S) Compared with Ca^{2+}–Oxygen
Distances in Calcium Ch-4S[a,b]

Donor atom	CaCh-4S	YbCh-4S	YbCh-6S	YbCh
O-6"	264	247	257	253
O-6'	290	253	253	271
S-O	255	190	395[c]	—

[a] From Balt et al. (1983), with permission.
[b] Ch-4S = chondroitin 4-sulfate; Ch-6S = chondroitin 6-sulfate; Ch = chondroitin.
[c] Value is too large for coordination.

the two groups occupied the same site. Furthermore, both hemin and Lu^{3+}-mesoporphyrin IX quenched globin fluorescence in the same manner. Further progress in NMR studies of these substances has been impeded by problems of denaturation of the Ln^{3+}-substituted myoglobins at higher concentrations (Horrocks, 1982).

5.6 Other Substances

Bratt and Hogenkamp (1982), having previously assigned the ^{13}C resonances of cyanocobalamin (vitamin B_{12}), examined the influence of Gd^{3+} upon its ^{13}C- and ^{31}P-NMR spectra. Gd^{3+} was shown to bind to the phosphate moiety, broadening the ^{31}P resonance, as well as the ^{13}C resonances of the ribose ring. However, it did not affect the ^{13}C resonances of 5,6-dimethylbenzimidazole or the corrin ring. When a monocarboxylic acid substituent was introduced into the molecule, the effect of Gd^{3+} was quite different, indicating additional association of Gd^{3+} with the carboxyl substituents.

The ionophore A23187 is an antibiotic which has been widely used in studies of Ca^{2+} metabolism. It shows a marked specificity for divalent cations in the approximate order $Mn^{2+} > Ca^{2+} > Mg^{2+} > Sr^{2+} > Ba^{2+}$. From differential spectral titration, La^{3+} appeared to form two types of complex with A23187 (Pfeiffer et al., 1974). The first equivalence point, at a La^{3+}/A23187 ratio of 0.66, could be explained by the formation of a 2:3 complex, or a mixture of 1:1 and 2:1 complexes, or some higher-order aggregates. La^{3+} also affected the fluorescence of A23187. At a concentration corresponding to the equivalence point noted by spectral titration, La^{3+} had no effect on the position of the emission peak but lowered the fluorescent emission by 58%. At higher concentration, La^{3+}

shifted the emission peak to a higher wavelength, with only a small further decrease in relative intensity. Further studies by Shastri *et al.* (1987) showed that A23187 transported Ln^{3+} into phospholipid vesicles. The 2:1 complex was identified as the transporting species. Intravesicular La^{3+} reduced the rate of transport of Pr^{3+}, Nd^{3+}, and Eu^{3+}.

Lasalocid A (X537A), an antibiotic from *Streptomyces lasaliensis*, transports metal cations and biogenic amines across membranes. It can transport Pr^{3+} across phospholipid bilayers (Section 6.2.1). Pr^{3+} was found to quench the fluorescence of methanolic solutions of X537A (Chen and Springer, 1978) and to alter its CD spectrum, without alteration in the absorption bands of Pr^{3+} (Alpha and Brady, 1973; Chen and Springer, 1978). Shifts in the ^{1}H- and ^{13}C-NMR spectra were attributable to binding of Pr^{3+} to the salicylate moiety of the ionophore, with the binding being rapid on the NMR time scale (Chen and Springer, 1978). By titrating the shifts with Pr^{3+}, the authors suggested the formation of Pr $(X537A)^{2+}$, $Pr(X537A)_2^+$, and $Pr(X537A)_3$ complexes with stepwise formation constants $K_1 = 10^7$, $K_2 = 10^6$, and $K_3 = 10^5 M^{-1}$. From the linewidth data, the authors calculated the average lifetime of X537A and Pr^{3+} binding to be $3.9 \mu s$. According to Shastri *et al.* (1987), it is the 2:1 complex which transports Ln^{3+} across membranes of phospholipid vesicles.

Richardson and Gupta (1981) used electronic absorption and emission techniques to study the interaction of a variety of lanthanides with X537A in methanol. They observed a nonlinear increase in the Nd^{3+} absorption band which leveled off at a X537A:Nd^{3+} ratio of 2, suggesting the formation of 1:1 and 2:1 complexes. Studies of Tb^{3+} luminescence enhancement by X537A revealed circularly polarized emissions, implying the interaction of Tb^{3+} with at least three donor groups. Several lanthanides quenched the intrinsic fluorescence of X537A. The authors suggested that Tb^{3+} has a coordination number of 6 upon binding the first X537A molecule, while the second X537A binds as a bidentate ligand through its salicylate residue. Binding of Tb^{3+} to salicylate alone showed different fluorescent characteristics, suggesting that the interaction of Tb^{3+} with X537A cannot be simply explained by attachment to the salicylate residue alone. Hanna *et al.* (1983) showed by NMR spectroscopy that in dimethylformamide or chloroform solution, Gd^{3+} bound solely to the carboxylate group of X537A.

Lanthanides have also been shown to bind to high-density lipoproteins (Assmann *et al.*, 1974), bilirubin (Velapoldi and Menis, 1971), melanin (Sarna *et al.*, 1976), and amphetamines (Smith *et al.*, 1976). The interactions of the lanthanides with phospholipids are discussed in Section 6.2.1.

5.7 Summary

Lanthanides interact with a variety of biological substances, especially if they provide carboxyl or phosphate groups as ligands. However, weak binding to uncharged compounds also occurs through coordination to carbonyl or hydroxyl oxygens. Spectroscopic evidence suggests that lanthanides interact with certain sugars and nucleosides in this way.

Purine and pyrimidine nucleotides have much higher affinities for Ln^{3+} ions, with the K_d values for nucleotide triphosphates (NTP) being below $1\,\mu M$. Affinity increases as the size of the Ln^{3+} ions decreases. Tb^{3+} luminescence is considerably enhanced upon binding to xanthine or guanine nucleotides. Other nucleotides promote Tb^{3+} luminescence either weakly or not at all. These differences are not attributable to different affinities for Tb^{3+} but reflect the nature of the substituents in the purine or pyrimidine ring systems. For given base, the ability to enhance Tb^{3+} luminescence increases in the order NMP < NDP < NTP < polynucleotide.

NMR spectroscopy confirms coordination of Ln^{3+} ions to two phosphate oxygen atoms on nucleotide monophosphates. With NTP, the Ln^{3+} ion coordinates with the α and β phosphoryl groups. At pH values of 1–2, NTP form 1:1 complexes with Ln^{3+} ions. At neutral pH values and above, 1:2 and possibly 1:3 Ln^{3+}-ATP complexes are formed. Conformational information can be gained from the quantitative analysis of alterations in the NMR spectra of nucleotides in response to paramagnetic Ln^{3+} ions.

Nucleic acids have high affinities for lanthanides, with K_d values as low as $10^{-8}\,M$ in some cases. Whereas single-stranded (ss) molecules are potent enhancers of Tb^{3+} luminescence, double-stranded (ds) polynucleotides have, at best, only weak activity. The ability of Tb^{3+} luminescence to detect selectively regions of single-strandedness is of great experimental value. It can be used to monitor the melting and annealing of DNA and to detect subtle conformational changes. This difference between ds and ss polynucleotides is thought to reflect difference in the way in which Ln^{3+} ions coordinate with each form. Whereas Ln^{3+} ions can only coordinate with the phosphate groups of ds molecules, they gain access to the bases of ss polynucleotides, coordinating with N-7 and the C-6 carbonyl oxygen atoms of guanine and xanthine. In the latter configuration, energy transfer from the purine bases to Tb^{3+} is facilitated.

Lanthanides bind to DNA until a Ln^{3+}:DNA phosphate ratio of 1:3 is reached, at which point the DNA precipitates. As histones mask some of these binding sites, Tb^{3+} luminescence can be used to estimate the amount of naked DNA in chromatin. Lanthanides induce a number of conformational changes in DNA, including the adaptation of the Z configuration. They also lower the melting temperature of DNA.

Lanthanides have a particularly high affinity for tRNA, where they occupy four or five high-affinity Mg^{2+} sites. Studies of the interactions between Ln^{3+} ions and tRNA molecules have been aided by the presence of an unusual base close to certain of their binding sites. In yeast tRNA, this is the Y base; in *E. coli* tRNA, it is 4-thiouracil. Selective excitation at the absorption maxima of these bases allows selective luminescence enhancement of the adjacent Eu^{3+} or Tb^{3+} ion. While low concentrations of Eu^{3+} substitute for Mg^{2+} in thermally stabilizing tRNA, high concentrations of Eu^{3+} not only destabilize the molecule, but also catalyze its hydrolysis. DNA is not hydrolyzed under these conditions.

Although uncharged sugars have weak affinities for Ln^{3+} ions, those with carboxyl substituents bind lanthanides more strongly. Sugar phosphates also sequester lanthanides, but the addition of a sulfate group does not greatly increase the molecule's affinity for Ln^{3+} ions.

Other molecules of biochemical interest which interact with lanthanides include porphyrins, vitamin B_{12}, the Ca^{2+} ionophores A23187 and X537A, bilirubin, high-density lipoproteins, melanin, and amphetamines. The interaction of lanthanides with phospholipids is discussed in Section 6.2.1.

References

Alpha, S. R., and Brady, A. H., 1973. Optical activity and conformation of the cation carrier X537A, *J. Am. Chem. Soc.* 95:7043–7049.

Angyal, S. J., 1972. Complexes of carbohydrates with metal cations. I. Determination of the extent of complexing by NMR spectroscopy, *Aust. J. Chem.* 25:1957–1966.

Anthonsen, T., Larsen, B., and Smidsrod, O., 1972. NMR-studies of the interaction of metal ions with poly(1,4-hexuronates). I. Chelation of europium ions by D-galacturonic acid, *Acta Chem. Scand.* 26:2988–2989.

Anthonsen, T., Larsen, B., and Smidsrod, O., 1973. NMR-studies of the interaction of metal ions with poly(1,4-hexuronates). II. The binding of europium ions to sodium methyl α-D-galactopyranosiduronate, *Acta Chem. Scand.* 27:2671–2673.

Arquilla, M., Thompson, L. M., Pearlman, L. F., and Simpkins, H., 1983. Effect of platinum antitumor agents on DNA and RNA investigated by terbium fluorescence, *Cancer Res.* 43:1211–1216.

Assmann, G., Sokoloski, E. A., and Brewer, H. B., 1974. ^{31}P nuclear magnetic resonance spectroscopy of native and recombined lipoproteins, *Proc. Natl. Acad. Sci. USA* 71: 549–553.

Balt, S., DeBolster, M. W., and Visser-Luirink, G., 1983. A ^{13}C-n.m.r. study of the binding of ytterbium (III) to chondroitin sulphate and chondroitin, *Carbohydr. Res.* 121:1–11.

Barela, T. D., Burchett, S., and Kizer, D. E., 1975. Terbium binding to ribosomes and ribosomal RNA, *Biochemistry* 14:4887–4892.

Barry, C. D., North, A. C. T., Glasel, J. A., Williams, R. J. P., and Xavier, A. V., 1971. Quantitative determinations of mononucleotide conformations in solution using lanthanide ion shift and broadening NMR probes, *Nature* 232:236–245.

Barry, C. D., Glasel, J. A., North, A. C. T., Williams, R. J. P., and Xavier, A. V., 1972. The quantitative conformations of some dinucleoside phosphates in solution, *Biochim. Biophys. Acta* 262:101–107.

Barry, C. D., Glasel, J. A., Williams, R. J. P., and Xavier, A. V., 1974a. Quantitative determination of conformations of flexible molecules in solution using lanthanide ions as nuclear magnetic resonance probes: application to adenosine-5'-monophosphate, *J. Mol. Biol.* 84:471–490.

Barry, C. D., Martin, D. R., Williams, R. J. P., and Xavier, A. V., 1974b. Quantitative determination of the conformation of cyclic 3',5'-adenosine monophosphate in solution using lanthanide ions as nuclear magnetic resonance probes, *J. Mol. Biol.* 84:491–502.

Bayley, P., and Debenham, P., 1974. The effect of lanthanide ions on the conformation of adenine mononucleotides and dinucleotides, *Eur. J. Biochem.* 43:561–568.

Beattie, J. K., and Kelso, M. T., 1981. Equilibrium and dynamics of the binding of calcium ion to sorbitol (D-glucitol), *Aust. J. Chem.* 34:2563–2568.

Birdsall, B., Birdsall, N. J. M., Feeney, J., and Thornton, J., 1975. A nuclear magnetic resonance investigation of the conformation of nicotinamide mononucleotide in aqueous solution, *J. Am. Chem. Soc.* 97:2845–2850.

Bratt, G. T., and Hogenkamp, H. P., 1982. The interaction of cyanocolbamin and some of its analogs with manganese(II) and gadolinium(III), *Arch. Biochem. Biophys.* 218:225–232.

Burns, V. W., 1985. Heavy-atom effects on energy transfer from polynucleotides to terbium(III), *Biopolymers* 24:1293–1300.

Chargaff, E., Vischer, E., Doninger, R., Green, C., and Misani, F., 1949. The composition of the desoxypentose nucleic acids of thymus and spleen, *J. Biol. Chem.* 177:405–416.

Chen, S. T., and Springer, C. S., 1978. Interaction of antibiotic lasalocid A (X537A) with praseodymium(III) in methanol, *Bioinorg. Chem.* 9:101–122.

Daman, M. E., and Dill, K., 1982. ^{13}C-n.m.r.-spectral study of the binding of Gd^{3+} to glycophorin, *Carbohydr. Res.* 111:205–214.

Davis, S. A., and Richardson, F. S., 1980. Circularly polarized luminescence induced by terbium–nucleoside interactions in aqueous solution, *J. Inorg. Nucl. Chem.* 42:1793–1795.

Dill, K., Daman, M. E., Batstone-Cunningham, R. L., Lacombe, J. M., and Pavia, A. A., 1983a. ^{13}C-n.m.r.-spectral study of the mode of binding of Gd^{3+} to various glycopeptides, *Carbohydr. Res.* 123:123–135.

Dill, K., Daman, M. E., Batstone-Cunningham, R. L., Denarie, M., and Pavia, A. A., 1983b. ^{13}C-n.m.r.-spectral study of the mode of binding of Gd^{3+} and Mn^{2+} to a tri-O-D-galactosylated hexapeptide, *Carbohydr. Res.* 123:137–144.

Dobson, C. M., Geraldes, C. F. G. C., Ratcliffe, G., and Williams, R. J. P., 1978. Nuclear-magnetic-resonance studies of 5'-ribonucleotide and 5'-deoxyribonucleotide conformation in solution using the lanthanide probe method, *Eur. J. Biochem.* 88:259–266.

Draper, D. E., 1985. On the coordination properties of Eu^{3+} bound to tRNA, *Biophys. Chem.* 21:91–101.

Eads, C. D., Mulqueen, P., Horrocks, W. D., and Villafranca, J. J., 1984. Characterization of ATP complexes with lanthanide(III) ions, *J. Biol. Chem.* 259:9379–9383.

Ellis, K. J., and Morrison, J. F., 1974. The interaction of europium(III) ion with nucleotides, *Biochim. Biophys. Acta* 362:201–208.

Fazakerley, G. V., and Reid, D. G., 1979. Determination of the interaction of ADP and dADP with copper(II), manganese(II) and lanthanide(III) ions by nuclear-magnetic-resonance spectroscopy, *Eur. J. Biochem.* 93:535–563.

Fazakerley, G. V., Russell, J. C., and Wolfe, M. A., 1977a. Determination of the *syn–anti* equilibrium of some purine 3′:5′-nucleotides by nuclear-magnetic relaxation perturbation in the presence of a lanthanide-ion probe, *Eur. J. Biochem.* 76:601–605.

Fazakerley, G. V., Linder, P. W., and Reid, D. G., 1977b. Determination of the solution conformation of dephospho coenzyme A by nuclear-magnetic-resonance spectroscopy with lanthanide probes. A method for analysis when more than one complex species is present, *Eur. J. Biochem.* 81:507–514.

Formoso, C., 1973. Fluorescence of nucleic acid–terbium(III) complexes, *Biochem. Biophys. Res. Commun.* 53:1084–1087.

Galea, J., Beccaria, R., Ferroni, G., and Belaich, J. P., 1978. Thermodynamic studies on formation of europium(III)–adenine nucleotide complexes, *Electrochim. Acta* 23:647–652.

Geraldes, C. F. G. C., 1979. Nuclear magnetic resonance study of the solution conformation of adenine mononucleotides using the lanthanide probe method, *J. Magn. Reson.* 36: 89–98.

Geraldes, C. F. G. C., and Williams, R. J. P., 1978. Nucleotide torsional flexibility in solution and the use of lanthanides as nuclear-magnetic-resonance conformational probes. The case of adenosine-5′-monophosphate, *Eur. J. Biochem.* 85:463–470.

Gersanovski, D., Colson, P., Houssier, C., and Fredericq, E., 1985. Terbium^{3+} as a probe of nucleic acids structure. Does it alter the DNA conformation in solution? *Biochim. Biophys. Acta* 824:313–323.

Gross, D. S., and Simpkins, H., 1981. Evidence for two-site binding in the terbium(III)–nucleic acid interaction, *J. Biol. Chem.* 256:9593–9598.

Gross, D. S., Rice, S. W., and Simpkins, H., 1981. Influence of inorganic cations and histone proteins on the terbium(III)–nucleic acid interaction, *Biochim. Biophys. Acta* 656:167–176.

Gross, D. S., Simpkins, H., Bubienko, E, and Borer, P. N., 1982. Proton magnetic resonance analysis of terbium ion–nucleic acid complexes: further evidence for two-site binding to polynucleotides, *Arch. Biochem. Biophys.* 219:401–410.

Gutman, M., and Levy, M. A., 1983. Fluorescence decay time measurements of Eu^{3+}–ATP–enzyme complexes. Replacement of the metal hydration water by active site ligands, *J. Biol. Chem.* 258:12132–12134.

Haertlé, T., Kretschmer, E., and Augustyniak, J., 1980. Tb^{3+} as a marker in studies of tRNA interactions, in *Biological Implications of Protein–Nucleic Acid Interactions* (J. Augustyniak, ed.), A. Mickiewicz University Press, Poznan, Poland, pp. 629–633.

Haertlé, T., Augustyniak, J., and Guschlbauer, W., 1981. Is Tb^{3+} fluorescence enhancement only due to binding to single strand polynucleotides? *Nucleic Acids Res.* 9:6191–6197.

Hanna, D. A., Yeh, C., Shaw, J., and Everett, G. W., 1983. Gadolinium(III) and manganese(II) binding by a polyether ionophore. Influence of cation charge and solvent polarity on the binding sites of lasalocid A (X-537A), *Biochemistry* 22:5619–5626.

Holbrook, S. R., Sussman, J. L., Warrant, R. W., Church, G. M., and Kim, S. H., 1977. RNA–ligand interactions: (1) magnesium binding sites in yeast tRNA^{Phe}, *Nucleic Acids Res.* 4:2811–2820.

Horrocks, W. DeW., 1982. Lanthanide ion probes of biomolecular structure, in *Advances in Inorganic Biochemistry* (G. L. Eichhorn and L. G. Marzilli, eds.), Vol. 4, Elsevier, New York, pp. 201–261.

Horrocks, D. DeW., and Hove, E. G., 1978. Water soluble lanthanide porphyrins: shift reagents for aqueous solutions, *J. Am. Chem. Soc.* 100:4386–4392.

Horrocks, W. DeW., and Wong, C. P., 1976. Lanthanide porphyrin complexes. Evaluation of nuclear magnetic resonance dipolar probe and shift reagent capabilities, *J. Am. Chem. Soc.* 98:7157–7162.

Horrocks, W. DeW., Venteicher, R. F., Spilburg, C. A., and Vallee, B. L., 1975. Lanthanide porphyrin probes of heme proteins. Insertion of ytterbium (III) mesoporphyrin IX into apomyoglobin, *Biochem. Biophys. Res. Commun.* 64:317–322.

Houssier, C., Maquet, M. N., and Fredericq, E., 1983. Denaturation level of DNA–Pt complexes evidenced by Tb^{3+} fluorescence enhancement and electric dichroism, *Biochim. Biophys. Acta* 739:312–316.

Inagaki, F., Tasumi, M., and Miyazawa, T., 1978. Structures and populations of conformers of nucleoside monophosphates in aqueous solution. I. General methods of conformational search with lanthanide-ion probes and spin-coupling constants and application to uridine-5'-monophosphate, *Biopolymers* 17:267–289.

Izatt, R. M., Christensen, J. J., and Rytting, J. H., 1971. Sites and thermodynamic quantities associated with proton and metal ion interactions with ribonucleic acid, deoxyribonucleic acid and their constituent bases, nucleosides and nucleotides, *Chem. Rev.* 71: 439–481.

Izumi, K., 1980. Carbon-13 NMR spectra of sodium D-gluco- and D-galactopyranuronates in the presence of lanthanide ions, *Agric. Biol. Chem.* 44:1623–1631.

Jack, A., Ladner, J. E., Rhodes, D., Brown, R. S., and Klug, A., 1977. A crystallographic study of metal-binding to yeast phenylalanine transfer RNA, *J. Mol. Biol.* 111:315–328.

Jones, C. R., and Kearns, D. R., 1974. Investigation of the structure of yeast tRNA[Phe] by nuclear magnetic resonance: paramagnetic rare earth ion probes of structure, *Proc. Natl. Acad. Sci. USA* 71:4237–4240.

Kayne, M. S., and Cohn, M., 1972. Cation requirements of isoleucyl-RNA synthetase from *Escherichia coli, Biochem. Biophys. Res. Commun.* 46:1285–1291.

Kayne, M. S., and Cohn, M., 1974. Enhancement of Tb(III) and Eu(III) fluorescence in complexes with *Escherichia coli* tRNA, *Biochemistry* 13:4159–4165.

Kearns, D. R., and Bolton, P. H., 1978. Proton probes of the tertiary structure of transfer RNA molecules, in *Biomolecular Structure and Function* (B. Pullman, ed.), Academic Press, New York, pp. 493–516.

Kieboom, A. P. G., Sinnema, A., Van der Toorn, J. M., and Van Bekkum, H., 1977. ^{13}C NMR study of the complex formation of sorbitol (glucitol) with multivalent cations in aqueous solution using lanthanide(III) nitrates as shift reagents, *Recl. Trav. Chim. Pays-Bas* 96:35–37.

Lavallee, D. K., and Zeltmann, A. H., 1974. Conformation of cyclic β-adenosine 3',5'-phosphate in solution using the lanthanide shift technique. *J. Am. Chem. Soc.* 96: 5552–5556.

Lee, C. Y, and Raszka, M. J., 1975. Determination of solution structure of diphosphopyridine coenzymes with paramagnetic shift and broadening reagents, *J. Magn. Reson.* 17: 151–160.

Martin, R. B., 1983. Structural chemistry of calcium: lanthanides as probes, in *Calcium in Biology* (T. G. Spiro, ed.), Wiley, New York, pp. 237–270.

Morley, P. J., Martin, R. B., and Boatman, S., 1981. Characterization of excitation spectra for Tb^{3+} luminescence from nucleic acids: calcium binding environs in icosahedral viruses, *Biochem. Biophys. Res. Commun.* 101:1123–1130.

Morrison, J. F., and Cleland, W. W., 1983. Lanthanide–adenosine 5'-triphosphate complexes: determination of their dissociation constants and mechanism of action as inhibitors of yeast hexokinase, *Biochemistry* 22:5507–5513.

Pavlick, D., and Formoso, C., 1978. Lanthanide fluorescence studies of transfer RNA_f^{Met} conformation, *Biochemistry* 17:1537–1540.

Pfeiffer, D. R., Reed, P. W., and Lardy, H. A., 1974. Ultraviolet and fluorescent spectral properties of the divalent cation ionophore A23187 and its metal ion complexes, *Biochemistry* 13:4007–4014.

Richardson, F. S., and Gupta, A. D., 1981. Spectroscopic studies on the interaction of the antibiotic lasalocid A (X537A) with lanthanide(III) ions in methanol, *J. Am. Chem. Soc.* 103:5716–5725.

Ringer, D. P., Burchett, S., and Kizer, D. E., 1978. Use of Tb(III) fluorescence enhancement to selectively monitor DNA and RNA guanine residues and their alteration by chemical modification, *Biochemistry* 17:4818–4824.

Ringer, D. P., Howell, B. A., and Kizer, D. E., 1980. Use of terbium fluorescence enhancement as a new probe for assessing the single-strand content of DNA, *Anal. Biochem.* 103:337–342.

Ringer, D. P., Etheredge, J. L., and Kizer, D. E., 1985. The influence of DNA sequence on terbium(III) fluorescence enhancement by DNA, *J. Inorg. Biochem.* 24:137–145.

Robertus, J. D., Ladner, J. E., Finch, J. T., Rhodes, D., Brown, R. S., Clark, B. F. C., and Klug, A., 1974. Structure of yeast phenylalanine tRNA at 3 Å resolution, *Nature* 250: 546–551.

Robins, M. J., MacCoss, M., and Wilson, J. S., 1977. Nucleic acid related compounds 27. "Virtual coupling" of the anomeric proton of cyclic 2'-deoxynucleoside 3',5'-monophosphates. Reassessment of conformation using praseodymium shifts and assignment of H-2', 2'' signals by biomimetic deuteration at ϵ-2', *J. Am. Chem. Soc.* 99:4660–4666.

Rordorf, B. F., and Kearns, D. R., 1976. Effect of europium(III) on the thermal denaturation and cleavage of transfer ribonucleic acids, *Biopolymers* 15:1491–1504.

Sarna, T., Hyde, J. S., and Swartz, H. M., 1976. Ion-exchange in melanin: an electron spin resonance study with lanthanide probes, *Science* 192:1132–134.

Shastri, B. P., Sankaram, M. B., and Easwaran, K. R., 1987. Carboxylic ionophore (lasalocid A and A23187)-mediated lanthanide ion transport across phospholipid vesicles, *Biochemistry* 26:4925–4930.

Shimizu, T., Mims, W. B., Peisach, J., and Davis, J. L., 1979. Analysis of the electron spin echo decay envelope for Nd^{3+}:ATP complexes, *J. Chem. Phys.* 76:2249–2254.

Shimizu, T., Mims, W. B., Davis, J. L., and Peisach, J., 1983. Studies of the coordination of rare earths and transition metal nucleotide complexes by an electron spin echo method, *Biochim. Biophys. Acta* 757:29–39.

Simpkins, H., and Pearlman, L. F., 1984. The binding of actinomycin D and adriamycin to supercoiled DNA, single-stranded DNA and polynucleotides, *Biochim. Biophys. Acta* 783:293–300.

Simpkins, H., Pearlman, L. F., and Thompson, L. M., 1984. Effects of adriamycin on supercoiled DNA and calf thymus nucleosomes studied with fluorescent probes, *Cancer Res.* 44:613–618.

Smith, R. V., Erhardt, P. W., Rusterholz, D. B., and Barfknecht, C. F., 1976. NMR study of amphetamines using europium shift reagents, *J. Pharm. Sci.* 65:412–417.

Srivastava, T. S., 1980. Gadolinium(III) myoglobin: interaction of gadolinium(III) mesoporphyrin IX with apomyoglobin, *Curr. Sci.* 49:429–430.

Stokke, T., and Steen, H. B., 1985. Neither adriamycin nor actinomycin D displaces Tb^{3+} from DNA, *Biochim. Biophys. Acta* 825:416–418.

Stout, C. D., Mizuno, H., Rao, S. T., Swaminathan, P., Rubin, J., Brenan, T., and Sunderalingham, M., 1978. Crystal and molecular structure of yeast phenylalanine transfer RNA. Structure determination, difference Fourier refinement, molecular conformation, metal and solvent binding, *Acta Crystallogr.* B34:1529–1544.

Suddarth, F. L., Quigley, G. J., McPherson, A., Sneden, D., Kim, J. J., Kim, S. H., and Rich, A., 1974. Three-dimensional structure of yeast phenylalanine transfer RNA at 3.0 Å resolution, *Nature* 248:20–24.

Tanswell, P., Thornton, J. M., Korda, A. V., and Williams, R. J. P., 1975. Quantitative determination of the conformation of ATP in aqueous solution using the lanthanide cations as nuclear-magnetic-resonance probes, *Eur. J. Biochem.* 57:135–145.

Thompson, L. M., Arquilla, M., and Simpkins, H., 1982. The interaction of platinum complexes with nucleosomes investigated with fluorescent probes, *Biochim. Biophys. Acta* 698:173–182.

Topal, M. D., and Fresco, J. R., 1980. Fluorescence of terbium ion–nucleic acid complexes: a sensitive specific probe for unpaved residues in nucleic acids, *Biochemistry* 19:5531–5537.

Velapoldi, R. A., and Menis, O., 1971. Formation and stabilities of free bilirubin complexes with transition and rare-earth elements, *Clin. Chem.* 17:1165–1170.

Wolfson, J. M, and Kearns, D. R., 1975. Europium as a fluorescent probe of transfer RNA, *Biochemistry* 14:1436–1444.

Yokoyama, S., Inagaki, F., and Miyazawa, T., 1981. Advanced nuclear magnetic resonance lanthanide probe analyses of short-range conformational interrelations controlling ribonucleic acid structures, *Biochemistry* 20:2981–2988.

Yonuschot, G., and Mushrush, G. W., 1975. Terbium as a fluorescent probe for DNA and chromatin, *Biochemistry* 14:1677–1678.

Yonuschot, G., Robey, G., Mushrush, G. W., Helman, D., and Van de Woude, G., 1978. Measurement of binding of terbium to DNA, *Bioinorg. Chem.* 8:397–404.

Interactions of Lanthanides with Tissues, Cells, and Cellular Organelles

6.1 Introduction

Over 75 years ago, Mines (1910, 1911) described the inhibitory effect of lanthanide ions on the beating of perfused frogs' hearts. Shortly thereafter, similar responses were noted for skeletal muscle (Hober and Spaeth, 1914). However, the modern experimental use of lanthanides in cellular biochemistry and physiology originates with the paper of Lettvin *et al.* (1964). Drawing attention to the similar ionic radii of Ca^{2+} and La^{3+}, these authors suggested that La^{3+} ions should occupy Ca^{2+}-binding sites on nerve axons but, being of higher valency, should bind much more strongly than Ca^{2+}. This observation anticipated the more detailed comparisons of Ca^{2+} and Ln^{3+} ions by bioinorganic chemists (Section 2.5). Lettvin *et al.* (1964) proposed that La^{3+} ions would function as nerve-blocking agents. The subsequent confirmation of this prediction (Takata *et al.*, 1966) led the way to the widespread use of Ln^{3+} ions in studies of cellular Ca^{2+} fluxes, especially with regard to their role in excitation and other stimulus-coupled responses. Much of the development of this area of lanthanide research stems from the investigations of Van Breeman, Weiss, and their colleagues on muscle cells. Reviews of various aspects of this field have been written by Weiss (1974), Mikkelsen (1976), and Dos Remedios (1981).

In this chapter, we shall examine the effects of lanthanides on cells, tissues, organs, and organelles under *in vitro* conditions; their effects *in vivo* are discussed in Chapter 8. According to present thinking, the physiological properties of the lanthanides can be largely explained on the basis of their attachment to the outside of the cell membrane, with resulting disturbances in the cellular transport of metal ions. In this perspective, this chapter describes sequentially the interactions of lanthanides with the cell surface, the resulting effects on transmembrane ion fluxes, and the consequences for cellular physiology.

6.2 Membrane Interactions of Lanthanide Ions

6.2.1 Binding of Lanthanides to Artificial Phospholipid Membranes

As most of the evidence suggests that the primary site of interaction of Ln^{3+} ions with living cells is at the external surface, the binding of lanthanides to membranes merits close attention. As biological membranes are so complex, artificial membranes have been widely used to study the interaction between phospholipids, the major components of cellular membranes, and Ln^{3+} ions. The first example of such investigations is that of Rojas *et al.* (1966), who demonstrated the rapid adsorption of $^{147}Pm^{3+}$ on phosphatidylserine monolayers. La^{3+} ions and Ca^{2+} ions competed with Pm^{3+} for binding sites. By masking negative charges on membranes comprised of cholesterol and phospholipid, La^{3+} ions reversed their polarity from cation permselectivity to anion permselectivity (Van Breeman, 1968). From the consequent changes in surface potential, Barton (1968) determined relative binding strengths of $Ce^{3+} > La^{3+} \gg Ca^{2+}$. Competition was observed between Ca^{2+} and Ln^{3+} ions.

Most studies of the interactions between lanthanides and artificial membranes have employed ^{1}H- or ^{31}P-NMR spectroscopy. As the major phospholipid in biological membranes is phosphatidylcholine (lecithin), bilayer vesicles of this compound have usually been selected as model membranes. Although Ln^{3+} ions form 1:1 complexes with the small molecules that make up the phosphatidylcholine polar group, they form 2:1 complexes with vesicular phosphatidylcholine. This stoichiometry reflects the chelation of one Ln^{3+} ion by two phosphate groups (Hauser *et al.*, 1977). Phosphate groups are the only ligands for lanthanides on these membranes (Hauser *et al.*, 1975).

As shown in Table 6-1, estimates of the binding constants of various lanthanides vary by over two orders of magnitude. One reason for this is the experimental difficulty of determining these values. Nearly all calculations require various assumptions which alter the final value. Cooperativity is suggested by the way in which the apparent affinities of various lanthanides for phosphatidylcholine vesicles depend upon the quantity of bound Ln^{3+} ions (Hauser *et al.*, 1977). Chrzeszczyk *et al.* (1981) have suggested the existence of interconvertible high-affinity, "relaxed" (R) sites and low-affinity, "tense" (T) sites. Although the R/T ratio is normally

Table 6-1. Affinities of Lanthanide Ions for Phosphatidylcholine
Bilayer Membranes

Lanthanide ion	Approximate K_a (M^{-1})	Comment	Reference
La	120		Akutsu and Seelig, 1981
Tb	565	Values obtained by	Grasdalen $et\,al.$, 1977
	715	different methods	
Various	30		Hauser $et\,al.$, 1975
			Hauser, 1976
Nd	120–1000	Calculated range,	Levine $et\,al.$, 1973
Eu	284–9400	given different	
		theoretical	
		concentrations of	
		binding sites	
Pr	2	Low-affinity, T-site	Chrzeszczyk $et\,al.$,
	3000	High-affinity, R-site	1981

0.14 at 52°C, addition of Pr^{3+} ions favors the conversion of T-sites to R-sites. Conformational changes (q.v.) induced by Ln^{3+} ions are thought to underlie this phenomenon.

Binding of lanthanides is independent of pH between pH 3 and pH 7 but is enhanced by various anions in the order $Cl^- < Br^- < NO^- < SCN^- < I^- < ClO_4^-$. The nature of the counterion has relatively little effect (Hauser $et\,al.$, 1977). According to Huang $et\,al.$ (1974), the molecular packing of the polar head groups is not flexible enough to permit every phosphate group to interact with a Ln^{3+} ion. They calculated that only 7% of the phosphate groups on the outer surface of the bilayer vesicles of phosphatidylcholine could serve as strong binding sites for Pr^{3+} ions. Mutual repulsion between adjacent Ln^{3+} ions could also contribute to this effect (Levine $et\,al.$, 1973). The greater binding of lanthanides as the salt concentration increases may result from steric changes. In particular, the polar head groups possess a more extended conformation, thus becoming more accessible to Ln^{3+} ions.

Bentz $et\,al.$ (1988) report a K_a of 10^5 M^{-1} for the binding of La^{3+} to phosphatidylserine liposomes at pH 7.4. They suggest that La^{3+} chelates phosphatidylserine at two sites, one of which is the primary amino group.

The affinity of Ca^{2+} for phosphatidylcholine has also been difficult to measure. Most authors now agree that the interaction is a weak one, with affinities 5 to 10 times less than those of Ln^{3+} ions (Grasdalen $et\,al.$, 1977; Akutsu and Seelig, 1981). However, Ca^{2+} and Eu^{3+} promote the association of prothrombin with phospholipid to similar degrees (Section 4.5.1).

As reflected in results of the binding studies described above, the attachment of Ln^{3+} ions to these artificial membranes alters their conformation (Brown and Seelig, 1977). This has been confirmed in a variety of ways. McIntosh (1980), for example, noted an increase of about 6 Å in the width of the bilayer after adding lanthanides. This observation complements other data detailing the conformation of the polar head groups in these membranes. Whereas the head groups normally lie parallel to the surface of the bilayer, in the presence of lanthanides they adopt a more extended configuration, becoming oriented perpendicularly to the surface (Hauser et al., 1976; Lichtenberg et al., 1979). Electron paramagnetic resonance (EPR) studies confirm an increase in the ordering of phospholipid molecules after adding lanthanide ions (Butler et al., 1970), which raises the transition temperature (Levine et al., 1973; Chowdhry et al., 1984). Increasing the rigidity of the polar head groups in this way alters the characteristics of the channels that the antibiotic alamethicin forms in bilayer membranes (Gogelein et al., 1981). Lanthanides also change the curvature of the vesicles (McIntosh, 1980). As certain membrane-associated enzymes are sensitive to this parameter, their response to lanthanides might reflect such a change. Phospholipase (Section 4.18) is a possible example.

Bystrov et al. (1971, 1972) made the valuable dicovery that paramagnetic lanthanides could be used in conjunction with NMR spectroscopy to distinguish between the inner and outer polar head groups in bilayer vesicles. This was quickly confirmed independently (e.g., Kostelnik and Castellano, 1972; Andrews et al., 1973; Huang et al., 1974) and has been put to very good experimental use. It has been particularly valuable in studies of the transport of metal ions across lipid bilayers by ionophores (e.g., Hunt et al., 1978; Donis et al., 1981; Grandjean and Laszlo, 1982; Bartsch et al., 1983; Deleers et al., 1983). Certain anesthetics displace Ln^{3+} ions from phospholipid bilayers (e.g., Fernandez and Cerbon, 1973).

Lanthanides promote the fusion of phospholipid vesicles (Blioch et al., 1968; Bentz et al., 1988). However, high concentrations of La^{3+} disrupt the fusion of unilamellar phosphatidylserine liposomes, possibly by altering the charge on the surface of the vesicles. Whereas the fusion of phosphatidylserine vesicles by La^{3+} occurs without leakage of intravesicular contents, fusion in response to Ca^{2+} is leaky. Much higher concentrations of Ca^{2+} than La^{3+} are required for fusion to occur (Bentz et al., 1988).

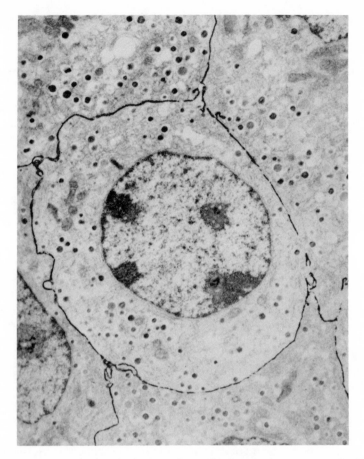

Figure 6-1. Electron micrograph of pancreatic β-cells after 60-min exposure to 2 mM La^{3+}. Magnification 600×. From Flatt *et al.* (1980a), with permission.

6.2.2 Binding of Lanthanides to Plasmalemmae

It is well established that lanthanides have a high affinity for the external surfaces of cells. Electron microscopy (EM), for example, provides visual evidence of this for Ln^{3+} ions and colloids (Section 3.7). An example is shown in Fig. 6-1. Biochemical analysis confirms the interaction of lanthanides with artificial phospholipid bilayers (Section 6.2.1), intact cells (q.v.), and a variety of compounds purified from plasmalemmae (Chapters 4 and 5). Furthermore, physiological studies have docu-

Table 6-2. Effects of Membrane Modifications on Lanthanide Binding

Type of cell	Treatment	Effect on binding[a]	Reference
Ehrlich ascites	Chloroform–methanol	0	Levinson et al.,
	Neuraminidase	0	1972a
	Pronase	—	
HeLa	Neuraminidase	—	Boyd et al., 1976
Cardiac myoblasts	Neuraminidase	—	Frank et al., 1977
Embryonic chick; various cell types	Trypsin, pronase, DNAase, α-amylase, neuraminidase	0	Lesseps, 1967
	Phospholipase C	—	

[a] 0: No effect on binding of Ln^{3+} ions to membrane.
—: Inhibition of binding of Ln^{3+} ions to membrane.

mented the relationship of such interactions to changes in cellular function (Section 6.4). However, given the common view that it is the interaction of Ln^{3+} ions with the external surface of the cell (Section 6.2.3) that determines their subsequent physiological effects, surprisingly little effort has been directed toward the rigorous biochemical characterization of the binding sites involved.

The major constituents of the plasmalemma are lipid, protein, and polysaccharide. Lanthanides have no affinity for neutral lipids but, as discussed in the last section, bind weakly to artificial phospholipid bilayers. Many proteins (Chapter 4) have much greater affinities than phospholipids for Ln^{3+} ions, while the carboxyl groups of sialic acid moieties of glycoproteins are also good ligands. On this simplified basis, one might expect cell surfaces to possess more than one class of Ln^{3+}-binding site, differing in affinity and binding capacity. This is indeed the case (q.v.), but the specific receptor molecules have not been well characterized.

One approach to the identification of surface receptors for Ln^{3+} ions has been to measure the effects of specific treatments on the abilities of membranes to sequester lanthanides. In most cases, this has involved the use of organic solvents to remove lipids, neuraminidase to remove sialic acid residues, or proteinases to remove proteins. The results (Table 6-2) do not permit general conclusions to be drawn. In view of the profound effects of lanthanides on transmembrane Ca^{2+} fluxes (Section 6.3), the outer surfaces of Ca^{2+} channels are likely to provide high-affinity binding sites for Ln^{3+} ions. El-Fakahany et al. (1983), in their electron microscopic and energy dispersive analysis of X rays (EDAX) studies of lanthanide binding to neuroblastoma cells, reported that La^{3+} ions associated with the membrane in randomly distributed clusters or "hot spots." The authors suggested that these confined areas of the cell membrane may be Ca^{2+} channels surrounding receptors. This view was supported by evidence that muscarine acetylcholine receptor agonists and, to a lesser degree, antagonists were able to displace La^{3+} ions from the cell membrane. There is *in vitro* evidence of the binding of Ln^{3+} ions to purified acetylcholine receptors (Section 4.25), insulin receptors (Section 4.25), and other membrane proteins such as adenylate cyclase (Section 4.25) and glycophorin (Section 5.4).

Luminescence studies (Section 3.3) have identified aromatic amino acids in the vicinity of Tb^{3+}-binding sites on a variety of cells. With transformed pituitary cells, Trp served as the energy donor (Canada, 1983), while Tyr was identified as the donor on lymphocytes (Mikkelsen, 1976), brush border membranes (Ohyashiki *et al.*, 1985), and erythrocyte ghosts (Mikkelsen and Wallach, 1974). However, while detecting the presence of aromatic amino acids within a few angstroms of the sequestered Tb^{3+} ions (Section 3.3), luminescence methods alone cannot determine the identity of the actual ligand.

In agreement with EM studies, quantitative biochemical data show that appreciable amounts of lanthanides can be sequestered on the surfaces of cells. Levinson *et al.* (1972a), for example, measured rapid binding of up to 6.6×10^5 La^{3+} ions per Ehrlich ascites tumor cell. Equivalent to 13×10^6 ions/μm^2 of cell surface, this figure was independent of cell number. It is very difficult to explain the finding of these authors that bound $^{140}La^{3+}$ could not be displaced by Ca^{2+}, Hg^{2+}, or nonradioactive La^{3+}. The authors rejected the possibility of transport and intracellular trapping of La^{3+} ions on the grounds that maximum La^{3+} binding occurred within 35 s and was independent of temperature. Whether these putative proteinaceous sites (Table 6-2) undergo a conformational shift upon binding La^{3+} which renders the sequestered ion resistant to displacement by Ca^{2+} or additional La^{3+} was not discussed. As the binding curve was sigmoidal, cooperativity or multiple binding sites may have been involved. HeLa cells only sequester lanthanides at a particular stage of the cell cycle, early in G_1 (Fiskin *et al.*, 1980). Binding of La^{3+} ions is increased by prednisone (Boyd *et al.*, 1976).

Most analyses have detected at least two classes of Ln^{3+}-binding site on the surfaces of intact cells (Table 6-3). The higher-affinity sites are generally less abundant and have K_d values in the micromolar range. Ca^{2+} ions usually compete for these sites, but have K_d values in the millimolar range. As expected for ionic interactions, the K_d values increase with ionic strength. Low-affinity sites have often been difficult to study kinetically and are poorly characterized.

Table 6-3. Characteristics of Lanthanide Binding Sites on Cell Surfaces

Cell or membrane	Binding sites	Reference
Pituitary	High affinity, K_d = 11 μM Low affinity, K_d = 560 μM K_d for Ca^{2+} = 28 mM	Canada, 1983
Intestinal	>1 class High affinity, K_d = 12.5 μM	Ohyashiki et al., 1985
Erythrocytes	>1 class based on Tb^{3+} luminescence	Mikkelsen and Wallach, 1974
Skeletal muscle	"Fast" and "slow" binding sites	Grinvald and Yaari, 1978
Synaptosomes	High affinity, K_d = 0.6 μM Low affinity, K_d = 27 μM	Tapia et al., 1985
Mouse synaptosomes	High affinity, K_d = 0.87 μM Low affinity, K_d = 74 μM	Madeira and Antunes- Madeira, 1973
Rat synaptosomes	K_d for La^{3+} = 0.2 μM K_d for Y^{3+} = 0.7 μM	Nachshen, 1984
Insect synaptosomes	Only 1 class, K_d = 4 μM	Breer and Jeserich, 1981
Axonal membranes	At least 2 classes (i) K_d = 2.2 μM (ii) K_d = 6.9 μM K_d for Ca^{2+} = 1.8 mM	Deschenes et al., 1981
Pancreatic islets	At least 2 classes for Tm^{3+}	Flatt et al., 1981

Several physicochemical changes accompany the binding of Ln^{3+} ions to cell surfaces. One is an increase in the membrane potential. In Ehrlich ascites tumor cells, the resting membrane potential increases from 8.3 ± 0.5 mV to 56 ± 5.1 mV after the addition of 1 mM La^{3+} (Smith et al., 1972). Ca^{2+} produces a smaller increase in membrane potential which decays with a half-life of 6 s. That produced by La^{3+} is stable for at least 10 min. The increased electropositivity of the membrane alters the electrophoretic behavior of Ehrlich ascites cells (Smith, 1976), erythrocytes, and platelets (Kosztolanyi et al., 1977). Smith (1976) calculated that at concentrations of La^{3+} above 0.5 mM, Ehrlich ascites cells assumed a net positive charge. This effect is also seen with phospholipid vesicles (Section 6.2.1).

A second effect of Ln^{3+} ions is to increase the specific membrane resistance. In Ehrlich ascites cells, the control resistance of 30.8 Ω/cm² increases to 76.9 Ω/cm² with 1 mM La^{3+} and 253.2 Ω/cm² with 20 mM Ca^{2+} (Smith et al., 1972). Similar responses occur with guinea pig papillary muscle (Haas, 1975). Finally, as with model membranes (Section 6.2.1), there is an increase in membrane rigidity. This has been shown for the membranes of *Bacillus subtilis* (Ehrstrohm et al., 1973), synaptosomes

(Uyesaka $et\,al.$, 1976), purified heart plasmalemmae (Gordon $et\,al.$, 1978), and intact platelet membranes (Sauerheber $et\,al.$, 1980). Uyesaka $et\,al.$ (1976) showed that Ca^{2+} could not mimic La^{3+} or Ce^{3+} ions in increasing the rigidity of synaptosomal membranes. They consequently concluded that whereas Ca^{2+} bound only to proteins in the membranes, Ln^{3+} ions bound both to these proteins and to the phospholipid bilayer. Lanthanides have a greater effect on the fluidity of the plasmalemmae of heart cells than on that of platelet membranes, possibly because the former have a higher content of phospholipid.

The interactions of lanthanides with the membranes of intracellular organelles are discussed in Section 6.5.

6.2.3 Impermeability of Membranes to Lanthanides

One of the perceived experimental advantages of using Ln^{3+} ions in studies of cellular metabolism is their apparent inability to penetrate the cellular membrane. This important assumption is worth considering further, as most conclusions drawn from such studies rest upon it. Furthermore, there are sporadic literature reports which contradict the notion of lanthanide impenetrability.

The direct evidence that lanthanide ions cannot enter the cytoplasm of living cells is provided by EM and by EDAX analysis of Ln^{3+}-treated cells. Lanthanides are electron-dense, heavy metals which show up well in the electron microscope. Countless electron micrographs, involving several different species and many different types of cell, have confirmed that La^{3+} ions bind to the external surface of the plasmalemma but are normally unable to penetrate into the cell. Indeed, lanthanum is now a standard marker for the extracellular compartment (Section 3.7). A non-exhaustive list of representative examples is given in Table 6-4, and a representative electron micrograph is shown in Fig. 6-1. Nevertheless, contradictory reports do exist, especially those reporting intracellular deposits of lanthanides in muscle cells (e.g., Weihe $et\,al.$, 1977; Lesson and Higgs, 1982; Dunbar, 1982). Much less use has been made of lanthanides other than La^{3+} in EM. However, Gd^{3+} stains only the periphery of skeletal nerve fibers (Hambly and Dos Remedios, 1977), while La^{3+}, Sm^{3+}, and Tm^{3+} are all entirely restricted to the plasma membrane of pancreatic β-cells (Flatt $et\,al.$, 1980b).

EM does, however, provide examples of the entry of lanthanide aggregates into cells by phagocytosis or pinocytosis (e.g., Briggs $et\,al.$, 1975; Tuchweber $et\,al.$, 1976; Squier and Rooney, 1976; Strum, 1977). In addition, there is indisputable EM evidence of intracellular penetration of

Table 6-4. Electron Microscopic Studies Confirming the
Impermeability of the Plasmalemma to Lanthanum

Type of cell or tissue	Reference
Myocardial cells perfused *in situ*	Frank and Rich, 1983
	Frank *et al.*, 1977
	Martinez-Palomo *et al.*, 1973
	Hatae, 1982
	Burton *et al.*, 1981
	Shine, 1973
	Bockman *et al.*, 1973
Myocardial cells *in vitro*	Langer and Frank, 1972
Fibroblasts	Langer and Frank, 1972
Skeletal muscle fibers	Henkart and Hagiwara, 1976
	Sperelakis *et al.*, 1973
	Kidokoro *et al.*, 1974
Neuroblastoma cells	El-Fakahany *et al.* 1983
Oral epithelial cells	Squier and Rooney, 1976
Brain	Brightman and Reese, 1969
Bladder epithelium	Strum, 1977
Endothelial cells	Frank *et al.*, 1977
Isolated pancreatic acinar cells	Wakasugi *et al.*, 1981
Fragments of pancreas	Chandler and Williams, 1974
Isolated pancreatic β-cells	Flatt *et al.*, 1980a
Chondrocytes	Morris and Appleton, 1984
Isolated adrenal medullas	Bresnahan *et al.*, 1980
Isolated adrenal cortical cells	Haksar *et al.*, 1976
Gall bladder	Machen *et al.*, 1972
Plant protoplasts	Taylor and Hall, 1979

lanthanides into dead or damaged cells. This has been shown, for example, with glutaraldehyde-damaged skeletal muscle (Hambly and Dos Remedios, 1977), the degenerating midgut cells of an insect (Humbert, 1979), structurally damaged barnacle muscle (Henkart and Hagiwara, 1976), ATP-depleted or stored red cells (Szasz *et al.*, 1978), and cardiac cells damaged by ischemia (Burton *et al.*, 1981) or by neuraminidase treatment (Frank and Rich, 1983). Intracellular deposits have also been observed following extended incubation of cells with La^{3+} ions *in vitro;* these are usually ascribed to a time-dependent loss of cellular integrity. This point is worth remembering, as most of the evidence of an extracellular accumulation of lanthanides has come from experiments where cells were incubated with Ln^{3+} ions for relatively short times of two hours or less.

EDAX has additionally confirmed the plasmalemmal association of Ln^{3+} ions with cultured neuroblastoma cells (El-Fakahany *et al.*, 1983), cross-striated muscle and smooth muscle (Swales and Gardner, 1985), and epithelial cells (Squier and Edie, 1983). However, neither EM nor EDAX is a particularly sensitive technique, and the foregoing results certainly do not exclude the possible cellular penetration of a small quantity of lanthanum. As the cytosolic concentration of free Ca^{2+} is very low, the intracellular presence of a small quantity of Ln^{3+} ions could have dramatic physiological consequences.

Given a reliable method to distinguish surface binding from internal uptake, tracer studies with radioactive lanthanides should provide a more sensitive monitor of the possible cellular uptake of Ln^{3+} ions. By such means, Van Breeman et al. (1977) detected the probable intracellular accumulation of $^{170}Tm^{3+}$ by muscle cells. However, according to Weiss and Goodman (1976), the apparent uptake of $^{169}Pm^{3+}$ by smooth muscle is a surface phenomenon. Autoradiography should be of value in such studies. Indeed, reference is sometimes made to the autoradiographic studies of Laszlo et al. (1952), discussed in Section 7.5.1, which show that $^{140}La^{3+}$ fails to penetrate the cells of various visceral organs following intraperitoneal (i.p.) injection into mice. Given the strong tendency of lanthanides to precipitate when administered in this way, such a result is unsurprising and of little relevance to controlled, in vitro, experimental conditions. In view of the high intracellular concentrations of certain phosphorylated metabolites with known peak assignments, ^{31}P-NMR spectroscopy might serve to monitor the possible cellular influx of paramagnetic lanthanides.

The other type of evidence which supports the notion that Ln^{3+} ions do not penetrate the cell is indirect and involves data which can be most simply explained by lanthanide impenetrability. One example of this sort of evidence is the finding that binding of $^{140}La^{3+}$ to cells is extremely rapid, readily reversible, and independent of temperature (Grinvald and Yaari, 1978; Szasz et al., 1978). Other evidence is of the general type where added Ln^{3+} ions produce an effect which cannot be mimicked, or which goes in the other direction, when Ln^{3+} ions are deliberately introduced into the cell. Tb^{3+} ions, for instance, normally inhibit platelet aggregation, but they stimulate it when transported into the cell by the ionophore A23187 (Del Principe et al., 1984). Likewise, externally applied La^{3+} ions induce maturation in Xenopus oocytes, but fail to do so if microinjected into the cells (Schorderet-Slatkine et al., 1976). Similarly, Tb^{3+} ions inhibit the efflux of K^+ when added to intact resealed erythrocyte ghosts (Szasz et al., 1978; Wood and Mueller, 1985) but activate it when experimentally introduced at the inner surface of the erythrocyte membrane (Wood and Mueller, 1985). Several other examples of this type exist (e.g., Hambly and Dos Remedios, 1977; Cartmill and Dos Remedios, 1980).

Claims of providing direct evidence for the entry of Ln^{3+} ions into intact, living cells rest largely upon one type of experiment, and it is flawed. This experiment involves treating the cells with Ln^{3+} ions, breaking open the cells, fractionating the various cellular compartments, and measuring their lanthanide content. Such an approach ignores the fact that Ln^{3+} ions form purely ionic, and hence readily reversible, chemical associations with biological ligands (Section 2.3). Regardless of the localization of the Ln^{3+} ions prior to disruption of the cell, the sequestered lanthanides will redistribute during fractionation among the competing ligands in the homogenate. Thus, the final apparent subcellular localization of the Ln^{3+} ions will not depend upon their original binding sites, but upon their relative affinities for all the ligands to which they were exposed during subsequent homogenization and fractionation. Such studies are further weakened in some cases (e.g., Chiari et al., 1980) by the use of perfusing solutions which contain phosphate or bicarbonate.

Indirect evidence of the cellular uptake of lanthanide ions comes in various forms. Hodgson et al. (1972), for example, have calculated that smooth muscle sequesters more $^{140}La^{3+}$ than can be explained by binding to the external surface alone. Axelrod and Klein (1974) have interpreted the changes in luminescence of Eu^{3+} and Tb^{3+} ions upon binding to nerve axons as indicating slow, continual uptake into the axoplasm. An intracellular site of action has been invoked to explain the inhibition by Gd^{3+} of the Ca^{2+}-independent release of vasopressin by neurohypophysis induced by cold (Muscholl et al., 1985). Failure of attempts to remove Ln^{3+} ions from cells, or to reverse their physiological effects by high concentrations of Ca^{2+} or EGTA, is also sometimes used as evidence supporting the internalization of lanthanides. Examples include the La^{3+}-insensitive sequestration of $^{140}La^{3+}$ by Ehrlich ascites tumor cells (Levinson et al., 1972a) and the inability of Ca^{2+} to reduce the inhibition by Gd^{3+} of electrically evoked vasopressin release (Muscholl et al., 1985). It should be noted that evidence of the latter kind only applies if the Ln^{3+} ions work through a Ca^{2+}-dependent mechanism. Prolonged treatment of guinea pig smooth muscle with $2\,mM$ or $10\,mM$ La^{3+} produces a small increase in tension which has been interpreted as reflecting a slow entry of La^{3+} ions into the cells (Mayer et al., 1972; Fig. 6-5). In addition, La^{3+} ions increase the contraction amplitude and resting tension of heart muscle under certain conditions (Mezon and Bailey, 1975; Ravens, 1975). Rozza et al. (1975) were able to elicit contractions in Ca^{2+}-free, K^+-depolarized preparations of rat uterine smooth muscle, which they interpreted as a result of La^{3+} influx. However, as their bathing solution contained millimolar concentrations of phosphate, sulfate, and bicarbonate, the relevance of their data is obscure.

The notion that the plasmalemma normally excludes Ln^{3+} ions is supported by studies with artificial lipid unilamellar and bilamellar vesicles. NMR spectroscopy reveals that lanthanides do not penetrate phosphatidylcholine liposomes (Bystrov et al., 1971; Fernandez et al., 1973; Lawaczeck et al., 1976; Hunt, 1980). An interesting effect of chain length upon permeability was noted for lecithin bilayers (Hauser and Barratt, 1973). Nd^{3+} ions readily penetrated bilayers where the chain length was $C = 10–16$ but failed to enter vesicles formed from lecithin with C-16 or from natural egg lecithin. The permeability of artificial membranes can be increased by incorporating agents such as phytol, α-tocopherol, phytic acid (Cushley and Forrest, 1977), rhodopsin in the presence of light (O'Brien et al., 1977), glycophorin (Gerritsen et al., 1979), or certain other agents discussed by Ting et al. (1981).

In summary, the bulk of the available evidence supports the view that Ln^{3+} ions generally cannot penetrate the outer membranes of healthy living cells. However, the evidence is far from complete and certainly cannot exclude the possible penetration of lanthanides under certain experimental conditions, especially if incubation times are prolonged. Investigations in which lanthanide impenetrability is a key assumption would be strengthened by independent evidence that this is indeed so for the relevant set of experimental conditions. Lanthanides do not appear to be able to enter bacteria, fungi, or algae (Section 7.3).

6.3 Influence of Lanthanides on Transmembrane Fluxes of Metal Ions

6.3.1 General Considerations

From the evidence discussed in the last section, we seem secure in supposing that, for the most part, the primary site of interaction of lanthanides with living cells is at the external surface of the cell. Although the precise biochemical nature of the binding sites is unclear, at least some of these sites also accept Ca^{2+} ions. As the K_d values of Ln^{3+} ions for such sites are generally lower than those of Ca^{2+} ions, we might expect lanthanides to displace Ca^{2+} from the plasmalemma. Evidence that this is indeed so will be discussed in the following section. Indeed, "the lanthanum method" (Section 3.9) is routinely used to discriminate between intracellular Ca^{2+} and surface-bound Ca^{2+}, while reference is frequently made to the "lanthanum-resistant" and "lanthanum-displaceable" pools of Ca^{2+} as synonyms for these two reservoirs of Ca^{2+}. In the great majority of cases, the attachment of Ln^{3+} ions to the surface of the cell inhibits the uptake of Ca^{2+} ions. In many types of cell, Ln^{3+} ions also inhibit the efflux of Ca^{2+}, although the data here are less consistent, and the effect is generally much weaker than the inhibition of Ca^{2+} influx. Before examining the data directly, it is worth mentioning the different ways in which Ca^{2+} ions enter and leave living cells. Various aspects of these processes have been reviewed by Kostyuk (1981), Hagiwara and Byerly (1981a,b), Tsien (1983), and Evans (1988).

Although cells normally maintain cytosolic Ca^{2+} concentrations of around 100 nM, the extracellular concentration of free Ca^{2+} ions is generally about 1–2 mM. Such a large concentration gradient is maintained by a limiting membrane which, in its resting state, is almost impermeable to Ca^{2+}. Leakage is dealt with by various pumps which require ATP and, in excitable cells, by a Na^+–Ca^{2+} exchanger (q.v.). However, the controlled entry of Ca^{2+} is fundamental to many cellular processes, particularly those involving physiological responses to hormones or other stimuli. During this type of stimulus-coupled response, the cytosolic concentration of free Ca^{2+} may increase manyfold. In some cells, this reflects the release of Ca^{2+} from intracellular stores. In others, it results from the influx of Ca^{2+} from the extracellular environment.

There are at least four routes through which Ca^{2+} penetrates the cell membrane. One is via specific channels through which Ca^{2+} ions pass. These channels are gated such that Ca^{2+} transport is low in the resting state but high following activation by electrical stimulation or by the occupation of a membrane receptor by an agonist. The former types of channel are known as "voltage-operated channels" and the latter as "receptor-operated channels." Of these, the voltage-operated channels in excitable tissues have received by far the most attention. Both types of channel can be blocked by Ln^{3+} ions. Inhibition of Ca^{2+} transport is probably not a general result of membrane stabilization, reduction in surface charge, or nonspecific interactions with the various components of the membrane (Nelson *et al.*, 1984). Unlike organic blockers of calcium channels, Ln^{3+} ions also displace Ca^{2+} ions from the cell surface. This is seen experimentally as an abrupt discharge of Ca^{2+} by the cell and should not be confused with efflux due to the transport of intracellular Ca^{2+}.

In excitable cells, Ca^{2+} ions also enter and leave by a Na^+–Ca^{2+} exchanger, which can operate in either direction. Most authors (e.g., Reeves and Sutko, 1979; Gill *et al.*, 1981) report the sensitivity of this exchange to Ln^{3+} ions. With the Na^+–Ca^{2+} exchanger of cardiac sarcolemmal vesicles, La^{3+} appears to be a stronger inhibitor than Y^{3+} (Bers *et al.*, 1980), although Katzung *et al.* (1973) found this transporter to resist inhibition by lanthanides. Calcium can also cross cellular membranes via a Ca^{2+}–Ca^{2+} exchanger and through slow passive "leakage" down the concentration gradient. Both of these are sensitive to lanthanides. Indeed, Ln^{3+} ions will block the spontaneous passage of Ca^{2+} ions through artificial membranes comprising phospholipid and cholesterol (Van Breeman, 1969).

Cytosolic fluxes in Ca^{2+} levels occur very rapidly and are usually recorded as "spikes" of sudden increase in concentration. These can be monitored with certain luminescent indicators or detected electrically as an inward Ca^{2+} current. Because the Ca^{2+} current is usually smaller and slower than the Na^+ current, its discovery was considerably delayed. However, it has now been detected in a large number of tissues (Reuter, 1973; Hagiwara and Byerly, 1981a,b).

Removal of Ca^{2+} from the cytoplasm involves ATP-dependent "pumps" which transfer Ca^{2+} from the cytoplasm into mitochondria, the endoplasmic reticulum, or across the plasmalemma and into the extracellular space. Ca^{2+} ions can also be transported out of the cell by the Na^+–Ca^{2+} exchanger. In certain cells, Ca^{2+} efflux is strongly inhibited by Ln^{3+} ions; in others, it is unaffected (Sections 6.3.2 and 6.3.3). Although, for the chemical reasons discussed in Section 2.5, the effects of Ln^{3+} are largely considered in relation to their effects on Ca^{2+} metabolism, perturbations in the cellular handling of monovalent cations are sometimes reported for lanthanide-treated cells. Likely explanations are interference by Ln^{3+} ions with the Na^+–Ca^{2+} exchanger and with a Ca^{2+}-activated K^+ channel which ejects K^+ from the cell. La^{3+} ions also inhibit the Na^+/K^+-ATPase of rat heart sarcolemma *in vitro* (Takeo *et al.*, 1979). Whether La^{3+} ions have access to this enzyme *in vivo* is unclear.

As the metal-ion transporters in cellular membranes are under such intensive scrutiny, we can expect modifications and additions to the foregoing summary.

6.3.2 Lanthanides and Cation Fluxes in Muscles and Nerves

Langer and Frank (1972) extensively characterized Ca^{2+} exchange in cultured rat heart cells. They found that 75% of the total cellular Ca^{2+} was exchangeable, with 43% of the exchange occurring during a rapid phase with a half-time of 1.15 min. The remainder exchanged with a half-time of 19.2 min. Half millimolar La^{3+} displaced about half of the rapidly exchanging Ca^{2+}, but none of the slowly exchanging Ca^{2+}. Kinetic studies revealed that, overall, La^{3+} displaced 10–20% of the exchangeable Ca^{2+} and then strongly inhibited further Ca^{2+} influx and efflux. Langer and Frank (1972) identified the La^{3+}-displaceable pool with superficially bound Ca^{2+}. Experiments with embryonic chicken heart (Shigenobu and Sperelakis, 1972) and frog ventricular strips (Stefanou and Wooster, 1976) have confirmed that La^{3+} blocks Ca^{2+} transport both into and out of the cell. However, Katzung *et al.* (1973) reported that because of its inability to block the Na^+-Ca^{2+} exchanger, La^{3+} inhibited only Ca^{2+} influx. Most other authors have disputed this. Indeed, Gill *et al.* (1981) found the Na^+-Ca^{2+} exchanger to be 100 times more sensitive than the ATPase-dependent system to La^{3+} ions. According to Reeves and Sutko (1979), 50% inhibition of the exchanger occurs with $10\,\mu M$ La^{3+}. By inhibiting the influx of Ca^{2+} into muscles and nerves, La^{3+} ions abolish the "Ca spike" measured electrically (Hagiwara and Byerly, 1981a,b).

Following treatment of rat myocardial cells with neuraminidase, La^{3+} displaced about 83% of the cellular Ca^{2+}, suggesting that removal of sialic residues permitted La^{3+} to gain access to intracellular Ca^{2+}. EM confirmed this conclusion. The amount of Ca^{2+} displaced by La^{3+} was also increased by ouabain, which gives a positive inotropic response (Nayler, 1973). As well as inhibiting the slow inward current in calf cardiac Purkinje fibers, $0.5\,mM$ La^{3+} decreased the slow outward current due to K^+ and increased the net outward plateau current. This was mediated by changes in membrane surface charge which stabilized the inward Na^+ current (Kass and Tsien, 1975).

Because of the paucity of sarcoplasmic reticulum (SR) in smooth muscle cells, the pool of Ca^{2+} attached to the outer sarcolemmal surface has received much attention as an important cellular reservoir of Ca^{2+} ions. As this pool is particularly sensitive to Ln^{3+} ions, lanthanides have been put to particularly good use in studies of Ca^{2+} fluxes in smooth muscle. Indeed, the "lanthanum method" (Section 3.9) was originally devised for experiments with smooth muscle. In general, the results have been qualitatively similar to those described for cardiac muscle. There is a rapid, apparently increased efflux of Ca^{2+} due to its displacement from the external surface. Cellular uptake of Ca^{2+} is inhibited strongly, while cellular efflux is less strongly inhibited.

Millimolar concentrations of La^{3+} ions block the uptake of $^{45}Ca^{2+}$ by intestinal (Weiss and Goodman, 1969; Mayer *et al.*, 1972), uterine (Goodman and Weiss, 1971a), and vascular (Goodman and Weiss, 1971b) smooth muscle. However, efflux is more weakly inhibited, such that 10 mM La^{3+} depletes cells of virtually all their $^{45}Ca^{2+}$ in 20 min (Weiss, 1977). Complete inhibition of the efflux of Ca^{2+} from smooth muscle can only be achieved with a combination of $80\,mM$ La^{3+} and cooling to $0.5°C$ (Karaki and Weiss, 1979). The order of effectiveness of $1.5\,mM$ concentrations of Ln^{3+} ions in inhibiting the uptake of $^{45}Ca^{2+}$ by vascular smooth muscle is $Nd^{3+} > La^{3+} \approx Eu^{3+} \gg Lu^{3+}$ (Weiss and Goodman, 1975). Other metal cations in smooth muscle cells are also sensitive to La^{3+}

ions. In the taenia coli of guinea pigs, $5\,mM$ La^{3+} caused a marked loss of K^+, Na^+, and Mg^{2+} within 1 h. The major effect was a large reduction in Na^+ uptake. Shrinkage, due to a reduction in the tissues' extracellular space, also occurred, possibly reflecting changes in the surface charge (Burton and Godfraind, 1974; Brading and Widdicombe, 1977).

La^{3+} ions inhibit the uptake of $^{45}Ca^{2+}$ into skeletal muscle of frog sartorius muscle, rectus abdominus muscle (Weiss, 1970) and toe muscle (Shetty and Frank, 1985), while they suppress the "Ca spike" of barnacle muscle (Hagiwara and Byerly, 1981a). Ca^{2+} efflux is also blocked, although in insects, this may reflect intracellular penetration by La^{3+} ions (Dunbar, 1982). There appears to be no change in Na^+ or K^+ content of sartorius muscle as a result of lanthanum treatment, but Na^+ levels fall, while K^+ concentrations rise, in the rectus abdominus muscle (Weiss, 1970).

La^{3+} ions also inhibit both the uptake and release of Ca^{2+} by nervous tissue. Inhibition of the basal uptake of $^{45}Ca^{2+}$ by brain slices (Weiss and Wheeler, 1978), neurohypophysis (Russell and Thorn, 1974), synaptic tissue (Miledi, 1971), and rabbit vagus nerve (Kalix, 1971) have been reported along with inhibition of Ca^{2+} entry through voltage-sensitive channels (Miledi, 1971; Hagiwara and Byerly, 1981a,b) and the Na^+-Ca^{2+} exchanger (Bers et al., 1980). As a result of the last of these, Na^+ efflux from squid axons is also inhibited (Baker et al., 1969). In squid axons (Starzak and Starzak, 1978) and Xenopus nerve fibers (Arhem, 1980), Ln^{3+} ions also decrease the rates of opening of delayed (K^+) and early (Na^+) channels and decrease leakage currents. Transmembrane potentials increased. Van Breeman and DeWeer (1970) showed that La^{3+} ions decrease Ca^{2+} efflux from the giant axons of squids which have been previously injected with $^{45}Ca^{2+}$. This probably reflects inhibition of the ATP-dependent Ca^{2+} pump that has been demonstrated in sarcolemmal and synaptic plasma membranes (Schellenberg and Swanson, 1981; Gill et al., 1981).

Using guinea pig brain synaptosomes, Gill et al. (1981) measured half-maximal inhibition of Ca^{2+} transport via the Na^+-Ca^{2+} exchanger system with $2\,\mu M$ La^{3+}. Similar inhibition of the ATPase-dependent transport system required about $200\,\mu M$ La^{3+}. More detailed kinetic analyses by Rahaminoff and Spanier (1984) showed that La^{3+}, Pr^{3+}, or Tb^{3+} equally reduced Na^+-driven Ca^{2+} influx into rat synaptic plasma membrane vesicles. Inhibition was competitive, with K_i values for the various lanthanides in the range of $2-3\,\mu M$. Externally applied Ln^{3+} ions also inhibited Ca^{2+} efflux from membrane vesicles via the Na^+-Ca^{2+} exchanger working in the reverse direction. However, when La^{3+} was incorporated into the vesicles, there was no inhibition of Na^+-driven Ca^{2+} uptake or efflux. Similarly, externally applied La^{3+} ions were poor inhibitors of Na^+-gradient-driven Ca^{2+} efflux when added to inside-out vesicles. The authors speculated that the Na^+–Ca^{2+} exchanger may have two Ln^{3+}-sensitive sites, one of which competitively binds Ca^{2+} and the other of which is a regulatory site. La^{3+} ions also inhibited ATP-dependent Ca^{2+} pumping, but much more weakly than they inhibited the Na^+–Ca^{2+} exchanger.

Nachshen (1984) has resolved the voltage-dependent entry of Ca^{2+} into rat brain synaptosomes into "fast" (1–2 s) and "slow" (10 s) components which probably represent different fast and slow channels. Concentrations of Y^{3+} and Ln^{3+} below 1 μM strongly inhibited Ca^{2+} transport through the fast channels. This was not due to changes in the transmembrane potential or the negative surface potential but probably represented direct action on the Ca^{2+} channels themselves. As inhibition occurred rapidly and was quickly reversed, it probably reflected binding of lanthanides to the outside surface of the channels. Inhibition was competitive, with K_i values of 0.2 μM for Y^{3+} and 0.7 μM for La^{3+}. No K^+-stimulated influx of $^{141}Ce^{3+}$ could be detected. Inhibition of Ca^{2+} influx through the K^+-stimulated slow channels by Ln^{3+} ions was much weaker but showed the same interesting increase with increasing ionic radius.

6.3.3 Lanthanides and Cation Fluxes in Nonexcitable Cells

Because they lack elaborate intracellular Ca^{2+} sinks, erythrocytes depend heavily on the Ca^{2+}-ATPase of their cell surface to maintain low cytosolic concentrations of Ca^{2+}. As a result, the ATP-dependent Ca^{2+} "pump" of intact erythrocyte ghosts has been widely studied. Szasz *et al.* (1978) noted that 0.25 mM La^{3+} completely blocked the efflux of $^{45}Ca^{2+}$ from preloaded cells. Various other lanthanides ranging from Pr^{3+} to Lu^{3+} were equally effective inhibitors (Sarkadi *et al.*, 1977). In Mg^{2+}-depleted cells, La^{3+} ions also blocked Ca^{2+}–Ca^{2+} exchange and, at concentrations above 0.25 mM La^{3+}, passive Ca^{2+} leakage down the concentration gradient. Because of this, rates of Ca^{2+} accumulation were greatly increased in the presence of 0.25 mM La^{3+}, although higher concentrations were inhibitory. However, the increased Ca^{2+} uptake by Mg^{2+}-depleted cells was strongly inhibited by 0.2 mM La^{3+}. Nevertheless, propranolol-induced $^{45}Ca^{2+}$ uptake was unaffected by La^{3+}. The erythrocyte is thus unusual in that it is Ca^{2+} efflux, rather than influx, that is more strongly inhibited by La^{3+} ions.

La^{3+} ions also prevented the efflux of K^+ produced by either high concentrations of intracellular Ca^{2+} or by propranolol treatment of Mg^{2+}-

depleted cells. However, when cells were preloaded with Ca^{2+} by A23187, the rate of rapid K^+ transport was unaffected by external La^{3+} ions (Szasz *et al.*, 1978). Further work by Wood and Mueller (1985) confirmed that Tb^{3+} ions applied externally to resealed erythrocyte ghost membranes inhibited the net efflux of K^+ via the Ca^{2+}-activated channel. However, internalized Tb^{3+} ions could activate this channel as well as Ca^{2+}. Externally applied Tb^{3+} ions were also able to inhibit K^+ efflux following experimental activation of the channel by partial trypsinization.

The lanthanum method (Section 3.9) has been vigorously applied to the study of Ca^{2+} fluxes in the pancreas. La^{3+} ions inhibit the rate of basal and carbachol-stimulated uptake and efflux of $^{45}Ca^{2+}$ by fragments of pancreas (Heisler and Grondin, 1973). While K^+ concentrations are not greatly affected by La^{3+}, the Na^+ content is lowered, although this can be accounted for by the reduced volume of the extracellular space (Chandler and Williams, 1974). Isolated pancreatic islets react in a similar fashion, with the efflux of Ca^{2+} being more weakly inhibited than its influx (Hellman *et al.*, 1976).

In isolated pancreatic β-cells, 0.5 mM La^{3+} displaces Ca^{2+} from superficial binding sites and inhibits both the uptake and efflux of Ca^{2+} under conditions of both high and low glucose (Flatt *et al.*, 1980b). La^{3+}, at concentrations between 0.8 mM and 2 mM, also inhibits both basal Ca^{2+} efflux and the accelerated efflux of Ca^{2+} due to glucose (Fig. 6-2). In addition, La^{3+} interferes with fluxes of univalent ions, decreasing the cellular content of Na^+ (Brading and Widdicombe, 1977). A similar effect on Na^+ has been reported for Ehrlich ascites cells (q.v.).

Pancreatic acinar cells respond rather differently to La^{3+}. According to Korc (1983), 0.1 mM La^{3+} does not affect the basal rate of $^{45}Ca^{2+}$ influx into isolated pancreatic acini but inhibits the accelerated influx due to cholecystokinin octapeptide. It produces only a marginal inhibition of basal or stimulated Ca^{2+} efflux. However, studies by Wakasugi *et al.* (1981) on isolated pancreatic acinar cells showed enhanced $^{45}Ca^{2+}$ uptake at 0.1 and 0.3 mM La^{3+}, but reduced uptake at 10 mM La^{3+}. Intermediate concentrations of 2 mM and 5 mM produced a rapid release of Ca^{2+} from the surface of the cells, without further uptake. The net cellular accumulation of $^{45}Ca^{2+}$ was increased by 0.1 and 0.3 mM La^{3+} but decreased by 1, 2, or 5 mM La^{3+}. Addition of carbamylcholine to the cells produced an abrupt release of $^{45}Ca^{2+}$ from the cell surface, followed by its rapid reuptake. La^{3+}, at a concentration of 0.3 mM, had no effect on this release, whereas 1, 2, and 5 mM La^{3+} prevented release; 5 mM La^{3+} also prevented reuptake.

According to Claret-Berthon *et al.* (1977), La^{3+} ions rapidly displace superficially bound Ca^{2+} from perfused rat livers and inhibit Ca^{2+} efflux.

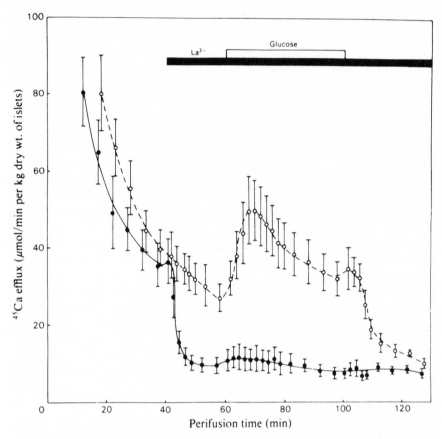

Figure 6-2. Effects of La^{3+} on glucose stimulation of $^{45}Ca^{2+}$ efflux in isolated pancreatic islets. $\bigcirc$, Control islets; $\bullet$, $2\,mM$ La^{3+}. From Flatt *et al.* (1980a), with permission.

However, in contrast to nearly all other types of cell yet investigated, hepatocytes apparently take up more $^{45}Ca^{2+}$ when treated with 0.2–1 mM La^{3+}, with half-maximal stimulation occurring at 0.7 mM La^{3+} (Parker and Barritt, 1981). La^{3+} ions increased both the initial rate of $^{45}Ca^{2+}$ uptake and the plateau level of uptake. Kinetic analyses revealed the existence of two intracellular, exchangeable pools of Ca^{2+}, both of which were affected by La^{3+}. Uptake of $^{45}Ca^{2+}$ into both these pools at 1°C by cells previously maintained in the nominal absence of Ca^{2+} was also enhanced by La^{3+}. However, these results should be interpreted with caution, as the incubation medium contained phosphate, sulfate, and bicarbonate. In view of this, it is interesting that Hinnen *et al.* (1979), who also used phosphate in their incubation buffer, recorded an enhancement of about 30% in the uptake of $^{45}Ca^{2+}$ by Ehrlich ascites tumor cells in the presence of La^{3+} or Tb^{3+} ions. La^{3+} ions also drastically affected the cellular concentrations of monovalent ions in these cells. Levinson *et al.* (1972a) reported that 1 mM La^{3+} caused an 87% reduction in K^+ content, 79% reduction in Cl^- content, and 21% reduction in Na^+ content within 5 min. Concomitant with this was an increase in the recorded membrane potential. As the authors pointed out, the loss of Na^+ occurred against the concentration gradient. Furthermore, La^{3+}-treated cells excluded trypan blue and did not lose protein, suggesting that generalized membrane breakdown was not to blame. The changes also occurred at 0.5°C. The authors concluded that La^{3+}-treated cells became permeable when subjected to mechanical stress (Smith *et al.*, 1972; Levinson *et al.*, 1972a,b). In later work (Smith, 1976), it was reported that La^{3+} did not affect the rate of K^+ efflux but inhibited Na^+ efflux. Presumably these cells were not mechanically stressed.

Although La^{3+} ions inhibit the efflux of Ca^{2+} ions from the adrenal medulla, they may have the odd effect in this tissue of promoting Ca^{2+} influx through an atypical route (Bresnahan *et al.*, 1980). However, isolated chromaffin cells respond to 50 μM Gd^{3+} in a more usual way, with $^{45}Ca^{2+}$ uptake and efflux and $Ca^{2+}-Ca^{2+}$ exchange reduced (Bourne and Trifaro, 1982).

In parotid glands, La^{3+} ions block Ca^{2+} entry and inhibit the sustained efflux of $^{86}Rb^+$ and K^+ in response to carbachol or norepinephrine (Keryer and Rossignol, 1978; Marier *et al.*, 1978). However, in the absence of these stimuli, 1 mM La^{3+} alone stimulates K^+ efflux. Like Ca^{2+} ions, La^{3+} ions impair the turnover of phosphatidylinositol in these glands.

Inhibition of Ca^{2+} fluxes has also been reported for duodenal segments (Pento, 1978), thyroid slices (Pento, 1977), neutrophils (Boucek and Snyderman, 1976), basophils (Beaven *et al.*, 1984), and platelets (Robblee and Shepro, 1976). In corn roots, La^{3+} and Ca^{2+} both inhibit K^+ uptake (Nagahashi *et al.*, 1974). Lanthanides also have important effects on the transport of ions across skin and bladder. These are discussed in Section 6.4.12.

6.4 Effects of Lanthanides on Cellular Metabolism

6.4.1 General Considerations

The foregoing section has provided numerous examples of the ways in which lanthanides interfere with movements of calcium into and out of resting and stimulated cells. Because of the importance of Ca^{2+} both to the basal metabolism of cells and to their stimulus-coupled responses, lanthanides strongly influence cellular physiology. This is particularly true of changes in cellular physiology provoked by specific, external stimuli; a vast body of evidence now implicates cytosolic fluxes in the concentration of Ca^{2+} in the triggering mechanisms of a wide range of cellular responses. These include muscle contraction, neurotransmitter release, hormonal responses, cell division, and a wide range of secretory events. The literature on this subject is enormous, and it increases rapidly.

The "trigger" calcium may enter the cytosol through specific voltage-operated or receptor-operated channels as discussed in Section 6.3.1. Alternatively, calcium may be released intracellularly from the endoplasmic reticulum. This mechanism involves stimulated phospholipid metabolism, with the production of inositol triphosphate as an intracellular mediator of this release. In general terms, La^{3+} should impair those cellular responses which depend upon influx of Ca^{2+} from the outside but not affect responses which are triggered by the intracellular liberation of Ca^{2+}. Indeed, La^{3+} has been used experimentally in this way to distinguish between these two sources of Ca^{2+}. However, while there is experimental evidence in support of this generalization, it is not quite this straightforward. For example, La^{3+} inhibits the release of insulin and amylase from the pancreas even though the trigger Ca^{2+} is released intracellularly (Section 6.4.9). It has been suggested that lanthanides interfere with the exocytolic machinery in these cases. An additional complication involves that fraction of the Ca^{2+} released into the cytoplasm from intracellular stores which is pumped out of the cell as the cytosolic Ca^{2+} concentration returns to the resting level. Extracellular Ca^{2+} is subsequently required to top up the intracellular reservoir. Thus, several rounds of cellular activation in the presence of La^{3+} can deplete the intracellular Ca^{2+} pool and prevent further responses.

6.4.2 Cardiac Tissue

Among the first physiological experiments conducted with lanthanides were those of Mines (1910, 1911), showing that $1-10\,\mu M$ concentrations of La^{3+}, Ce^{3+}, Y^{3+}, Pr^{3+}, Nd^{3+}, Sm^{3+}, Tm^{3+}, Dy^{3+}, and "didymium" (now known to be a mixture of Nd^{3+} and Pr^{3+}; Section 1.2) reversibly and quickly produced diastole arrest in frogs' hearts. One of Mines' original myograms is shown in Fig. 6-3. This was seen when the heart was perfused with Ringer solution, despite the presence of bicarbonate, but did not occur when Ln^{3+} ions were injected into the blood-

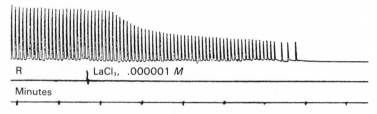

R   LaCl₃,   .000001 M

Minutes

Figure 6-3. Myogram showing the arrest of frog heart by La^{3+}. From Mines (1910), with the publisher's permission.

stream. From the discussion in Section 7.5.2, it is clear that free Ln^{3+} ions would probably not have been available to the cardiac tissue under *in vivo* conditions.

Fawzi and McNeill (1985) produced 50% inhibition of the contraction of adult guinea pig hearts at $0.19 \pm 0.01 \,\mu M$ La^{3+}, although higher concentrations were needed to inhibit the positive inotropic effect of isoproterenol. Dose–response curves for this effect gave evidence of two inhibitory components acting as if there were high- and low-affinity La^{3+} sites with apparent K_a values, in the presence of $1\,mM$ Ca^{2+}, of $(2.29 \pm 0.4) \times 10^7 M^{-1}$ and $(6.31 \pm 1.10) \times 10^5 M^{-1}$. Low-affinity sites accounted for an inhibition in contractile force of 23%, and the high-affinity sites an additional 64% (Wong *et al.*, 1976).

Certain cardiac responses to perfused La^{3+} depend upon age (George and Jarmakani,1983). Although La^{3+} quickly inhibited the developed tension in both adult and newborn rabbit ventricular portions, it increased the resting tension in the newborn only. The explanation is unknown, but these results may reflect maturational changes in membrane structure. La^{3+} also decreased the rate of tension development. Sanborn and Langer (1970) found 5–$20\,\mu M$ La^{3+} to be a potent uncoupler of excitation and contraction, which did not alter the action potential in arterially perfused, lapine intraventricular septa. Cartmill and Dos Remedios (1980) noted that Tm-EGTA was far less effective than other Ln-EGTA complexes in inhibiting the twitch response of toad atrial septum. Although they drew attention to the closeness of the ionic radii of Ca^{2+} and Tm^{3+}, the discussion in Section 2.5 (Table 2-12) advises caution in interpreting these sorts of results.

Most authors explain the inhibition of cardiac muscle contraction by lanthanides by reference to displacement of superficially bound Ca^{2+}. However, Ravens (1975) considered this explanation too simple for the complicated phenomena observed in his experiments. The contraction amplitude of guinea pig cardiac muscle initially decreased, but then in-

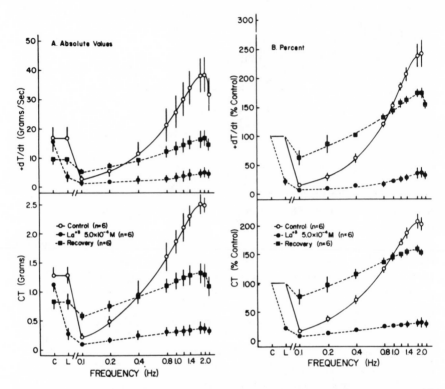

Figure 6-4. Frequency–force responses of atrial muscle after recovery from treatment with La³⁺. From Durrett and Adams (1980), with permission.

creased, following exposure to 50–100 μM La³⁺. In addition, the resting membrane potential was strongly reduced, while the contractile response proved irreversible. Durrett and Adams (1980) confirmed that 50 and 500 μM concentrations of La³⁺, which reduced contractility by about 50% and 90%, respectively, were irreversible inhibitors in isolated guinea pig atria. Recovery never exceeded about 45% of the control value. However, if La³⁺-treated, washed hearts were subjected to low-frequency electrical stimulation, the contractile response was greater than control. However, the responses to high frequencies were still attenuated (Fig. 6-4). Like Ravens (1975), these authors concluded that an explanation based simply on reversible effects on Ca²⁺ uptake was not sufficient to explain such phenomena. In particular, they suggested that irreversible damage to the myofibers may occur (Durrett and Adams, 1980). Other strange results were found by Shine (1973). Although 200 μM La³⁺ produced an abrupt decline in the twitch tension of isolated, perfused intraventricular septum of rabbits' hearts to 8.1 ± 2.9% of control, it only weakly inhibited contractures caused by 20 mM quinine. Also, when the twitch tension had been abolished by perfusion in the absence of added Ca^{2+}, 200 μM La^{3+} inhibited the re-established twitch tension much more weakly. This concentration of La^{3+} ions had no effect on the membrane resistance or internal longitudinal resistance of cardiac Purkinje fibers (Pressler et al., 1982).

Odd late effects were also reported by Kawata et al. (1983). Using trabecular strips of bullfrog ventricle and atrial strips of papillary muscle from adult guinea pigs, they showed a dose-dependent, rapid inhibition of twitch tension by La^{3+}. However, on prolonged incubation, the twitch tension slowly recovered back to about 20% of the control level. Returning the tissue to normal Ringer solution produced a very large after-potentiation. La^{3+} (100 μM) dramatically increased the resting tension. After reperfusing normal Ringer solution, there was a transient increase in resting tension followed by a slow recovery. Inhibition of twitch tension by La^{3+} was accompanied by a decrease in action potential, reflecting a decrease in the slow inward current. There was no change in the resting potential during La^{3+} perfusion or washout. Paradoxically, contractions produced by sodium-free, high-potassium solutions were markedly augmented by 0.2 mM La^{3+}, although the twitch was completely suppressed. Both the rate of rise and height of contracture were considerably augmented during La^{3+} washout. The authors presented evidence to suggest that the permeability of the cell membrane to Ca^{2+} was transiently increased after La^{3+} washout. The positive inotropic action of catecholamine was strongly and irreversibly inhibited by La^{3+} ions. Neuraminidase treatment did not affect the excitation or twitch tension, but the inhibitory effect of La^{3+} on twitch tension was markedly reduced.

Addition of La^{3+} (0.1–4 mM) to cultures of neonatal rat ventricular myocardiocytes diminished contraction frequency and strength. At the same time, the membrane potential and overshoot were reduced, and the action potentials tended to be prolonged. Complete inhibition of spontaneous contraction was always accompanied by membrane depolarization and complete absence of action potentials. These effects were reversible (Kitzes and Berns, 1979). Embryonic chick myocardioblasts in culture were 10-fold more sensitive to La^{3+} ions than was intact ventricular tissue. Amplitude contraction was reduced by concentrations of La^{3+} which had no effect on the rate of contraction (Barry et al., 1978). When the Na^+ channels of chick embryonic hearts were inactivated by tetrodoxin or high K^+, catecholamines still induced slow contractions via divalent cation channels. La^{3+} ions abolished these contractions (Shigenobu and Sperelakis, 1972).

Lanthanides not only suppress the basal contraction and isoproterenol inotropy of perfused rat hearts, but they also inhibit isoproterenol-induced glycogenolysis (Bockman *et al.*, 1973). Glycogen breakdown involves a Ca^{2+}-dependent activation of phosphorylase b to phosphorylase a. Isoproterenol increases the percentage of phosphorylase present as the active (a) form. Although La^{3+} ions did not alter the basal level of active phosphorylase, they inhibited the isoproterenol-induced increase in phosphorylase a. An analogous response has been found with the fat glands of cockroaches (McClure and Steele, 1981). In a possibly related phenomenon, $50 \mu M$ La^{3+} increased the basal uptake of glucose by guinea pig atria. However, sugar transport stimulated by insulin or other treatments was not strongly affected by La^{3+}. At a concentration of $1 mM$, La^{3+} had no effect on basal or stimulated transport, and intracellular Na^+ and K^+ levels were unaffected (Bihler *et al.*, 1980).

6.4.3 Smooth Muscle

La^{3+} has been particularly helpful in investigations of Ca^{2+}-related functions in smooth muscle, as the ultrastructure of this tissue shows few well-defined intracellular reservoirs. Most studies on the effects of Ln^{3+} ions on smooth muscle refer back to the work of Weiss and Goodman (1969), who showed that $1.8 mM$ La^{3+} inhibited the contraction of guinea pig ileal smooth muscle in response to either acetylcholine or $80 mM$ K^+. Inhibition was only reversed after resting the muscles in La^{3+}-free Tyrode's solution for at least two hours. La^{3+} ions also inhibited both the resting muscle tone and the increase in tone that followed immersion in $36 mM$ Ca^{2+} for $2 h$. Further investigation (Goodman and Weiss, 1971a,b) of the dose–response relationship showed that contraction of ileal muscle in response to high K^+ or acetylcholine was about 50% inhibited by 0.9 μM La^{3+}. However, in rat uterine strips, La^{3+} inhibited contractions induced by high K^+ about 100 times more strongly than those induced by acetylcholine. A similar dichotomy was found for rabbit aortic smooth muscle, where La^{3+} inhibited contractions induced by high K^+ much more strongly than those due to norepinephrine (Van Breeman, 1969; Goodman and Weiss, 1971a,b). Contractions evoked by histamine were most resistant to La^{3+} ions (Goodman and Weiss, 1971a), while Turrin and Oga (1977) even reported an enhanced maximum response of taenia coli to histamine in the presence of $10 \mu M$ Sm^{3+}. Inhibition of contraction by La^{3+} has also been reported for the smooth muscle of human and rat stomach (Hava and Hurwitz, 1973), rabbit vein (Collins *et al.*, 1972), rat myometrium (Hodgson and Daniel, 1973), and guinea pig vas deferens (Magaribuchi *et al.*, 1977). Inhibition of rabbit thoracic aorta contraction in response to phenylephrine (Deth and Lynch, 1981) was dose dependent in the $0.1–10\,mM$ La^{3+} range, and inhibition was quicker at higher doses. The effect of $1\,mM$ La^{3+} was only slowly reversed upon washing, with a half-recovery time of 75 min. With guinea pig stomach muscle, La^{3+} inhibited contraction due to $PGF_{2\alpha}$ much more strongly than that due to high K^+; 50% inhibition of the former occurred with $0.6\,\mu M$ La^{3+} (Ishizawa et al., 1980).

Although Mayer et al. (1972) confirmed that $50\,\mu M–10\,mM$ La^{3+} rapidly inhibited tension development by the taenia coli of guinea pigs, an odd late effect was observed; unusual late effects on cardiac muscle were mentioned in the last section. After incubation of taenia coli muscle with $50\,\mu M\,La^{3+}$ for about 2–3 h, there occurred a very large increase in tension without any superimposed phasic contractions (Fig. 6–5). This reached a maximum at 4–5 h and then decreased. The increase in tension corresponded with the delayed increase in cellular permeability to $^{45}Ca^{2+}$ mentioned in Section 6.3.2, while the channel blocker D600 inhibited this response. The authors drew attention to the gradual and small, but apparently real, increase in tension that occurred with millimolar concentrations of La^{3+} (Fig. 6-5), suggesting that it was caused by slow intracellular penetration of La^{3+} as a result of prolonged incubation at relatively high doses. However, Triggle et al. (1975) failed to produce a contractile response in ileal smooth muscle by adding Tm^{3+} in the presence of the ionophore A23187. Mayer et al. (1972) also noted that $10\,mM$ La^{3+} inhibited the immediate increase in tension produced by $154\,mM$ K^+ but permitted a reduced, delayed response peaking 1.5–2 h later.

Apart from Ce^{3+}, the other Ln^{3+} ions are better inhibitors than La^{3+} of the phasic and tonic components of the mechanical responses of guinea pig ileal longitudinal muscle (Triggle and Triggle, 1976). Tm^{3+} ions were most inhibitory, with ED_{50} values around $10^{-6}M$; the unusually high ED_{50} value of around $10^{-3}M$ for Ce^{3+} ions remains unexplained. At a concentration of $2.5 \times 10^{-6}M$, Tm^{3+} ions inhibited contraction by 90%, yet failed to inhibit $^{45}Ca^{2+}$ uptake; calcium uptake was only blocked by $1\,mM\,Tm^{3+}$. As described in Section 6.4.9, Tm^{3+} also blocks insulin secretion by pancreatic β-cells at concentrations which do not interfere with Ca^{2+} fluxes. Contraction of vas deferens in response to norepinephrine or high K^+ was equally inhibited by Tm^{3+} or La^{3+}. Complete inhibition required $1\,mM\,Ln^{3+}$. However, suprainhibitory doses often provoked a slow, small contractile response in the absence of any other stimulus (Swamy et al., 1976). The authors suggested that Ln^{3+} ions act by generally stabilizing the membrane rather than blocking specific Ca^{2+}-binding sites. Lanthanum ions also inhibited the contraction of vas deferens in response to cooling, but this was followed by a slow re-establishment of

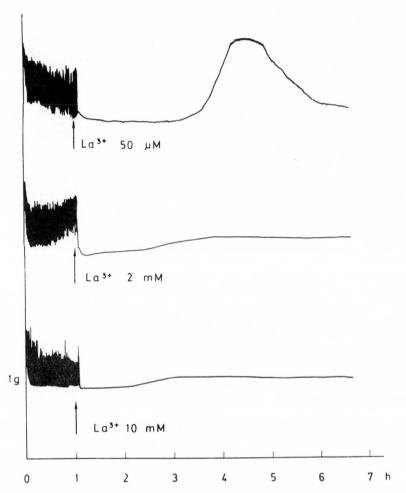

Figure 6-5. Effect of La^{3+} on tension development by guinea pig taenia coli. From Mayer *et al.* (1972), with permission.

tension. Upon rewarming, a phasic contraction occurred. Sunano (1984) explained this as an initial blocking of Ca^{2+}, followed by slow influx of La^{3+} into the cell.

One interesting observation to come from such studies was a paradoxical contractile response to La^{3+} by the aortic smooth muscle of spontaneously hypertensive, but not normal, rats (Goldberg and Triggle, 1977). Similar effects were noted for smooth muscle of the carotid (Bohr, 1974) and femoral (Hansen and Bohr, 1975) artery. These results have been explained on the basis of abnormal Ca^{2+} metabolism by the membranes of spontaneously hypertensive rats, although pH changes may also be involved. However, Laher and Triggle (1984) found that the aortic smooth muscle of certain strains of normal rats also contracted in response to 5 mM La^{3+}. In Wistar–Kyoto rats, for instance, this response was over 40% of that caused by epinephrine, although the response of the spontaneously hypertensive rats was greater. These responses to La^{3+} increased with age from 4 weeks to 16 weeks in all the strains of rat that were tested. However, the La^{3+} response was absent from adult rats treated with antihypertensive drugs. The authors suggested that the contractile response to La^{3+} was a secondary response to elevated blood pressure and provided evidence that genetic factors, independent of blood pressure, played a major role in determining the response to lanthanum. One complicating factor in the interpretation of these data was the use of a Krebs buffer which contained phosphate, bicarbonate, and sulfate. When injected intravenously into experimental animals, lanthanides have a transient hypotensive effect (Section 8.5).

6.4.4 Skeletal Muscle

In 1914, Hober and Spaeth reported the inhibiton of skeletal muscle contraction by Ln^{3+} ions, an observation which has been repeated many times since. In frog sartorius muscle, 1 mM La^{3+} inhibited tension responses induced by 80 mM K^+, which opens voltage-operated channels, but not by caffeine, which releases intracellular Ca^{2+} (Weiss, 1970). However, under hypertonic conditions, La^{3+} gains access to the terminal cisternae of the sarcoplasmic reticulum, and caffeine contracture is also inhibited (Sperelakis et al., 1973). With the frog rectus abdominus muscle (Weiss, 1973), tension responses to 80 mM K^+, acetylcholine, and nicotine were suppressed by La^{3+} ions.

Tension development in the anterior byssus retractor muscle of mussels in response to serotonin and dopamine was also inhibited by La^{3+} ions (Muneoka and Twarog, 1977). Electron microscopy showed that La^{3+} altered the arrangement of the gap junctions in these tissues (Brink et al., 1979).

In isolated frog muscle fibers, low concentrations (50–300 μM) of La^{3+} increased the latent period and the rate of tension development, potentiated peak twitch amplitude, and prolonged the relaxation period of electrically stimulated fibers. There was no change in resting potential, but a reduction in the rate of rise, amplitude, and rate of fall of action potential. At concentrations of La^{3+} above 0.5 mM, the muscle became

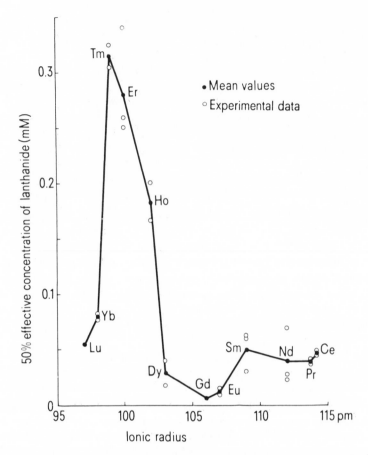

Figure 6-6. Inhibition of twitch tension in toad skeletal muscle as a function of the ionic radius of lanthanides. From Hambly and Dos Remedios (1977), with permission.

electrically inexcitable (Andersson and Edman, 1974a). Bowen (1972) found no major differences between a range of different Ln^{3+} ions and Y^{3+} in their potencies as inhibitors of the twitch tension of frog sartorius muscles. However, the ability of Ln^{3+} ions to inhibit twitch tension in toad skeletal muscle fibers changes with ionic radius (Fig. 6-6). Hambly and Dos Remedios (1977) discussed the possibility that the weaker inhibition produced by Tm^{3+} reflects the fact that its ionic radius at a coordination number of 8 is close to that of Ca^{2+} at a coordination number of 6 (Section 2.5; Table 2-1). The same group also found Tm^{3+} to be an unusually weak inhibitor of cardiac muscle contraction (Section 6.4.2). However, it is the strongest inhibitor of smooth muscle contraction (Section 6.4.3).

Addition of La^{3+} (10–100 μM) to cultures of embryonic chick skeletal muscle cells inhibits accumulation of muscle-specific creatine kinase without inhibiting cell fusion or increases in cAMP (Morris, 1980). However, higher doses of La^{3+} (0.3–1 mM) do inhibit myoblast fusion (Entwistle et al., 1988). These results are a little difficult to assess because of the chemical complexity of the culture medium and serum, with which La^{3+} undergoes all sorts of complex interactions (Section 2.8), and the extended incubation period of 3 days.

6.4.5 Nervous Tissue

The prediction by Lettvin et al. (1964) that La^{3+} ions should inhibit cationic conductance in nerves was first tested with lobster giant axons (Takata et al., 1966). At a concentration of 11 mM, La^{3+} ions reversibly reduced the rate of rise, rate of fall, and height of the peak of the action potential. Conductances for Na^+ and K^+ were reduced, thus blocking the conductance of the nerve. High concentrations of Ca^{2+} have a similar effect; however, La^{3+} is about 20 times as effective as Ca^{2+}.

Ca^{2+} enters the presynaptic terminal on depolarization, leading to neurotransmitter release. This phenomenon is most easily studied in pinched-off nerve endings known as synaptosomes. Most, but not all, authors find La^{3+} ions to inhibit the uptake of Ca^{2+} by the synaptosomes (see Section 6.3.2), thereby blocking electrically stimulated neurotransmitter release. However, La^{3+} ions provoked a large stimulation in spontaneous release of γ-aminobutyric acid (GABA), glycine, glutamine, and aspartate from synaptosomes (Osborne and Bradford, 1975). The reason for this is unclear, but there is evidence that under certain conditions La^{3+} can increase the influx of Ca^{2+} into synaptosomes (Section 6.3.2). Tapia et al. (1985), using synaptosomes isolated from mouse brain, studied the release of the neurotransmitter GABA in response to high K^+. Inhibitions produced by 1, 2.5, and 50 μM La^{3+} were about 0, 50, and 80%, respectively. When Na^+ was removed and the medium made iso-osmotic with sucrose, both 2.5 μM and 50 μM La^{3+} gave almost complete inhibition of GABA secretion. However, in this medium, unlike the Na^+-containing medium, these concentrations of La^{3+} ions evoked spontaneous release of GABA. Spontaneous release of [^{14}C]glutamate was also stimulated under these conditions. As the high-affinity sites for La^{3+} had a K_d value of 2.3 μM, it is these sites which were probably involved in the effects of 2.5 μM La^{3+}. Spontaneous release of GABA by La^{3+} in the absence of Na^+ may be related to the large, spontaneous release of acetylcholine evoked by millimolar La^{3+} in frog neuromuscular junction (Heuser and Miledi, 1971; Miledi *et al.*, 1980). La^{3+} ions prevented the electrically stimulated release of neurotransmitter amino acids from spinal-medullary synaptosomes of rats (Osborne and Bradford, 1975) and from squid giant synapse (Miledi, 1971). In chick brain synaptosomes, La^{3+} ions depressed the activities of neural ATPase and acetylcholinesterase (Basu *et al.*, 1984). Some of these effects may also occur *in vivo*. After i.p. injection of a 250-mg/kg dose of La^{3+} into chicks, their isolated brain synaptosomes took up [^{14}C]glutamate at a reduced rate. There was also a marked decrease in its release from preloaded vesicles in the presence of high K^+ or 1.2 mM Ca^{2+}. Inhibition of uptake was uncompetitive. These results are a little hard to interpret, as the brain does not accumulate much La^{3+} following i.p. injection (Section 7.5.1). Furthermore, any La^{3+} which had become metabolically associated with brain tissue might be expected to have been stripped from the synaptosomes by their suspending salt solution, which contained phosphate and malonic acid and had the relatively high pH of 7.65.

La^{3+}, Pr^{3+} Er^{3+}, and Y^{3+} have an interesting dual effect at neuromuscular synapses. They abolish the induction of transmitter release by nerve impulses, thereby abolishing evoked endplate potentials, but they provoke very rapid and large increases in miniature endplate potentials (mepps) due to spontaneous release of transmitter (Blioch *et al.*, 1968; DeBassio *et al.*, 1970; Heuser and Miledi, 1971; Bowen, 1972; Alnaes and Rahamimoff, 1974; Metral *et al.*, 1978). La^{3+} ions also heighten the response of endplates in muscle to acetylcholine, increasing the amplitude, rise time, and half-fall time of the mepps (Parsons *et al.*, 1971).

In nerve–sartorius muscle preparations of frog, millimolar concentrations of La^{3+} ions rapidly abolished the release of transmitter. With 0.1 mM La^{3+}, inhibition was readily reversed, whereas muscles did not recover from the effect of 1 or 2 mM La^{3+} for at least 6 h. These concentrations of La^{3+} ions did not block nerve impulses, and presynaptic spikes were still detectable, indicating that impulses were still invading the nerve endings. Although La^{3+} ions inhibited impulse-induced release of transmitter, they provoked a large spontaneous discharge of transmitter which evoked a huge increase in mepp generation in the absence of added Ca^{2+}. This duality has been observed for synaptosomes in Na^+-free medium. At 1 mM La^{3+}, the mepps were generated so rapidly as to be uncountable (Table 6-5). Thus, the increase in mepp generation was at least 10^4-fold. The rate remained uncountably high for at least 1 h at 4°C before slowly subsiding from about 2 h onward, until most endplates were devoid of mepps after 16 h. At room temperature, the decay occurred

Table 6-5. Effect of La^{3+} on the Number of Miniature Endplate Potentials per Second in Frog Neuromuscular Junctions[a]

Fiber	Mepps/s			
	After 4 h in 0 Ca^+ 2 mM Mg^{2+}	After 45 min in 1 μM La^{3+}	After 30 min in 10 μM La^{3+}	After 1 to 4 min in 1 mM La^{3+}
1	0.01	0.03	0.11	>100
2	0.07	0.06	0.15	>100
3	0.03	0.03	0.42	>100
4	0.03	0.08	0.13	>100
5	0.05	0.08	0.18	>100
6	0.07	0.05	0.12	>100
7	0.08	0.06	0.08	>100
8	0.05	0.02	0.05	>100
9	0.02	0.13	0.17	>100
Mean	0.047	0.060	0.157	>100

[a] From Heuser and Miledi (1971), with permission.

within 10 h. Electron microscopy suggested that the mepps decay because all the synaptic vesicles have been used up. La^{3+} ions had an analogous action on squid synapses. Kriebel and Gross (1974) identified a class of small, spontaneous mepps at the frog nerve–muscle junction, with an amplitude about one-seventh that of normal. After La^{3+} (1 mM) treatment, the percentage representation of these small mepps increased (Kriebel and Florey, 1983).

The same dual effect occurs in insect neuromuscular synapses (Washio and Miyamoto, 1983), where there is evidence of a postsynaptic effect of La^{3+} ions. The excitatory response produced by glutamate, a transmitter in this tissue, was inhibited by 0.1 mM La^{3+}. The glutamate potential decreased 50% within 2 min of application of La^{3+} ions but recovered completely 10 min after washing. The spontaneous release of transmitter provoked by La^{3+} was suppressed by high Ca^{2+}, an effect also noted by DeBassio et al. (1970). Kinetic analysis suggested that La^{3+} ions and glutamate did not compete for the same binding sites.

In spite of complete depletion of vesicles, in La^{3+}-treated nerve terminals, the acetylcholine content of the terminals increased above control levels (Miledi et al., 1980). This newly formed acetylcholine was present in the cytoplasm, but it was not within vesicles. However, unlike Heuser and Miledi (1971), Miledi et al. (1980) found no loss of vesicles at 4°C. There was an increase in the percentage of small mepps during La^{3+} treatment, while La^{3+} ions decreased the size of the normal mepps. An initial increase in the size of mepps could be ascribed to a postsynaptic action of La^{3+}. The mean amplitude of small mepps was not affected by La^{3+}.

Because of their ability to provoke spontaneous transmitter release, La^{3+} ions have been tested as a possible agent with which to overcome the inhibition of neuromuscular transmission caused by tetanus toxin (Mellanby and Thompson, 1981). Using neuromuscular junctions of gold-fish fin muscle, Mellanby and Thompson confirmed the ability of $2\,mM$ La^{3+} ions to provoke mepps. After a delay of up to $2\,h$, La^{3+} ions were also able to provoke mepps in junctions where tetanus toxin had abolished them. However, they were unable to produce frequencies as high as those they had produced in control junctions. Similar effects were found with the rat soleus muscle (Bevan and Wendor, 1984). Unlike normal junctions, the nerve endings of tetanus-blocked terminals were not depleted of syn-aptic vesicles by La^{3+}. Instead, vesicles lined up in a row just inside the terminal membrane, both at synaptic and nonsynaptic positions (Mellanby et al., 1988).

Spontaneous discharge of the neurohypophysial hormones oxytocin, vasopressin, and neurophysin from isolated neurohypophyses was also increased by La^{3+} ions. However, La^{3+} inhibited the enhanced secretion provoked by high K^+ or electrical stimulation. La^{3+} ions reduced the release of lactate dehydrogenase, suggesting greater membrane integrity (Matthews et al., 1973; Russell and Thorn, 1974).

Muscholl et al. (1985) observed that incorporation of Gd^{3+} ions into the homogenization buffer during the formation of synaptosomes from neurohypophyses inhibited the exocytolic release of vasopressin induced by cold treatment. However, Gd^{3+} had no effect on the spontaneous release of vasopressin from isolated neurointermediate lobes but produced a dose-dependent inhibition of discharge provoked electrically, by 60 $mM\,K^+$, by cold, or by the ionophore X537A. The authors drew attention to the biphasic inhibition of electrically provoked discharge, one inhibitory response occurring at $30-300\,\mu M$, and an additional one at $1-3\,mM\,Gd^{3+}$. As discussed in Section 6.3, Nachshen and Blaustein (1980) provided evidence of functionally separate "fast" and "slow" voltage-operated Ca^{2+} channels in synaptosomal membranes, with the former being more sensitive to inhibition by Ln^{3+} ions. The biphasic response to Gd^{3+} may reflect this. However, Muscholl et al. (1985) speculated about an intra-synaptosomal site of action for Gd^{3+} ions.

Lanthanum inhibition of the depressant effect that norepinephrine has on cerebral cortical neurons (Yarbrough e al., 1974) may occur through a Ca^{2+}-independent mechanism. Norepinephrine mediates its effects on these cells through cAMP. Nathanson et al. (1976) demonstrated that La^{3+}

ions strongly inhibited the adenylate cyclase activity of homogenates of this tissue, with 50% inhibition at about $2 \mu M$ La^{3+}. As adenylate cyclase is membrane bound, the authors speculated that externally bound La^{3+} ions had access to this enzyme in intact cells. La^{3+} ions had no effect on the firing induced by γ-aminobutyric acid, which does not work through cAMP. In other tissues, La^{3+} ions have been found to have little effect (Fawzi and McNeill, 1985) or an inhibitory effect (Wong et al., 1972; Haksar et al., 1976; Flatt et al., 1980a) on cAMP production. However, in platelets, La^{3+} ions provoked a remarkable increase in cAMP accumulation (Best et al., 1980).

6.4.6 Platelets

Millimolar concentrations of La^{3+} ions inhibit the secretion of dense granule constituents in response to thrombin (Holmsen et al., 1971; Robblee and Shepro, 1976). This is assumed to reflect antagonism of Ca^{2+} handling at the cell surface, but the actual mechanism may be more complicated than this. Best et al. (1980) noted that $0.1–0.5 \, mM \, La^{3+}$ markedly increased the basal and prostaglandin E_1-stimulated production of cAMP. For some reason, this stimulation was not reversed by the addition of Ca^{2+} in concentrations as high as $10 \, mM$. This effect of La^{3+} is in marked contrast to that found in adrenocortical cells (Haksar et al., 1976) and Purkinje cells (Nathanson et al., 1976), where the production of cAMP is inhibited. As platelet adenylate cyclase, like the Purkinje cell enzyme, was inhibited by La^{3+} ions, it is highly unlikely that La^{3+} penetrates the platelet plasma membrane. La^{3+} ions also prevented the inhibition of cAMP accumulation in platelets exposed to ADP, but not that produced by epinephrine. Sauerheber et al. (1980) suggested that La^{3+} ions inhibit certain platelet responses by decreasing lipid fluidity. As a result of binding to the external surface, La^{3+} ions reduce the surface charge of platelets and alter their electrophoretic mobility (Kosztolanyi et al., 1977).

At concentrations which increased cAMP levels, La^{3+} inhibited 5-hydroxytryptamine secretion by washed platelets in response to collagen, thrombin, and sodium arachidonate (Best et al., 1980). La^{3+} $(1 \, mM)$ also inhibited thrombin-induced release of serotonin, without altering the basal release (Zilberman et al., 1982). However, Worner and Brossmer (1976) recorded release of serotonin in response to $1 \, mM \, La^{3+}$, Nd^{3+}, or Tm^{3+}, although $10 \, mM$ concentrations were inhibitory. These authors also identified concentrations of La^{3+} which potentiated release of serotonin in response to collagen or thrombin. They concluded that this reflected interaction of La^{3+} with the activator molecules rather than with the platelets.

Millimolar concentrations of Tb^{3+} were unable to substitute for Ca^{2+} during platelet aggregation and inhibited aggregation induced by ADP (Holmsen *et al.*, 1971; Del Principe *et al.*, 1984). Inhibition was stronger for Tm^{3+} than for Nd^{3+} and weakest for La^{3+} (Worner and Brossmer, 1976). However, aggregation was strongly induced by Tb^{3+} in the presence of A23187, although, under these conditions, Tb^{3+} inhibited the secretion of ATP. Thus, Tb^{3+} presumably does not normally penetrate the limiting membrane of platelets. High concentrations of Ln^{3+} ions cause nonspecific clumping of platelets.

6.4.7 Mast Cells

Mast cells have been among the most widely studied model systems with which to study Ca^{2+}-dependent, stimulus-coupled secretion. The rise in cytosolic Ca^{2+} can occur by recruitment from intracellular or extracellular locations, depending upon the experimental conditions and nature of the secretagogue. Foreman and Mongar (1973) first reported that low concentrations ($\sim 100\,\mu M$) of La^{3+} ions reversibly and strongly inhibited anaphylactic histamine secretion. At concentrations of around $10\,\mu M$, La^{3+} provoked a spontaneous release of histamine. Pearce and White (1981) examined this phenomenon in greater detail, describing interesting variations dependent upon the concentration and nature of the Ln^{3+} ion and also upon the type of agonist.

With allergen as an agonist, Pr^{3+}, Nd^{3+}, Dy^{3+}, and Lu^{3+} inhibited histamine secretion to an approximately equal degree in a dose-dependent manner over the range $10^{-9}-10^{-3}\,M$. La^{3+} gave an anomalous, triphasic dose–response behavior (Fig. 6-7). With anti-IgE as an activator, only Lu^{3+} showed a straightforward dose–response relationship, while the other Ln^{3+} displayed odd bi- or triphasic behavior (Fig. 6-7). Indeed, at 10^{-5} and $10^{-4}\,M$, Nd^{3+}, Lu^{3+}, and La^{3+} stimulated histamine release. Histamine release in response to concanavalin A was inhibited by 10^{-4} or $10^{-6}\,M\,Ln^{3+}$ in the order $La^{3+} < Pr^{3+} < Nd^{3+}$, Sm^{3+}, Eu^{3+}, Dy^{3+}, Lu^{3+}. The relative effects of various Ln^{3+} ions on activation by peptide 401 depended on whether or not Ca^{2+} was present in the extracellular medium (Fig. 6-8). The inhibitory effects of Lu^{3+} and La^{3+}, but not Dy^{3+}, were reversed by washing the cells with buffer. It is interesting to note that the expected antagonism between Ca^{2+} and Ln^{3+} ions was only seen for $La^{3+}-Dy^{3+}$. With Er^{3+}, Yb^{3+}, and Lu^{3+}, inhibition was greater in the presence of Ca^{2+}. Histamine release induced by dextran, which is

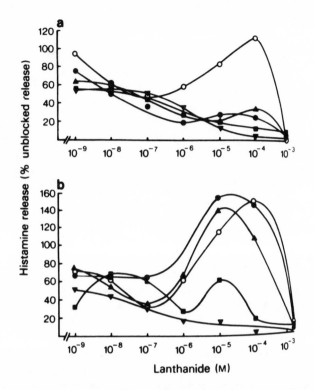

Figure 6-7. Effect of various Ln^{3+} ions on histamine release from mast cells induced by (a) allergen and (b) anti-IgE. ○, La^{3+}; ▲, Pr^{3+}; ●, Nd^{3+}; ■, Dy^{3+}; ▼, Lu^{3+}. From Pearce and White (1981), with permission.

totally dependent upon the presence of phosphatidylserine and extracellular Ca^{2+}, was also inhibited by Ln^{3+} ions, although this may have been due to interaction of Ln^{3+} ions with the phosphatidylserine. Whereas inhibition in the presence of extracellular Ca^{2+} was immediate, in the absence of Ca^{2+} the degree of inhibition by La^{3+} concentrations of $<10^{-4} M$ increased with preincubation times of up to 20 min. Ouabain had little effect on the dose–response curve for inhibition by La^{3+} (Frossard et al., 1983).

Lanthanides induced histamine release when transported into mast cells with A23187 (Amellal and Landry, 1983), strongly suggesting that the inhibitory effects described above were mediated via interaction at the external cell surface. In the concentration range $10^{-5}-10^{-3} M$, Tb^{3+} ions were slightly more effective than Ca^{2+} at provoking histamine release in the presence of A23187; but, unlike Ca^{2+}, Tb^{3+} at concentrations above

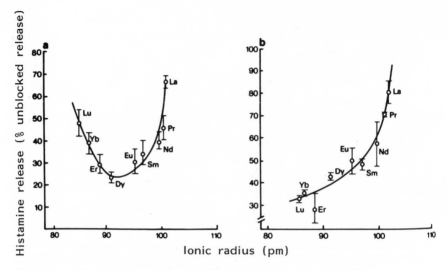

Figure 6-8. Effect of ionic radius on inhibition by $10^{-6}M$ Ln^{3+} ions of histamine release produced by peptide 401 in (a) the presence or (b) the absence of 1 mM Ca^{2+}. From Pearce and White (1981), with permission.

1 mM were inhibitory. Qualitatively similar results were obtained with La^{3+} ions. The effects of low concentrations of Tb^{3+} and Ca^{2+} were additive. As Ln^{3+}-dependent stimulation of histamine release was sensitive to inhibitors of calmodulin function, Amellal and Landry (1983) suggested that Ln^{3+} ions activate calmodulin-dependent reactions in the same way that Ca^{2+} ions do. Such activation by Ln^{3+} has been shown biochemically (Section 4.4.3).

At concentrations of 1–10 μM, which inhibit the Ca^{2+} "spike," La^{3+} inhibited histamine release from a rat basophilic cell line in response to aggregated ovalbumin or concanavalin A (Beaven et al., 1984).

6.4.8 Adrenal Tissue

The adrenal medulla contains chromaffin cells which secrete the catecholamines epinephrine, otherwise known as adenaline, and norepinephrine, under the control of the sympathetic nervous system. The adrenal cortex is responsible for synthesizing and secreting steroids under the control of adrenocorticotropic hormone (ACTH) and angiotensin II. Lanthanides affect secretory processes in both the cortex and the medulla.

Gd^{3+} had no effect on the spontaneous release of norepinephrine in cultured chromaffin cells but inhibited responses to high K^+ or acetylcholine (Bourne and Trifaro, 1982). However, perfusion of the unstimulated, isolated, bovine adrenals with a solution containing $0.5\,mM\,La^{3+}$ provoked a substantial release of catecholamine over an 8-min period. A second round of exposure to the same concentration of La^{3+} provoked very little release (Borowitz, 1972) and strongly inhibited catecholamine release in response to acetylcholine or high K^+. It was suggested that the first exposure of the medulla to La^{3+} displaced superficially bound Ca^{2+}, which entered the cell, provoking granule release. With La^{3+} now occupying these surface sites, further exocytosis was inhibited.

As the stimulation of catecholamine release by La^{3+} ions required Ca^{2+} ions but was insensitive to verapamil, H^+, or, of course, lanthanides, Ng *et al.* (1982) proposed that Ca^{2+} entered the chromaffin cells by an unusual route in the presence of La^{3+} ions. The potencies of various lanthanides in provoking catecholamine release increased with ionic radius in the sequence $Nd^{3+} < Pr^{3+} < Ce^{3+} < La^{3+}$; the smaller lanthanides Eu^{3+}, Gd^{3+}, Dy^{3+}, and Yb^{3+} were very weak in this respect (Borowitz and Noller, 1977). Nevertheless, Cohen and Gutman (1979) failed to elicit catecholamine release from isolated rat adrenals with $1\,mM\,La^{3+}$. Although La^{3+} ions inhibited secretion in response to K^+ or acetylcholine, they did not modulate the activation caused by agents such as tyramine, salbutamol, or theophylline, which do not require external Ca^{2+} ions.

The initial burst of amine release in response to Ln^{3+} ions may also occur *in vivo*. Arvela (1979) noted that rat adrenals contained 24% less catecholamine after intravenous (i.v.) injection with a 2-mg/kg dose of Ce^{3+} (Section 8.7.2). Certain of the pharmacologic effects of lanthanides might be explained on this basis (Chapter 8).

Although La^{3+} ions did not alter basal secretion of aldosterone from isolated glomerular cells, they abolished its production in response both to the physiological activators angiotensin II and ACTH and to the experimental activator ouabain (Schiffrin *et al.*, 1981). These findings were essentially confirmed by Shima *et al.* (1978) and Fakunding and Catt (1980). However, both of these groups also recorded inhibition of the basal level of aldosterone production.

The notion that La^{3+} ions block aldosterone production by abolishing the influx of Ca^{2+} is supported by the work of Haksar *et al.* (1976). They found micromolar concentrations of La^{3+} to inhibit both steroidogenesis and increases in cAMP within isolated adrenal cortex cells after treatment with ACTH. However, steroidogenesis stimulated by dibutyryl cAMP or glucose was not sensitive to La^{3+}.

6.4.9 Pancreas

Treatment of isolated pancreatic acini with the hormone cholecystokinin (CCK) stimulates incorporation of [^{3}H]phenylalanine into pancreatic proteins, a process which is linked to the increased synthesis and secretion of digestive enzymes by the pancreas. Following a 70-min preincubation, concentrations of La^{3+} ions above $10^{-4}M$ reduced the basal rate of assimilation of [^{3}H]phenylalanine by diabetic rat acini over a 15-min period. This appeared to reflect inhibition of amino acid transport, rather than protein synthesis as such (Bieger *et al.*, 1975, 1977; Korc, 1983). At a concentration of $10^{-4}M$, La^{3+} ions inhibited neither amino acid uptake nor oxidative phosphorylation but strongly inhibited the stimulation of protein synthesis induced by the hormones CCK and carbachol, both of which are thought to act through Ca^{2+}-dependent mechanisms. In contrast, protein synthesis stimulated by insulin, whose mode of action is Ca^{2+} independent, was not inhibited by La^{3+} ions. The Ca^{2+}-flux experiments discussed in Secton 6.3.3 revealed that La^{3+} ions inhibited ^{45}Ca^{2+} influx without affecting efflux.

The release of α-amylase by pancreatic acini in response to CCK, bethanechol (Chandler and Williams, 1974), or carbonylcholine (Bieger *et al.*, 1975) was also inhibited by La^{3+} ions. Basal release of α-amylase was inhibited by concentrations of La^{3+} above 1 mM but was stimulated by 0.1 mM La^{3+}. Experiments by Oberdisse *et al.* (1976) suggest that this can also happen *in vivo*. These authors injected a 10-mg/kg dose of Pr^{3+} intravenously into rats and noted a strong consequent reduction in both the basal and carbachol-stimulated rates of secretion of trypsin and α-amylase. However, in contrast to the *in vitro* results, they found that protein synthesis was not strongly inhibited. Instead, they suggested interference with the secretory mechanism itself, linking this to EM evidence of structural damage to the Golgi.

At a concentration of 0.5 mM, La^{3+} inhibited both basal and glucose-stimulated secretion of insulin by isolated pancreatic islets (Flatt *et al.*, 1980a,b). Under certain conditions, it was possible to detect inhibition with as little as 30 μM La^{3+}. After washing away the lanthanide, the islets quickly recovered, producing a marked rebound effect of spontaneous insulin release (Fig. 6-9) both in the presence and the absence of glucose. This was not associated with increases in cellular cAMP, and 1 mM La^{3+} was without effect on the rate of glucose exudation. This rebound was inhibited by re-exposure to La^{3+} ions or to epinephrine but was potentiated by 3-isobutyl-1-methylxanthine, an inhibitor of phosphodiesterase. The authors suggested that the rebound effect was due to accumulation of cytoplasmic Ca^{2+} in the presence of La^{3+} ions. It should be noted,

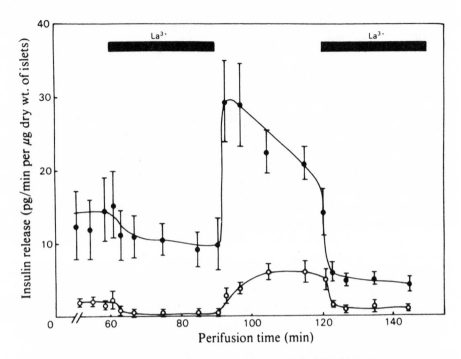

Figure 6-9. Dynamics of insulin release in response to 0.5 m*M* La^{3+} in the presence or absence of glucose during perifusion in Ca^{2+}-deficient medium. Experiments were performed in the presence (●) or absence (○) perifusion of 20 m*M* glucose. From Flatt *et al.* (1989a), with permission.

however, that Lorenz *et al.* (1979) found no secondary stimulation of spontaneous insulin release after removal of La^{3+} ions, instead noting a persistent suppression of glucose-stimulated insulin secretion. The inhibitory effects of La^{3+} ions and epinephrine were additive.

Intravenous injection of lanthanides causes a transient increase in plasma concentration of insulin. This is followed by a secondary decline to very low plasma levels which do not respond to glucose (Section 8.7.4).

The ^{45}Ca^{2+}-flux experiments described in Section 6.3.3 suggested that La^{3+} ions inhibit insulin secretion by blocking Ca^{2+} influx and efflux. However, Tm^{3+} ions were able to inhibit insulin release at concentrations where Ca^{2+} fluxes were unaffected. Furthermore, removal of Tm^{3+} did not produce the secondary stimulation of spontaneous insulin release seen with La^{3+}. Flatt *et al.* (1981) drew attention to the possible existence on the β-cell surface of regulatory sites at which Tm^{3+} modulates insulin secretion independently of Ca^{2+} transport. These could be the same sites which are responsible for suppressing insulin secretion at high concentrations of Ca^{2+} (Flatt *et al.*, 1980b).

6.4.10 Erythrocytes

We have already seen in Section 6.3.3 how La^{3+} ions bind to the erythrocyte surface and inhibit active Ca^{2+} efflux. This interaction has several consequences for the cell. By inhibiting Ca^{2+} efflux, La^{3+} ions maintained the spherical shape of spherochinocytotic erythrocytes. However, La^{3+} ions did not affect the shape of fresh human erythrocytes (Szasz *et al.*, 1978) nor alter the change to a spheroid shape upon ATP depletion.

At a concentration of $26.9\,\mu M$, La^{3+} ions protected chick erythrocytes from hypotonic hemolysis (Basu *et al.*, 1983) although their morphologies were greatly altered. The dose response of protection was biphasic. Erythrocytes recovered from chicks which had been injected intraperitoneally with La^{3+} were also protected from hypotonic lysis. In rats, moderate doses of lanthanides decreased the osmotic fragility of erythrocytes, while high doses rendered the cells more osmotically sensitive (Godin and Frohlich, 1981). Under *in vitro* conditions, concentrations of La^{3+} ions above $300\,\mu M$ caused erythrocytes to clump while increasing hemolysis (Sarkadi *et al.*, 1977). However, at a concentration of $10\,mM$, lanthanides have been reported to induce erythrocyte agglutination and promote cell fusion *in vitro* (Majumdar *et al.*, 1980). Enhanced fusion of artificial membranes in the presence of lanthanides has been described (Blioch *et al.*, 1968; Bentz *et al.*, 1988; Section 6.2.1).

6.4.11 Polymorphonuclear Leukocytes

At a concentration of $5\,mM$, Ho^{3+} ions inhibited the ability of polymorphonuclear (PMN) leukocytes to internalize *Staphylococcus aureus* by about 42% in 30 min. Ho^{3+} ions also changed the morphologies of the nuclei and prevented the characteristic shape change that accompanies contact of the cells with erythrocytes (Mircevova *et al.*, 1984). Lanthanides inhibit phagocytosis by the Kupffer cells of the liver *in vivo* (Section 8.10). It has been suggested that displacement of superficial Ca^{2+} from the outer membranes of these cells prevents attachment of the particles to the cell surface.

Lanthanides inhibit PMN and macrophage chemotaxis much more strongly, without changing the cellular morphology (Boucek and Snyderman, 1976). La^{3+} ions also suppress motility of PMN in the absence of an attractant. Aggregation induced by the chemoattractant formyl-Met-Leu-Phe (fMLP) was strongly inhibited by $10^{-6}-10^{-5}\,M\,La^{3+}$ (O'Flaherty *et al.*, 1978), concentrations at which chemotaxis was much more weakly inhibited. However, $10^{-5}-10^{-3}\,M\,La^{3+}$ spontaneously aggregated the cells in a Ca^{2+}-independent manner. This aggregation phenomenon has been observed with erythrocytes and other types of cells. It may reflect the ability of Ln^{3+} ions to act nonspecifically as a bridge between anionic sites on adjacent cells.

In the absence of Ca^{2+}, $10^{-5}-10^{-7}\,M\,La^{3+}$ had little effect on the release of lysozyme in response to fMLP. However, 10^{-3} and $10^{-4}\,M$ La^{3+} enhanced release. In the presence of $1.7\,mM\,Ca^{2+}$, La^{3+} produced a modest, biphasic, inhibition of the release of lysozyme and β-glucuronidase, without increasing the leakage of lactate dehydrogenase. The greater sensitivity of chemotaxis and aggregation to La^{3+} may result from their greater reliance upon external sources of Ca^{2+}.

6.4.12 Amphibian Bladder and Skin

The wall of the urinary bladder, which serves as a semipermeable barrier between the blood and urine, is the site of important transmembrane fluxes of water, H^+, and Na^+. Each of these is under hormonal control. In amphibians, the skin is responsible for similar physiological functions. As with other membranes which selectively transport metabolites from one side to the other, they exhibit polarity. The manner in which La^{3+} ions affect their metabolism depends upon the side of the membrane to which they are applied.

About 20 min after the serosal application of 1 or $1.5\,mM\,La^{3+}$ to isolated toad bladder, there commenced a steady decline in the transmembrane surface potential that continued for up to 3 h. While the potential difference decreased, the short-circuit current (scc) increased, indicating reduced resistance across the bladder (Strum, 1977). As the scc is carried by Na^+, its increase reflects enhanced transport of sodium across the wall of the bladder. Mucosal addition of La^{3+} also caused the scc to increase by about 35% in 30 min, after which time it slowly decreased, reaching a level 25% above controls by 1 h. The potential difference reflected these changes, decreasing 25% at 30 min, but recovering to only 8% below normal after 1 h. Although Sabatini and Arruda (1984) confirmed the mucosal effects of La^{3+} ions in toad bladder, they found that mucosal addition of 0.5 or $1\,mM\,La^{3+}$ to turtle bladder decreased the scc by about a half. This inhibition was not greatly pH dependent in the pH 5.4–7.4 range.

Other workers (e.g., Hille *et al.*, 1975; Hardy *et al.*, 1979) have confirmed that La^{3+} ions increase basal rates of Na^+ transport across the toad bladder. However, La^{3+} ions inhibit the stimulated transport of Na^+ which follows exposure to vasopressin or cAMP (Strum, 1977). After 2-h equilibration with La^{3+} ions, the response to cAMP or vasopressin was completely eliminated. Mucosal equilibration with $1.5\,mM\,La^{3+}$ for 1 h produced a delayed but much weaker inhibition of Na^+ transport in response to vasopressin.

Serosal addition of 1 or $1.5\,mM\,La^{3+}$ greatly and rapidly inhibited the flow of water across the toad bladder in response to vasopressin, antidiuretic hormone, or hypertonicity (Strum, 1977; Hardy *et al.*, 1979). According to Hardy *et al.* (1979), serosal application of $5\,mM\,La^{3+}$ increased the basal rate of transepithelial water flux in the toad bladder. However, others (Bourguet *et al.*, 1981; Wietzerbin *et al.*, 1974) reported that mucosal $10\,mM\,La^{3+}$ had little effect on water transport by the resting frog bladder, although it progressively abolished the increase in flux due to repeated application of oxytocin. Reversal of the response after removal of the hormone was also impaired.

Mucosal La^{3+} (1 mM) rapidly and reversibly inhibited H^+ secretion by turtle and toad bladder (Sabatini and Arruda, 1984). Inhibition by lanthanum was nearly eliminated by lowering the mucosal pH from 7.4 to 6.4. Serosal addition of $10\,mM\ La^{3+}$ to turtle bladder caused a slow decrease in H^+ secretion to a value 60% of that of controls after 2 h. Lower concentrations of La^{3+} ions had no effect. This response was only partially reversible.

The manner through which La^{3+} ions produce their effects on bladder is not clear. Strum (1977) reported that serosal La^{3+} ions affected the morphology of the epithelial cells. After 1 h, the granular cells showed the greatest changes and appeared to have swollen, producing fragmentation of the rough endoplasmic reticulum. However, the mitochondria appeared normal. After 2 h, nearly all the epithelial cells showed extensive deleterious alterations, which included vacuolation, separation from the basement lamina, and disruption of the desmosomes. However, the tight junctions (Section 3.7) remained intact, suggesting that La^{3+} ions stabilize these structures. Bladders exposed to La^{3+} ions on the mucosal side for 30 min to 1 h showed evidence of having phagocytosed lanthanum, as large heterophagic vacuoles containing deposits of lanthanum were observed.

In isolated toad bladders, high concentrations of Ca^{2+} ions, like La^{3+}, inhibited vasopressin-induced water transport more effectively than they inhibited vasopressin-induced Na^+ transport. Bourguet *et al.* (1981) have correlated structural changes in frogs' bladders with the antienergizing effect that La^{3+} ions have on oxytocin-induced water transport. Unlike Strum (1977), these authors found that 2 h exposure of the mucosal surface of frogs' bladders to $10 \, mM \, La^{3+}$ did not greatly alter morphology. However, also in contradiction to Strum (1977), they suggested that La^{3+} ions interfered with the integrity of tight junctions. They proposed that La^{3+} perturbed the fusion and turnover of cell surface aggregates which are important in the reaction to oxytocin. Sabatini and Arruda (1984) linked the action of La^{3+} ions to a transient, enhanced release of Ca^{2+} and a delayed, inhibited efflux of $^{45}Ca^{2+}$ after La^{3+} treatment. Modest inhibition of Ca^{2+} uptake was observed. Lanthanum also reduces the Na^+-dependent increase in oxygen uptake by toad bladders in the presence or absence of antidiuretic hormone (Cuthbert and Wong, 1971).

Amphibian skin is also an organ which transports water, Na^+, and H^+. Lanthanum has been shown to enhance Na^+ transport and to decrease water transport in response to vasopressin or serosal hypertonicity in frog skin (Wietzerbin et al., 1977). Mucosal La^{3+} ions increased the scc of frog skin, but repetitive addition caused the scc to fall (Goudeau et al., 1979). The increase in Na^+ transport in toad bladder and frog skin may be related to the masking of negative sites on the mucosal surface. It has been calculated that addition of $0.5 \, mM \, La^{3+}$ to the apical membrane of the frog skin decreased the surface potential by 24–28 mV (Hille et al., 1975). A mechanism involving charge masking is supported by the ability of other polyvalent cations to stimulate Na^+ transport.

6.4.13 Other Tissues and Cells

The remaining cellular and physiological effects of Ln^{3+} ions are summarized in Table 6-6. As we have seen in the previous discussions, many, but by no means all of them, involve inhibition of Ca^{2+}-requiring functions.

Although $0.1 \, mM \, La^{3+}$ had only a slight inhibitory effect on fluid secretion by insect salivary glands in response to 5-HT (5-Hydroxy-tryptamine or serotonin; Berridge et al., 1975), it strongly decreased the membrane potential in a manner that was completely reversed by $20 \, mM \, Ca^{2+}$. However, La^{3+} ions did not block the initial rapid depolarization that followed 5-HT stimulation. Fluid secretion was not inhibited by La^{3+} ions because it involved the release of Ca^{2+} from intracellular stores (Berridge et al., 1975).

The effects of Ln^{3+} ions on cell division are interesting. Hepler (1985) reported that $0.1 \, mM \, La^{3+}$ extended the metaphase of plant stamen hair

Table 6-6. Effects of Lanthanides on Other Tissues and Cells

Tissue or cell	Physiological parameter	Effect of Ln^{3+}	Reference
Frog esophageal mucosa	Pepsinogen secretion:		
	Basal level	($10\ mM\ La^{3+}$)	Fong, 1985
	Carbachol, cAMP stimulation	No effect	
		Inhibition	
Insect salivary gland	Fluid secretion	Slight inhibition by $0.1\ mM\ La^{3+}$	Berridge et al., 1975
Cockroach fat body	Activation of glycogen	Inhibited by $1.8\ mM\ La^{3+}$	McClure and Steele, 1981
	phosphorylase by trehalogen		
Marine sponge cells	Aggregation by a specific	All Ln^{3+} produce aggregation	Rice and Humphreys, 1983
	aggregation factor		
Plant stamen hair cells	Cell division	$0.1\ mM\ La^{3+}$ arrested	Hepler, 1985
3T3, 3T6 cells	DNA synthesis	Various Ln^{3+} stimulate at $5\ \mu M$	Smith and Smith, 1984
Xenopus oocytes	Meiotic maturation	La^{3+} acts as an initiator	Schorderet-Slatkine et al., 1976
Spermatazoa	Motility	Inhibition	Saling, 1982
	Binding to ova	Inhibition	
	Binding to zona pellucida	Unaffected	
Snake renal tubules	Transport of aminohippurate	$2\ mM\ La^{3+}$ inhibits	Dantzler and Brokl, 1984a
	Transport of urate	No effect	Dantzler and Brokl, 1984b
Mouse stomach	Fluid secretion induced by	Inhibition	Greenberg et al., 1982
	enterotoxin or 8-Br-cGMP		

Heart	"Calcium paradox"	Inhibition	Hunt and Willis, 1983
Mammary gland	Ejection of milk by oxytocin	Inhibition	Lawson and Schmidt, 1972
Gall bladder	Permeability to ions	Reduced cation / Increased anion	Machen et al., 1972
Adipocytes	Lipolysis	No effect	Schimmel, 1978
Parotid glands	Protein secretion		
	In response to carbachol	Inhibition	Keryer and Rossignol, 1978
	In response to norepinephrine	No effect	
	Phosphatidylinositol turnover	Inhibition	
Kidney	Renin release	Inhibition	Logan et al., 1977
	Renal vasoconstriction	Inhibition	
Thyroid	Calcitonin release	Inhibition	Pento, 1977
Amoeba	Movement	Inhibition	Hawkes and Holberton, 1973
	Pinocytosis	Induction	Josefsson and Hansson, 1976
Brush border	Glucose transport	Stimulation	Stevens and Kneer, 1988
	Proline transport	Stimulation	
Various	Binding and anti-viral effect of interferon	Enhanced	Sedmak et al., 1986

Table 6-7. Effects of La^{3+} on DNA Synthesis by
Cultured Murine Fibroblasts[a]

| Additions to medium | [³H]Thymidine incorporation (cpm $\times$ 10^{-3}/culture) | |
	Swiss 3T3	Swiss 3T6
None	4.4	8.7
La^{3+}	9.6	122.4
Al^{3+}	7.4	105.1
Al^{3+} + La^{3+}	7.2	122.1
Insulin	20.3	108.5
Insulin + La^{3+}	244.5	526.8
Insulin + Al^{3+}	210.8	528.6
Insulin + La^{3+} + Al^{3+}	197.9	556.8
Fetal bovine serum, 10%	456.4	597.6

[a] From Smith and Smith (1984), with permission.

cells. In some cases, metaphase was completely arrested. There was no change in the rate of cell plate initiation or in the rate of chromosome motion. However, micromolar concentrations of various Ln^{3+} ions accelerated another stage in the cell cycle, namely, the $G_0/G_1 \rightarrow S$ transition (Smith and Smith, 1984). As little as $1 \mu M$ La^{3+} had some stimulatory effect. In 3T3 fibroblasts, $5 \mu M$ La^{3+} acted synergistically with insulin or colchicine in stimulating DNA synthesis. In 3T6 fibroblasts, it had a direct effect of its own, which was nevertheless stimulated by insulin (Table 6-7). Although Ln^{3+} ions slowly precipitated in the culture medium, stimulation of DNA synthesis was independent of this effect, as a 2-h exposure of the cells to La^{3+} ions in buffer was sufficient for its induction. The mode of action is unclear, although it is significant that La^{3+} only slightly stimulated DNA synthesis in response to TPA (12-0-Tetradecanoylphorbol-13-acetate), which acts directly upon protein kinase C.

These findings may be related to the ability of La^{3+} and Tb^{3+} (0.01–1 mM) to induce the anchorage-independent growth of a preneoplastic, murine epidermal cell line. Ln^{3+} proved as effective as TPA in this regard, and were even able to induce such growth in a variant line which resisted induction by TPA. Unlike TPA, Ln^{3+} failed to activate protein kinase C in intact cells (Smith et al., 1986). Anchorage-independent growth is usually associated with malignant transformation. Although subcutaneously injected pellets of metallic lanthanides have been reported to cause neoplasms (Talbot et al., 1965; Ball et al., 1970), and $Y(NO_3)_3$ in the drinking water produces malignant tumors in rats (Schroeder and

Mitchener, 1971), the bulk of the evidence from animal studies suggests that lanthanides are not strong carcinogens (e.g., Ball and Van Gelder, 1966; MacDonald et al., 1950; Section 8.5). The therapeutic use of lanthanides as anti-cancer agents has been proposed (Section 9.5). Lanthanides inhibit DNA synthesis in lymphocytes (Yamage and Evans, 1989). They also reduce "patching" and "capping" (Horenstein and Emery, 1983), and inhibit increases in cAMP following treatment with isoprotererol (Borst and Conolly, 1988; Hui and Yu, 1988).

Whether these effects on DNA synthesis are related to the induction of meiotic maturation in *Xenopus* oocytes is unknown (Schorderet-Slatkine *et al.*, 1976). Induction by La^{3+} ions mimicked that of the natural inducer, progesterone, in that maturation was not abortive, and a similar range of new proteins was induced. When microinjected into the oocytes, La^{3+} ions were ineffective. Exposure of preimplantation murine embryos to 1 mM La^{3+} dramatically increases their ability to develop into blastocysts (Abramczuk, 1985). In addition, lanthanum parthenogenetically activates murine oocytes (Whittingham, 1980).

Various Ln^{3+} ions increase the efficiency with which muscarine receptor agonists elicit cGMP responses (El-Fakahany *et al.*, 1984). At the same time, they decrease the affinity of antagonists for these receptors. At a concentration of 3 mM, Tb^{3+} increased basal cGMP levels in these cells. Both 1 mM and 3 mM Tb^{3+} increased the effectiveness with which low concentrations of carbonylcholine increased cGMP levels, but there was a modest, dose-dependent inhibition at maximal carbonylcholine concentrations. The kinetics of cGMP production were also altered. At a concentration of 1 mM, Tb^{3+} increased the K_d for antagonists, without affecting the K_d for agonists. Under these conditions, the number of receptors on the cell surface did not appear to change. The authors suggested that the muscarine receptors in this cell line are linked to ROC (Receptor Operated Channel). As La^{3+} ions inhibit soluble preparations of guanylate cyclase, the effects on neuroblastoma cells are likely to be mediated at the cell surface.

Ln^{3+} inhibits the induction of ornithine decarboxylase in skin cells (Verma and Bontwell, 1981), the granulosa cells of the ovary (Veldhuis, 1982) and osteoblasts (Van Leeuwen *et al.*, 1988).

6.5 Cellular Organelles

6.5.1 Mitochondria

Lanthanides inhibit both energy-dependent transport (Mela, 1968) and energy-independent binding (Scarpa and Azzi, 1968) of Ca^{2+} by mi-

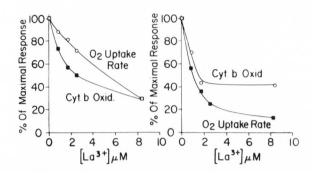

Figure 6-10. Effect of La^{3+} on respiratory rate and extent of cytochrome b oxidation induced by adding 850 μM Ca^{2+} to rat liver mitochondria. Left: In the presence of succinate and roterone; right: in the presence of glutamate, malate, and malonate. From Mela (1969a), with permission.

tochondria. In so doing, they inhibit Ca^{2+}-dependent activation of respiration, oxidation of cytochromes (Fig. 6-10), increases in intramitochondrial pH (Mela, 1969b), swelling (Selwyn *et al.*, 1970), and increases in acetyl-coenzyme A permeability (Benjamin *et al.*, 1983), without affecting oxidative phosphorylation or monovalent cation accumulation (Mela, 1969a,b). However, the nature of the La^{3+}-sensitive Ca^{2+} transporting sites, the type of inhibition involved, and whether mitochondria inwardly transport La^{3+} ions remain unresolved. As well as inhibiting the mitochondrial accumulation of Ca^{2+}, lanthanides reduce the uptake of Sr^{2+} and Ba^{2+}. High concentrations also suppress the uptake of Mn^{2+}, although enhancement occurs at low concentrations of lanthanides (Vainio *et al.*, 1970).

According to Mela (1969b), the inhibition of mitochondrial Ca^{2+} transport by Pr^{3+} was noncompetitive, with a K_i of 0.05 μM. However, both Reed and Bygrave (1974b) and Scarpa and Azzone (1970) have reported competitive inhibition kinetics, although providing a similar K_i value of 0.02 μM (Reed and Bygrave, 1974b). All these authors agree that the inhibitory effects of Ln^{3+} ions decline rapidly with time. Ca^{2+}-dependent oxygen consumption was strongly inhibited by the addition of 3.1 μM La^{3+} seconds before the introduction of Ca^{2+}, whereas a 5-min preincubation with La^{3+} ions had no effect on subsequent Ca^{2+}-stimulated oxygen consumption. La^{3+} ions had little effect on the basal rate of oxygen consumption during this period (Reed and Bygrave, 1974a). Similarly, the inhibition by La^{3+} of cytochrome oxidation disappeared in 3–4 min (Mela, 1969b); the rate of recovery depended upon the initial concentration of La^{3+} ions. Both sets of authors considered the transport of La^{3+} ions into the mitochondria as a possible explanation of the transient nature of the inhibition. Whether or not transport occurs is disputed (Reed and Bygrave, 1974a; Piccinini $et\,al.$, 1975; Hashimoto and Rottenberg, 1983). Neutron activation analysis of mitochondria has confirmed a small, but apparently real, accumulation of lanthanum in the mitochondrial matrix. However, studies using Tb^{3+} luminescence have led to the opposite conclusion (Hashimoto and Rottenberg, 1983). Although transport is an energy-dependent process, there was no difference in the Tb^{3+}-binding capacity or dissociation constant between energized and de-energized mitochondria. It was suggested that the stimulation of mitochondrial succinate oxidation in response to $75\,\mu M\,La^{3+}$, which had been used to support the notion of La^{3+} transport (Reed and Bygrave, 1974a), in fact reflected a nonspecific salt effect (Hashimoto and Rottenberg, 1983). The matter remains unresolved.

La^{3+} transport has also been invoked (Villani $et\,al.$, 1975) to explain the enhanced efflux of Ca^{2+} from the mitochondrion in response to La^{3+} ions (Carafoli and Rossi, 1971). The pool of exchangeable Ca^{2+} in mitochondria has been shown to increse from $2.1 \pm 0.3\%$ of the total calcium of control mitochondria to $9.6 \pm 1.2\%$ of the total calcium of mitochondria treated with $100\,\mu M\,La^{3+}$ (Villani $et\,al.$, 1975). However, the total calcium content of whole mitochondria was not affected by La^{3+}.

The La^{3+}-binding sites on mitochondria have not been clearly identified. Reed and Bygrave (1974b) reported a K_d value of $10\,\mu M$ with a B_{max} of $27-30$ nmol La^{3+}/mg protein. This K_d value is well above the K_i for Ca^{2+} transport reported in the same paper, suggesting that the mechanism of inhibition of Ca^{2+} transport may be complicated. The authors suggested that the Ca^{2+} carrier had a number of Ca^{2+}-binding sites but that the binding of La^{3+} to only one of these sites could inhibit transport. They also equated these La^{3+}-binding sites with the low-affinity Ca^{2+}-binding sites identified by Reynafarje and Lehninger (1969) as having a K_d of $42\,\mu M$ and B_{max} of 29 nmol La^{3+}/mg protein. This implies that the high-affinity binding sites are not inhibited by La^{3+}. This idea is supported by the observation that $100\,\mu M\,La^{3+}$, in the presence of $40\,\mu M\,Ca^{2+}$, displaces only a small portion of the Ca^{2+} bound by the mitochondrial membrane (Villani $et\,al.$, 1975). Nevertheless, Lehninger and Carafoli (1971) did not detect a class of mitochondrial Ca^{2+}-binding sites which were resistant to lanthanides. Hashimoto and Rottenberg (1983) were able to identify three classes of Tb^{3+}-binding site on the basis of K_d values and the half-life of the luminescence decay. These were: high-affinity ($K_d = 1.5\,\mu M$), fast-decay ($t_{1/2} \approx 0.18$ ms) sites (20% of total); high-affinity ($K_d = 1.5\,\mu M$), slow-decay ($t_{1/2} \approx 0.52$ ms) sites (10% of total); low-affinity ($K_d = 25\,\mu M$), slow-decay ($t_{1/2} \approx 0.52$ ms) sites (70% of total).

From the K_d values, it would seem that the low-affinity sites identified here correspond to the sites measured by Reed and Bygrave (1974b). The high-affinity sites may be the Ca^{2+} carrier of the inner mitochondrial membrane, which has a K_d for La^{3+} of $0.83\ \mu M$ (Lehninger and Carafoli, 1971). The binding sites on the inner membrane revealed by Tb^{3+} luminescence are unaffected by inhibitors of respiration or of oxidative phosphorylation. Luminescence titration studies suggest a pK_a of around 6 for the Tb^{3+}-binding ligands (Mikkelsen and Wallach, 1976).

The slow decay of Tb^{3+} luminescence occurred by transfer of energy to suitable chromophores, probably the heme groups of cytochromes. All the cytochromes, apart from cytochrome c, are buried in the membrane core, 20–50 Å from the surface. However, as cytochrome c depletion did not affect luminescence, it was suggested (Hashimoto and Rottenberg, 1983) that Tb^{3+} was brought into closer association with the other cytochromes in the membrane core. From the excitation spectrum of the luminescence of Tb^{3+} when bound to submitochondrial particles, it appeared that the donor chromophores for luminescence included Phe and Tyr residues.

As Tb^{3+} binds ionically, its luminescence could be used to estimate the surface potential. Measurements with $5\ \mu M\ Tb^{3+}$ suggested a positive local surface potential in intact mitochondria but a negative potential in submitochondrial particles. This was interpreted as reflecting two different types of binding site, possibly the matrix and cytoplasmic sides of the Ca^{2+} carrier (Hashimoto and Rottenberg, 1983). Niggli *et al.* (1978) have confirmed that the inhibition of Ca^{2+} transport by La^{3+} ions is not due to nonspecific surface charge effects.

The ability to inhibit mitochondrial Ca^{2+} transport is shared by most, if not all, lanthanides. Mela (1969a,b) showed that Ho^{3+}, Pr^{3+}, Ce^{3+} and Ce^{4+} ions all had similar potencies, inhibiting by 50% at about 0.07 nmol Ln^{3+}/mg protein. La^{3+} ions inhibited a little less strongly, producing 50% inhibition at 0.2 nmol/mg. Reed and Bygrave (1974b) found La^{3+} and Nd^{3+} to be indistinguishable in their ability to inhibit Ca^{2+} transport. However, the ability to inhibit Ca^{2+} binding to the outer surface of mitochondria in the absence of transport does change across the lanthanide series (Tew, 1977). Whereas the ability to inhibit respiration-independent Ca^{2+} binding increased with ionic radius, inhibition of respiration-dependent Ca^{2+} binding was strongest for those Ln^{3+} whose ionic radii are closest to that of Ca^{2+} (Fig. 6-11). However, Crompton *et al.* (1979) found that the ability of Ln^{3+} ions to inhibit Ca^{2+} influx increased with decreasing ionic radius, whereas the larger Ln^{3+} ions were better inhibitors of Ca^{2+} efflux. If both sets of experimental results are valid, they suggest that the binding of Ln^{3+} ions to mitochondria and their interference with Ca^{2+} fluxes are dissociable events.

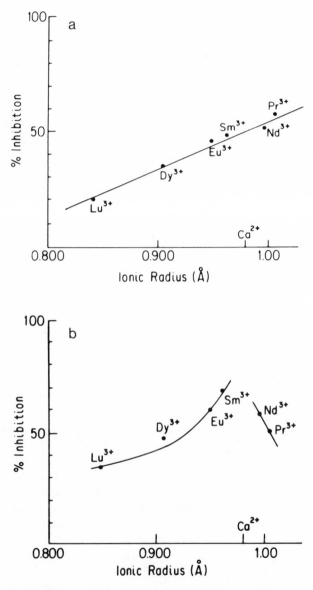

Figure 6-11. (a) Lanthanide inhibition of (a) respiration-independent and (b) respiration-dependent Ca^{2+} binding. From Tew (1977), with permission.

6.5.2 Endoplasmic Reticulum

In contrast to the situation with mitochondria, the accumulation of Ca^{2+} by the endoplasmic reticulum (ER) is relatively insensitive to La^{3+} ions. Indeed, early experiments (Entman *et al.*, 1969) failed to detect any effect on Ca^{2+} fluxes or on the Ca^{2+}-ATPase activity of canine cardiac microsomes. Later work suggested that the $10^{-4}M$ and $10^{-5}M$ concentrations of La^{3+} used by these authors were subinhibitory. Furthermore, as with mitochondria, the inhibition of Ca^{2+} transport into ER by La^{3+} ions declines with time (Krasnow, 1972). As Entman *et al.* (1969) preincubated the microsomes with La^{3+} ions for 20 min, they would have missed the early inhibitory phase. The results of many of these experiments are a little hard to interpret, as millimolar concentrations of ATP are included in the reaction. Not only does ATP have a high affinity for Ln^{3+} ions (Section 5.1; Table 5-3), but the phosphate released as the ATP hydrolyzes will precipitate lanthanides. Indeed, the latter event may explain why inhibition declines with time. An alternative explanation involves the transport of La^{3+} ions into the microsomes (Krasnow, 1972). However, EM studies have failed to detect such uptake (Dos Remedios, 1977). Low concentrations of lanthanides have no effect on basal rates of Ca^{2+} efflux from ER (Krasnow, 1977), although millimolar amounts are inhibitory (Chiesi and Inesi, 1979).

The affinity of La^{3+} ($K_d = 6.5\,\mu M$) for the high-affinity Ca^{2+}-binding sites of rabbit skeletal muscle sarcoplasmic reticulum is lower than that of Ca^{2+} ($K_d = 1.3\,\mu M$). La^{3+} also binds to the low-affinity Ca^{2+} sites with a K_d value of $6.5\,\mu M$, whereas K_d for Ca^{2+} is $32\,\mu M$. Gd^{3+} and Yb^{3+} are also able to displace Ca^{2+} competitively from the sarcoplasmic reticulum (Chevallier and Butow, 1971). The K_d for Gd^{3+} ions is also in the micromolar range, with about 100 nmol/mg of these high-affinity sites existing on the ER. More than 300 nmol/mg of the low-affinity Gd^{3+} sites, which did not bind Ca^{2+} strongly, were also detected. According to Krasnow (1977), binding of Gd^{3+} to the ER was greater in the presence of ATP. However, Barry *et al.* (1979) noted that ATP provoked the release of 87% of $^{153}Gd^{3+}$ bound to the surface of sarcoplasmic reticulum from rabbit muscle. The poor ability of Ln^{3+} ions to compete for Ca^{2+} on the ER is reflected in the high concentrations that need to be added before inhibition of Ca^{2+} uptake is observed (Krasnow, 1972; Batra, 1973). At such concentrations, microsomal Ca^{2+}-ATPase is also inhibited. Inhibition of Ca^{2+} uptake has been reported to be noncompetitive (Krasnow, 1972), with 50% inhibition occurring with $125\,\mu M\ La^{3+}$; Gd^{3+} is a slightly weaker inhibitor.

In addition to using the Ca^{2+}-ATPase-driven reaction, microsomes can accumulate Ca^{2+} via a $Na^{+}–Ca^{2+}$ exchanger (Section 6.3.1). In rat brain microsomes, this form of Ca^{2+} uptake was only inhibited when the concentration of La^{3+} ions exceeded $10\,\mu M$ (Schellenberg and Swanson, 1981), although 50% inhibition by $10\,\mu M\,La^{3+}$ was reported for heart sarcolemmal vesicles (Reeves and Sutko, 1979). La^{3+} uncompetitively inhibits the Na^{+}/K^{+}-ATPase of guinea pig heart microsomes (Nayler and Harris, 1976).

6.5.3 Other Organelles

Lanthanides bind to the outer surface of chromaffin granules, decreasing the negative surface charge (Matthews *et al.*, 1972), promoting aggregation and producing structural changes in the core (Morris and Schober, 1977). Terbium luminescence studies suggest a K_d for Tb^{3+} of about $15\,\mu M$, with competitive inhibition by Ca^{2+} and Mg^{2+} (Morris and Schober, 1977). Luminescence was maximum with an excitation wavelength of 285 nm; endogenous tryptophan fluorescence was concomitantly diminished.

Upon addition of Tb^{3+} to preparations of spinach chloroplast membranes, the luminescence of both the Tb^{3+} and chlorophyll is altered (Mills and Hind, 1978).

6.6 Summary

Most physiological responses of isolated cells, tissues, and organs to lanthanides can be explained in terms of the interactions between Ln^{3+} ions and the outer surface of their plasmalemmae. Lanthanides have a high affinity for cellular membranes, with K_d values in the micromolar range. In addition, there are usually one or more classes of lower-affinity sites, which are often present in greater number. Although lanthanides can bind to membrane proteins, to the sialic acid residues of membrane glycoproteins, and to phospholipid bilayers, it is unclear which of these serve as lanthanide receptors on intact cells. No consensus has emerged from the few types of cell yet examined in this regard (Table 6-2). Physiological studies suggest that the outer surfaces of Ca^{2+} channels should provide high-affinity binding sites for Ln^{3+} ions.

Despite their avid binding to the cell surface, Ln^{3+} ions are unable to penetrate artificial phospholipid bilayers and probably cannot enter healthy cells. However, there are sufficient numbers of dissenting reports to warrant caution in the latter case; the possibility that lanthanides enter specific types of cell under certain conditions cannot be excluded. When present in aggregates, lanthanides are internalized through phagocytosis.

Lanthanides stabilize phospholipid membranes, increasing rigidity, making the surface charge increasingly positive, and, at higher concentration, promoting aggregation and membrane fusion. With biological membranes, these effects are supplemented by increases in the membrane potential and membrane resistance, and perturbations in transmembrane Ca^{2+} fluxes. It is commonly observed that addition of lanthanides to populations of cells produces a rapid, apparent efflux of calcium as Ca^{2+} ions are competitively displaced from the external cell surface. This is followed by strong inhibition of Ca^{2+} influx. In many types of cells, lanthanide ions block all known routes of Ca^{2+} entry into the cell, including both voltage- and receptor-operated channels, the $Ca^{2+}-Na^+$ exchanger, the $Ca^{2+}-Ca^{2+}$ exchanger, and even the passive leakage of Ca^{2+} ions down the concentration gradient. The efflux of Ca^{2+} is generally more resistant to La^{3+} ions. Exceptions to this general scheme do exist. In erythrocytes, for instance, it is the efflux of Ca^{2+} which is particularly sensitive to lanthanides. With the pancreas, physiologically active doses of Tm^{3+} ions do not alter rates of Ca^{2+} transport. Furthermore, it has been reported that lanthanides increase the uptake of Ca^{2+} by hepatocytes. Nevertheless, very high doses of lanthanides can effectively stop all exchange of Ca^{2+} between many types of cells and their external media. This property forms the basis for the "lanthanum method" used in measuring Ca^{2+} fluxes (Section 3.9).

Because lanthanides usually block the cellular uptake of Ca^{2+}, they generally inhibit those physiological processes which depend upon Ca^{2+} influx. Most important in this regard are numerous types of stimulus-coupled cellular response. Thus, lanthanides block the transmission of nervous impulses, prevent the contraction of smooth, skeletal, and cardiac muscle, depress reticuloendothelial function, and inhibit a variety of hormonal responses. In general, those physiological processes which depend upon the release of intracellular calcium or which are mediated by an alternative second messenger such as a cyclic nucleotide are unaffected by lanthanides. Indeed, Ln^{3+} ions are sometimes used experimentally to distinguish between these sorts of mechanisms. However, in certain cells, lanthanides alter the concentrations of cyclic nucleotides. Furthermore, several rounds of cellular stimulation in the presence of lanthanides can deplete intracellular reservoirs of calcium.

Despite suppressing a wide variety of stimulus-coupled, exocytic events, there exist instances where lanthanides provoke them. Sometimes lanthanides increase the basal level of activity while suppressing that due to the presence of agonists. Examples include the release of serotonin from platelets, catecholamines from the adrenal medulla, and acetylcholine and other neurotransmitters from nerve terminals. In many such cases, this effect is strongly concentration dependent, while in others it depends upon the ionic radius of the lanthanide. Lanthanides also promote DNA synthesis in cultured fibroblasts, but not lymphocytes, and stimulate the early stages of embryogenesis.

Mitochondrial membranes have high affinities for lanthanides, possessing at least two classes of binding site. Lanthanides inhibit the uptake of Ca^{2+} by mitochondria, thereby preventing a number of physiological responses coupled to this process. The inhibitory effects of Ln^{3+} ions disappear rapidly, but it is still unclear whether this reflects the internalization of Ln^{3+} ions by mitochondria. There is a much greater likelihood that lanthanides can penetrate the ER, an organelle whose functions are relatively resistant to lanthanides.

References

Abramczuk, J. W., 1985. The effects of lanthanum chloride on pregnancy in mice and on preimplantation mouse embryos *in vitro, Toxicology* 34:315–320.

Akutsu, H., and Seelig, J., 1981. Interaction of metal ions with phosphatidylcholine bilayer membranes, *Biochemistry* 10:7366–7373.

Alnaes, E., and Rahaminoff, R., 1974. Dual action or praseodymium (Pr^{3+}) on transmitter release at the frog neuromuscular synapse, *Nature* 247:478–479.

Amellal, M., and Landry, Y., 1983. Lanthanides are transported by ionophore A23187 and mimic calcium in the histamine secretion process, *Br. J. Pharmacol.* 80:365–370.

Andersson, K. E., and Edman, K. A. P., 1974a. Effects of lanthanum on the coupling between membrane excitation and contraction of isolated frog muscle fibres, *Acta Physiol. Scand.* 90:113–123.

Andersson, K. E., and Edman, D. P. P., 1974b. Effects of lanthanum on potassium contractures of isolated twitch muscle fibres of the frog, *Acta Physiol. Scand.* 90:124–131.

Andrews, S. B., Faller, J. W., Billiam, J. M., and Barrnett, R. J., 1973. Lanthanide ion-induced isotrophic shifts and broadening for nuclear magnetic resonance structural analysis of model membranes, *Proc. Natl. Acad. Sci. USA* 70:1814–1818.

Arhem, P., 1980. Effect of rubidium, caesium, strontium, barium and lanthanum on ionic currents in myelinated nerve fibres from *Xenopus laevis, Acta Physiol. Scand.* 108: 7–16.

Arvela, P., 1979. Toxicity of rare-earths, *Prog. Pharmacol.* 2:71–114.

Axelrod, D., and Klein, M. P., 1974. Fluorescence Ca^{2+} analogs in nerve: rare-earth ions, *Biochem. Biophys. Res. Commun.* 57:927–933.

Baker, P. F., 1978. The regulation of intracellular calcium in giant axons of *Loligo* and *myxicola, Ann. N.Y. Acad. Sci.* 307:250–268.

Baker, P. F., Blaustein, M. P., Hodgkin, A. L., and Steinhart, R. S., 1969. Influence of calcium on sodium efflux in squid axons, *J. Physiol.* 200:431–458.

Ball, R. A., and Van Gelder, G., 1966. Chronic toxicity of gadolinium oxide for mice following exposure by inhalation, *Arch. Environ. Health* 13:601–608.

Ball, R. A., Van Gelder, G., and Green, J. W., 1970. Neoplastic sequelae following subcutaneous implantation of mice with rare earth metals, *Proc. Soc. Exp. Biol. Med.* 135:426–430.

Barry, K. J., Bloomquist, E., and Mikkelsen, R., 1979. The interaction of lanthanides with isolated sarcoplasmic reticulum vesicles from rabbit skeletal muscle, *Arch. Int. Physiol. Biochim.* 87:493–499.

Barry, W. H., Goldminz, D., Kimball, T., and Fitzgerald, J. W., 1978. Influence of cell dissociation and culture of chick embryo ventricle on inotropic responses to calcium and lanthanum, *J. Mol. Cell. Cardiol.* 10:967–979.

Barton, P. G., 1968. The influence of surface charge density of phosphatides on the binding of some cations, *J. Biol. Chem.* 243:3884–3890.

Bartsch, R. A., Grandjean, J., and Laszlo, P., 1983. A synthetic crown ether carboxylic acid ionophore displays synergistic transport of Pr^{3+} in conjunction with lasalocid, *Biochem. Biophys. Res. Commun.* 117:340–343.

Basu, A., Bhattacharyya, A., Chakrabarty, K., and Chatterje, G. C., 1983. Stabilization of chick erythrocyte membrane by lanthanum, *Indian J. Exp. Biol.* 21:1–4.

Basu, A., Chakrabarty, K., Haldar, S., Addya, S., and Chatterje, G. C., 1984. The effects of lanthanum chloride administration in newborn chicks on glutamate uptake and release by brain synaptosomes, *Toxicol. Lett.* 20:303–308.

Batra, S., 1973. The effects of zinc and lanthanum on calcium uptake by mitochondria and fragmented sarcoplasmic reticulum of frog skeletal muscle, *J. Cell. Physiol.* 82:245–256.

Beaven, M. A., Rogers, J., Moore, J. P., Hesketh, R., Smith, G. A., and Metcalfe, J. C., 1984. The mechanism of the calcium signal and correlation with histamine release in 2H3 cells, *J. Biol. Chem.* 259:7129–7136.

Benjamin, A. M., Murthy, C. R. K., and Quastel, J. H., 1983. Calcium-dependent release of acetyl-coenzyme A from liver mitochondria, *Can. J. Physiol. Pharmacol.* 61:154–158.

Bentz, J., Alford, D., Cohen, J., and Düzgünes, N., 1988. La^{3+}-induced fusion of phosphatidylserine liposomes. Close approach, intermembrane intermediates, and the electrostatic surface potential, *J. Biophys.* 53:593–607.

Berridge, M. J., Oschman, J. L., and Wall, B. J., 1975. Intracellular calcium reservoirs in *Calliphora* salivary glands, in *Calcium Transport in Contraction and Secretion* (E. Carafoli, ed.), North-Holland, Amsterdam, pp. 131–138.

Bers, D. M., Philipson, K. D., and Nishimoto, A. Y., 1980. Sodium–calcium exchange and sidedness of isolated cardiac sarcolemmal vesicles, *Biochim. Biophys. Acta* 601:358–371.

Best, L. C., Bone, E. A., Jones, P. B., and Russell, R. G., 1980. Lanthanum stimulates the accumulation of cyclic AMP and inhibits secretion and thromboxane B2 formation in human platelets, *Biochim. Biophys. Acta* 632:336–342.

Bevan, S., and Werdon, L. M. B., 1984. A study of the action of tetanus toxin at rat soleus neuromuscular junctions, *J. Physiol.* 348:1–17.

Bieger, W., Seybold, J, and Kern, H. F., 1975. Studies on intracellular transport of secretory proteins in the rat exocrine pancreas. III. Effect of cobalt, lanthanum and antimycin A, *Virchows Arch.* 368:329–345.

Bieger, W., Peter, S., Volkl, A., and Kern, H. F., 1977. Amino acid transport in the rat exocrine pancreas. II. Inhibition by lanthanum and tetracaine, *Cell Tissue Res.* 180: 45–62.

Bihler, I., Hoeschen, L. E., and Sawh, P. C., 1980. Effect of heavy metals and lanthanum on sugar transport in isolated guinea pig left atria, *Can. J. Physiol. Pharmacol.* 58: 1184–1188.

Blioch, Z. L., Glagoleva, I. M., Liberman, E. A., and Nenashev, V. A., 1968. A study of the mechanism of quantal transmitter release at a chemical synapse, *J. Physiol.* 199: 11–35.

Bockman, E. L., Rubrio, R., and Berne, R. M., 1973. Effect of lanthanum on isoproterenol-induced activation of myocardial phosphorylase, *Am. J. Physiol.* 225:438–443.

Bohr, D. F., 1974. Reactivity of vascular smooth muscle from normal and hypertensive rats: effects of several cations, *Fed. Proc.* 33:127–132.

Borowitz, J. L., 1972. Effect of lanthanum on catecholamine release from adrenal medulla, *Life Sci.* 11:959–964.

Borowitz, J. L., and Noller, A., 1977. Adrenal catecholamine release by trivalent metallic cations, *Arch. Int. Pharmacodyn. Ther.* 230:150–155.

Borst, S., and Conolly, M., 1988. Calcium dependence of beta-adrenoceptor mediated cyclic AMP accumulation in human lymphocytes. *Life Sci.* 43:1021–1029.

Boucek, M. M., and Snyderman, R., 1976. Calcium influx requirement for human neutrophil chemotaxis: inhibition by lanthanum chloride, *Science* 193:905–907.

Bourguet, J., Chevalier, J., and Gobin, R., 1981. Ultrastructural studies on the mode of action of antidiuretic hormone: the inhibitory effect of lanthanum, *Ann. N.Y. Acad. Sci.* 372:131–143.

Bourne, G. W., and Trifaro, J. M., 1982. The gadolinium ion: a potent blocker of calcium channels and catecholamine release from cultured chromaffin cells, *Neuroscience* 7:1615–1622.

Bowen, J. M., 1972. Effects of rare earths and yttrium on striated muscle and the neuromuscular junction, *Can. J. Physiol. Pharmacol.* 50:603–611.

Boyd, K. S., Melnykovych, G., and Fiskin, A. M., 1976. A replica method for visualizing the distributon of lanthanum-binding sites over HeLa surfaces, *Eur. J. Cell. Biol.* 14:91–101.

Brading, A. F., and Widdicombe, J. H., 1977. The use of lanthanum to estimate the numbers of extracellular cation-exchange sites in the guinea pig's taenia coli, and its effects on transmembrane monovalent ion movements, *J. Physiol.* 266:255–273.

Breer, H., and Jeserich, G., 1981. Calcium binding sites of synaptosomes from insect nervous system as probed by trivalent terbium ions, *J. Neurochem.* 37:276–282.

Bresnahan, S. J., Baugh, L. E., and Borowitz, J. L., 1980. Mechanisms of La^{3+}-induced adrenal catecholamine release, *Res. Commun. Chem. Pathol. Pharmacol.* 28:229–244.

Briggs, R. T., Drath, D. B., Karnovsky, M. L., and Karnovsky, M. J., 1975. Localization of NADH oxidase on the surface of human polymorphonuclear leukocytes by a new cytochemical method, *J. Cell Biol.* 67:566–586.

Brightman, M. W., and Reese, T. S., 1969. Junctions between intimately apposed cell membranes in the vertebrate brain, *J. Cell. Biol.* 40:648–677.

Brink, P. R., Kensler, R. W., and Dewey, M. M., 1979. The effect of lanthanum on the nexus of the anterior byssus reactor muscle of *Mytilus edulis* L., *Am. J. Anat.* 154:11–26.

Brown, M. F., and Sellig, J., 1977. Ion-induced changes in head group conformation of lecithin bilayers, *Nature* 169:721–723.

Burton, J, and Godfraind, T., 1974. Sodium-calcium sites in smooth muscle and their accessibility to lanthanum, *J. Physiol.* 241:287–298.

Burton, K. P., Hagler, H. K., Willerson, J. T., and Buja, L. M., 1981. Abnormal lanthanum accumulation due to ischemia in isolated myocardium: effect of chlorpromazine, *Am. J. Physiol.* 241:H714–723.

Butler, K. W., Dugas, H., Smith, I. C. P., and Schneider, H., 1970. Cation-induced organization changes in a lipid bilayer model membrane, *Biochem. Biophys. Res. Commun.* 40:770–776.

Bystrov, V. F., Dubrovina, N. I., Barsukov, L. I., and Bergelson, L. D., 1971. NMR differentiation of the internal and external phospholipid membrane surfaces using paramagnetic Mn^{2+} and Eu^{3+} ions, *Chem. Phys. Lipids* 6:343–350.

Bystrov, V. F., Shapiro, Y. E., Viktorov, A. V., Barsukov, L. I., and Bergelson, L. D.,

1972. ^{31}P-NMR signals from inner and outer surfaces of phospholipid membranes, *FEBS Lett.* 25:337–338.

Canada, R. G., 1983. Terbium binding to neoplastic GH3 pituitary cells, *Biochem. Biophys. Res. Commun.* 111:135–142.

Carafoli, E., and Rossi, C. S., 1971. Calcium transport in mitochondria, *Adv. Cytopharmacol.* 1:209–227.

Cartmill, J. A., and Dos Remedios, C. G., 1980. Ionic radius specificity of cardiac muscle, *J. Mol. Cell. Cardiol.* 12:219–223.

Chandler, D. E., and Williams, J. A., 1974. Pancreatic acinar cells: effects of lanthanum ions on amylase release and calcium ion fluxes, *J. Physiol.* 243:831–846.

Chevallier, J, and Butow, R. A., 1971. Calcium binding to the sarcoplasmic reticulum of rabbit skeletal muscle, *Biochemistry* 10:2733–2737.

Chiari, M. C., Crespic, V. C., Favalli, L., and Piccinini, F., 1980. Uptake and intracellular distribution of lanthanum, *Arch. Int. Physiol. Biochem.* 88:225–230.

Chiesi, M., and Inesi, G., 1979. The use of quench reagents for resolution of single transport cycles in sarcoplasmic reticulum, *J. Biol. Chem.* 254:10370–10377.

Chowdhry, B. Z., Lipka, G., Dalziel, A. W., and Sturtevant, J. M., 1984. Effect of lanthanum ions on the phase transitions of lecithin bilayers, *Biophys. J.* 45:633–635.

Chrzeszczyk, A., Wishnia, A., and Springer, C. S., 1981. Evidence for cooperative effects in the binding of polyvalent metal ions to pure phosphatidylcholine bilayer vesicle surfaces, *Biochim. Biophys. Acta* 648:28–48.

Claret-Berthon, B., Claret, M., and Mazet, J. L., 1977. Fluxes and distribution of calcium in rat liver cells: kinetic analysis and identification of pools, *J. Physiol.* 272:529–552.

Cohen, J., and Gutman, Y., 1979. Effects of verapamil, dantrolene and lanthanum on catecholamine release from rat adrenal medulla, *Br. J. Pharmacol.* 65:641–645.

Collins, G. A., Sutter, M. C., and Teiser, J. C., 1972. Calcium and contraction in the rabbit anterior mesenteric-portal vein, *Can. J. Physiol. Pharmacol.* 50:289–299.

Crompton, M., Heid, I., Baschera, C., and Carafoli, E., 1979. The resolution of calcium fluxes in heart and liver mitochondria using the lanthanide series, *FEBS Lett.* 104: 352–354.

Cushley, R. J, and Forrest, B. J., 1977. Structure and stability of vitamin E–lecithin and phytanic acid–lecithin bilayers studied by ^{13}C and ^{31}P nuclear magnetic resonance, *Can. J. Chem.* 55:220–226.

Cuthbert, A. W., and Wong, P. Y. D., 1971. The effect of metal ions and antidiuretic hormone on oxygen consumption in toad bladder, *J. Physiol.* 219:39–56.

Dantzler, W. H., and Brokl, O. H., 1984a. Effects of low [Ca^{2+}] and La^{3+} on PAH transport by isolated perfused renal tubules, *Am. J. Physiol.* 246:F175–187.

Dantzler, W. H., and Brokl, O. H., 1984b. Lack of effect of low [Ca^{2+}], La^{3+} and pyrazinoate on urate transport by isolated, perfused snake renal tubules, *Pflügers Arch.* 401:262–265.

DeBassio, W. A., Schnitzler, R. M., and Parsons, R. C., 1970. Influence of lanthanum on transmitter release at the neuromuscular junction, *J. Neurobiol.* 2:263–278.

Deleers, M., Gelbcke, M., and Malaisse, W. J., 1983. Gliclazide and glibenclamide-mediated transport of Pr^{3+} across an artificial lipid membrane, *Arch. Int. Phamacodyn. Ther.* 262:313–314.

Del Principe, D., Menichelli, A., DeMatteis, W., and Agro, A. F., 1984. Interaction of terbium with human platelets, *Biochem. Biophys. Res. Commun.* 122:311–318.

Deschenes, R. J., Mautner, H. G., and Marquis, J. K., 1981. Local anesthetics noncompetitively inhibit terbium binding to the exterior surface of nerve membrane vesicles, *Biochim. Biophys. Acta* 649:515–520.

Deth, R., and Lynch, C., 1981. Inhibition of alpha-receptor-induced Ca^{2+} release and Ca^{2+} influx by Mn^{2+} and La^{3+}, *Eur. J. Pharmacol.* 71:1–11.

Donis, J., Grandjean, J., Grosjean, A., and Laszlo, P., 1981. ^{31}P NMR spectroscopic study of Pr^{3+} transport by etheromycin and by synthetic ionophores, *Biochem. Biophys. Res. Commun.* 102:690–696.

Dos Remedios, C. G., 1977. Lanthanide ions and skeletal muscle sarcoplasmic reticulum. I. Gadolinium localization by electron microscopy, *J. Biochem.* 81:703–708.

Dos Remedios, C. G., 1981. Lanthanide ion probes of calcium-binding sites on cellular membranes, *Cell Calcium* 2:29–51.

Dunbar, S. J., 1982. The use of lanthanum as a tool to study calcium fluxes in an insect visceral muscle, *Comp. Biochem. Physiol.* 72A:199–204.

Durrett, L. R., and Adams, H. R., 1980. A comparison of the influence of La^{3+}, D600 and gentamicin on frequency–force relationships in isolated myocardium, *Eur. J. Pharmacol.* 66:315–325.

Ehrstrom, M., Eriksson, G., Israelachuili, J., and Ehrenberg, A., 1973. The effects of some cations and anions on spin labeled cytoplasmic membranes of *Bacillus subtilis*, *Biochem. Biophys. Res. Commun.* 55:396–402.

El-Fakahany, E., Lopez, J. R., and Richelson, E., 1983. Lanthanum binding to murine neuroblastoma cells, *J. Neurochem.* 40:1687–1691.

El-Fakahany, E. E., Pfenning, M., and Richelson, E., 1984. Kinetic effects of terbium on muscarine acetylcholine receptors of murine neuroblastoma cells, *J. Neurochem.* 42:863–869.

Entman, M. L., Hansen, J. L., and Cook, J. W., 1969. Calcium metabolism in cardiac microsomes incubated with lanthanum ion, *Biochem. Biophys. Res. Commun.* 35:258–264.

Entwistle, A., Zalin, R. J., Bevan, S., and Warner, A. E., 1988. The control of chick myoblast fusion by ion channels operated by prostaglandins and acetylcholine, *J. Cell Biol.* 106:1694–1702.

Evans, C. H., 1988. Alkaline earths, transition metals and lanthanides, in *Calcium in Drug Actions* (P. F. Baker, ed.), Springer-Verlag, Heidelberg, pp. 527–546.

Fakunding, J. L., and Catt, K. J., 1980. Dependence of aldosterone stimulation in adrenal glomerulosa cells on calcium uptake: effects of lanthanum and verapamil, *Endocrinology* 107:1345–1353.

Fawzi, A. B., and McNeill, J. H., 1985. Effect of lanthanum on the inotropic response of isoproterenol: role of the superficially bound calcium, *Can. J. Physiol. Pharmacol.* 63:1106–1112.

Fernandez, M. J,. and Cerbon, J., 1973. The importance of the hydrophobic interactions of local anesthetics in the displacement of polyvalent cations from artificial lipid membranes, *Biochim. Biophys. Acta* 298:8–14.

Fernandez, M. S., Celis, H., and Montal, M., 1973. Proton magnetic resonance detection of ionophore mediated transport of praesodymium ions across phospholipid membranes, *Biochim. Biophys. Acta* 323:600–605.

Fiskin, A. M., Melnykovych, G., and Peterson, G., 1980. Transient expression of lanthanum binding sites over surfaces of HeLa cells, *Eur. J. Cell Biol.* 21:151–159.

Flatt, P. R., Boquist, L., and Hellman, B., 1980a. Calcium and pancreatic beta-cell function. The mechanism of insulin secretion studied with the aid of lanthanum, *Biochem. J.* 190:371–372.

Flatt, P. R., Berggren, P. O., Gylfe, E., and Hellman, B., 1980b. Calcium and pancreatic beta-cell function. IX. Demonstration of lanthanide-induced inhibition of insulin secretion independent of modifications in transmembrane Ca^{2+} fluxes, *Endocrinology* 107: 1007–1013.

Flatt, P. R., Gylfe, E., and Hellman, B., 1981. Thulium binding to the pancreatic beta-cell membrane, *Endocrinology* 108:2258–2263.

Fong, J. C., 1985. Effects of ionophore A23187 and lanthanum on pepsinogen secretion from frog esophageal mucosa *in vitro, Biochim. Biophys. Acta* 814:356–362.

Foreman, J. C., and Mongar, J. L., 1973. The action of lanthanum and manganese on anaphylactic histamine secretion, *Br. J. Pharmacol.* 48:527–537.

Frank, J. S., and Rich, T. L., 1983. Calcium depletion and repletion in rat heart: age-dependent changes in the sarcolemma, *Am. J. Physiol.* 245:H343–353.

Frank, J. S., Langer, G. A., Nudd, L. M., and Seraydarian, K., 1977. The myocardial cell surface, its histochemistry and the effect of sialic acid and calcium removal on its structure and cellular ionic exchange, *Circ. Res.* 41:702–714.

Frossard, N., Amellal, M., and Landry, Y., 1983. Sodium-potassium ATPase, calcium and immunological histamine release, *Biochem. Pharmacol.* 32:3259–3262.

George, B. L., and Jarmakani, J. M., 1983. The effects of lanthanum and manganese on excitation–contraction coupling in the newborn rabbit heart, *Develop. Pharmacol. Ther.* 6:33–44.

Gerritsen, W. J, Van Zoelen, E. J. J., Verkleij, A. J., DeKruijff, B., and Van Deenen, L. L. M., 1979. A ^{13}C NMR method for determination of the transbilayer distribution of phosphatidylcholine in large, unilamellar, protein-free and protein-containing vesicles, *Biochim. Biophys. Acta* 551:248–259.

Gill, D. L., Grollman, E. F., and Kohn, L. D., 1981. Calcium transport mechanisms in vesicles from guinea pig brain synaptosomes, *J. Biol. Chem.* 256:184–192.

Godfraind, T., 1976. Calcium exchange in vascular smooth muscle, action of noradrenaline and lanthanum, *J. Physiol.* 260:21–35.

Godin, D. V., and Frohlich, J., 1981. Erythrocyte alterations in praseodymium-induced lecithin:cholesterol acyltransferase (LCAT) deficiency in the rate: comparison with familial LCAT deficiency in man, *Res. Commun. Chem. Pathol. Pharm.* 31:555–566.

Gogelein, H., DeSmedt, H., Van Driessche, W., and Borghgraef, R., 1981, The effect of lanthanum on alamethicin channels in blacklipid bilayers, *Biochim. Biophys. Acta* 640: 185–194.

Goldberg, M. T., and Triggle, C. R., 1977. An analysis of the action of lanthanum on aortic tissue from normotensive and spontaneously hypertensive rats, *Can. J. Physiol. Pharmacol.* 55:1084–1090.

Goodman, F. R., and Weiss, G. B., 1971a. Dissociation by lanthanum of smooth muscle responses to potassium and acetylcholine, *Am. J. Physiol.* 220:759–766.

Goodman, F. R., and Weiss, G. B., 1971b. Effects of lanthanum on ^{45}Ca movements and on contractions induced by norepinephrine, histamine and potassium in vascular smooth muscle, *J. Pharmacol. Exp. Ther.* 177:414–425.

Gordon, L. M., Sauerheber, R. D, and Esgate, J. A., 1978. Spin label studies on rat liver and heart plasma membranes: effects of temperature, calcium and lanthanum on membrane fluidity, *J. Supramol. Struct.* 9:299–326.

Goudeau, H., Wietzerbin, J., and Gary-Bobo, C. M., 1979. Effects of mucosal lanthanum on electrical parameters of isolated frog skin. Mechanism of action, *Pflügers Arch.* 379: 71–80.

Grandjean, J., and Laszlo, P., 1982. Complexation of Pr^{3+} by two different ionophores enhances markedly its transport role, *Biochem. Biophys. Res. Commun.* 104:1293–1297.

Grasdalen, H, Eriksson, L. E. G., Westman, J., and Ehrenberg, A., 1977. Surface potential effects on metal ion binding to phosphatidyl choline membranes. ^{31}P NMR study of lanthanide and calcium ion binding to egg-yolk lecithin vesicles, *Biochim. Biophys. Acta* 469:151–162.

Greenberg, R. N., Murad, F., and Guerrant, R. L., 1982. Lanthanum chloride inhibition of the secretory response to *Escherichia coli* heat-stable enterotoxin, *Infect. Immun.* 35: 483–488.

Grinvald, A., and Yaari, Y., 1978. Utilization of fluorescent lanthanide ions for the study of cation binding to extracellular sites in frog skeletal muscle, *Life Sci.* 22:1573–1584.

Haas, A., 1975. Effects of lanthanum, calcium and barium on the resting membrane resistance of guinea-pig papillary muscles. *Naunyn-Schmiedeberg's Arch. Pharmacol.* 290:207–220.

Hagiwara, S., and Byerly, L., 1981a. Membrane biophysics of calcium currents, *Fed. Proc.* 40:2220–2225.

Hagiwara, S., and Byerly, L., 1981b. Calcium channel, *Annu. Rev. Neurosci.* 4:69–125.

Haksar, A., Maudsley, D. V., Peron, F. G., and Bedigian, E., 1976. Lanthanum: inhibition of ACTH-stimulated cyclic AMP and corticosterone synthesis in isolated rat adrenocortical cells, *J. Cell Biol.* 68:142–153.

Hambly, B. D., and Dos Remedios, C. G., 1977. Responses of skeletal muscle fibres to lanthanide ions. Dependence of the twitch response on ionic radii, *Experientia* 33: 1042–1044.

Hansen, T. R., and Bohr, D. F., 1975. Hypertension, transmural pressure and vascular smooth muscle response in rats, *Circ. Res.* 36:590–598.

Hardy, M. A., Balsam, P., and Bourgoigne, J. J., 1979. Reversible inhibition by lanthanum of the hydrosmotic responses to serosal hypertonicity in toad urinary bladder, *J. Membr. Biol.* 48:13–19.

Hashimoto, K., and Rottenberg, H., 1983. Surface potential in rat liver mitochondria: terbium ions as a phosphorescent probe for surface potential, *Biochemistry* 22:5738–5745.

Hatae, J., 1982. Effects of lanthanum on the electrical and mechanical activities in the frog ventricle, *Jpn. J. Physiol.* 32:609–625.

Hauser, H., 1976. The conformation of the polar group of lecithin and lysolecithin, *J. Colloid Interface Sci.* 55:85–93.

Hauser, H., and Barratt, M. D., 1973. Effect of chain length on the stability of lecithin bilayers, *Biochem. Biophys. Res. Commun.* 53:379–405.

Hauser, H., Phillips, M. C., Levine, B. A., and Williams, R. J. P., 1975. Ion binding to phospholipids. Interactions of calcium and lanthanide ions with phosphatidylcholine (lecithin), *Eur. J. Biochem.* 58:133–144.

Hauser, H., Phillips, M. C., Levine, B. A., and Williams, R. J. P., 1976. Conformation of the lecithin polar group in charged vesicles, *Nature* 261:390–394.

Hauser, H., Hinckley, C. C., Krebs, J., Levine, B. A., Phillips, M. C., and Williams, R. J. P., 1977. The interaction of ions with phosphatidylcholine bilayers, *Biochim. Biophys. Acta* 468:364–377.

Hava, M., and Hurwitz, A., 1973. The relaxing effect of aluminum and lanthanum on rat and human gastric smooth muscle *in vitro*, *Eur. J. Pharmacol.* 22:156–161.

Hawkes, R. B., and Holberton, D. V., 1973. A calcium-sensitive inhibition of amoeboid movement, *J. Cell Physiol.* 81:365–370.

Heisler, S., and Grondin, G., 1973. Effect of lanthanum on ^{45}Ca flux and secretion by protein from rat exocrine pancreas, *Life Sci.* 13:783–794.

Hellman, B., Sehlin, J., and Taljedal, I. B., 1976. Effects of glucose on ^{45}Ca^{2+} uptake by pancreatic islets as studied with the lanthanum method, *J. Physiol.* 254:639–656.

Henkart, M., and Hagiwara, S., 1976. Localization of calcium binding sites associated with the calcium spike in barnacle muscle, *J. Membr. Biol.* 27:1–20.

Hepler, P. K., 1985. Calcium restriction prolongs metaphase in dividing *Tradescantia* stamen hair cells, *J. Cell Biol.* 100:1363–1368.

Heuser, J., and Miledi, R., 1971. Effect of lanthanum ions on function and structure of frog neuromuscular junctions, *Proc. Roy. Soc. B* 179:247–260.

Hille, B., Woodhull, A. M., and Shapiro, B. J., 1975. Negative surface charge near sodium channels of nerve: divalent ions, monovalent ions and pH, *Phil. Trans. Roy. Soc. B* 270:301–318.

Hinnen, R., Miyamoto, H., and Racker, E., 1979. Ca^{2+} translocation in Ehrlich ascites tumor cells, *J. Membr. Biol.* 49:309–324.

Hober, R., and Spaeth, R. A., 1914. Über den Einfluss seltener Erden auf die Kontraktilitat des Muskels, *Arch. Ges. Physiol.* 159:433–456.

Hodgson, B. J., and Daniel, E. E., 1973. Studies concerning the source of calcium for contraction of rat myometrium, *Can. J. Physiol. Pharmacol.* 51:914–932.

Hodgson, B. J., Kidwai, A. M., and Daniel, E. E., 1972. Uptake of lanthanum by smooth muscle, *Can. J. Physiol. Pharmacol.* 50:730–733.

Holmsen, H., Whaun, J., and Day, H. J., 1971. Inhibition by lanthanum ions of ADP-induced platelet aggregation, *Experientia* 27:451–453.

Horenstein, A. L., and Emergy, A. E. H., 1983. Human lymphocyte capping and the effect of lanthanum in Duchenne muscular dystrophy, *Res. Commun. Chem. Pathol. Pharmacol.* 41:303–312.

Huang, C. H., Sipe, J., Chow, S. T., and Martin, R. B., 1974. Differential interaction of cholesterol with phosphatidylcholine on the inner and outer surfaces of lipid bilayer vesicles, *Proc. Natl. Acad. Sci. USA* 71:359–362.

Hui, K. K., and Yu, J. L., 1988. The effects of calcium channel blockers on isoproterenol-induced cyclic adenosine 3',5'-monophosphate generation in intact human lymphocytes, *Life Sci.* 42:2037–2045.

Humbert, W., 1979. Intracellular and intramitochondrial binding of lanthanum in dark degenerating midgut cells of a collembolan (insect), *Histochemie* 59:117–128.

Hunt, G. R. A., 1980. Transport of Pr^{3+} across phospholipid membrane by lipophilic β-diketone, *Chem. Phys. Lipids* 27:353–364.

Hunt, G. R. A., Tipping, L. R. H., and Belmont, M. R., 1978. Rate-determining processes in the transport of Pr^{3+} ions by the ionophore A23187 across phospholipid vesicular membranes. A ^{1}H-NMR and theoretical study, *Biophys. Chem.* 8:341–355.

Hunt, W. G., and Willis, R. J, 1983. Inhibition by lanthanum of the calcium paradox phenomenon in rat heart, *Aust. J. Exp. Biol. Med. Sci.* 61:313–320.

Ishizawa, M., Nitta, H., and Miyazaki, E., 1980. Inhibitory action of lanthanum (La) on $PGF_{2\alpha}$-induced contraction of guinea-pig stomach muscle, *Prostaglandins* 19:407–413.

Josefsson, J. O., and Hansson, S. E., 1976. Effects of lanthanum on pinocytosis induced by cations in *Amoeba proteus*, *Acta Physiol. Scand.* 96:443–445.

Kalix, P., 1971. Uptake and release of calcium in rabbit vagus nerve, *Pflügers Arch. Ges. Physiol.* 326:1–14.

Karaki, H., and Weiss, G. B., 1979. Alterations in high and low affinity binding of ^{45}Ca in rabbit aortic smooth muscle by norepinephrine and potassium after exposure to lanthanum and low temperature, *J. Pharmacol. Exp. Ther.* 211:86–92.

Kass, R. S., and Tsien, R. W., 1975. Multiple effects of calcium antagonists on plateau currents in cardiac Purkinje fibres, *J. Gen. Physiol.* 66:169–192.

Katzung, B. G., Reuter, H., and Porzig, H., 1973. Lanthanum inhibits Ca inward current, but not Na–Ca exchange in cardiac muscles, *Experientia* 29:1073–1075.

Kawata, H., Ohba, M., Hatae, J., and Kishi, M., 1983. Paradoxical after-potentiation of the myocardial contractability by lanthanum, *Jpn. J. Physiol.* 33:1–17.

Keryer, G., and Rossignol, B., 1978. Lanthanum as a tool to study the role of phosphatidylinositol in the calcium transport in rat parotid glands upon cholinergic stimulation, *Eur. J. Biochem.* 85:77–83.

Kidokoro, Y., Hagiwara, S., and Henkart, M. P., 1974. Electrical properties of obliquely striated muscle fibres of *Anodonta glochidium*, *J. Comp. Physiol.* 90:321–338.

Kitzes, M. C., and Berns, M. W., 1979. Electrical activity of rat myocardial cells in cultures: La^{3+}-induced alterations, *Am. J. Physiol.* 237:C87–C95.

Korc, M., 1983. Effect of lanthanum on pancreatic protein synthesis in streptozotocin-diabetic rats, *Am. J. Physiol.* 244:G321–326.

Kostelnik, R. J., and Castellano, S. M., 1972. 250-MHz proton magnetic resonance spectrum of a sonicated lecithin dispersion in water: the effect of ferricyanide, manganese(II), europium(III) and gadolinium(III) ions on the choline methyl resonance, *J. Magn. Reson.* 7:219–223.

Kostyuk, P. G., 1981. Calcium channels in the neuronal membrane, *Biochim. Biophys. Acta* 650:128–150.

Kosztolanyi, G., Jobst, K., and Kadas, I., 1977. Effect of lanthanum ion on platelet electrophoretic mobility, *Haematologia* 11:155–161.

Krasnow, N., 1972. Effects of lanthanum and gadolinium ions on cardiac sarcoplasmic reticulum, *Biochim. Biophys. Acta* 282:187–194.

Krasnow, N., 1977. Lanthanide binding to cardiac and skeletal muscle microsomes. Effects of adenosine triphosphate, cations and ionophores, *Arch. Biochem. Biophys.* 181:322–330.

Kriebel, M. E., and Florey, E., 1983. Effect of lanthanum ions on the amplitude distributions of miniature endplate potentials and on synaptic vesicles in frog neuromuscular junctions, *Neuroscience* 9:535–547.

Kriebel, M. E., and Gross, C. E., 1974. Multimodal distribution of frog miniature endplate potentials in adult, denervated and tadpole leg muscle, *J. Gen. Physiol.* 64:85–103.

Laher, I., and Triggle, C., 1984. Blood pressure, lanthanum- and norepinephrine-induced mechanical response in thoracic aortic tissue, *Hypertension* 6:700–708.

Langer, G. A., and Frank, J. S., 1972. Lanthanum in heart cell culture. Effect on calcium exchange correlated with its localization, *J. Cell Biol.* 54:441–445.

Laszlo, D., Ekstein, D. M., Lewin, R., and Stern, K. G., 1952. Biological studies on stable and radioactive rare earth compounds. I. On the distribution of lanthanum in the mammalian organism, *J. Natl. Cancer Inst.* 13:559–572.

Lawaczeck, R., Kainosho, M., and Chan, S. I., 1976. The formation and annealing of structural defects in lipid bilayer vesicles, *Biochim. Biophys. Acta* 443:313–330.

Lawson, D. M., and Schmidt, G. H., 1972. Effects of lanthanum on ^{45}Ca movements and oxytocin-induced milk ejection in mammary tissue, *Proc. Soc. Exp. Biol. Med.* 140: 481–484.

Lehninger, A. L., and Carafoli, E., 1971. The interaction of lanthanum with mitochondria in relation to respiration-coupled Ca^{2+} transport, *Arch. Biochem. Biophys.* 143:506–515.

Lesseps, R. J., 1967. The removal of phospholipase C of a layer of lanthanum-staining material external to the cell membrane in embryonic chick cells, *J. Cell Biol.* 34:173–183.

Lesson, T. S., and Higgs, G. W., 1982. Lanthanum as an intracellular stain for electron microscopy, *Histochem. J.* 14:553–560.

Lettvin, J. Y., Pickard, W. F., McGulloch, W. F., and Pitts, W. S., 1964. A theory of passive ion flux through axon membranes, *Nature* 202:1338–1339.

Levine, Y. K., Lee, A. G., Birdsall, N. J. M., Metcalfe, J. C., and Robinson, J. D., 1973. The interaction of paramagnetic ions and spin labels with lecithin bilayers, *Biochim. Biophys. Acta* 291:592–607.

Levinson, C., Mikiten, T. M., and Smith, T. C., 1972a. Lanthanum-induced alterations in cellular electrolytes and membrane potential in Ehrlich ascites tumor cells, *J. Cell Physiol.* 79:299–308.

Levinson, C., Smith, T. C., and Mikiten, T. M., 1972b. Lanthanum-induced alterations of cellular electrolytes in Ehrlich ascites tumor cells. A new view, *J. Cell. Physiol.* 80: 149–154.

Lichtenberg, D., Amselsen, S., and Tamir, I., 1979. Dependence of the conformation of the polar head groups of phosphatidylcholine on its packing in bilayers. Nuclear magnetic resonance studies on the effect of the binding of lanthanide ions, *Biochemistry* 18:4169–4172.

Logan, A. G., Tenyi, I., Peart, W. S., Breathnach, A. S., and Martin, B. G. H., 1977. The effect of lanthanum on renin secretion and renal vasoconstriction, *Proc. Roy. Soc. B* 195:327–342.

Lorenz, R., Sharp, R., and Burr, I. M., 1979. Effects of calcium, lanthanum and bicarbonate ion on epinephrine modification of insulin release *in vitro, Diabetes* 28:52–55.

MacDonald, N. W., Nusbaum, R. E., Alexander, G. V., Ezmirnan, F., Spain, P., and Rounds, D. E., 1952. The skeletal deposition of yttrium, *J. Biol. Chem.* 195:837–842.

Machen, T. E., Erlij, D., and Wooding, F. B. P., 1972. Permeable junctional complexes. The movement of lanthanum across rabbit gallbladder and intestine, *J. Cell Biol.* 54: 302–312.

Madeira, V. M. C., and Antunes-Madeira, M. C., 1973. Interaction of Ca^{2+} and Mg^{2+} with synaptic plasma membranes, *Biochim. Biophys. Acta* 323:396–407.

Magaribuchi, T., Nakajima, H., and Kiyomoto, A., 1977. Effects of diltiazem and lanthanum on the potassium contracture of isolated guinea pig smooth muscle, *Jpn. J. Pharmacol.* 27:333–339.

Majumdar, S., Baker, R. F., and Kalra, V. K., 1980. Fusion of human erythrocytes induced by uranyl acetate and rare earth metals, *Biochim. Biophys. Acta* 598:411–416.

Marier, S. H., Putney, J. W., and Van de Walle, C. M., 1978. Control of calcium channels by membrane receptors in the rat parotid gland, *J. Physiol.* 279:141–151.

Martinez-Palomo, A., Benitez, D., and Alanis, J., 1973. Selective deposition of lanthanum in mammalian cardiac cell membranes. Ultrastructural and electrophysiological evidence, *J. Cell Biol.* 58:1–10.

Matthews, E. K., Evans, R. J., and Dean, P. M., 1972. The ionogenic nature of the secretory-granule membrane. Electrokinetic properties of isolated chromaffin granules, *Biochem. J.* 130:825–832.

Matthews, E. K., Legros, J. J., Grau, J. D., Nordmann, J. J., and Dreifuss, J. J., 1973. Release of neurohypophysial hormones by exocytosis, *Nature* 241:86–88.

Mayer, C. J., Van Breeman, C., and Casteels, R., 1972. The action of lanthanum and D600 on the calcium exchange in the smooth muscle cells of the guinea-pig taenia coli, *Pflügers Arch. Ges. Physiol.* 337:333–350.

McClure, J. B., and Steele, J. E., 1981. The role of extracellular calcium in hormonal activation of glycogen phosphorylase in cockroach fat body, *Insect Biochem.* 11:605–613.

McIntosh, T. J., 1980. Differences in hydrocarbon tilt between hydrated phosphatidylethanolamine and phosphatidyl choline bilayers. A molecular packing model. *Biophys. J.* 29:237–246.

Mela, L., 1968. Interactions of La^{3+} and local anesthetic drugs with mitochondrial Ca^{2+} and Mn^{2+} uptake, *Arch. Biochem. Biophys.* 123:286–293.

Mela, L., 1969a. Reaction of lanthanides with mitochondrial membranes, *Ann. N.Y. Acad. Sci.* 147:824–828.

Mela, L., 1969b. Inhibition and activation of calcium transport in mitochondria. Effects of lanthanides and local anesthetic drugs, *Biochemistry* 8:2481–2486.

Mellanby, J., and Thompson, P. A., 1981. The interaction of tetanus toxin and lanthanum at the neuromuscular junction in the gold fish, *Toxicon* 19:547–554.

Mellanby, J., Beaumont, M. A., and Thompson, P. A., 1988. The effect of lanthanum on nerve terminals in goldfish muscle after paralysis with tetanus toxin, *Neuroscience* 25: 1095–1106.

Metral, S., Bonneton, C., Hort-Legrand, C., and Reynes, J., 1978. Dual action of erbium on transmitter release at the frog neuromuscular synapse, *Nature* 271:773–775.

Mezon, B., and Bailey, L. E., 1975. Prevention of relaxation by lanthanum in the kitten heart, *J. Mol. Cell. Cardiol.* 7:417–425.

Mikkelsen, R. B., 1976. Lanthanides as calcium probes in biomembranes, in *Biological Membranes* (D. Chapman and D. F. H. Wallach, eds.), Vol. 3, Academic Press, New York, pp. 153–190.

Mikkelsen, R. B., and Wallach, D. F. H., 1974. High affinity calcium binding sites on erythrocyte membrane proteins. Use of lanthanides as fluorescent probes, *Biochim. Biophys. Acta* 363:211–218.

Mikkelsen, R. B., and Wallach, D. F. H., 1976. Binding of fluorescent lanthanides to rat liver mitochondrial membranes and calcium ion-binding proteins, *Biochem. Biophys. Acta* 433:674–683.

Miledi, R., 1971. Lanthanum ions abolish the 'calcium response' of nerve terminals, *Nature* 229:410–411.

Miledi, R., Molenaar, P. C., and Polak, R. L., 1980. The effect of lanthanum ions on acetylcholine in frog muscle, *J. Physiol.* 309:199–214.

Mills, J. D., and Hind, G., 1978. Use of the fluorescent lanthanide Tb^{3+} as a probe for cation-binding sites associated with isolated chloroplast thylakoid membranes, *Phytochem. Photobiol.* 28:67–73.

Mines, G. R., 1910. The action of beryllium, lanthanum, yttrium and cerium on the frog's heart, *J. Physiol.* 40:327–345.

Mines, G. R., 1911. The action of tri-valent ions on living cells and on colloidal systems. II. Simple and complex kations, *J. Physiol.* 42:309–331.

Mircevova, L., Viktora, L., and Hermanova, E., 1984. Inhibition of phagocytosis of polymorphonuclear leucocytes by adenosine and $HoCl_3$ in vitro, *Med. Biol.* 62:326–330.

Morris, D. C., and Appleton, J., 1984. The effects of lanthanum on the ultrastructure of hypertrophic chondrocytes and the localization of lanthanum precipitates in condylar cartilages of rats fed on normal and rachitogenic diets, *J. Histochem. Cytochem.* 32: 239–247.

Morris, G. E., 1980. The use of lanthanum and cytochalasin B to study calcium effects on skeletal muscle differentiation in vitro, *J. Cell. Physiol.* 105:431–438.

Morris, S. J., and Schober, R., 1977. Demonstration of binding sites for divalent and trivalent ions on the outer surface of chromaffin-granule membranes, *Eur. J. Biochem.* 75:1–12.

Muneoka, Y., and Twarog, B. M., 1977. Lanthanum block of contraction and relaxation in response to serotonin and dopamine in molluscan catch muscle, *J. Pharmacol. Exp. Ther.* 202:601–609.

Muscholl, E., Racke, K., and Traut, A., 1985. Gadolinium ions inhibit exocytolic vasopressin release from the rat neurohypophysis, *J. Physiol.* 367:419–434.

Nachshen, D. A., 1984. Selectivity of the Ca binding site in synaptosome Ca channels. Inhibition of Ca influx by multivalent metal cations, *J. Gen. Physiol.* 83:941–967.

Nachshen, D. A., and Blaustein, M. P., 1980. Some properties of potassium-stimulated calcium influx in presynaptic nerve endings, *J. Gen. Physiol.* 76:709–728.

Nagahashi, G., Thomson, W. W., and Leonard, R. T., 1974. The casparian strip as a barrier to the movement of lanthanum in corn roots, *Science* 183:670–671.

Nathanson, J. A., Freedman, R., and Hoffer, B. J., 1976. Lanthanum inhibits brain adenylate cyclase and blocks noradrenergic depression of Purkinje cell discharge independent of calcium, *Nature* 261:330–332.

Nayler, W. G., 1973. An effect of ouabain on the superficially-located stores of calcium in cardiac muscle cells, *J. Mol. Cell. Cardiol.* 5:101–111.

Nayler, W. G., and Harris, J. P., 1976. Inhibition by lanthanum of the $Na^+ + K^+$ activated, ouabain-sensitive adenosinetriphosphatase enzyme, *J. Mol. Cell. Cardiol.* 8:811–822.

Nelson, M. T., French, R. J., and Kreuger, B. K., 1984. Voltage-dependent calcium channels from brain incorporated into planar lipid bilayers, *Nature* 308:77–80.

Ng, D., Shanbaky, N. M., and Borowitz, J. L., 1982. Novel calcium channels for lanthanide ion-induced adrenal catecholamine release, *Res. Commun. Pathol. Pharmacol.* 37: 259–265.

Niggli, V., Gazzotti, P., and Carafoli, E., 1978. Experiments on the mechanism of the inhibition of mitochondrial Ca^{2+} transport by La^{3+} and ruthenium red, *Experientia* 34: 1136–1137.

Oberdisse, E., Arvela, P., and Merke, H.-J., 1976. Effects of lanthanons on rat exocrine pancreas *in vivo, Acta Physiol. Scand. Suppl.* 440:137.

O'Brien, D. F., Zumbulyadis, N., Michaels, F. M., and Ott, R. A., 1977. Light-regulated permeability of rhodopsin:egg phosphatidylcholine recombinant membrane, *Proc. Natl. Acad. Sci. USA* 74:5222–5226.

O'Flaherty, J. T., Showell, H. J., Becker, E. L., and Ward, P. A., 1978. Substances which aggregate neutrophils. Mechanism of action, *Am. J. Pathol.* 92:155–166.

Ohyashiki, T., Chiba, K., and Mohri, T., 1979. Terbium as a fluorescent probe of the nature of Ca^{2+}-binding sites of rat intestinal mucosal membranes, *J. Biochem.* 86:1479–1485.

Ohyashiki, T., Ohtsuka, T., and Mohri, T., 1985. Characterization of interaction between Tb^{3+} and porcine intestinal brush-border membranes, *Biochim. Biophys. Acta* 817: 181–186.

Osborne, R. H., and Bradford, H. F., 1975. The influence of sodium, potassium and lanthanum on amino acid release from spinal-medullary synaptosomes, *J. Neurochem.* 25: 35–41.

Parker, J. C., and Barritt, G. J., 1981. Evidence that lanthanum ions stimulate calcium inflow to isolated hepatocytes, *Biochem. J.* 200:109–114.

Parsons, R. L., Johnson, E. W., and Lambert, D. H., 1971. Effects of lanthanum and calcium on chronically denervated muscle fibers, *Am. J. Physiol.* 220:401–405.

Pearce, F. L., and White, J. R., 1981. Effect of lanthanide ions on histamine secretion from rat peritoneal mast cells, *Br. J. Pharmacol.* 72:341–347.

Pento, J. T., 1977. Lanthanum inhibition of calcitonin secretion and calcium uptake in porcine thyroid slices, *Mol. Cell. Endocrinol.* 9:223–226.

Pento, J. T., 1978. Influence of lanthanum on calcium transport and retention in the rat duodenum, *Nutr. Metab.* 22:362–367.

Piccinini, F., Meloni, S., Chiarra, A., and Villani, F. P., 1975. Uptake of lanthanum by mitochondria, *Pharmacol. Res. Commun.* 7:429–435.

Pressler, M. L., Elharrar, V., and Bailey, J. C., 1982. Effects of extracellular calcium ions, verapamil and lanthanum on active and passive properties of canine cardiac Purkinje fibers, *Circ. Res.* 51:637–651.

Rahaminoff, H., and Spanier, R., 1984. The asymmetric effect of lanthanides on Na^+-gradient-dependent Ca^{2+} transport in synaptic plasma membrane vesicles, *Biochim. Biophys. Acta* 773:279–289.

Ravens, U., 1975. The effects of lanthanum on electrical and mechanical events in mammalian cardiac muscle, *Naunyn-Schmiedeberg's Arch. Pharmacol.* 288:133–146.

Reed, K. C., and Bygrave, F. L., 1974a. Accumulation of La by rat liver mitochondria, *Biochem. J.* 138:239–252.

Reed, K. C., and Bygrave, F. L., 1974b. The inhibition of mitochondrial calcium transport by lanthanides and ruthenium red, *Biochem. J.* 140:143–155.

Reeves, J. P., and Sutko, J. L., 1979. Sodium–calcium ion exchange in cardiac membrane vesicles, *Proc. Natl. Acad. Sci. USA* 76:590–594.

Reuter, H., 1973. Divalent cations as charge carriers in excitable membranes, *Prog. Biophys. Mol. Biol.* 26:1–43.

Reynafarje, B., and Lehninger, A. L., 1969. High affinity and low affinity binding of Ca^{2+} by rat liver mitochondria, *J. Biol. Chem.* 244:584–593.

Rice, D. J., and Humphreys, T., 1983. Two Ca^{2+} functions are demonstrated by the substitution of specific divalent and lanthanide cations for the Ca^{2+} required by the aggregation factor complex from the marine sponge, *Microciona prolifera*, *J. Biol. Chem.* 258:6394–6399.

Robblee, L. S., and Shepro, D., 1976. The effect of external calcium and lanthanum on platelet calcium content and on the release reaction, *Biochim. Biophys. Acta* 436:448–459.

Rojas, E., Lettvin, J. Y., and Pickard, W. F., 1966. A demonstration of ion-exchange phenomena in phospholipid monomolecular films, *Nature* 209:886–887.

Rozza, A., Favalli, L., Chiari, M. C., and Piccinini, F., 1975. Diphasic action of lanthanum in rat uterus, *Pharmacol. Res. Commun.* 7:171–180.

Russell, J. T., and Thorn, N. A., 1974. Calcium and stimulus-secretion coupling in the neurohypophysis. II. Effects of lanthanum, a verapamil analogue (D600) and prenylamine on 45-calcium transport and vasopressin release in isolated rat neurohypophyses, *Acta Endocrinol.* 76:471–487.

Sabatini, S., and Arruda, J. A., 1984. Effect of lanthanum on urinary acidification and sodium transport by the turtle and toad bladder, *Mineral Electrolyte Metals* 10:12–20.

Saling, P. M., 1982. Development of the ability to bind to zonae pellucidae during epididymal maturation: reversible immobilization of mouse spermatazoa by lanthanum, *Biol. Reprod.* 26:429–436.

Sanborn, W. G., and Langer, G. A., 1970. Specific uncoupling of excitation and contraction in mammalian cardiac tissue by lanthanum, *J. Gen. Physiol.* 5:191–217.

Sarkadi, B., Szasz, I., Gerloczy, A., and Grardos, G., 1977. Transport parameters and stoichiometry of active calcium ion extrusion in intact human red cells, *Biochim. Biophys. Acta* 464:93–107.

Sauerheber, R. D., Zimmerman, T. S., Esgate, J. A., Vanderlaan, W. P., and Gordon, L. M., 1980. Effects of calcium, lanthanum and temperature on the fluidity of spin-labeled human platelets, *J. Membr. Biol.* 52:201–219.

Scarpa, A., and Azzi, A., 1968. Cation binding to submitochondrial particles, *Biochim. Biophys. Acta* 150:473–481.

Scarpa, A., and Azzone, G. F., 1970. The mechanism of ion translocation in mitochondria. 4. Coupling of K^+ efflux with Ca^{2+} uptake, *Eur. J. Biochem.* 12:328–335.

Schellenberg, G. D., and Swanson, P. D., 1981. Sodium-dependent and calcium-dependent calcium transport by rat brain microsomes, *Biochim. Biophys. Acta* 648:13–27.

Schiffrin, E. L., Lis, M., Gutkowska, J., and Genest, J., 1981. Role of Ca^{2+} in response of adrenal glomerulosa cells to angiotensin II, ACTH, K^+ and ouabain, *Am. J. Physiol.* 241:E42–46.

Schimmel, R. J., 1978. Calcium antagonists and lipolysis in isolated rat epididymal adipocytes: effects of tetracaine, manganese, cobaltous and lanthanum ions and D600, *Horm. Metab. Res.* 10:128–134.

Schorderet-Slatkine, S., Schorderet, M., and Baulieu, E., 1976. Initiation of meiotic maturation in *Xenopus laevis* oocytes by lanthanum, *Nature* 262:289–290.

Schroeder, H. A., and Mitchener, M., 1971. Scandium, chromium (VI), gallium, yttrium,
rhodium, palladium, indium in mice: effects on growth and lifespan, *J. Nutr.* 101:
1431–1438.

Sedmak, J. J., MacDonald, H. S., and Kushnaryov, V. M., 1986. Lanthanide ion enhancement of interferon binding to cells, *Biochem. Biophys. Res Commun.* 137:480–485.

Selwyn, M. J., Dawson, A. P., and Dunnett, S. J., 1970. Calcium transport in mitochondria,
FEBS Lett. 10:1–5.

Shetty, S. S., and Frank, G. B., 1985. Displacement of the surface membrane bound calcium
of the skeletal muscle fibres of the frog: effects of lanthanum and opioids, *J. Pharmacol.
Exp. Ther.* 234:233–238.

Shigenobu, K., and Sperelakis, N., 1972. Calcium current channels induced by catecholamines in chick embryonic hearts whose fast sodium channels are blocked by tetrodoxin
or elevated potassium, *Circ. Res.* 31:932–952.

Shima, S., Kawashima, Y., and Hirai, M., 1978. Studies on cyclic nucleotides in the adrenal
gland. VII. Effects of angiotensin on adenosine 3′,5′-monophosphate and steroidogenesis in the adrenal cortex, *Endocrinology* 103:1361–1367.

Shine, K. I., 1973. Contractile calcium in rabbit myocardium: observations with nicotine
and quinine, *Am. J. Physiol.* 224:1024–1031.

Smith, J. B., and Smith, L., 1984. Initiation of DNA synthesis in quiescent Swiss 3T3 and
3T6 cells by lanthanum, *Biosci. Rep.* 4:777–782.

Smith, T. C., 1976. The effect of lanthanum on electrophoretic mobility and passive cation
movements of the Ehrlich ascites tumor cell, *J. Cell. Physiol.* 87:47–52.

Smith, T. C., Mikiten, T. M., and Levinson, C., 1972. The effect of multivalent cations on
the membrane potential of the Ehrlich ascites tumor cell, *J. Cell. Physiol.* 79:117–126.

Sperelakis, N., Valle, R., Orozco, R., Martinez-Palomo, A., and Rubio, R., 1973. Electro-
mechanical uncoupling of frog skeletal muscles by possible changes in sarcoplasmic
reticulum content, *Am. J. Physiol.* 225:793–800.

Squier, C., and Edie, J., 1983. Localization of lanthanum tracer in oral epithelium using
transmission electron microscopy and the electron microprobe, *Histochem. J.* 15:
1123–1130.

Squier, C. A., and Rooney, L., 1976. The permeability of keratinized and nonkeratinized
oral epithelium to lanthanum *in vivo*, *J. Ultrastruct. Res.* 54:286–295.

Starzak, M. E., and Starzak, R. J., 1978. The compensation of potential changes produced
by trivalent erbium ion in squid giant axon with applied potentials, *Biophys. J.* 24:
555–560.

Stefanou, J., and Wooster, M. J., 1976. Effects of lanthanum and low sodium on calcium
movements and mechanical activity in frog ventricular strips, *Proc. Physiol. Soc.* 260:
23P–24P.

Stevens, B. R., and Kneer, C., 1988. Lanthanide-stimulated glucose and proline transport
across rabbit intestinal brush-border membranes, *Biochim. Biophys. Acta* 942:205–208.

Strum, J. M., 1977. Lanthanum "staining" of the lateral and basal membranes of the
mitochondria-rich cell in toad bladder epipthelium, *J. Ultrastruct. Res.* 59:126–139.

Sunano, S., 1984. The effects of Ca antagonists, manganese and lanthanum on cooling-
induced contracture of depolarized vas deferens, *Jpn. J. Pharmacol.* 34:51–56.

Swales, L. S., and Gardner, D. R., 1985. X-ray analysis of *Lymnaea stagnalis* muscle fibres
does not suggest that the sarcolemma is permeable to lanthanum ions, *J. Cell Sci.* 75:
181–194.

Swamy, V. C., Triggle, C. R., and Triggle, D. J., 1976. The effects of lanthanum and thulium
on the mechanical responses of rat vas deferens, *J. Physiol.* 254:55–62.

Szasz, I., Sarkadi, B., Schubert, A., and Gardos, G., 1978. Effects of lanthanum on calcium-dependent phenomena in human red cells, *Biochim. Biophys. Acta* 512:331–340.

Takata, M., Pickard, W. F., Lettvin, J. Y, and Moore, J. W., 1966. Ionic conductance changes in lobster axon membrane when lanthanum is substituted for calcium, *J. Gen. Physiol.* 50:461–471.

Takeo, S., Duke, P., Tamm, G. M., Singal, P. K., and Dhalla, N. S., 1979. Effects of lanthanum on the heart sarcolemmal ATPase and calcium binding activities, *Can. J. Physiol. Pharmacol.* 57:496–503.

Talbot, R. B., Davison, F. C., Green, J. W., Reece, W. O., and Van Gelder, G., 1965. Effects of subcutaneous injection of rare earth metals, United States Atomic Energy Department Report 1170.

Tapia, R., Arias, C., and Morales, E., 1985. Binding of lanthanum ions and ruthenium red to synaptosomes and its effects on neurotransmitter release, *J. Neurochem.* 45:1464–1470.

Taylor, A. R. D., and Hall, J. L., 1979. An ultrastructural comparison of lanthanum and silicotungstic acid/chromic acid as plasma membrane stains of isolated protoplasts, *Plant Sci. Lett.* 14:139–144.

Tew, W. P., 1977. Use of coulombic interactions of the lanthanide series to identify two classes of Ca^{2+} binding sites in mitochondria, *Biochem. Biophys. Res. Commun.* 78:624–630.

Ting, D. Z., Hagan, P. S., Chan, S. I., Doll, J. D., and Springer, C. S., 1981. Nuclear magnetic resonance studies of cation transport across vesicle bilayer membranes, *Biophys. J.* 34:189–216.

Triggle, C. R., and Triggle, D. J., 1976. An analysis of the action of cations of the lanthanide series on the mechanical responses of guinea-pig ileal longitudinal muscle, *J. Physiol.* 254:39–54.

Triggle, C. R., Grant, W. F., and Triggle, D. J., 1975. Intestinal smooth muscle contraction and the effects of cadmium and A23187, *J. Pharmacol. Exp. Ther.* 194:182–190.

Tsien, R. W., 1983. Calcium channels in excitable cell membranes, *Annu. Rev. Physiol.* 45:341–358.

Tuchweber, B., Trost, R., Salas, M., and Sieck, W., 1976. Effect of praseodymium nitrate on hepatocytes and Kupffer cells in the rat, *Can. J. Physiol. Pharmacol.* 54:898–906.

Turrin, M. Q. A., and Oga, S., 1977. Effects of samarium (Sm^{3+}) on the contractility of isolated guinea-pig ventral taenia coli to histamine, *Jpn. J. Pharmacol.* 27:592–595.

Uyesaka, N., Kamino, K., Ogawa, M., Inouye, A., and Machida, K., 1976. Lanthanum and some other cation-induced changes in fluidity of synaptosomal membrane studied with mitroxide stearate spin labels, *J. Membr. Biol.* 27:283–295.

Vainio, H., Mela, L., and Chance, B., 1970. Energy dependent bivalent cation translocation in rat liver mitochondria, *Eur. J. Biochem.* 12:387–391.

Van Breeman, C., 1968. Permselectivity of a porous phospholipid-cholesterol artificial membrane. Calcium and lanthanum effects, *Biochem. Biophys. Res. Commun.* 32:977–983.

Van Breeman, C., 1969. Blockade of membrane calcium fluxes by lanthanum in relation to vascular smooth muscle contractility, *Arch. Int. Physiol. Biochem.* 77:710–716.

Van Breeman, C., and Deth, R., 1976. La^{3+} and excitation contraction coupling in vascular smooth muscle, in *Ionic Actions on Vascular Smooth Muscle* (E. Betz, ed.), Springer, Berlin, pp. 26–33.

Van Breeman, C., and De Weer, P., 1970. Lanthanum inhibition of ^{45}Ca efflux from the squid giant axon, *Nature* 226:760–761.

Van Breeman, C., Farinas, B. R., Gerba, P., and McNaughton, E. D., 1972. Excitation–contraction coupling in rabbit aorta: studied by the lanthanum method for measuring cellular calcium influx, *Circ. Res.* 30:44–54.

Van Breeman, C., Hwang, O., and Siegel, B., 1977. The lanthanum method, in *Excita-

tion–Contraction Coupling in Smooth Muscle (R. Casteels, T. Godfraind, and J. C. Ruegg, eds.), Elsevier/North-Holland, Amsterdam, pp. 243–252.

Van Leeuwen, J-P. T. M., Bos, M. P., and Herrmann-Erlee, M. P. M., 1988. Involvement of cAMP and calcium in the induction of ornithine decarboxylase activity in an osteoblast cell line, *J. Cell. Physiol.* 135:488–494.

Veldhuis, J. D., 1982. Regulation of ovarian ornithine decarboxylase. Role of calcium ions in enzyme induction in isolated swine granulosa cells in vitro, *Biochim. Biophys. Acta* 720:211–216.

Verma, A. K., and Boutwell, R. K., 1981. Intracellular calcium and skin tumor promotion: calcium regulation of the induction of epidermal ornithine decarboxylase activity by the tumor promoter 12-*O*-tetradecanoylphorbol-13-acetate, *Biochem. Biophys. Res. Commun.* 101:375–383.

Villani, F., Piccinini, F., Chiarra, A., and Brambilla, G., 1975. Effect of lanthanum on calcium exchangeability in mitochondria, *Biochem. Pharmacol.* 24:1349–1351.

Vogel, W., 1974. Calcium and lanthanum effects at the nodal membrane, *Pflügers Arch.* 350:25–39.

Wakasugi, H., Stolze, H., Haase, W., and Schulz, I., 1981. Effect of La^{3+} on secretagogue-induced Ca^{2+} fluxes in rat isolated pancreatic acinar cells, *Am. J. Physiol.* 240:G281–289.

Washio, H., and Miyamoto, T., 1983. Effect of lanthanum ions on neuromuscular transmission in insects, *J. Exp. Biol.* 107:405–414.

Weihe, E., Hartschuh, W., Metz, J., and Bruhl, U., 1977. The use of ionic lanthanum as a diffusion tracer and as a marker of calcium binding sites, *Cell Tissue Res.* 178:285–302.

Weiss, G. B., 1970. On the site of action of lanthanum in frog sartorius muscle, *J. Pharmacol. Exp. Ther.* 174:517–528.

Weiss, G. B., 1973. Inhibition by lanthanum of some calcium related actions in the frog rectus abdominus muscle, *J. Pharmacol. Exp. Ther.* 185:551–559.

Weiss, G. B., 1974. Cellular pharmacology of lanthanum, *Annu. Rev. Pharmacol.* 14:343–354.

Weiss, G. B., 1977. Approaches to delineation of differing calcium binding sites in smooth muscle, in *Excitation-Contraction Coupling in Smooth Muscle* (R. Casteels, T. Godfraind, and J. C. Ruegg, eds.), Elsevier/North-Holland, Amsterdam, pp. 253–260.

Weiss, G. B., 1982. Calcium and excitation contraction coupling in vascular smooth muscle, *Can. J. Physiol. Pharmacol.* 60:483–488.

Weiss, G. B., and Goodman, F. R., 1969. Effects of lanthanum on contraction, calcium distribution and Ca^{2+} movements in intestinal smooth muscle, *J. Pharmacol. Exp. Ther.* 169:46–55.

Weiss, G. B., and Goodman, F. R., 1975. Interaction between several rare earth ions and calcium ion in vascular smooth muscle, *J. Pharm. Exp. Ther.* 195:557–564.

Weiss, G. B., and Goodman, F. R., 1976. Distribution of a lanthanide (^{147}Pm) in vascular smooth muscle, *J. Pharm. Exp. Ther.* 198:366–374.

Weiss, G. B., and Wheeler, E. S., 1978. Inhibition of ^{45}Ca movements by lowered temperature or lanthanum in rat brain slices, *Arch. Int. Pharmacodyn. Ther.* 233:4–20.

Whittingham, D. G., 1980. Parthogenesis in mammals, in *Oxford Reviews of Reproductive Biology* (C. A. Finn, ed.), Vol. 2, Clarendon Press, Oxford, p. 205.

Wietzerbin, J., Lange, Y., and Gary-Bobo, C. M., 1974. Lanthanum inhibition of the action of oxytocin on the water permeability of the frog urinary bladder: effect on the serosa and the apical membrane, *J. Membr. Biol.* 17:27–40.

Wietzerbin, J., Goudeau, H., and Gary-Bobo, C. M., 1977. Influence of membrane polarization and hormonal stimulation on the action of lanthanum on frog skin sodium permeability, *Pflügers Arch.* 370:145–153.

Wong, P. Y. D., Bedwani, J. R., and Cuthbert, A. W., 1972. Hormone action and the levels of cyclic AMP and prostaglandins in the toad bladder, *Nature New Biol.* 238:27–31.

Wong, P. Y. D., Hwang, J. C., and Yeung, C. H., 1976. Interaction between lanthanum and calcium in isolated guinea pig heart, *Eur. J. Pharmacol.* 36:253–256.

Wood, P. G. and Mueller, H., 1985. The effects of terbium(III) on the Ca-activated K channel found in the resealed human erythrocyte membrane, *Eur. J. Biochem.* 146:65–69.

Worner, P., and Brossmer, R., 1976. Aggregation and release reaction of washed platelets stimulated by lanthanum, *Life Sci.* 19:661–676.

Yamage, M., and Evans, C. H., 1989. Suppression of mitogen- and antigen-induced lymphocyte proliferation by lanthanides, *Experientia* (in press).

Yarbrough, C. G., Lake, N., and Phillis, J. W., 1974. Calcium antagonism and its effect on the inhibitory actions of biogenic amines on cerebral cortical neurones, *Brain Res.* 67: 77–88.

Zilberman, Y., Gutman, Y., and Koren, R., 1982. The effect of verapamil, lanthanum and local anesthetics on serotonin release from rabbit platelets, *Biochim. Biophys. Acta* 691:106–114.

The Occurrence and Metabolism of Lanthanides

7

7.1 Introduction

At first sight, it might seem strange to spend a chapter discussing the metabolism of substances with no known metabolic role. Nevertheless, there exists an extensive literature on lanthanide metabolism. Studies of this subject have been largely prompted by toxicological considerations (Chapter 8), with particular concern about human exposure to radioactive lanthanides formed as a result of nuclear reactions. For obvious reasons, investigations with mammals have predominated, although the possible passage of radioactive lanthanides through the food chain has led to the inclusion of other organisms in such studies. In areas where the ambient concentration of lanthanides in the soil is high, their assimilation is being monitored to obtain an indication of the likely distribution of the actinides, elements whose metabolic behavior shows similarities to that of the lanthanides (Durbin, 1962; Eisenbud *et al.*, 1984). An additional motive for studying lanthanide metabolism has been medical (Chapter 9). Possible medical applications of the lanthanides have been sought for over a century, a process which has necessarily entailed consideration of their metabolism.

The results of these investigations are discussed in this chapter. However, the *in vitro* interactions of lanthanides with organs, cells, and organelles isolated from metazoans are not addressed, as these were extensively described in Chapter 6. Likewise, only passing mention is made of the metabolism of lanthanides by tumor-bearing animals; this is dealt with more appropriately in Section 9.6.2.

Table 7-1. Geological Abundance of the Lanthanides

Element	Average abundance in earth's crust (ppm)	Composition[a] of monazite (%)	Composition[a] of gadolinite (%)
La	18.0	23.0	1.0
Ce	46.0	48.0	2.0
Pr	5.5	6.0	2.0
Nd	24.0	20.0	5.0
Sm	6.5	2.0	5.0
Eu	1.1	0.02	<0.1
Gd	6.4	0.5	5.0
Tb	0.9	0.04	0.05
Dy	4.5	0.08	6.0
Ho	1.2	0.01	1.0
Er	2.5	0.03	4.0
Tm	0.2	0.003	0.6
Yb	2.7	0.005	4.0
Lu	0.8	0.003	0.6
Y	28.0	0.08	63.0

[a] Figures for the ores are representative only, as the precise composition varies from source to source.

7.2 Sources and Biological Distribution

Under natural conditions, lanthanides become available, via the groundwater, through leaching from mineral deposits. Certain lanthanides are detectable at low levels in higher organisms, suggesting that they have some ability to travel up through the food chain, although inhalation is also a route to their biological fixation. In the latter context, it should be noted that crustal weathering releases lanthanides into the atmosphere, where their concentrations in aerosols reflect the composition of local rocks (Sugimae, 1980).

The amounts of the various lanthanides which are made available in the groundwater depend, in the first instance, upon their abundance in the soil. The approximate average abundance of the lanthanides in the earth's crust is given in Table 7-1. Members of the lanthanide series obey the Oddo–Harkins rule that elements of even atomic number are more abundant than their neighbors of odd atomic number. The source ores tend to be enriched in either the heavy or the light lanthanides. Monazite is an important ore, which contains about 50% by weight rare earth elements (Table 7-1). Concentrations of monazite are found in Australia, the USSR, South Africa, and the U.S. Other important ores are gadolinite

(Table 7-1); xenotime, a phosphate containing 50–60% Y and a predominance of heavier lanthanides; bastnaesite, a fluorocarbonate of La and Ce, with smaller amounts of Nd and Pr; and fergusonite, euxenite, samarskite, and blomstrandine, which are rich in the heavy lanthanides.

The People's Republic of China is thought to contain among the world's largest deposits of lanthanides, the main ones being at Baiyuneba, in Inner Mongolia (Baiyuneba means "Mount of Riches" in Mongolian). Another large deposit of interest occurs in Brazil, in the state of Minas Gerais, where there is an ore containing 100,000 tons of lanthanide elements (Wedow, 1967). Interesting studies of the biological assimilation of these lanthanides are under way (Eisenbud et al., 1984).

Concentrations of lanthanides in the groundwater are much lower than those in the soil through which they percolate. In a study in Virginia (Robinson et al., 1958), the soil concentrations of lanthanides were about 100 ppm, whereas in the water of two wells in the same area, concentrations were in the ppb to undetectable range. In the Brazilian study mentioned above, the concentrations of lanthanides in the soil were measured as follows: Ce (514–944 μg/g dry soil), La (79–405 μg/g), Nd (27–156 μg/g), and Sm (12–27 μg/g). However, the mean concentration of lanthanides in the water was only 2 μg/liter. Such low concentrations probably reflect the insolubility of lanthanide phosphates and oxides which form most of their ores. From such data, Eisenbud et al. (1984) calculated a mobilization rate of La by groundwater of only 1.6×10^{-9}/y. Concentrations of 0.002–8.03 μg/liter are found in certain Japanese hot springs (Ikeda and Takahashi, 1978). Seawater contains 0.3–0.4 ppb of each of Ce, La, and Y in predominantly particulate form (Revelle and Schaefer, 1957).

The occurrence of trace amounts of lanthanides in plants has been known for over a century (Cossa, 1870). Some studies have suggested that the concentrations of lanthanides in plants passively reflect concentrations in the water (Borneman-Starinkevitch et al., 1941). However, Milton et al. (1944) provided evidence that some trees can discriminate against lanthanides. Thus, despite finding lanthanide concentrations of about 600 ppm in the leaves of chestnut trees in Virginia, these authors failed to detect any lanthanides in the leaves of adjacent oak trees. Nevertheless, hickory trees may possess a remarkable capacity for concentrating lanthanides. Robinson et al. (1938) reported levels of 300–2300 ppm in the leaves of hickory trees in the eastern U.S.; Y, La, Ce, and Nd accounted for up to 91% of these lanthanides (Robinson et al., 1958). Other plants that appear able to concentrate lanthanides in this way are certain mosses (Shacklette, 1965) and ferns (Erametsa and Haukka, 1970). It has been suggested that such plants could be used in prospecting for lanthanide deposits. Although the mechanism of this accumulation is unknown, the ability of lanthanides to form extremely stable complexes with suitable ligands and to form a variety of insoluble precipitates (Section 2.3) would seem to provide ample scope for tissue retention. However, Bowen (1956) disputed the assumption that hickory trees actively concentrate lanthanides, claiming instead that they merely reflect the lanthanide content of their environment. According to Bowen's unpublished data, hickory trees maintained experimentally in water did not greatly concentrate added Ce, Y, or Pr. Furthermore, uptake of lanthanides only occurred at a particular stage in the development of the leaf, with older leaves showing no specific uptake. Evidence that many plants are able to discriminate against lanthanides was obtained by Laul and Weimer (1982), who found concentrations of lanthanides in a variety of crops to be only 10^{-4}–10^{-5} that of their concentrations in soil. However, bean seedlings are apparently able to take up and to translocate Y^{3+} (Rediske and Selders, 1954). Corn roots, on the other hand, transport La^{3+} only as far as the Casparian strip, which separates the cortex from the vascular tissue (Nagahashi *et al.*, 1974). In the Brazilian study described earlier, mean lanthanide concentrations in various crops ranged from under 1 ppb for Sm in beans and corn to 2.46 ppm for Ce in couvé (Linsalata *et al.*, 1986a). Comparison of the lanthanide concentrations in these crops with that of the soil provided ratios in the range of 10^{-3}–10^{-5}, suggesting poor uptake. Plants which can take up lanthanides may do so either without fractionating the individual members of the series or with selective enrichment of the lighter elements. However, cerium may be underrepresented in some cases (e.g., Robinson *et al.*, 1958) due to its ability to exist in the Ce(IV) state, which is not well assimilated.

Some representative values for lanthanide abundances in plants are given in Table 7-2. Several plants are available as Standard Reference Materials (SRM) from the U.S. National Bureau of Standards (NBS). Orchard leaves (SRM number 1571) was the first biological specimen to be included as an SRM; its lanthanide concentrations are shown in Table 7-3.

Several types of algae (e.g., *Rhodymenia*) sequester Y^{3+} from seawater, whereas others (e.g., *Carteria*) discriminate against it in the presence of Sr^{2+} (Spooner, 1949; Rice, 1956; Boroughs *et al.*, 1957). Apparent assimilation by algae may represent passive surface absorption rather than true uptake. Fish may constitute a barrier to the progression of lanthanides through the food chain (Boroughs *et al.*, 1956).

Trace amounts of lanthanides are often detected in mammals (Tables 7-3, 7-4). Levels of Yb in the eyes of laboratory mice were about 10 times those in other organs (Samochocka *et al.*, 1984a,b; Table 7-5). The reason for this is obscure, although lanthanides in the environment may have

Table 7-2. Concentrations of Lanthanides in Various Plants[a]

Plant	Tissue	Lanthanide concentration[b] (ppm)	Reference
Fern	Leaf	0.7–98	Erametsa and Haukka, 1970
Chestnut tree	Leaf	~600	Milton et al., 1944
Oak tree	Leaf	u.d.[c]	Milton et al., 1944
Hickory tree	Leaf	300–2300	Robinson et al., 1938
Various crops	Root, leaf, fruit	4–690	Robinson, 1943
Spinach (NBS SRM 1570)	Leaf	0.02–0.37	Gladney, 1980; NBS
Tomato (NBS SRM 1573)	Leaf	0.02–0.9	Gladney, 1980; NBS
Pine tree (NBS SRM 1575)	Needle	0.006–0.4	Gladney, 1980; NBS
Brewer's yeast (NBS SRM 1569)		2.3–7.1	Gladney, 1980
Kale	Leaf	0.06 (La)	Nadkarni and Morrison, 1978

[a] See also Table 7-3.
[b] Concentrations are on a dry weight basis.
[c] u.d.: Undetectable.

easier access to the eye than to other organs. The corneal stroma has a strong affinity for lanthanides (Grant and Kern, 1957; Section 8.5). However, in mice, the greatest amounts of Yb were associated with the retina and sclera (Table 7-5). Lenses of human eyes accumulate La (Swanson and Truesdale, 1971) as they age, higher levels being found in cataractous tissue (Table 7-6). However, Sihvonen (1972) detected no age- or sex-related differences between the concentrations of various lanthanides in several human organs.

The results of Leddicotte and his colleagues (Table 7-4), who measured lower concentrations of lanthanides in bone than in kidney, are surprising. Metabolic studies (Section 7.5) show the skeleton to be the main long-term reservoir of lanthanides, while very low levels of lanthanides accumulate in the kidneys. According to the Russian literature cited by Sihvonen (1972), La and Ce have been detected in human kidney stones.

The NBS provides a standard preparation of bovine liver (SRM number 1577), whose lanthanide concentrations are given in Table 7-3. Inspection of Tables 7-2, 7-3, and 7-4 reveals that plants generally have higher concentrations of lanthanides than animals. This presumably reflects the poor transfer of lanthanides through the food chain.

Table 7-3. Concentrations of Lanthanides in NBS Standard
Reference Materials[a]

Element	Unit	Concentration	
		Orchard leaves (SRM 1571)	Bovine liver (SRM 1577)
La	ppm	1.2 ± 0.4	0.028 ± 0.023
Ce	ppm	1 ± 0.1	0.046
Pr	ppb	60	4.6
Nd	ppb	400	170
Sm	ppb	110 ± 20	1.4
Eu	ppb	90 ± 110	0.2–3,100
Gd	ppb	1.64–100	<1
Tb	ppb	1.2–80	<1.6
Dy	ppb	81	2.4
Ho	ppb	20	<0.9
Er	ppb	30	<0.5
Tm	ppb	7	<0.3
Yb	ppb	27 ± 11	0.5–830
Lu	ppb	6.1–10	<0.1
Y	ppm	0.48	<1

[a] Data from Gladney (1980), who compiled these values from an exhaustive review of the primary literature.

Examination of human spleens in Finland (Erametsa and Sihvonen, 1971) has provided the values shown in Table 7-7. Individual variations were very large, and the spleens which accumulated high amounts of one lanthanide did not necessarily contain high levels of another. Differences between the spleens of alcoholic and nonalcoholic individuals are interesting but unexplained. It is possible that the liver, normally a main early site of accumulation of lanthanides, is damaged in alcoholics, decreasing its ability to sequester these elements. Such a condition might lead to spillover into the spleen, an organ rich in cells of the reticuloendothelial system, which are important in clearing lanthanides from the circulation (Section 7.5). This suggestion receives support from the observation of Trnovec et al. (1974) that the hepatotoxin, CCl_4, decreases the uptake of $^{144}Ce^{3+}$ by the liver, with a corresponding increase in its uptake by the spleen.

Lanthanide levels may also change in other diseases. Wester (1965) noted greater concentrations of lanthanides in infarcted cardiac tissue than in normal tissue, while there is a dramatic increase in the levels of lanthanides in the synovial fluid of rheumatoid joints (Esposito et al., 1986a; Section 9.9; Table 9-5). Esposito et al. (1986b) have suggested that

Table 7-4. Concentrations of Lanthanides in Organs of Normal Animals[a]

Animal	Organ	Lanthanide concentration (ppb)	Reference
Rabbit	Liver, bone blood	u.d.[b]	Kramsch et al., 1980
Mouse	Various organs[c]	120–2,100	Samochocka et al., 1984
Human	Eye[d]	u.d.–620,000	Swanson and Truesdale, 1971
	Bone	500	Brooksbank and Leddicotte, 1953
	Kidney	10,300	Leddicotte and Tipton, 1958
	Kidney	0.1	Gerhardsson et al., 1984
	Spleen[e]	420–12,400	Erametsa and Sihvonen, 1971
	Heart[f]	u.d.–2.5	Webster, 1965
	Larynx[g]	0.6–94.6	Esposito et al., 1986b
	Lung[h]	0.46–70.6	Sabbioni et al., 1982b
	Lung[i]	4.5	Gerhardsson et al., 1984
	Liver	5.5	Gerhardsson et al., 1984
	Lymph nodes[h]	0.7–106	Sabbioni et al., 1982b
	Blood	u.d.–2.2	Sabbioni et al., 1982b
	Plasma[j]	0.16–45.1	Esposito et al., 1986a
	Synovial fluid[j]	u.d.	Esposito et al., 1986a
	Urine	u.d.–2.7	Sabbioni et al., 1982b
	Erythrocyte[k]	4.3	Esposito et al., 1986b
	Various organs	u.d.–220,000	Sihvonen, 1972

[a] See also Tables 7-3, 7-5, 7-6, 7-7, 7-8, 9-5.
[b] u.d.: Undetectable.
[c] Enriched in eye (Table 7-5).
[d] Increases with age and with cataracts (Table 7-6).
[e] Higher values in alcoholics (Table 7-7).
[f] Higher values in infarcted tissue.
[g] Lower in malignant tissue.
[h] Higher values in rare earth pneumoconiosis (Table 7-8).
[i] Higher values in smelter workers.
[j] Higher values in rheumatoid arthritis (Table 9-5) and inflammation.
[k] Lower in laryngeal cancer. Increased in inflammation.

La may serve as a marker for diagnosis or prognosis in cancer. Concentrations of La in malignant laryngeal tissue were lower than normal. Furthermore, no La could be detected in erythrocyte lysates from these patients. Erythrocytes from healthy donors contained 14.8 ppb La, a value which increased to 33.2 ppb during inflammation. Plasma La was twelve times normal in patients with laryngeal carcinoma, while the normal level doubled during inflammation.

Piccinini et al. (1975) found lanthanide levels of 137 ng/mg protein in mitochondria. Mitochondrial membranes contained 75 ng/mg protein, while no La could be detected in the matrix.

Certain activities of man have resulted in greater environmental and tissue accumulation of lanthanides. For example, radioactive Ce from the

Table 7-5. Concentrations of Yb in Various
Organs of Laboratory Mice[a]

Organ	Concentration (ppm)
Liver	0.23 ± 0.04
Kidney	0.17 ± 0.03
Heart	0.12 ± 0.03
Spleen	0.30 ± 0.05
Pancreas	0.32 ± 0.05
Brain	0.29 ± 0.06
Eye	
Lens + cornea	0.13 ± 0.03
Retina + sclera	3.40 ± 0.05
Vitreous body	0.002
Total eye	2.1 ± 0.3

[a] From Samochocka *et al.* (1984a), with permission.

fallout after nuclear explosions has been recovered from the lungs and lymph nodes of residents of Vienna (Liebscher *et al.*, 1961). Cerium-144 and cerium-141 have been detected in the atmosphere (Matsunami *et al.*, 1974) and in the sea (e.g., Yamoto, 1984) as a result of human activity. Enormously high concentrations of certain lanthanides occur in the lungs

Table 7-6. Age-Related Changes in the
Concentrations of La in Normal and
Cataractous Human Lens[a]

Age (years)	Abundance of La (μg/g dry wt.) in components of human lens[b]		
	Capsule and epithelium	Cortex	Nucleus
Normals			
0–5	<9	<3	<2
10–20	<25	<10	<2
50–60	400	280	170
70–85	620	410	270
Cataractous			
40–55	490	400	360
60–75	710	600	440
80+	820	600	440

[a] Taken from Swanson and Truesdale (1971), with permission.
[b] $n \geq 6$; SD < 5%.

Table 7-7. Concentrations of Lanthanides and Yttrium in Human Spleen[a]

	Concentration in ashed tissue (ppm)[b]				
	La	Ce	Sm	Eu	Y
Alcoholics	17.7 ± 40.5	2.6 ± 4.8	76.0 ± 61.8	7.9 ± 11.5	1.5 ± 2.7
(n = 11)	(0.1–137.5)	(0.1–16.5)	(0–194.0)	(0–35.0)	(0–9.3)
Nonalcoholics	12.4 ± 20.7	5.5 ± 13.0	0.42 ± 0.9	0.5 ± 1.0	1.3 ± 2.3
(n = 11)	(0.2–68.9)	(0.1–44.0)	(0–2.9)	(0–3.1)	(0–8.0)

[a] Gd, Tb, Ho, Tm, Lu: undetectable in all specimens; Dy, Yb: detectable in only one specimen; Pr, Nd, Er: detectable in only three specimens.
[b] Recalculated, with permission, from Erametsa and Sihvonen (1971). Values are means ± SD, with ranges given in parentheses; in calculating the means, undetectable levels have been assigned zero value.

of industrial workers suffering from rare earth pneumoconiosis as a result of exposure to lanthanide fumes (Sabbioni et al., 1982b; Table 7-8; Section 8.5).

Workers in a Swedish copper smelter also exhibited higher pulmonary concentrations of La. However, despite the presence of La in cigarette smoke, there was no difference in pulmonary concentrations between smokers and nonsmokers (Gerhardsson et al., 1984). The burning of coal, which contains about 10 ppm La, Ce, and Nd, releases lanthanides into the atmosphere (Sabbioni et al., 1982a). Concentrations of 2–9 ng/m^3 have been detected in the atmospheres of various cities (Olmez and Aras, 1977). Fertilizers also contain lanthanides, which leach La and Ce into the soil at a rate of about 14.9 g/ha per year (Maas, 1980). The Chinese have recently begun treating crops with lanthanide fertilizers to improve yield (Guo, 1985; Section 8.3).

High concentrations of Ce, Pr, and Sm have been measured in samples taken beside the M25 motorway around London. Levels fall sharply with distance from the motorway, and appear to be related to motor vehicle activity. The source of the lanthanides is unclear, but it may be linked to the use of diesel fuel (Ward, 1988).

7.3 Microbial Interactions with the Lanthanides

Several studies make reference to the accumulation of lanthanides by bacteria, algae, fungi, and fungal spores (Richards and Troutman, 1940; Spooner, 1949; Miller et al., 1953; Rice, 1956; Boroughs et al., 1957; Miller, 1959; Johnson and Kyker, 1961, 1966; Sobek and Talburt, 1968). However, it is unclear whether this reflects the transport and intracellular

Table 7-8. Lanthanide Concentrations in a Patient with Rare Earth Pneumoconiosis[a,b]

	Concentration (ppb; wet weight)							
	Lung		Lymph node		Blood		Urine	
Lanthanide	Patient	Control	Patient	Control	Patient	Control	Patient	Control
La	45,600	16.6	2,310	28	N.D.[c]	0.16	0.15	0.13
Ce	166,500	70.6	4,903	93	<0.8	2.2	1.8	0.43
Nd	57,700	46.2	2,375	106	3.75	<1.5	18	2.7
Sm	4,550	2.5	321	3.7	N.D.	0.09	0.085	0.05
Eu	87.5	1.23	4.8	4.4	0.62	0.5	0.01	<0.003
Tb	230	1.75	1.61	0.7	<0.005	0.006	0.05	<0.01
Yb	252	3.5	3.08	9.7	<0.09	<0.03	0.006	0.0015
Lu	25	0.46	0.5	1.3	N.D.	<0.009	0.001	<0.00015

[a] From Sabbioni et al. (1982b), with permission.
[b] Patient was a worker in a lithographic laboratory. Control values are averages of 11 autopsy samples of lymph node and lung and 3 control individuals who donated blood and urine.
[c] N.D.: Not determined.

accumulation of lanthanides by these microorganisms or passive attachment to the cell surface. Such evidence as there is favors the latter. For instance, lanthanum treatment of *E. coli* prior to electron microscopy stains only the cell wall (Cassone and Garaci, 1974), while uptake of $^{140}La^{3+}$ by *Streptococcus faecalis* is independent of temperature (Wurm, 1951). With *Saccharomyces cerevisiae*, accumulation of radioactivity from 1 mM $^{144}Ce^{3+}$-citrate was independent of temperature, unaffected by metabolic inhibitors, and reversed by acidification or the addition of EDTA (Johnson and Kyker, 1966). Furthermore, dead cells accumulated almost as much ^{144}Ce as live cells, all of which strongly suggest binding to the outer surface of the cell. Bowen and Rubinson (1951) have reported the uptake of La^{3+} by *Candida albicans*. However, the results of their later work led Bowen (1956) to conclude that this was purely a surface phenomenon and that lanthanide ions only incorporated onto the capsule of the cell. The same conclusion applied to the apparent uptake of lanthanides by plankton (Bowen, 1956). Nevertheless, considerable concentration of lanthanides can occur at the surfaces of plankton. Concentration factors of 500 for filamentous algae and 1000 for phytoplankton have been measured in the Columbia River (Krumholz and Foster, 1957).

Incorporation of millimolar concentrations of lanthanum into the culture media of bacteria and fungi inhibits their growth (Chapter 8). However, certain fungi detoxify lanthanides by incorporating them into insoluble oxalate crystals (Talburt and Johnson, 1967).

7.4 Metabolism of Lanthanides by Invertebrates and Fish

Data on the metabolism of lanthanides by nonmammalian higher organisms, other than birds, are few. Insect larvae in the Columbia River concentrate radioactive lanthanides by a factor of 200 (Krumholz and Foster, 1957). Brine shrimp (*Artemia salini*) also apparently accumulate Y^{3+} (Boroughs *et al.*, 1958), but at least one type of fish (*Tilapia mossambica*) does so to only a limited degree. Of the 1.3% of an intragastric dose of $^{90}Y^{3+}$ remaining in the fish after 14 days, 40% was recovered from the visceral organs, 30% from the muscles, 20% from the skeleton, and smaller amounts from the integument and gills (Boroughs *et al.*, 1956). Nevertheless, Krumholz and Foster (1957) reported a lanthanide concentration factor of 100 for fish in the Columbia River. Freshwater mollusks accumulate $^{144}Ce^{3+}$ (Polikarpov, 1958).

Larvae of *Drosophila repleta* do not assimilate $^{140}La^{3+}$ (Bowen, 1951). If $^{140}La^{3+}$ is smuggled into the larvae as its parent isotope $^{140}Ba^{2+}$, it remains within the body without evidence of transport or excretion.

7.5 Metabolism of Lanthanides by Mammals and Birds

7.5.1 Metabolism of Soluble Lanthanide Salts after Extravascular Administration

Nearly all authors agree that almost no absorption of lanthanides occurs through the gastrointestinal (G.I.) tract of adult vertebrates. Hamilton (1948), for instance, could measure a retention of only 0.3% of a dose of $^{140}LaCl_3$ fed orally to rats 4 days earlier. In subsequent experiments (Hamilton, 1949), values of <0.5% for Pr^{3+} and <0.05% for Pm^{3+}, Ce^{3+}, and Y^{3+} were recorded. In agreement with this, about 90% of the radioactivity from a single oral dose of $^{91}YCl_3$ was found in the feces of rats after 2.5–3 days, with almost complete elimination after a week (Norris *et al.*, 1956). After daily feeding of $^{91}YCl_3$ for 6 months, the retention was under 0.4% of the total administered dose. Similarly low assimilation of orally administered lanthanides has been found with chickens (Mraz *et al.*, 1964) and quail (Robinson *et al.*, 1978). Graul and Hundeshagen (1959), in contrast, reported uptake of 51% of a dose of $^{90}Y^{3+}$ following intragastric introduction into guinea pigs. The reason for their anomalously high uptake value is unknown, analysis of the paper being hindered by the absence of a "methods" section. However, the data in this report are also odd in suggesting an unusually high assimilation of yttrium by the pituitary.

Rabinowitz *et al.* (1988) reported a substantial uptake of ^{140}La by adult rats allowed to ingest $^{140}LaCl_3$ in their drinking water. Lungs, liver, kidney, bone, and teeth accumulated much of the label. Most of the ^{140}La detected in the teeth was associated with the outer layers, suggesting surface adsorption from the drinking water. However, the transfer to the teeth of i.p. injected Y^{3+} (Thomassen and Leicester, 1964) and i.v. injected Ln^{3+} (Ewaldsson and Magnusson, 1964a,b; Table 7-13; Fig. 7-5) has been detected. The use of lanthanides in the prevention of caries has been suggested (Section 9.3.1).

Experiments are under way to monitor the ingestion levels of lanthanides in an area of Brazil where their concentration in the soil is very high (Section 7.1). Fecal concentrations of lanthanides are as follows: Ce ($11–378\,\mu g/g$ dry weight), La ($4.1–63.1\,\mu g/g$), and Sm ($0.24–5.15\,\mu g/g$). In accordance with the very low absorption of lanthanides from the G.I. tract, these values reflect the relative abundance of these lanthanides in the soil. Fecal concentrations for a control population in New York were: Ce ($0.4–5.4\,\mu g/g$ dry weight), La ($0.2–1.5\,\mu g/g$), and Sm ($0.02–0.28\,\mu g/g$) (Linsalata *et al.*, 1986b).

Little is known of the metabolic fate of the trace amounts of lanthanides which are absorbed from the digestive system. The earliest data on this were provided by Baehr and Wessler (1909). Without the aid of sophisticated analytical techniques, they recorded the presence of cerium in the liver of a dog fed solid $Ce(NO_3)_3$. Norris *et al.* (1956) found that most of the ^{90}Y absorbed following oral administration to rats ended up in the skeleton. Lower amounts were recovered from the liver and kidney. However, Kramsch *et al.* (1980) could detect no lanthanum in the bones of rabbits which had been fed 40 mg/kg every day for 8 weeks. This discrepancy probably reflects the greater affinity that yttrium and the heavier lanthanides have for bone (q.v.). The lighter lanthanides tend to accumulate instead in the liver, and, indeed, an average of 33 μg of La was measured in the whole liver of such rabbits (Kramsch *et al.*, 1980). These authors also measured a concentration of 1.7 μg La/100 ml in the blood of monkeys 4 h after a meal containing 40 mg $LaCl_3$/kg. This fell to undetectable levels 12 h after feeding. According to Sihvonen (1972), rats fed $LnCl_3$ accumulate lanthanides in both the pancreas and the liver, with lower amounts in the spleen.

Poor absorption from the G.I. tract explains the low toxicity of orally administered lanthanides (Section 8.4; Table 8-2), while suggesting their use as digesta markers in mammals and as aids to computed tomography and to nuclear magnetic resonance imaging of the G.I. tract (Section 9.6).

Newborn animals have a much greater ability than adults to assimilate lanthanides from the G.I. tract (Inaba and Lengemann, 1972; Naharin *et al.*, 1974; Eisele *et al.*, 1980). About 91% of the $^{144}Ce^{3+}$ assimilated from the G.I. tract by newborn pigs was present in the skeleton (Mraz and Eisele, 1977). The reason for the difference between neonates and older animals is unclear, but it also exists for other heavy metals (Jugo, 1977). Inaba and Lengemann (1972) concluded that it may involve retention of lanthanide by the epithelial cells of the intestinal mucosa. This in turn might reflect the greater nonspecific phagocytic activity of these cells in suckling animals. If pinocytosis is inhibited by injection of hydrocortisone, the assimilation of $^{144}Ce^{3+}$ by newborn rats falls to adult levels (Shiraishi and Ichikawa, 1972).

In view of the poor oral uptake of lanthanides by adult mammals, nearly all metabolic studies have involved parenteral administration. One complication common to the interpretation of all these studies is the tendency of lanthanides to form insoluble precipitates of $Ln(OH)_3$ at physiological pH values. In addition, they form poorly understood "radiocolloids" at pH values well below those at which they precipitate (Schweitzer and Jackson, 1952; Section 2.4). Furthermore, the relatively high physiological concentrations of phosphate and carbonate promote the formation of insoluble precipitates of lanthanide phosphates and carbonates. These factors, and the presence of a variety of organic ligands at different locations in the body with various affinities for lanthanides, render the detailed biochemical analysis of lanthanide metabolism very difficult.

An additional complication is the effect of "carrier." As this factor is common to all the metabolic studies discussed from here on, it is worth discussing it at this juncture. "Carrier" is the nonradioactive lanthanide existing in a radioactive preparation of the same lanthanide. It has the effect of reducing the specific activity of the preparation and increasing its overall concentration of lanthanide. As the solubility constants of many of the salts that lanthanides form $in\,vivo$ are very low (Section 2.3), the increase in concentration upon adding carrier can change the lanthanides from soluble ions to insoluble precipitates after injection. As ions, Ln^{3+} have the capacity to bind to soluble carrier ligands and thereby travel efficiently from the injection site. As precipitates, lanthanides are much more likely to remain close to this site. Formation of particles targets the lanthanides for internalization by cells of the reticuloendothelial system. The Kupffer cells of the liver constitute the major site of accumulation of such cells, although sizable numbers are found in the spleen and, to a lesser degree, in organs such as the lung. In addition, macrophages, monocytes, and polymorphonuclear leukocytes are phagocytes which circulate in the blood and accumulate at sites of inflammation.

In general terms, then, the addition of carrier will reduce the assimilation of lanthanides from their site of application and increase their uptake by phagocytic cells. A good example of this phenomenon has been provided by Kyker (1956), who injected different doses of Y^{3+} into the peritoneal cavities of rats. As the dose per animal increased from 2.2×10^{-10} g to 10^{-4} g and then to 10^{-3} g, the percentage of the injected material which remained intraperitoneal rose from 33.3% to 75% and then to 95%, respectively. Similar observations were found with mice (Lewin $et\,al.$, 1954) and humans (Andrews, 1956), bearing ascites tumors. The importance of gravimetric dose was also shown by analysis of the yttrium content of the femurs of rats, which had received intraperitoneal (i.p.) injections of YCl_3 at doses of 1–60 mg/kg every other day for 5 months (MacDonald $et\,al.$, 1952). Deposition of yttrium in bone was not linear with dose. The authors speculated that after the bone had accumulated 150–200 ppm Y, further deposition became difficult as a second, slower process of yttrium capture was utilized. As might be predicted from the close chemical similarity between the members of the lanthanide series (Chapter 2), one lanthanide can act as a carrier for another (Kyker, 1956).

Effects of carrier notwithstanding, most researchers have found that, when injected intraperitoneally, lanthanides tend to remain within the

Table 7-9. Distribution of [140]La One Week after i.p.
Administration of [140]LaCl$_3$ in Mice[a]

Organ	Recovery of radioactivity per organ (% of injected dose)
Abdominal organs	
Liver	17.8
G.I. tract	46.9
Spleen	3.0
Kidney	0.7
Total abdominal	68.4
Extra-abdominal organs	
Bone	0.06
Muscle	0.10
Lung	0.04
Heart	0.02
Total extra-abdominal	0.22

[a] Taken from Laszlo *et al.* (1952), with permission.

abdominal cavity and to coat the surface of abdominal organs. This property has been explored in the experimental radiotherapy of ascites tumors (Section 9.5). Laszlo *et al.* (1952), for instance, recovered over 68% of the radioactivity in a dose of [140]LaCl$_3$ from the abdomens of mice injected intraperitoneally 7 days earlier (Table 7-9). In agreement with this, Kyker (1956) was able to recover 48–86% of [166]Ho^{3+} from the abdomens of mice 3 days following i.p. injection. At least some of the high abdominal retention of i.p.-injected lanthanides appears to be due to precipitation, as Graca *et al.* (1957) recovered caseous precipitates of Lu from the peritoneal cavities of mice and guinea pigs. Greater transport of lanthanides from sites of i.p. injection occurs in the presence of chelators (Section 7.5.4).

The fraction of lanthanides transported to other tissues following i.p. administration is small. One week after i.p. injection of [140]LaCl$_3$ into mice, Laszlo *et al.* (1952) recovered only 0.22% of the administered radioactivity from the bone, muscle, lung, and heart combined (Table 7-9). Autoradiography confirmed that the extensive association of [140]La with the abdominal organs was a surface phenomenon only. Similarly small values have been reported for the transport of i.p.-injected ^{90}YCl$_3$ to plasma, liver, and spleen in humans (Andrews, 1956). What little excretion of ^{90}Y^{3+} took place over a period of 6 days occurred primarily through the urine. Transport of i.p.-injected Y^{3+} to the teeth has been recorded (Thomassen and Leicester, 1964). The presence of ascites tumor cells did not affect the overall distribution of i.p.-injected $^{90}Y^{3+}$ or $^{140}La^{3+}$, although there was considerable uptake by the tumors (Lewin et al., 1953, 1954). Transfer of radioactivity to the skeleton of puppies 1–4 days after i.p. injection of $^{91}YCl_3$, $^{144}CeCl_3$, or $^{170}TmCl_3$ has been detected autoradiographically (Jowsey et al., 1958). However, the percentage of the applied dose that this represented was not given.

Although bone and liver are nearly always identified as the major sites to which intraperitoneally injected lanthanides are transferred, exceptions exist. Graul and Hundeshagen (1959) noted, in guinea pigs, an unusually high uptake of yttrium by the pancreas, which accounted for 50% of a dose of $^{90}YCl_3$ administered i.p. 4 days earlier. Bone, in contrast, only accumulated 25–28% of the dose during the first 8 days. Genital organs were also labeled, with the specific activity of the ovaries being 2.5 times greater than that of the testes. As generally observed with lanthanides, there was no transfer of label from pregnant mice to the embryos.

Rates of absorption and excretion of lanthanides from other extravascular sites of injection are equally low. Hamilton (1948) injected $^{88}YCl_3$ intramuscularly (i.m.) into rats. After 24 h, 99% of the radioactivity was still at the site of injection. This fell to 36.1% after 4 days, 19.7% after 16 days, and 6.3% after 64 days. Similarly, 6 days after a subcutaneous (s.c.) injection of $^{140}LaCl_3$ into mice, 54% of the radioactivity remained where it had been injected (Laszlo et al., 1952). Only 0.5% of the dose traveled to the liver in this period. Similar studies with $^{90}Y^{3+}$ gave equivalent results. When the label was injected into subcutaneous tumors, there was even greater retention of radioactivity at the site of injection (Goldie and West, 1956).

Contamination through wounds is one of the commonest routes by which industrial workers take up lanthanides. When applied to wounds in experimental animals, $^{144}Ce^{3+}$ tends to remain at the wound site. The small amount that is absorbed behaves as if it had been intravenously injected (Section 7.5.2) (Takada, 1978). However, in recent clinical trials (Section 9.3.2), evaluating $Ce(NO_3)_3$ as a topical antiseptic, no systemic absorption of Ce^{3+} through burn wounds has been detected.

Radioactive $^{90}YCl_3$ has been introduced intrapleurally into human cancer patients with bilateral effusions and ascites (Laszlo, 1956). Less than 1% was excreted in 16 days. Radioactivity remained very low in the plasma, bone, and liver during this period. However, radioactivity could be recovered by lavage of the injected pleural cavity. Hamilton (1948) found that $^{144}CeCl_3$ solution introduced into the lungs themselves was cleared much more quickly, with a half-life of about 14 days. Over 80%

Table 7-10. Localization of Free [169]Yb in the Tissues of a Rabbit Injected 28 Days Earlier into the Right Knee Joint[a]

Tissue	Dry weight (mg)	[169]Yb (cpm $\times$ 10^{-3})	cpm/mg dry weight
Right patellar tendon and synovium[b]	48.0	19.30	400.3[c]
Ear cartilage	32.0	0.16	5.0
Hyaline cartilage—tibia	7.0	0.91	126.4
Hyaline cartilage—femur	4.0	0.96	239.6
Patella (cartilage and bone)	149.0	12.70	85.5
Medial meniscus	14.1	4.50	322.3[c]
Lateral meniscus	21.4	7.40	347.7[c]
Synovium and subsynovial fat	49.5	121.70	2460.4
Kidney	68.0	0.54	7.9
Liver	66.5	0.17	2.6
Bone marrow (red)	42.6	0.49	11.4

[a] From McCarty et al. (1979), with permission.
[b] Similar tissue from the left (control) knee showed virtually no radioactivity.
[c] Small pieces of synovium were probably included in these samples.

of intrapleurally administered ^{90}YCl$_3$ remained in the pleural space of rats and rabbits 1 week after injection (Lahr et al., 1955).

Radioactivity from ^{177}LuCl$_3$ injected into the mammary glands of dogs also remained at the injection site for at least 21 days, with some transfer to the lymph nodes draining the site (Kyker et al., 1956). The authors explained this by the formation of colloids incorporating ^{177}Lu which were taken up by the reticuloendothelial cells of the lymphatic system. In these experiments, there was no transfer to the heart, lung, kidney, spleen, liver, sternum, or rib. Such efficient localization of the β-emitting ^{177}Lu suggested utility in the radiotherapy of breast tumors (Section 9.5). An analogous conclusion was reached by Jasmin et al. (1977) after demonstrating retention of ^{144}Ce^{3+} close to the injection site in the maxillary sinus of rats. After injection of ^{90}YCl$_3$ into the prostate, about one-third of the dose was excreted in the urine after 48 h. About one-third remained in the prostate, and 5% traveled to the liver (Cooper et al., 1956).

Clearance of ^{169}YbCl$_3$ from the knee joints of rabbits following intra-articular injection is biphasic (McCarty et al., 1979). These authors demonstrated a rapidly cleared fraction, representing about 35% of the injected dose, which disappeared from the joint with a half-life of less than a day. The remainder was removed with a half-life of 20–30 days. Surprisingly, little transfer of radioactivity to other organs occurred (Table 7-10); it seems that most of the Yb^{3+} cleared from the joint was efficiently excreted in the urine. This is not the case with lanthanides administered through

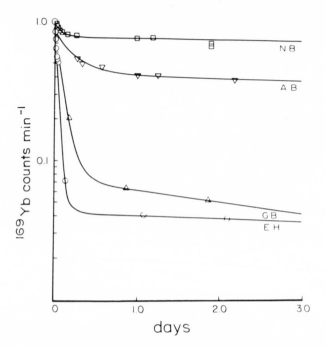

Figure 7-1. Clearance of ^{169}YbCl$_3$ from the knees of four different arthritic patients. From McCarty *et al.* (1979), with permission.

other routes. It is possible, though speculative, to suggest that lanthanides leave the joint in tight association with a soluble physiological chelator which is only found intra-articularly. Synovial fluid, for instance, has an unusually high concentration of hyaluronic acid, which forms tight complexes with Ln^{3+} ions. Clearance of ^{169}YbCl$_3$ from the knee joints of four arthritic patients was also biphasic, but of variable rate from patient to patient (Fig. 7-1). The variability may reflect the different arthritic conditions of the knees. In these patients, over 99% of the radioactivity removed from the knee joints appeared in the urine within 72 h.

Particles of ^{144}CeCl$_3$ of median diameter 0.83 μm are cleared from the lungs of hamsters following inhalation, according to a three-component equation. There is an early rapid component which leads to removal of 80% of the dose after a week. This component reflects movement of the label from the respiratory tract and lung to the G.I. tract and subsequent fecal excretion. The residual ^{144}Ce then travels from the lung to the liver and skeleton, which are responsible for the long-term retention of radioactivity (e.g., Sturbaum *et al.*, 1970; Cuddihy *et al.*, 1976). Radioactivity from intrapleurally applied $^{90}YCl_3$ solutions is poorly transported from the injection site (Laszlo, 1956). Cuddihy *et al.* (1976) have provided a detailed mathematical model of the distribution of $^{144}CeCl_3$ in beagles after inhalation.

7.5.2 Metabolism of Soluble Lanthanide Salts after Intravenous Injection

The key matter to be addressed when considering the metabolism of intravenously injected lanthanide ions is that of their chemical form in the blood. Unfortunately, this is still poorly understood. The possibilities are enormous. Not only would one predict, from purely chemical considerations (Chapter 2), the formation of radiocolloids at low lanthanide concentrations and of insoluble hydroxides, phosphates, and carbonates at higher concentrations, but one would also expect extensive interaction with ligands in the blood. All of these may well occur, depending on the conditions.

By far the greater proportion of i.v.-injected lanthanides recovered from whole blood is found in the plasma, with low cellular association (Baltrukiewicz *et al.*, 1975). Although lanthanides precipitate in neutral aqueous solutions which contain phosphate and bicarbonate, no visible precipitate forms when moderate pharmacologic doses of lanthanides are added to blood or serum *in vitro* (Kyker, 1962; Kanapilly, 1980). Furthermore, blood–Ln^{3+} mixtures are stable to centrifugation and dialysis (Kyker, 1962; Rosoff *et al.*, 1958). Kanapilly (1980) has made careful *in vitro* studies of these matters, with the following results.

Although almost insoluble in a synthetic ultrafiltrate of blood, lanthanides are almost completely soluble in dog serum. Indeed, they are more soluble in serum than in distilled water. Phosphate is the most potent precipitant of lanthanides from simple salt solutions. Carbonate precipitates lanthanides from saline solution only in the absence of citrate, while the formation of insoluble hydroxides is not observed at physiological pH and ionic strength (Kanapilly, 1980). This finding agrees with that of Schatzki and Newsome (1978), who showed freshly prepared, aqueous solutions of La^{3+} at pH 7.4 to be free from large aggregates and precipitates. Increasing the amount of carrier (Section 7.5.1) increased the degree of lanthanide precipitation from saline, synthetic ultrafiltrate, saline–bicarbonate, and saline–phosphate solutions.

All these data suggest that lanthanides do not simply fall out of solution as insoluble precipitates after i.v. injection, as chemical considerations alone might lead one to expect. Despite quite clear evidence of this, the majority of papers on the subject of lanthanides in the blood insist on

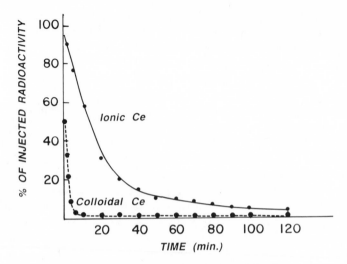

Figure 7-2. Clearance of radioactivity from the blood of rats following intravenous injection of $^{144}CeCl_3$ or $^{144}Ce(OH)_3$. From Aeberhardt *et al.* (1962), with permission.

commenting, incorrectly, that lanthanides form insoluble hydroxides and phosphates in blood after i.v. injection.

These *in vitro* observations probably reflect the state of affairs *in vivo*. Large deposits of lanthanides are not seen in the Kupffer cells of the liver until high doses are injected (Tuchweber *et al.*, 1976; Section 8.7). Autoradiography has confirmed that the uptake of physiologically active doses of lanthanide ions by the liver is mediated by the hepatocytes (Aeberhardt *et al.*, 1962). Uptake by the Kupffer cells only occurred after the injection of colloidal preparations. Further evidence that straightforward precipitation of lanthanides does not occur after i.v. injection is provided by the results of Aeberhardt *et al.* (1962). They compared the metabolism of $^{144}CeCl_3$ with colloidal suspensions of $^{144}Ce(OH)_3$ having a particle size of about $0.3\,\mu$m. As shown in Fig. 7-2, the radioactivity from $^{144}CeCl_3$ disappeared from the blood of rats at a much faster rate than did the colloidal radioactivity. Dobson *et al.* (1949) had earlier shown that "large" colloidal particles of Y or La are cleared from the bloodstream of rabbits with a half-life of 30 s to 1 min. Their "intermediate"-sized particles had half-lives in the blood of 30–80 min.

The probable explanation of all these effects is that, upon i.v. injection, Ln^{3+} ions form soluble complexes with ligands present in the plasma. Possible ligands in plasma include phospholipids, amino acids, nucleotides, proteins, and the numerous other chemicals in blood with oxygen donor atoms as ligands (Section 2.3). Reference to Chapter 4 alone reveals a number of serum proteins which interact with lanthanides. These include albumin, transferrin, immunoglobulins, and the several "Gla-proteins" involved in blood clotting. The fact that lanthanides have anticoagulant properties (Section 8.8) presumably infers their binding to the clotting proteins under physiological conditions. However, transferrin may not be a carrier for lanthanides *in vivo*. Its affinity for Fe^{3+} exceeds its affinity for Ln^{3+}, while spectroscopic studies have failed to detect the presence of transferrin-Ln^{3+} complexes in blood (Schomacker *et al.*, 1986). Pederson (1979) has reported preliminary evidence of fibrinogen-La^{3+} complexes in rabbit blood.

Under *in vitro* conditions, albumin and various globulins bind lanthanides (Rosoff and Spencer, 1979) and prevent their precipitation from an artificial serum ultrafiltrate (Kanapilly, 1980). Attempts have been made to separate Ln^{3+}–carrier complexes from blood following *in vitro* mixing or i.v. injection. Kawin (1953) subjected the blood and urine of rats to paper chromatography 10 min after intravenous injection of $^{91}Y^{3+}$. Whereas about 90% of the radioactivity in urine comigrated with free Y^{3+}, only 4% of plasma radioactivity did so. Instead, the bulk of the $^{91}Y^{3+}$ was associated with 10 ninhydrin-positive compounds of different R_f values. Similar experiments, where $^{91}Y^{3+}$ was incubated for 30 min *in vitro* with sheep plasma, revealed the presence of at least two yttrium carriers (Zilversmit and Hood, 1953). More detailed studied by Ekman *et al.* (1961) using rat serum mixed with lanthanides *in vitro* or *in vivo* showed albumin to be the major complexing agent. Several smaller carriers were also detected. Again, no difference existed between the results from *in vitro* and *in vivo* experiments. Although albumin has only a modest affinity for lanthanides (Section 4.21), it is likely to be a major serum carrier for these ions by virtue of its high concentration in blood. However, as the ability of plasma to bind lanthanides exceeds that predicted by its albumin content, additional carriers must exist (Rosoff *et al.*, 1958).

Published plasma half-lives of i.v.-injected lanthanides fall into two groups (Table 7-11). Some authors report half-lives in the range of 10–20 min. As the EDTA and DTPA chelates of Ln^{3+} ions share this half-life, it may mean that lanthanides form stable associations with small complexing agents in blood. The second group of reported half-lives fall in the range of 1–3 h (Table 7-11). These values are similar to those of the "smaller" particles of Dobson *et al.* (1949). In at least two of the experiments yielding the higher half-life, "carrier" lanthanide was present (see discussion of "carrier" in Section 7.5.1). This raises the possibility that the complexing ability of the plasma was exceeded and colloidal precipitates formed. In this context, it has been estimated that the total

Table 7-11. Rates of Removal of Lanthanides from Blood

Animal	Injected material	Half-life[a]	Reference
Rat	Ce^{3+}	13 min	Aeberhardt *et al.*, 1962
	$Ce(OH)_3$	1 min	Aeberhardt *et al.*, 1962
	(0.3 μm particles)		
Human	Y^{3+}	2 h	Hart *et al.*, 1955
	Y-EDTA	<10 min	Hart *et al.*, 1955
Cow	Y^{3+}	15 min	Hood and Comar, 1956
Monkey	Y^{3+}	10 min	Daigneault, 1963
Human	Pm^{3+}	10 min	Palmer *et al.*, 1970
Mouse	La^{3+}	1 h	Laszlo *et al.*, 1952
Rat	Yb^{3+}	1 h	Baltrukiewicz *et al.*, 1975
Rat	Gd-DTPA	20 min	Weinmann *et al.*, 1984b
Quail	La^{3+}, Gd^{3+}	3 h	Robinson *et al.*, 1978, 1981b
Rat	Y^{3+}-citrate	<5 min	Schubert *et al.*, 1950

[a] Several of these values were estimated by interpolation of graphs and tables. They are therefore approximate.

plasma of a 250-g rat is capable of complexing approximately 15 μg (~0.1 μmol) of Ln^{3+} ions (Durbin *et al.*, 1956b).

An alternative explanation to saturation of the blood's ability to complex Ln^{3+} ions is one based on the nature of the complexes formed at higher dose. The rate at which lanthanides are cleared from blood may well depend strongly upon the plasma half-life of the ligands to which they are attached. At high concentrations of Ln^{3+}, additional plasma molecules of extended half-life may complex these ions. Whatever the correct explanation, Kyker (1956) has confirmed that the rate of clearance of lanthanides from blood decreases with increasing dose.

Lanthanides are excreted from the body slowly after i.v. injection of free cations. Norris *et al.* (1956), for example, reported only 9–14% excretion of ^{144}Ce in 4 days following injection of $^{144}CeCl_3$ solutions into rats. Most authors report that the kidney provided the major early route of excretion, with the G.I. tract taking over after a day (e.g., Fig. 7-3). Fecal excretion occurs as a consequence of biliary clearance from the liver and direct transfer to the digestive system (Magnusson, 1963; Mahlum, 1967). Patterns of excretion for $^{140}La^{3+}$ (Hart *et al.*, 1955; Laszlo *et al.*, 1952) and $^{152-154}Eu^{3+}$ (Berke, 1968) agree quite well with those reported for Ce^{3+}. It seems likely that the early, urinary excretion of lanthanides reflects the fraction of the injected material which forms tight complexes with soluble ligands. Once this has cleared, the less mobile material is slowly excreted fecally. Dose is also a factor here. Less than 0.5% of an i.v. dose of 85 mg $^{90}YCl_3$ in humans was excreted in the urine

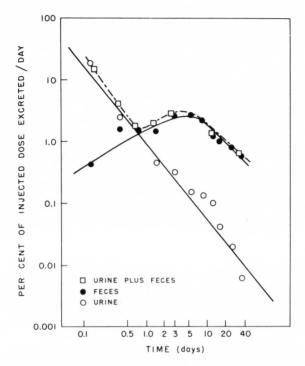

Figure 7-3. Excretion of radioactivity from mice following intravenous injection of ^{144}CeCl$_3$. From Norris *et al.* (1956), with permission.

over a 3-day period. However, at a dose of 4.2 μg, over 12% was recovered in the urine during the same time (Andrews, 1956).

Most authors identify the liver or the skeleton as the main sites of accumulation of i.v.-injected lanthanides. Laszlo *et al.* (1952) found the liver to be by far the most radioactive organ following i.v. injection of ^{140}LaCl$_3$ into mice, containing over 74% of the administered dose (Table 7-12). Radioactivity was slowly released from this organ, but even a week later it remained the major bodily reservoir of ^{140}La^{3+}. These results do not simply reflect the large size of the liver, as its uptake of La^{3+} per gram of tissue was also much greater than that of any other organ. The spleen accumulated appreciable amounts of ^{140}La^{3+}, but other organs, including bone, were poorly labeled. Similar organ distributions were found with i.v.-injected ^{169}Yb^{3+} (Baltrukiewicz *et al.*, 1975) and ^{144}Ce^{3+} (Mahlum, 1967). The results of Hart *et al.* (1955) concerning the distribution of i.v.-injected ^{90}YCl$_3$ in mice are in general agreement with these findings. After 40–45 h, about 77% of an i.v. dose of ^{90}YCl$_3$ was retained

Table 7-12. Distribution Pattern Following
Intravenous Injection of ^{140}LaCl$_3$ Solution
into Mice[a]

Organ	Recovery of radioactivity (% of total)			
	1 h	20 h	72 h	144 h
Liver	N.G.[b]	74.3	69.1	53.0
Spleen	N.G.	N.G.	5.5	4.8
Kidney	N.G.	0.6	0.7	0.5
Bone	N.G.	6.2	8.2	6.0
Carcass	N.G.	29.6	11.5	11.7
Blood	49.7	0	0	0

[a] From Laszlo et al. (1952), with permission.
[b] N.G.: Not given.

in mice, of which 91% was located in the liver. The liver was also found to be the main site of accumulation of ^{153}Gd^{3+}, retaining over 56% of the radioactivity 1 week after i.v. injection of ^{153}GdCl$_3$ into rats (Weinmann et al., 1984a). The spleen contained about 25% of the counts while the rest of the body, including the bones, retained about 16%.

Magnusson (1963) made detailed studies of the distribution of i.v.-injected lanthanide chlorides in rats. During the first 96 h, uptake by the liver followed the order Ce, Pm $\geq$ Ho > Tb > Yb > Y. Apart from the reversal of Ho and Tb, this sequence demonstrates that the lighter the lanthanide, the greater its affinity for liver, with Y behaving as a heavy lanthanide. A similar sequence was observed for the hepatic uptake of lanthanides from Ln^{3+}-citrate complexes (Section 7.5.3), but this trend has not been generally reported following i.v. injection of the chlorides. Regardless of the initial uptake of lanthanides, the radioactivities of the liver fell after 1–3 h. Excretion of lanthanides through both the bile and the wall of the G.I. tract was noted (Magnusson, 1963).

Hepatocarcinogens were not found to alter the ability of liver to sequester ^{144}Ce^{3+}, but they inhibited subsequent release of Ce from this organ (Mahlum, 1967). However, CCl$_4$ prevented the uptake of Ce by liver and prolonged its plasma half-life (Trnovec et al., 1974). Cortisone depresses the activity of the reticuloendothelium system and accordingly reduces by 50% the incorporation of ^{91}Y^{3+} into liver and spleen, while extending its dwell time in the plasma (Kawin, 1957). Nevertheless, activation of the reticuloendothelial system with zymosan did not affect the distribution of Y (Daigneault, 1963). Parathyroid hormone is also without effect (Kawin, 1957).

Not all studies of the metabolism of i.v.-injected lanthanides report such a low relative uptake by bone. Hamilton (1949), for instance, showed that the skeleton can sequester more than 30% of an i.v. dose of $^{140}LaCl_3$. In humans, $^{143}Pm^{3+}$ distributes almost equally between the liver and the skeleton. As is usually found, hepatic sequestration was very rapid, while fixation in the skeleton took several hours. However, unlike rodents, where the hepatic pool of lanthanides is relatively labile, the organ distribution of ^{143}Pm in humans remained stable for at least a year (Palmer et al., 1970). Detailed studies by Berke (1968) using $^{152-154}EuCl_3$ found the half-life for whole body clearance of Eu^{3+} by rats to be about 40 days. The body burden of the isotope stabilized at about 50% of the original dose after 8–9 weeks. Initially, $^{152-154}Eu^{3+}$ accumulated in both the liver and the skeleton. However, whereas the skeletal pool was metabolically stable, the liver rapidly lost its radioactivity (Fig. 7-4). Of the radioactivity released by the liver, two-thirds was excreted and one-third retained by the skeleton. Thus, most of the 50% of the Eu that was lost from the body during the first 40 days had come from the liver. By 252 days postinjection, the skeleton accounted for about 85% of the body's load of Eu. By 445 days, this had risen to nearly 100%. Release of Eu from bone could be resolved into an early, small, labile fraction of half-life 52 days, and a later, large, stable fraction of half-life 2.7 years.

The results of Aeberhardt et al. (1962) and Takada (1978) for $^{144}Ce^{3+}$ metabolism are in general qualitative agreement with those of Berke (1968) for $^{152-154}Eu^{3+}$, although in the former experiments the distribution of Ce initially favored the liver over the skeleton. Autoradiography demonstrated that hepatic ^{144}Ce was associated with the hepatocytes (Aeberhardt et al., 1962). After the first day, the liver released its $^{144}Ce^{3+}$, with the liver concentration of ^{144}Ce falling to very low levels in about 20 days. During this time, the skeletal concentration of ^{144}Ce increased, accounting for nearly 20% of the injected dose after 3 days and most of the body burden by 10–20 days.

The reason for the disagreement over the relative importance of the liver and bone in sequestering lanthanides following i.v. injection of a simple salt solution remains unknown. However, three possible resolutions come to mind. From the above discussion, it is clear that timing is important; skeletal uptake is very stable, while hepatic uptake is labile. A second factor is dosage. As discussed in Section 7.5.1, the presence of "carrier" lanthanide affects the rate of mobilization of intraperitoneally administered Ln^{3+} and the plasma half-life of intravenously administered lanthanides. The organ distribution of the injected material may also be

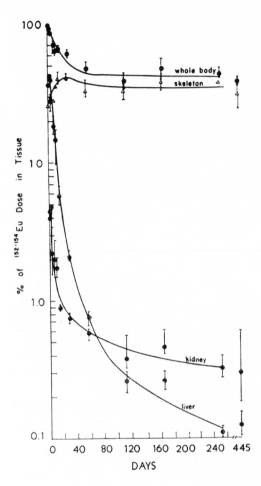

Figure 7-4. Organ distribution of radioactivity in rats following i.v. injection of $^{152-154}$ EuCl$_3$. From Berke (1968), with permission.

affected. In particular, the formation of particles will target the lanthanides to the reticuloendothelial system, the Kupffer cells of the liver being one of its major components. As most of the larger particles may be removed from the blood during their first pass through the liver (Dobson *et al.*, 1949), the liver could act to some degree as a governor, determining how much lanthanide remains available for other organs to assimilate. While on the subject of dosage, it should be pointed out that moderate doses of lanthanides depress the reticuloendothelial system (Lazar, 1973) and may thus inhibit sequestration by the Kupffer cells (Section 8.10). A third factor which may alter the bodily distribution of lanthanides is age. For example, the skeletal uptake of $^{90}Y^{3+}$ is greater in calves and young rabbits than in the adult animals (Hood and Comar, 1956; Rayner et al., 1953).

The mechanism of localization of lanthanides in bone has caused some controversy. At issue is whether the lanthanides associate with the bone mineral or the organic matrix of bone. Hamilton (1948) and Laszlo et al. (1952) presented autoradiograms showing concentrations of lanthanides beneath the endosteum, the periosteum, and the epiphysial plate, suggesting their association with the organic matrix. Copp et al. (1951) arrived at the same results for the deposition of $^{88}Y^{3+}$ and $^{144}Ce^{3+}$ in the bones of normal and rachitic rats. Thomasset et al. (1976) recorded deposition of $^{177}Lu^{3+}$ on the endosteal surfaces, on the metaphysis, and in the area of the blood vessels in the compact bone diaphysis. Dietary deficiency of phosphorus mobilized Ca, but not hydroxyproline or Lu. Calcium deficiency, on the other hand, raised the concentration of hydroxyproline and Lu in the urine. Thus, under these experimental conditions, Lu seems to be associated with the organic matrix rather than the mineral. However, Jowsey et al. (1958) injected neutral aqueous solutions of $^{90}YCl_3$, $^{144}CeCl_3$, or $^{170}TmCl_3$ intraperitoneally into calcium-deficient puppies and came to a different conclusion. The three isotopes were deposited on the highly mineralized, nongrowing surfaces in areas of resorption and inactivity, but never on the uncalcified osteoid tissue of growing surfaces. Furthermore, lanthanides were not eluted from bone by chemical removal of the organic fraction of bone but remained with the mineral fraction. These authors thus concluded that the lanthanides were sequestered by the inorganic mineral of bone and not by the organic matrix.

In vitro investigations have demonstrated that both mineral and osteoid have binding sites for lanthanides. For example, Mössbauer spectroscopic studies of the interaction between $^{161}Tb^{3+}$ and powdered bone suggest that lanthanides are absorbed onto the surface of the bone mineral (Rimbert et al., 1981). Ultrastructural studies, meanwhile, reveal an interaction of the organic matrix of bone with La^{3+} and Tb^{3+} (Bonucci et al., 1988). Our own experiments magnetizing bone and collagen with Er^{3+} (Evans and Tew, 1981; Section 3.8.3) and investigating the interactions of lanthanides with collagen (Drouven and Evans, 1986; Section 4.12) show that both the mineral and the osteoid can sequester lanthanides. However, the mineral is the more efficient of the two. What happens in vivo may depend upon blood flow, turnover rates, diffusion barriers, and, perhaps, particle size.

Although the liver and the skeleton almost invariably sequester the bulk of the injected, radioactive lanthanide, other organs take up some of the label. Several authors have detected an early and rather labile uptake of lanthanides by lung (Laszlo *et al.*, 1952; Norris *et al.*, 1956; Hood and Comar, 1956; Baltrukiewicz *et al.*, 1975). This is attributable to the presence of pulmonary reticuloendothelial cells. Laszlo *et al.* (1952) found that pulmonary sequestration of $^{140}La^{3+}$ increased sharply with the size of the injected dose. Uptake by spleen and kidney has also been recorded. Although i.v.-administered lanthanides do not accumulate in muscle or nervous tissue, small to moderate amounts have been found in the pituitary, thyroid, adrenal gland, pancreas, and ovaries (Hood and Comar, 1956; Daigneault, 1963; Ojemann *et al.*, 1961). Unusually high specific radioactivity occurred in the choroid plexus of cats' eyes 1 h after injection of $^{165}DyCl_3$ (Ojemann *et al.*, 1961). Lanthanide concentrations in the eye are given in Tables 7-5 and 7-6.

In the experimental results discussed so far, the individual organs had been dissected from the animals and their radioactivities measured. The presence of radioactive lanthanides in organs not selected for study would, of course, have been missed. To remedy this, Ewaldsson and Magnusson (1964a,b) made autoradiograms of sagittal sections of entire pregnant mice, following i.v. injection of $^{144}CeCl_3$, $^{147}PmCl_3$, $^{160}TbCl_3$, $^{166}HoCl_3$, or $^{169}YbCl_3$. In this way, every major site of lanthanide accumulation could be screened. Representative autoradiograms are shown in Fig. 7-5, while Table 7-13 summarizes the distributions of the lanthanides in various organs. As expected from the foregoing discussion, the skeleton and liver were strongly labeled. However, the qualitative pattern of labeling of the liver differed between the heavy and light lanthanides (Fig. 7-5). This may be relevant to the observation that light lanthanides produce a characteristic fatty liver, while the heavy lanthanides instead cause focal necrosis (Section 8.7.2). Large amounts of radioactivity were also detected in kidney and teeth. We have already noted that the teeth of rats accumulate La^{3+} and Y^{3+} from drinking water and following i.p. injection (Thomassen and Leicester, 1964; Rabinowitz *et al.*, 1988). The hyaline cartilage of the tracheal rings sequestered large amounts of each lanthanide except Yb^{3+}. Quite high labeling of the adrenals and ovaries was detected. In view of the sex dependency of the rare earth fatty liver (Section 8.7.2) and the effect of adrenalectomy on its development, this observation may also be of toxicologic relevance.

Autoradiographic evidence (Ewaldsson and Magnusson, 1964a) suggests only weak transfer of ^{160}Tb and ^{166}Ho to the 2-day prepartum fetuses following i.v. injection of the mother mice. No transfer was visible for ^{169}Yb-injected mice. Transfer of ^{144}Ce (Zylicz *et al.*, 1975) and ^{165}Dy (Ojemann *et al.*, 1961) to fetuses was barely detectable. Very low, but detectable, amounts of Ce were measured in the chlorioallantoic placenta and in the yolk sac (Zylicz *et al.*, 1975). Hood and Comar (1956) also concluded that a placental barrier to the transmission of ^{90}Y existed in

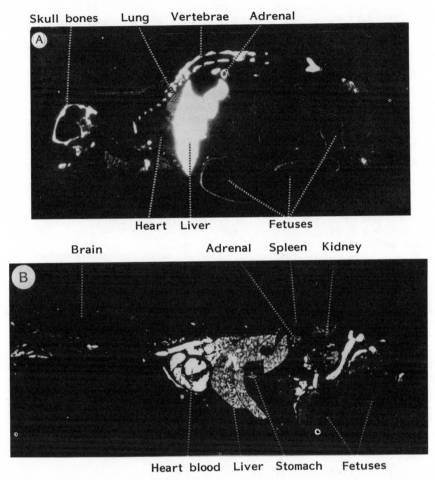

Figure 7-5. Autoradiographic localization of (A) ^{144}Ce and (B) ^{169}Yb in pregnant mice. From Ewaldsson and Magnusson (1964a,b), with permission. Lanthanide chlorides were administered intravenously 24 h before sectioning.

rats and cows. In agreement with this, Beno (1973) could detect no lanthanides in the amniotic fluid and very little in the fetuses and placentas of rats at any time after i.v. injection of the mother.

In view of the greater G.I. absorption of lanthanides by neonates (Section 7.5.1), the possible transfer of lanthanides from mothers to milk is of toxicological concern. In cows, secretion of ^{90}Y into milk was very small, with a cumulative dose after 112 days of less than 1.5% of the injected radioactivity (Hood and Comar, 1956). However, up to 10% of i.v.-injected ^{169}Yb^{3+} passed into the milk of rats. The transfer of ^{144}Ce^{3+} was much lower (Baltrukiewicz et al., 1976).

Table 7-13. Organ Distribution of Lanthanides after Intravenous Injection
of Chlorides into Pregnant Mice[a]

Organ	Ce	Pm	Tb	Ho	Yb
Skeleton	+ + +[b]	+ + +[b]	+ + +[c]	+ + +[c]	+ + +[c]
Liver	+ + +	+ + +	+ + +	+ + +	+ + +
Teeth[d]	+ + +	+ + +	+ +	N.G.[e]	+ +
Cartilage[f]	+ + +	+ + +	+ +	+ +	−
Kidney	+ + +	+ + +	+	+	+
Spleen[g]	+ +	+	+ +	+ +	+ +
Adrenal gland[h]	+ +	+	+ + +	+ + +	+ +
Ovary	+ +	+	+	+	+
Stomach[i]	+ +	+ +	+	+	+
Intestine	+	+	+	+	+
Mammary gland	+	+	±	±	±
Fetal membranes	+ +	+	+	−	−
Lung	+	−	±	−	±
Fetus	±	±	±	±	−
Myocardium	±	±	±	±	±
Skeletal muscle	−	−	±	±	±
Skin	−	−	±	±	±
Gall bladder	N.G.	N.G.	+	+	+
Urinary bladder	−	+	N.G.	N.G.	N.G.
Brain	−	−	N.G.	N.G.	N.G.

[a] Table compiled from data of Ewaldsson and Magnusson (1964a,b). Mice were i.v. injected
with aqueous solution of the chlorides of ^{144}Ce, ^{147}Pm, ^{160}Tb, ^{166}Ho, or ^{169}Yb before killing,
sagittal sectioning, and autoradiography. Uptake has been designated, by reference to the
authors' description of their autoradiograms, as: −, no radioactivity detected; ±, radioactivity
slight or questionable; +, radioactivity low to moderate; + +, radioactivity high; + + +,
radioactivity intense. Examples of autoradiograms are shown in Fig. 7-5.
[b] Most activity in endosteum and periosteum; activity low in bone marrow.
[c] Most activity in bone marrow; activity low in periosteum; absent from cortex.
[d] Activity mostly in enamel pulp; low in dentine.
[e] N.G.: Not given.
[f] Tracheal rings.
[g] Activity concentrated in the red pulp.
[h] Activity concentrated in cortex.
[i] Largely associated with excretion.

Robinson and co-workers have published a series of papers on the
metabolism of various lanthanides by quail (Robinson *et al.,* 1978, 1979,
1980, 1981a,b, 1984; Robinson and Wasnidge, 1982). The plasma half-life
of intravenously injected material was 3 h, considerably larger than that
reported for mammals (Table 7-11). The high Ca^{2+} concentration in the
blood of egg-laying females was probably not responsible for this long
half-life, as male quails gave identical results. Liver assimilated by far the
greatest amounts of the injected radioactive lanthanide. In females, there
was substantial uptake by the oocytes and ovaries. In males, uptake of
the testes was equal to the ovarian uptake in the females, but liver levels

were higher. Other sites of lanthanide accumulation were the intestine and cecum, these probably reflecting biliary excretion. Nearly no $^{140}La^{3+}$ was incorporated into the eggshell; nor did La inhibit the deposition of $^{47}Ca^{2+}$ at this site. However, about 25% of the injected radioactivity was found in the yolk, principally in association with phosvitin and lipovitellin. These proteins are derived from vitellin, which is synthesized in the liver and transported in the plasma to the yolk granules. Robinson *et al.* (1978) speculated that vitellogenin was a major carrier of Ln^{3+} ions in the blood and that lanthanides could be used as *in vivo* labels of plasma vitellogenin.

Each of several lanthanides tested was metabolized in a similar fashion by quail. As with mammals, the presence of "carrier" lanthanide affected the tissue distribution (Section 7.5.1). Increasing the dose of injected material had the effect of lowering its uptake by the oocytes and increasing its uptake by liver, spleen, and ovary. Transfer of lanthanides to the yolk was so effective that over 90% of the lanthanide injected into the mother was recovered in the eggs over an 8-day period. During this process, label which was initially sequestered by the liver was secondarily released to the oocytes. Upon hatching of these eggs, nearly all the label was found in the yolk sacs. Radioactivity fell by about 50% during the first 2 weeks after hatching as the yolk sac was resorbed but then remained stable into adulthood. There was almost no transfer of radioactive lanthanide from the yolk sac to other organs in the adult quail. Furthermore, very little radioactivity occurred in the eggs laid by these F_1 quail.

7.5.3 Effects of Chelators on Lanthanide Metabolism

Chelators are important in studies of lanthanide metabolism for two reasons. Firstly, they prevent the formation of colloids and precipitates, thereby permitting studies of the metabolism of an unambiguously soluble form of the lanthanide. Secondly, chelators are of potential medical use in removing lanthanides from the body following accidental exposure. The extent to which chelators fulfill these two functions depends upon their affinities for the various Ln^{3+} ions and, to some degree, their metabolic stability. EDTA, for instance, is not degraded by mammals (Foreman and Trigillo, 1954), whereas citrate is rapidly metabolized.

During the passage of Ln–chelate complexes through the body, physiological ligands will compete for the lanthanides. In addition, other metal ions will compete with the Ln^{3+} ions for the chelator. The extent to which exchange occurs is, of course, determined by the relative affinities of the chelator and the competing ligands for Ln^{3+} ions and endogenous metal ions. Citrate, for instance, has a relatively low affinity for Ln^{3+} ions (Table

2-8) and, as we shall see, gives them up to the liver and skeleton. At the other extreme, DTPA complexes of Ln^{3+} are extremely stable (Table 2-4) and pass intact through the body, leaving behind a minimal residue. The chemistry of lanthanide–ligand interactions has been extensively discussed in Section 2.3. In this section, we shall discuss the metabolism of progressively stronger lanthanide complexes, starting with acetate complexes, the weakest, and ending with DTPA complexes, the strongest of those whose metabolism has been investigated.

Incorporation of i.p.-injected Yb^{3+}-acetate (Table 2-5) into the eyes (Samochocka *et al.*, 1984a) and brain (Samochocka *et al.*, 1984b) of rodents has been reported. The total Yb concentration of the eyes rose from 2.1 ± 0.3 ppm in control mice to 7.8 ± 1 ppm in mice 2 h following i.p. injection of an unspecified volume of $3 \mu M$ Yb^{3+}-acetate. The lens, cornea, retina, and sclera all contributed to this increase. Examination of control mice (Section 7.1; Table 7-5) revealed an unusually high concentration of Yb present naturally in the eye. Relatively high levels have also been found in human eyes (Section 7.1; Table 7-6). In similar experiments with rats injected i.p. with 1 ml of a 31 mM solution of Yb^{3+}-acetate, the levels of Yb in the brain rose from 0.07 ± 0.05 ppm to 0.4 ± 0.02 ppm after 2 h. The highest concentrations were found in the cerebral cortex and striatum. These results are difficult to evaluate and to compare to the findings of other researchers. In their autoradiographic studies, Ewaldsson and Magnusson (1964b) detected no transfer of $^{169}YbCl_3$ to the eyes or brain of mice following i.v. injection 5 min to 24 h previously. However, Ojemann *et al.* (1961) found transport of ^{165}Dy to the choroid plexus of cats.

Citrate forms stronger complexes than acetate with lanthanides (Tables 2-5 and 2-8). Durbin *et al.* (1956a) published the classic paper on the distribution of Ln^{3+}-citrates in rats. Following i.m. injection, there was a rapid excretion of lanthanides, amounting to 40–60% of the initial dose (Fig. 7-6). Although lanthanides were excreted both by the kidney and by the G.I. tract, the fecal route predominated for the lighter lanthanides and the urinary route for the heavier lanthanides. Urinary excretion of all lanthanides fell to low levels after about a week. The major organ of initial accumulation of the light lanthanides was the liver (Fig. 7-7), while the heavier lanthanides tended to accumulate in the skeleton (Fig. 7-8). Retention by the skeleton was much more stable, as discussed in Section 7.5.2 for $EuCl_3$ (Berke, 1968). Skeletal half-lives of 2.5, 2.8, and 2.8 years were calculated for Eu, Tb, and Tm, respectively. Lanthanides released from the liver were presumably excreted via the bile into the G.I. tract, thereby accounting for the greater fecal excretion of the lighter lanthanides. Despite the ability of citrate to mobilize lanthanides from sites of

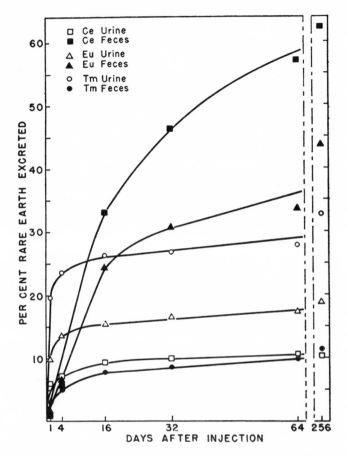

Figure 7-6. Cumulative excretion of [144]Ce, [152–154]Eu, and [170]Tm by rats following intramuscular injection of their citrate complexes. From Durbin *et al*. (1956a), with permission.

intramuscular (Durbin *et al*., 1956a) and subcutaneous (Stineman *et al*., 1978) injection, absorption of citrate complexes of [144]Ce, [152–154]Eu, [160]Tb, and [170]Tm from the G.I. tract of rats (Durbin *et al*., 1956a) and mice (Stineman *et al*., 1978) remained under 0.1% of the administered dose. Sturbaum *et al*. (1970) found [144]Ce^{3+}-citrate to be no more easily excreted than [144]CeCl$_3$ following i.p. administration to hamsters, with 60–80% of the dose remaining in the animals after 10 days. The presence of citrate did not alter the bodily distribution of [144]Ce. On the other hand, [177]Lu^{3+}-citrate was much more readily excreted following i.p. injection into mice, with 50% being eliminated within a day (Muller *et al*., 1978). This material

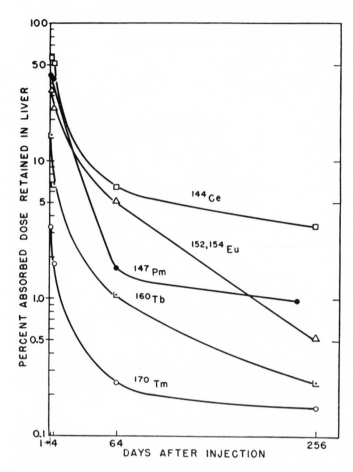

Figure 7-7. Liver retention of lanthanides by rats following intramuscular injection of their citrate complexes. From Durbin *et al.* (1956a), with permission.

showed the typical behavior of a heavy lanthanide–citrate complex, with the radioactivity localizing predominantly in the skeleton. The greater mobilization of i.p. Lu^{3+}-citrate than of Ce^{3+}-citrate may reflect the higher affinity of Lu^{3+} for citrate. Of interest is the increase in hepatic sequestration of Lu^{3+} as the amount of nonradioactive "carrier" (Section 7.5.1) was increased (Fig. 7-9).

Results obtained after i.v. injection of Ln^{3+}-citrate complexes were nearly identical to those found with i.m. injection (Durbin *et al.*, 1956b). In these studies, the different relative affinities of the light and heavy lanthanides for the liver and skeleton were apparent as early as 1 min

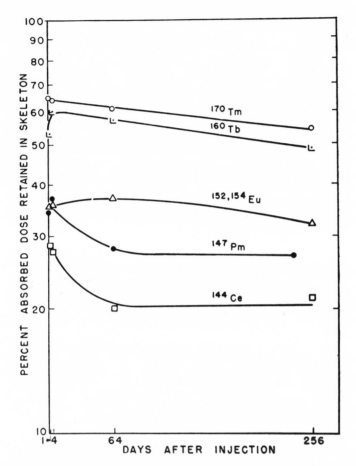

Figure 7-8. Skeletal retention of lanthanides by rats following intramuscular injection of their citrate complexes. From Durbin *et al.* (1956a), with permission.

after i.v. injection. In keeping with the suggestion (Section 7.5.2) that skeletal uptake is partly controlled by the extent of hepatic sequestration during the lanthanide's first transit through the liver, the lighter lanthanides left the blood more rapidly than the heavier lanthanides. They also form weaker complexes with citrate (Table 2-8), a property which may facilitate their exchange with ligands in the liver. Mathematical analysis of the removal of Ln^{3+}-citrate complexes from plasma yields three- or four-component functions which described the clearance curves.

Much research has been conducted with polyaminopolycarboxylate chelators such as NTA, EDTA, and DTPA (Section 2.3) which form

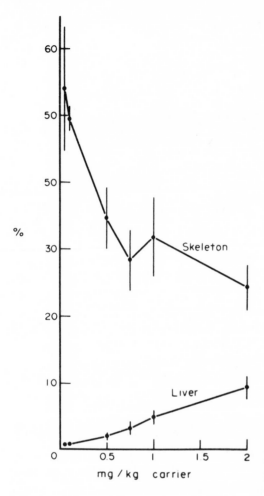

Figure 7-9. Effect of dose on the distribution of lutetium in mice. From Muller *et al.* (1978), with permission.

progressively stronger complexes with lanthanides (Table 2-4). These agents have been shown to reduce the binding of lanthanides to serum proteins (Rosoff and Spencer, 1979), with the binding decreasing as a function of the stability constant of the chelate (Rosoff *et al.*, 1958). Studies of the filterability of ^{91}Y chelates mixed with serum *in vitro* show that when the log K_1 is above about 18, the radioactivity remains filterable. This is presumably because the complex stays undissociated. Once the stability falls below this level, there is an abrupt decrease in filterability such that only 3–4% of the radioactivity of ^{91}Y-HEEDTA [HEEDTA = *N*-(2-hydroxyethyl)ethylenediaminetriacetic acid; log K_1 = 14.5] can be filtered

Table 7-14. Effect of Chelate Stability on Excretion
of Y from Mice[a,b]

Chelate	Stability constant (log K_1)[c]	Excretion of ^{90}Y from mice	
		24 h	96 h
Y-IPDTA	14–15	95.5	—
Y-EDTA	14.2 (18.09)	94.8	96.2
Y-EDTA-OL	10–11	75.4	72.2
Y-NTA	10.9 (11.48)	18.0	30.0
Y-EDG	8.6	13.5	23.5
Y-IDA	6.8	8.2	19.5
Y-Fe^{3-} Spec.	5.2	1.4	—
YCl$_3$	N.A.[d]	9.1	21.9

[a] From Hart (1956), with permission.
[b] The various forms of yttrium were injected intravenously.
[c] Values of log K_1 given in parentheses are those given in Table 2-4 and are probably more accurate than those from the original table.
[d] N.A.: Not applicable.

from bovine serum (Rosoff *et al.*, 1958). In other experiments, [153]Sm-NTA, [140]La-NTA, and [91]Y-NTA were injected intravenously and the distribution of the radioactivity monitored by electrophoresis of the serum. Whereas [140]La migrated with the γ-globulins, [153]Sm migrated with all the globulins, but [91]Y, which forms tighter complexes with NTA, did not migrate with any of the proteins (Rosoff and Spencer, 1975).

As the stability constant of the chelate increases, there is greater excretion and less bodily retention of the lanthanides (Kroll *et al.*, 1957). This is well illustrated by the data of Hart (1956) (Table 7-14). As comparisons between different lanthanide complexes of the same chelator show, the trend evident from Table 7-14 is not simply a reflection of nonspecific effects of the different organic chelators. For example, after i.v. injection of [140]La-EDTA, which has a smaller stability constant than ^{90}Y-EDTA, much less [140]La than [91]Y was excreted by human patients (Hart, 1956). Similarly, Foreman and Finnegan (1956) showed that the ability of EDTA to promote the excretion of lanthanides from rats increases with the atomic number of the lanthanide. Although the excretion of La-EDTA greatly exceeds excretion following administration of unchelated LaCl$_3$, it is still quite low. Far greater amounts of [140]La are excreted if additional EDTA is supplied (Hart and Laszlo, 1953), presumably by mass action effects. The effectiveness of the additional EDTA decreases as its administration is delayed. Evidently the sequestered [140]La becomes increasingly unavailable as time progresses. This has been also

Table 7-15. Effect of Chelation on the Tissue
Distribution of $^{90}Y^{3+}$ Injected Intravenously into Mice

Organ	Percentage of injected dose[b]		
	$^{90}YCl_3$	^{90}Y-NTA	^{90}Y-EDTA
Muscle	1.46	7.50	0.15
Bone	1.86	8.40	0.64
Lung	0.67	0.04	0.005
Heart	0.07	0.01	0.002
Liver	69.80	5.80	0.09
Spleen	0.07	0.55	0.003
Kidneys	0.42	0.94	0.05
G.I. tract	2.52	3.30	0.08
Total retention	77	27	1.0

[a] From Hart *et al.* (1955), with permission.
[b] Measurements were taken 40–45 h after injection.

reported for the removal of ^{143}Pm from humans by DTPA (Palmer *et al.*, 1970).

Coupled to the enhanced excretion rates in the presence of chelators are decreased plasma dwell times. These, too, are linked to the stability constants. Thus, 10 min following i.v. injection in humans, only 29% of injected ^{90}Y-EDTA remains in the blood; for $^{90}YCl_3$, this figure is 95% (Hart, 1956). Elimination of ^{169}Yb-EDTA from dog plasma is biphasic. In normal dogs, the fast phase has a half-life of about 5 min, while in nephrectomized dogs, the value is 7–8 min (Molnar *et al.*, 1975). There is little, if any, absorption of ^{169}Yb-EDTA from the stomach of guinea pigs.

Chelators not only promote urinary excretion of lanthanides as a function of stability constant, but they also alter the tissue distribution of the residual Ln (Table 7-15). Thus, whereas 91% of the residual $^{90}YCl_3$ was found in the liver of mice 40–45 h after i.v. injection, this fell to 22% for ^{90}Y-NTA and 9% for ^{90}Y-EDTA (Hart *et al.*, 1955). Table 7-15 also confirms the drastic fall in retention of $^{90}Y^{3+}$ produced by the chelators. Chelators progressively shift the main site of retained lanthanides from liver to bone. Relative uptake by muscle is also favored by chelation (Table 7-15; Hart *et al.*, 1955). A similar result was found for ^{140}La-EDTA (Laszlo, 1956). The excretion of ^{140}La-EDTA by mice amounted to 33% of the injected dose after 24 h (Laszlo, 1956). This is considerably higher than the 5% reported for humans over the same period (Hart, 1956).

DTPA forms extremely stable complexes with lanthanides (Table 2-4) which appear to permeate the extracellular compartment of mammals without decomposition. Clinical use has been made of Gd-DTPA for NMR

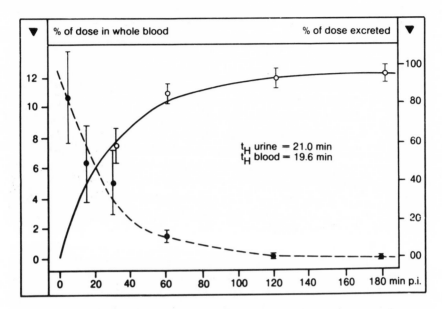

Figure 7-10. Blood and urine radioactivity following intravenous injection of [153]Gd-DTPA into rats. From Weinmann *et al.* (1984a), with permission.

imaging and [169]Yb-DTPA for tumor scanning (Section 9.6). About 99% of an i.v. dose of [169]Yb-DTPA is excreted from mice, rats, and man in a day (Hosain *et al.*, 1968; Baltrukiewicz *et al.*, 1975). During its passage through the body, the radioactivity initially labels the blood, lungs, and kidneys. Radiation levels quickly fall, with the kidney showing longest retention as the label is filtered by the glomeruli prior to urinary excretion. As the rate of clearance of [169]Yb-DTPA equals that of inulin, excretion can be completely accounted for by glomerular excretion. In tumor-bearing mice, high tumor/soft tissue ratios have been obtained, while clinical studies have confirmed the ability of [169]Yb-DTPA to detect brain tumors. These aspects are discussed in Chapter 9.

Weinmann *et al.* (1984a) confirmed the rapid urinary excretion of Gd-DTPA from rats. Half-times in the blood and urine were both about 20 min (Fig. 7-10). The plasma concentrations declined according to biexponential functions, explicable by a distribution phase of half-life 0.20 ± 0.13 h and a disposition and elimination phase of half-life 1.58 ± 0.13 h (Weinmann *et al.*, 1984b). NMR spectroscopic studies have confirmed the presence of high concentrations of Gd-DTPA in the kidney, whence the agent is rapidly excreted. Moderate amounts were detected in organs with a rich blood supply, such as heart, pancreas, and lung (Wolf and Fobben,

1984). Chromatographic analysis of human urine following i.v. administration of Gd-DTPA has confirmed that this agent does not dissociate during its passage through the body (Weinmann et al., 1984b).

Because of its high affinity for lanthanides, DTPA has received attention as a chelator with which to remove these metals from the body after accidential exposure. DTPA was effective in removing ^{143}Pm from humans, although the effect declined rapidly as the delay in administering the DTPA increased (Palmer et al., 1970). This decline was attributable to a marked decrease in the rate of urinary excretion, with fecal elimination remaining constant. DTPA releases more ^{143}Pm from the liver than from the skeleton. Indeed, in rats, Thomasset et al. (1976) were able to target ^{177}Lu to bone by first injecting ^{177}Lu^{3+}-citrate intravenously into rats and then administering DTPA twice a week. After 28 days, about 75% of the injected dose had been excreted. Nearly all of the radioactivity remaining in the animal was restricted to the skeleton. Urinary excretion of ^{177}Lu could then be monitored as an index of bone resorption. Better removal of lanthanides is accomplished by administering the DTPA in moderate doses at regular intervals than by giving large doses infrequently (Takada and Fujita, 1979). Encapsulation of the chelator in liposomes also aids removal (Blank et al., 1980). Chelation therapy is much less efficient in rats at 2 weeks of age than at 6 weeks of age (Kargacin et al., 1983).

7.5.4 Metabolism of Insoluble Lanthanide Particles

The metabolism aspect of insoluble lanthanide particles will be dealt with only briefly, as its characteristics are those of insoluble particles in general, rather than lanthanides in particular. Unless injected intravenously, parenterally administered particles tend to remain at the site of application over periods of several months. After intravenous injection, particles are removed from the circulation by cells of the reticuloendothelial system, particularly the Kupffer cells of the liver.

The main concern in connection with insoluble lanthanide particles has been the fate of radioactive particles inhaled into the lungs following atmospheric nuclear explosions or industrial exposure. The main radiolanthanides formed are ^{144}Ce and ^{90}Y, a daughter of ^{90}Sr. Because of its long half-life of 285 days and its relatively high fission yield, ^{144}Ce is of particular importance. Liebscher et al. (1961) found high concentrations of ^{144}Ce in the lungs and pulmonary lymph nodes of inhabitants of Vienna. The radioactivity was thought to have arisen during the nuclear test explosions of 1957 and 1958. After the recent Chernobyl disaster, we may all be unknowing participants in a similar unscheduled experiment. Transfer of inhaled lanthanides to the lymph nodes is also suggested by studies of workers exposed to lanthanide fumes and dusts (Sabbioni et al., 1982b). However, blood and urine levels of lanthanides were normal in these workers (Table 7-8).

Whereas particles of soluble lanthanide salts are quickly transported from the lung (Section 7.5.1), pulmonary retention of inhaled lanthanide particles is high. Particles containing ^{144}Ce combined with fused clay of diameter 1.6–2.1 μm had a half-life of 170 days in the lungs of beagles. Excretion was correspondingly slow. Of the body burden 200–300 days after inhalation, over 80% remained in the lung with small amounts in the lymph nodes, liver, and skeleton (Muggenburg et al., 1975).

Similar results have been obtained with other types of insolubilized cerium. For example, experiments with inhaled ^{144}Ce(OH)$_3$ particles 1.4 μm in diameter suggested a 130-day half-time in the lungs of rats (Thomas et al., 1972). This is a little slower than removal from beagle dogs inhaling ^{144}CeO$_2$, where less than 10% remained in the lung after a year (Stuart et al., 1964). Inhaled ^{144}Ce was excreted through the feces and urine. Initial excretion was predominantly fecal, possibly due to coughing up and swallowing the particles. The low urinary excretion probably reflects the insolubility of Ce(OH)$_3$. However, transfer of label from the lungs to the liver and skeleton was seen. A similar distribution has been noted for mice (Lundgren et al., 1980) and rats (Norris et al., 1956) after inhaling ^{144}CeO$_2$. There was a secondary release of ^{144}Ce from the liver, with a half-time of 165 days. However, skeletally sequestered radioactivity was less metabolically labile and was released more slowly with a half-time of 280 days. Similar differences between the retention of lanthanides by liver and bone were noted in previous sections of this chapter. The average whole body retention of ^{144}Ce was 230 days. Infection with influenza virus increases the half-life of ^{144}CeO$_2$ in the lung (Lundgren et al., 1978). Although soluble lanthanide particles are cleared from lungs much more quickly, the organ distribution of the mobilized material remains the same. Most of the nonpulmonary radioactivity is recovered from the liver and the skeleton regardless of whether its source was an insoluble or a soluble particle in the lung.

Suspensions of ^{144}Ce(OH)$_3$ particles with an average diameter of 0.3 μm disappeared from the blood of rats with a half-life of about a minute (Fig. 7-2; Aeberhardt et al., 1962). Over 85% was captured by the liver, with the Kupffer cells being autoradiographically identified as the cells responsible. As with i.v. injections of soluble ^{144}Ce^{3+}, the liver rapidly released its radioactivity, part of which was captured and retained by the skeleton. There was also a very early, transient uptake of colloidal ^{144}Ce(OH)$_3$ by the reticuloendothelial cells of the lung. This amounted to

10–13% of the injected material between 5 and 30 min after i.v. injection but fell to 0.3% after a day. Insoluble particles of Gd_2O_3, under 2 μm in diameter, can be detected by NMR techniques in the liver, spleen, and lung of rabbits following i.v. injection (Burnett et al., 1985). However, in contrast to the results of others (Aeberhardt et al., 1962; Seltzer et al., 1981), Burnett et al. (1985) did not observe uptake by the cells of the reticuloendothelial system. Instead, the Gd_2O_3 accumulated extracellularly as a "sludge" in the hepatic and splenic sinusoids. Extensive aggregation of the particles could have been to blame.

Suspensions of lanthanide particles, when parenterally administered outside the bloodstream, tend to remain at the injection site. This property has been used by certain investigators to try to localize therapeutic doses of radiation at specific sites. The long-term goal of such studies is the radiotherapy of tumors (Section 9.5). This approach either involves local irradiation of the tissue, as mentioned above, or irradiation of the lymph nodes (Walker, 1950) in an attempt to control metastases. Three days after injection of particles of $^{90}YF_3$ (30 nm diameter) directly into sarcoma 180 tumors in mice, 97% of the dose remained associated with the tumors. Most of the activity transported from the tumors was recovered in the liver. However, quite efficient transport of i.p. $^{90}YF_3$ to the liver and spleen occurred, with 6% being excreted (Mayer and Morton, 1956). Autoradiography confirmed that the isotope was distributed within, rather than upon, these organs. The presence of ascites tumors altered the distribution of i.p. $^{90}YF_3$ considerably. Excretion was reduced to 1%, and 42% of the administered amount was recovered from the ascites. A further 53% of the label remained on the serous cavities in the peritoneal cavity. These workers also injected suspensions of $^{90}YF_3$ into the hamstring muscle area of mice (Mayer and Morton, 1956). After 3 days, less than 0.1% of the injected material was excreted, while much less than 0.1% accumulated in any of the seven organs measured. Of these, the kidney was most radioactive, accumulating 18×10^{-5}%/g of the injected dose. The inguinal lymph nodes contained 1.1–3.8%/g of the injected label. The residuum remained where it had been injected.

When incorporated into crystals of calcium pyrophosphate, $^{169}Yb^{3+}$ was cleared from rabbits' knees much more slowly than were solutions of $^{169}YbCl_3$ (McCarty et al., 1979). After 35 days, most of the injected radioactivity was recovered from the synovium, presumably as a result of phagocytosis. With crystals the size of 10–20 μm, the half-life in the joint was about 7–12 days. For particles of 10–50 μm, this was 16 days, and for particles bigger than 50 μm, 84 days. In human arthritic patients, the half-life of crystals in the joint ranged from 32 to 99 days.

7.6 Summary

The lanthanides are more common than the term "rare earth" indicates. Their average abundance in the earth's crust is over 100 ppm, which makes them more abundant than elements such as lead and silver. In areas rich in lanthanide ores, their content in the soil can reach over 1 mg per gram of dry soil. Under natural conditions, lanthanides are made available to the food chain through leaching into the groundwater and through the formation of atmospheric dusts. Industrial exposure involves the inhalation of fumes and wound contamination. Lanthanides have a very limited ability to travel up through the food chain. Even in areas with high amounts of lanthanides in the soil, their concentration in the groundwater is low. Certain plants appear able to discriminate against lanthanides, while others assimilate relatively large amounts. However, very poor absorption from the G.I. tracts of vertebrates limits their assimilation of lanthanides from food. The amounts of lanthanides detected in mammalian organs reflect this, with concentrations being in the ppm to undetectable range. There is nearly no transfer of lanthanides from mother to fetus.

Ln^{3+} ions rapidly bind in large quantities to the external surfaces of microorganisms. As with metazoan cells (Chapter 6), lanthanides are probably not transported into bacteria, algae, or yeast. Among the reasons for the poor intestinal absorption of lanthanides are the formation of insoluble precipitates, the binding of lanthanides to digesta, and, perhaps, the inability of lanthanide ions to make use of the Ca^{2+}-transporting mechanisms in the intestine. Much more lanthanide is made available to the body by parenteral injection or inhalation. In general, systemic absorption of lanthanides after administration of simple, soluble salts increases in the order p.o. $\ll$ s.c. $<$ i.m. $<$ i.p. $<$ inhalation $\ll$ i.v. However, despite the huge range in efficiency of absorption, from less than 1% following oral administration to 100% for i.v. injection, the organ distribution of the physiologically available material remains qualitatively constant. The liver and the skeleton sequester most of the activity. Hepatic sequestration is much more labile than uptake by the skeleton. Some of the material released by the liver is recaptured by the skeleton, such that several months after administration, the skeleton often contains nearly the entire body load of lanthanide. This difference in organ retention may reflect the extracellular deposition of lanthanides in the matrix of bone, compared to the cellular association of hepatic lanthanides, which is likely to be more metabolically labile. Apart from the liver and skeleton, the spleen, lung, certain endocrine glands, and several other organs may sequester lanthanides.

Excretion occurs both fecally and in the urine. The former represents material excreted both from the liver, through the bile, and directly into the G.I. tract. The urinary component probably reflects the excretion of soluble lanthanide complexes. It is generally highest early after injection and falls off more rapidly than the fecal route. Rates of overall excretion fall with time, as the labile pools of sequestered lanthanides are depleted.

Although the lanthanides are chemically rather similar, there are some differences in their relative body distributions. The lighter lanthanides tend to accumulate predominantly in the liver, while the heavier lanthanides and yttrium prefer bone. This trend is most clearly observed following the administration of lanthanide–citrate complexes. These organ specificities may be important in the development of the rare earth fatty liver (Section 8.7.2), which is only produced by the lighter lanthanides. Lanthanide metabolism is affected by dose. The presence of nonradioactive "carrier" in the lanthanide preparation decreases the rates of removal from sites of extravascular injection and increases the dwell time in the blood after i.v. injection. Carrier tends to decrease excretion rates and to favor hepatic sequestration.

When lanthanides are introduced into the body as chelates, they become absorbed more completely, rates of excretion increase, and the distribution pattern may alter. The important parameter is the stability constant of the Ln–chelator complex. With very strong chelators such as DTPA, no exchange occurs between the complex and physiological ligands. As a result, the compound passes quickly through the body and is excreted unchanged in the urine. With complexes of lower stability, excretion is less complete, while the main effect on distribution is to reduce the fractional uptake by the liver and increase accumulation in the bone and other organs.

Insoluble particles which contain lanthanides are, like other particulate substances, internalized by phagocytic cells. When such particles are introduced at sites with access to phagocytes, such as the joint, blood, or in areas with a good supply of lymph, clearance occurs efficiently. However, if introduced at other sites, such as intramuscularly, removal is very slow.

References

Aeberhardt, A., Nizza, P., and Remy, J., 1962. Etude comparée du métabolisme du cérium 144 en fonction de son état physicochimique chez le rat, *Int. J. Radiat. Biol.* 5:217–246.
Andrews, G. A., 1956. The effect of dose on the mobilization of rare-earth radioisotopes.

II. Preliminary clinical studies, in *Rare Earths in Biochemical and Medical Research* (G. C. Kyker and E. B. Anderson, eds.), U.S. Atomic Energy Commission, Report ORINS-12, pp. 241–258.

Baehr, G., and Wessler, H., 1909. The use of cerium oxalate for the relief of vomiting: an experimental study of the effects of some salts of cerium, praseodymium, neodymium and thorium, *Arch. Intern. Med.* 2:517–531.

Baltrukiewicz, Z., Burakowski, T., and Pogorzelska-Lis, M., 1975. Distribution and retention of ^{169}YbCl$_3$, ^{169}YbEDTA and ^{169}YbDTPA injected intravenously in the rat, *Acta Physiol. Pol.* 26:205–212.

Baltrukiewicz, Z., Burakowski, T., and Derecki, J., 1976. Effects of ethylenediaminetetraacetic acid (EDTA) and diethylenetriaminepentaacetic acid (DTPA) derivatives on penetration of ytterbium-169 and cerium-144 into the rat offspring, *Acta Physiol. Pol.* 27: 175–181.

Beno, M., 1973. The fetal uptake of cerium-144–praseodymium-144 after injection of its chelates into pregnant rat, *Health Phys.* 25:575–580.

Berke, H. L., 1968. The metabolism of the rare earths. 1. The distribution and excretion of intravenous $^{152-154}$europium in the rat, *Health Phys.* 15:301–312.

Blank, M. L., Cress, E. A., Byrd, B. L., Washburn, L. C., and Snyder, F., 1980. Liposomal encapsulated Zn-DTPA for removing intracellular ^{169}Yb, *Health Phys.* 39:913–920.

Bonucci, E., Silvestrini, G., and DiGrezia, R., 1988. The ultrastructure of the organic phase associated with the inorganic substance in calcified tissues, *Clin. Orthop. Rel. Res.* 233:243–261.

Borneman-Starinkevitch, I. D., Borovick, S. A., and Borovsky, I. B., 1941. Rare earths in plants and soils, *Dokl. Akad. Nauk SSSR* 30:227–231.

Boroughs, H., Townsley, S. J., and Hiatt, R. W., 1956. The metabolism of radionuclides by marine organisms. II. The uptake accumulation and loss of yttrium-91 by marine fish, and the importance of short-lived radionuclides in the sea, *Biol. Bull.* 111:352–357.

Boroughs, H., Chipman, W. A., and Rice, T. R., 1957. Laboratory experiments on the uptake, accumulation and loss of radionuclides by marine organisms, in *The Effects of Atomic Radiation on Oceanography and Fisheries*, Publication No. 551, National Academy of Sciences, National Research Council, Washington, D.C., pp. 80–87.

Boroughs, H., Townsley, S. J., and Ego, W., 1958. The accumulation of ^{90}Y from an equilibrium mixture of ^{90}Sr-^{90}Y by *Artemia salina*, in Progress Report 1957–1958, Hawaii Marine Laboratory, U.S. Atomic Energy Commission, pp. 20–31.

Bowen, V. T., 1951. The uptake and distribution of barium^{140} and lanthanum^{140} in the larvae of *Drosophila repleta*, *J. Exp. Zool.* 118:509–525.

Bowen, V. T., 1956. Published discussion to paper of Laszlo (1956).

Bowen, V. T., and Rubinson, A. C., 1951. Uptake of lanthanum by a yeast, *Nature* 167: 1032.

Brooksbank, W. A., and Leddicotte, G. W., 1953. Ion-exchange separation of trace impurities, *J. Phys. Chem.* 57:819–825.

Burnett, K. R., Wolf, G. L., Schumacher, H. R., and Goldstein, E. J., 1985. Gadolinium oxide: a prototype agent for contrast enhanced imaging of the liver and spleen with magnetic resonance, *Magn. Reson. Imaging* 3:65–71.

Cassone, A., and Garaci, E., 1974. Lanthanum staining of the intermediate region of the cell wall in *Eschericia coli*, *Experientia* 30:1230–1232.

Cooper, J. A. D., Bulkley, G. J., and O'Conor, V. J., 1956. Intraprostatic injection of radioactive yttrium chloride in the dog, in *Rare Earths in Biochemical and Medical Research* (G. C. Kyker and E. B. Anderson, eds.), U.S. Atomic Energy Commission, Report ORINS-12, pp. 323–331.

Copp, D. H., Hamilton, J. G., Jones, D. C., Thompson, D. M., and Cramer, C., 1951. The effect of age and low phosphorus rickets on calcification and the deposition of certain radioactive metals in bones, U.S. Atomic Energy Commission UCRL-1464, 31 pages.

Cossa, A., 1870. Sulla diffusione del cerio, del lantano e del didimio, *Gazz. Chim. Ital.* 9: 118–140.

Cuddihy, R. G., Boecker, B. B., McClellan, R. O., and Kanapilly, G. M., 1976. [144]Ce in tissues of beagle dogs after inhalation of $CeCl_3$ with special emphasis on endocrine glands and reproductive organs, *Health Phys.* 30:53–59.

Daigneault, E. A., 1963. The distribution of intravenously administered yttrium chloride (carrier free) in the rhesus monkey, *Toxicol. Appl. Pharmacol.* 5:331–343.

Dobson, E. L., Gofman, J. W., Jones, H. B., Kelly, L. S., and Walker, B. S., 1949. Studies with colloids containing radioisotopes of yttrium, zirconium, columbium and lanthanum. II. The controlled selective localization of radioisotopes of yttrium, zirconium and columbium in the bone marrow, liver and spleen, *J. Lab. Clin. Med.* 34:305–312.

Drouven, B. J., and Evans, C. H., 1986. Collagen fibrillogenesis in the presence of lanthanides, *J. Biol. Chem.* 261:11792–11797.

Durbin, P. W., 1962. Distribution of transuranic elements in mammals, *Health Phys.* 8: 665–671.

Durbin, P. W., Williams, M. H., Gee, M., Newman, R., and Hamilton, J. G., 1956a. Metabolism of the lanthanons in the rat, *Proc. Soc. Exp. Biol. Med.* 91:78–85.

Durbin, P. W., Asling, C. W., Johnston, M. E., Hamilton, J. G., and Williams, M. H., 1956b. The metabolism of the lanthanons in the rat. II. Time studies on the tissue deposition of intravenously administered radioisotopes, in *Rare Earths in Biochemical and Medical Research* (G. C. Kyker and E. B. Anderson, eds.), U.S. Atomic Energy Commission, Report ORINS-12, pp. 171–192.

Eisele, G. R., Mraz, F. R., and Woody, M. C., 1980. Gastrointestinal uptake of [144]Ce in the neonatal mouse, rat and pig, *Health Phys.* 39:185–192.

Eisenbud, M., Krauskopf, K., Franca, E. P., Lei, W., Ballard, R., and Linsalata, P., 1984. Natural analogues for the transuranic actinide elements: an investigation in Minas Gerais, Brazil, *Environ. Geol. Water Sci.* 6:1–19.

Ekman, L., Valmet, E., and Aberg, B., 1961. Behaviour of yttrium-91 and some lanthanons towards serum proteins in paper electrophoresis, density gradient electrophoresis and gel filtration, *Int. J. Appl. Radiat. Isotop.* 12:32–41.

Erametsa, O., and Haukka, M., 1970. The occurrence of lanthanides in ferns, *Suom. Kemistilehti* 43:189–193.

Erametsa, O., and Sihvonen, M. L., 1971. Rare earths in the human body. II. Yttrium and lanthanides in the spleen, *Ann. Med. Exp. Fenn.* 49:35–37.

Esposito, M., Oddone, M., Accardo, S., and Cutolo, M., 1986a. Concentrations of lanthanides in plasma and synovial fluid in rheumatoid arthritis, *Clin. Chem.* 32:1598.

Esposito, M., Collecchi, P., Brera, S., Mora, E., Mazzucotelli, A., Cutolo, M., and Oddone, M., 1986b. Plasma and tissue levels of some lanthanide elements in malignant and nonmalignant tissues, *Sci. Total Environ.* 50:55–63.

Evans, C. H., and Tew, W. P., 1981. Isolation of biological materials by use of erbium(III)-induced magnetic susceptibilities, *Science* 213:653–654.

Ewaldsson, B., and Magnusson, G., 1964a. Distribution of radioterbium, radioholmium and radioyttrium in mice, *Acta Radiol. Ther. Phys. Biol.* 2:121–128.

Ewaldsson, B., and Magnusson, G., 1964b. Distribution of radiocerium and radio-promethium in mice. An autoradiographic study, *Acta Radiol.* 2:65–72.

Foreman, H., and Finnegan, C., 1956. Chelation in physiological systems, in *Rare Earths in Biochemical and Medical Research* (G. C. Kyker and E. B. Anderson, eds.), U.S. Atomic Energy Commission, Report ORINS-12, pp. 136–142.

Foreman, H. M., and Trigillo, T. T., 1954. The metabolism of ^{14}C-labelled ethylenediamine tetraacetic acid in human beings, *J. Lab. Clin. Med.* 43:566–571.

Gaudry, A., Maziere, B., Comar, D., and Nau, D., 1976. Multielement analysis of biological samples after intense neutron irradiation and fast chemical separation, *J. Radioanal. Chem.* 29:77–87.

Gerhardsson, L., Wester, P. O., Nordberg, G. F., and Brune, D., 1984. Chromium, cobalt and lanthanum in lung, liver and kidney tissue from deceased smelter workers, *Sci. Total Environ.* 37:233–246.

Gladney, E. S., 1980. Elemental concentrations in NBS biological and environmental standard reference materials, *Anal. Chim. Acta* 118:385–396.

Goldie, H., and West, H. D., 1956. Effect of peritumoral tissue infiltration with radioactive yttrium on growth and spread of malignant cells, *Cancer Res.* 16:484–489.

Graca, J. G., Garst, E. L., and Lowry, W. E., 1957. Comparative toxicity of stable rare earth compounds. 1. Effect of citrate complexing on stable rare earth chloride toxicity, *A.M.A. Arch. Ind. Health* 15:9–14.

Grant, W. M., and Kern, H. L., 1956. Cations and the cornea. Toxicity of metals to the stroma, *Am. J. Ophthalmol.* 42:167–181.

Graul, H. E., and Hundeshagen, H., 1959. Investigations of radio-yttrium (Y^{90}) metabolism. Studies of organ-distribution with special regard to the radioautographic method of demonstration and to paper electrophoresis under various experimental conditions, *Int. J. Appl. Radiat. Isotop.* 5:243–252.

Guo, B., 1985. Present and future of rare earth research in Chinese agriculture, *J. Chin. Rare Earth Soc.* 3:89–94.

Hamilton, J. G., 1948. The metabolic properties of the fission products and actinide elements, *Rev. Mod. Phys.* 20:718–728.

Hamilton, J. G., 1949. The metabolism of the radioactive elements created by nuclear fission, *N. Engl. J. Med.* 240:863–870.

Hart, H., 1956. Modification of distribution and excretion of rare earths by chelating agents, *in Rare Earths in Biochemical and Medical Research* (G. C. Kyker and E. D. Anderson, eds.), U.S. Atomic Energy Commission, Report ORINS-12, pp. 118–135.

Hart, H. E., and Laszlo, D., 1953. Modification of the distribution and excretion of radioisotopes by chelating agents, *Science* 118:24–25.

Hart, H. E., Greenberg, J., Lewis, R., Spencer, H., Stern, K. G., and Laszlo, D., 1955. Metabolism of lanthanum and yttrium chelates, *J. Lab. Clin. Med.* 46:182–192.

Hood, S. L., and Comar, C. L., 1956. Tissue distribution and placental transfer of yttrium 91 in rats and cattle, *in Rare Earths in Biochemical and Medical Research* (G. C. Kyker and E. B. Anderson, eds.), U.S. Atomic Energy Commission, Report ORINS-12, pp. 280–300.

Hosain, F., Reba, R. C., and Wagner, H. N., 1968. Ytterbium-169 diethylenetriaminepentaacetic acid complex. A new radiopharmaceutical for brain scanning, *Radiology* 91: 1199–1203.

Ikeda, N., and Takahashi, N., 1978. Neutron activation analysis of the rare earth elements in Nasu Hot Springs, *Radioisotopes* 27:300–305.

Inaba, J., and Lengemann, F. W., 1972. Intestinal uptake and whole body retention of ^{144}Ce by suckling rats, *Health Phys.* 22:169–175.

Jasmin, J. R., Brocheriou, C., Klein, B., Morin, M., Smadja-Joffe, F., Cernea, P., and

Jasmin, C., 1977. Induction in rats of paranasal sinus carcinomas with radioactive cerium chloride, *J. Nat. Cancer Inst.* 58:423–427.

Johnson, G. T., and Kyker, G. C., 1961. Fission product and cerium uptake by bacteria, yeasts and moulds, *J. Bacteriol.* 81:783–740.

Johnson, G. T., and Kyker, G. C., 1966. The mechanism of cerium uptake by *Saccharomyces cerevisiae, Mycologia* 58:91–99.

Jowsey, J., Rowland, R. E., and Marshall, J. H., 1958. The deposition of the rare earths in bone, *Radiat. Res.* 8:490–501.

Jugo, S., 1977. Metabolism of toxic heavy metals in growing organisms: a review, *Environ. Res.* 13:36–46.

Kanapilly, G. M., 1980. *In vitro* precipitation behavior of trivalent lanthanides, *Health Phys.* 39:343–346.

Kargacin, B., Kostial, K., and Landeka, M., 1983. The influence of age on the effectiveness of DTPA in reducing [141]Ce retention in rats, *Int. J. Radiat. Biol.* 44:363–366.

Kawin, B., 1953. Metabolism of radioyttrium, *Arch. Biochem. Biophys.* 45:230–231.

Kawin, B., 1957. Effects of cortisone acetate upon the distribution and excretion of radioyttrium, *Nature* 179:871–872.

Kramsch, D. M., Aspen, A. J., and Apstein, C. S., 1980. Suppression of experimental atherosclerosis by the Ca^{2+}-antagonist lanthanum, *J. Clin. Invest.* 65:967–981.

Kroll, H., Korman, S., Siegel, E., Hart, H. E., Rosoff, B., Spencer, H., and Laszlo, D., 1957. Excretion of yttrium and lanthanum chelates of cyclohexane 1,2-trans-diamine tetraacetic acid and diethylenetriamine pentaacetic acid in man, *Nature* 180:919–920.

Krumholz, L. A., and Foster, R. F., 1957. Accumulation and retention of radioactivity from fission products and other radiomaterials by fresh-water organisms, in *The Effects of Atomic Radiation on Oceanography and Fisheries,* Publication No. 551, National Academy of Sciences, National Research Council, Washington, D.C., pp. 88–95.

Kyker, G. C., 1956. The effect of dose on the mobilization or rare-earth radioisotopes. I. Animal studies, in *Rare Earths in Biochemical and Medical Research* (G. C. Kyker and E. B. Anderson, eds.), U.S. Atomic Energy Commission, Report ORINS-12, pp. 222–240.

Kyker, G. C., 1962. Rare earths, in *Mineral Metabolism* (C. L. Comar and F. Bonner, eds.), Vol. 8, Part B, Academic Press, New York, pp. 499–541.

Kyker, G. C., Christopherson, W. M., Berg, H. F., and Brucer, M., 1956. Selective irradiation of lymph nodes by radiolutecium (Lu[177]), *Cancer* 9:489–498.

Lahr, T. N., Olsen, R., Gleason, G. I., and Tabern, D. L., 1955. Animal distribution of colloids of Au[198], P[32] and Y[90]. An improved method of tissue assay for radioactivity, *J. Lab. Clin. Med.* 45:66–80.

Laszlo, D., 1956. Distribution of lanthanum and yttrium in the mammalian organism, in *Rare Earths in Biochemical and Medical Research* (G. C. Kyker and E. B. Anderson, eds.), U.S. Atomic Energy Commission, Report ORINS-12, pp. 193–221.

Laszlo, D., Ekstein, D. M., Lewin, R., and Stern, K. G., 1952. Biological studies on stable and radioactive rare earth compounds. I. On the distribution of lanthanum in the mammalian organism, *J. Natl. Cancer Inst.* 13:559–572.

Laul, J. C., and Weimer, W. C., 1982. Behavior of REE in geological and biological samples, in *The Rare Earths in Modern Science and Technology* (G. J. McCarthy, H. B. Silber, and J. J. Rhyne, eds.), Vol. 3, Plenum Press, New York, pp. 531–535.

Lazar, G., 1973. The reticuloendothelial-blocking effect of rare earth metals in rats, *J. Reticuloendothel. Soc.* 13:231–237.

Leddicotte, G. W., and Tipton, I. H., 1958. Personal communication cited in Kyker, 1962.

Lewin, R., Stern, K. G., Ekstein, D. M., Woidowsky, L., and Laszlo, D., 1953. Biological studies on stable and radioactive rare earth compounds. II. The effect of lanthanum on mice bearing Ehrlich ascites tumor, *J. Natl. Cancer Inst.* 14:45–56.

Lewin, R., Hart, H. E., Greenberg, J., Spencer, H., Stern, K. G., and Laszlo, D., 1954. Biological studies on stable and radioactive rare earth compounds. III. Distribution of radioactive yttrium in normal and ascites-tumor-bearing mice, and in cancer patients with serous effusions, *J. Natl. Cancer Inst.* 15:131–143.

Liebscher, K., Schonfeld, T., and Schaller, A., 1961. Concentration of inhaled Ce-144 in pulmonary lymph nodes of human beings, *Nature* 192:1308.

Linsalata, P., France, E. P., Campos, M. J., Lobao, N., Ballad, R., Lei, W., Ford, H., Morse, R. S., and Eisenbud, M., 1986a. Radium, thorium and the light rare earth elements in soils and vegetables grown in an area of high natural radioactivity, in *Environmental Research for Actinide Elements,* National Technical Information Service, Springfield, Va.

Linsalata, P., Eisenbud, M., and Franca, E. P., 1986b. Ingestion estimates of Th and the light rare earth elements based on measurements of human feces, *Health Phys.* 50: 163–167.

Lundgren, D. L., Hahn, F. F., Crain, C. R., and Sanchez, A., 1978. Effect of influenza virus infection on the pulmonary retention of inhaled [144]Ce and subsequent survival of mice, *Health Phys.* 34:557–567.

Lundgren, D. L., McClellan, R. O., Hahn, F. F., Newton, G. J., and Diel, J. H., 1980. Repeated inhalation exposure of mice to $^{144}CeO_2$. I. Retention and dosimetry, *Radiat. Res.* 82:106–122.

Maas, G., 1980. Estimation of the amounts of heavy metal deposited on Belgian farm soils by the application of chemical fertilizers, *Rev. Agric.* 33:329–336.

MacDonald, N. S., Nusbaum, R. E., Alexander, G. V., Ezmirlian, F., Spain, P., and Rounds, D. E., 1952. The skeletal deposition of yttrium, *J. Biol. Chem.* 195:837–841.

Magnusson, G., 1963. The behaviour of certain lanthanons in rats, *Acta Pharmacol. Toxicol.,* Suppl. 3, 20:1–95.

Mahlum, D. D., 1967. Influence of hepatocarcinogen feeding on the retention of cerium[144] by the liver, *Toxicol. Appl. Pharmacol.* 11:585–590.

Matsunami, T., Mizohata, A., and Mamuro, T., 1974. Isotopic ratios of radioruthenium and radiocerium in rain water at Osaka in relation to nuclear explosions during the period of late 1969–1972, *J. Radiat. Res.* 15:96–102.

Mayer, S. W., and Morton, M. E., 1956. Preparation and distribution of yttrium 90 fluoride, in *Rare Earths in Biochemical and Medical Research* (G. C. Kyker and E. B. Anderson, eds.), U.S. Atomic Energy Commission, Report ORINS-12, pp. 263–279.

McCarty, D. J., Palmer, D. W., and Halverson, P. B., 1979. Clearance of calcium pyrophosphate dihydrate crystals *in vivo.* I. Studies using [169]Yb labeled triclinic crystals, *Arthritis Rheum.* 22:718–727.

Miller, L. P., 1959. Factors influencing the uptake and toxicity of fungicides, *Trans. N.Y. Acad. Sci.* 21:442–445.

Miller, L. P., McCallan, S. E. A., and Weed, R. M., 1953. Accumulation of 2-heptadecyl-2-imidazoline, silver and cerium by fungus spores in mixed and consecutive treatments, *Contrib. Boyce Thompson Inst.* 17:283–298.

Milton, C., Murata, K. J., and Knechtel, M. K., 1944. Weinschenkite, yttrium phosphate dihydrate from Virginia, *Am. Mineral.* 29:92–107.

Molnar, Gy., Pal, I., Stutzel, M., Szilvasi, I., and Lengyel, M., 1975. Absorption, distribution and elimination of [169]Yb-EDTA in animal experiments. GFR estimates in rats, *Strahlentherapie* 74:142–151.

Mraz, F. R., and Eisele, G. R., 1977. Gastrointestinal absorption and distribution of ^{144}Ce in the suckling pig, *Health Phys.* 33:494–495.

Mraz, F. R., Wright, P. L., Ferguson, T. M., and Anderson, D. L., 1964. Fission product metabolism in hens and transference to eggs, *Health Phys.* 10:777–782.

Muggenburg, B. A., Mauderly, J. L., Boecker, B. B., Hahn, F. F., and McClellan, R. O., 1975. Prevention of radiation pneumonitis from inhaled cerium-144 by lung lavage in beagle dogs, *Am. Rev. Respir. Dis.* 111:795–802.

Muller, W. A., Linzner, U., and Schaffer, E. H., 1978. Organ distribution studies of lutetium-177 in mouse, *Int. J. Nucl. Med. Biol.* 5:29–31.

Nadkarni, R. A., and Morrison, G. H., 1978. Use of standard reference materials as multielement irradiation standards in neutron activation analysis. *J. Radioanal. Chem.* 43: 347–369.

Nagahashi, G., Thomson, W. W., and Leonard, R. T., 1974. The Casparian strip as a barrier to the movement of lanthanum in corn roots, *Science* 183:670–671.

Naharin, A., Lubin, E., and Feige, Y., 1974. Internal deposition of ingested cerium in suckling mice, *Health Phys.* 27:207–211.

Norris, W. P., Lisco, H., and Brues, A. M., 1956. The radiotoxicity of cerium and yttrium, in *Rare Earths in Biochemical and Medical Research* (G. C. Kyker and E. B. Anderson, eds.), U.S. Atomic Energy Commission, Report ORINS-12, pp. 102–115.

Ojemann, R. G., Brownell, G. L., and Sweet, W. H., 1961. Possible radiation therapy of cephalic neoplasms by perfusion of short-lived isotopes. II. Dysprosium-165. Metabolism in mouse and cat, *Neurochirurgia* 4:41–57.

Olmez, I., and Aras, N. K., 1977. Trace elements in the atmosphere, *J. Radioanal. Chem.* 37:671–676.

Palmer, H. E., Nelson, I. C., and Crook, G. H., 1970. The uptake, distribution and excretion of promethium in humans and the effect of DTPA on these parameters, *Health Phys.* 18:53–61.

Pederson, O. O., 1979. An electron microscopic study of the permeability of intraocular blood vessels using lanthanum as a traces *in vivo, Exp. Eye Res.* 29:61–69.

Piccinini, F., Meloni, S., Chiarra, A., and Villani, F. P., 1975. Uptake of lanthanum by mitochondria, *Pharmacol. Res. Commun.* 7:424–435.

Polikarpov, G. C., 1958. Accumulation of the radioisotope of cerium by freshwater molluscs, Translation 8347-1/0-sc.1. UKAEA, 1960.

Rabinowitz, J. L., Fernandez-Gavarron, F., and Brand, J. G., 1988. Tissue uptake and intracellular distribution of 140-lanthanum after oral intake by rat, *J. Toxicol. Environ. Health* 24:229–235.

Rayner, B., Tutt, M., and Vaughan, J., 1953. The deposition of ^{91}Y in rabbit bones, *Br. J. Exp. Pathol.* 34:138–145.

Rediske, J. H., and Selders, A. A., 1954. The uptake and translocation of yttrium by higher plants, *Am. J. Bot.* 41:238–242.

Revelle, R., and Schaefer, M. B., 1957. General considerations concerning the oceans as a receptacle for artificially radioactive materials, in *The Effects of Atomic Radiation on Oceanography and Fisheries,* Publication No. 551, National Academy of Sciences, National Research Council, Washington, D.C., pp. 1–25.

Rice, T. R., 1956. The accumulation and exchange of strontium by marine planktonic algae, *Limnol. Oceanogr.* 1:123–128.

Rimbert, J. N., Kellershohn, C., Dumas, F., Fortier, D., Maziere, M., and Hubert, C., 1981. Mössbauer spectroscopy and perturbed angular correlation studies of the rare- and alkaline-earth bone uptake, *Biochimie* 63:931–936.

Robinson, G. A., and Wasnidge, D. C., 1981. Comparison of the accumulation of ^{125}I and [144]Ce in the growing oocytes of the Japanese quail, *Poultry Sci.* 60:2195–2199.

Robinson, G. A., Wasnidge, D. C., and Floto, F., 1978. Distribution of [140]La and [47]Ca in female Japanese quail and in the eggs laid, *Poultry Sci.* 57:190–196.

Robinson, G. A., Martin, K. A., Wasnidge, D. C., and Floto, F., 1979. Lipovitellin and phosvitin as major [140]La-binding components of Japanese quail egg yolk *in vivo* and *in vitro*, *Poultry Sci.* 58:1361–1366.

Robinson, G. A., Wasnidge, D. C., and Floto, F., 1980. Radiolanthanides as markers for vitellogenin-derived proteins in the growing oocytes of the Japanese quail, *Poultry Sci.* 59:2312–2321.

Robinson, G. A., Wasnidge, D. C., Floto, F., and Templeton, G. A., 1981a. Distribution of [153]Gd in F_1 quail, *Poultry Sci.* 60:563–568.

Robinson, G. A., Wasnidge, D. C., Floto, F., and Gibbins, A. M., 1981b. Gadolinium as a useful radiolanthanide for long-term labeling of tissues in Japanese quail, *Poultry Sci.* 60:861–866.

Robinson, G. A., Wasnidge, D. C., and Floto, F., 1984. A comparison of the distributions of the actinides uranium and thorium with the lanthanide gadolinium in the tissues and eggs of Japanese quail: concentrations of uranium in feeds and foods, *Poultry Sci.* 63: 883–891.

Robinson, W. O., 1943. The occurrence of rare earths in plants and soils, *Soil Sci.* 56:1–6.

Robinson, W. O., Whetstone, R., and Scribner, B. F., 1938. The presence of rare earths in hickory leaves, *Science* 87:470–471.

Robinson, W. O., Bastron, H., and Murata, K. J., 1958. Biogeochemistry of the rare-earth elements with particular reference to hickory trees, *Geochim. Cosmochim. Acta* 14: 55–67.

Rosoff, B., and Spencer, H., 1975. Studies of electrophoretic binding of radioactive rare earths, *Health Phys.* 28:611–612.

Rosoff, B., and Spencer, H., 1979. Binding of rare earths to serum proteins and DNA, *Clin. Chim. Acta* 93:311–319.

Rosoff, B., Lewis, R., Hart, H. E., Williams, G. L., and Laszlo, D., 1958. Interaction of yttrium compounds with serum and serum constituents *in vitro*, *Arch. Biochem. Biophys.* 78:1–9.

Sabbioni, E., Goetz, L., and Bignoli, G., 1982a. Mobilization of heavy metals from fossil-fueled power plants. Potential ecological and biochemical implications. IV. Assessment of the European Communities situation, Report EUR 698/IV.

Sabbioni, E., Pietra, R., Gaglione, P., Vocaturo, G., Colombo, F., Zanoni, M., and Rodi, F., 1982b. Long-term occupational risk of rare-earth pneumoconiosis. A case report investigated by neutron activation analysis, *Sci. Total Environ.* 26:19–32.

Samochocka, K., Czauderna, M., Kalicki, A., Konecki, J., and Wolna, M., 1984a. The incorporation of selenium and ytterbium into the eyes of mice, *Int. J. Appl. Radiat. Isotop.* 35:1134–1135.

Samochocka, K., Czauderna, M., Konecki, J., and Wolna, M., 1984b. The incorporation of Yb and Se into the brains of rats and their effect on the Zn level, *Int. J. Appl. Radiat. Isotop.* 35:1136–1137.

Schatzki, P. F., and Newsome, A., 1975. Neutralized lanthanum solution, a largely non-colloidal ultra structural tracer, *Stain Technol.* 50:171–178.

Schomacker, K., Franke, W. G., Henke, E., Fromm, W. D., Maka, G., and Beyer, G. J., 1986. The influence of isotopic and nonisotopic carriers on the biodistribution and biokinetics of M^{n+}-citrate complexes, *Eur. J. Nucl. Med.* 11:345–349.

Schubert, J., Finkel, M. P., White, M. R., and Hirsch, G. M., 1950. Plutonium and yttrium content of the blood, liver and skeleton of the rat at different times after intravenous administration, *J. Biol. Chem.* 182:635–642.

Schweitzer, K. G., and Jackson, M., 1952. Radiocolloids, *J. Chem. Educ.* 29:513–522.

Seltzer, S. E., Adams, D. F., Davis, M. A., Hessel, S. J., Havron, A., Judy, P. F., Hurlburt, A. J., and Hollenbert, N. K., 1981. Hepatic contrast agents for CT: High atomic number particulate material, *J. Comput. Assist. Tomogr.* 5:370–373.

Shacklette, H. T., 1965. Geological Survey Bulletin 1198-D, U.S. Govt. Printing Office, Washington, D.C.

Shiraishi, Y., and Ichikawa, R., 1972. Absorption and retention of ^{144}Ce and ^{95}Zr–^{95}Nb in newborn juvenile and adult rats, *Health Phys.* 22:373–378.

Sihvonen, M. L., 1972. Accumulation of yttrium and lanthanoids in human and rat tissues as shown by mass spectrometric analysis and some experiments with rats, *Ann. Acad. Sci. Fenn. Ser. A* 168:1–62.

Sobek, J. M., and Talburt, D. E., 1968. Effects of the rare earth cerium on *Escherichia coli*, *J. Bacteriol.* 95:47–51.

Spooner, G. M., 1949. Observations on the absorption of radioactive strontium and yttrium by marine algae, *J. Mar. Biol. Assoc. U.K.* 28:587–625.

Stineman, C. H., Massaro, E. J., Lown, B. A., Morganti, J. B., and Al-Nakeeb, S., 1978. Cerium tissue/organ distribution and alterations in open field and exploratory behavior following acute exposure of the mouse to cerium (citrate), *J. Environ. Pathol. Toxicol.* 2:553–570.

Stuart, B. O., Casey, H. W., and Bair, W. J., 1964. Acute and chronic effects of inhaled ^{144}CeO$_2$ in dogs, *Health Phys.* 10:1203–1209.

Sturbaum, B., Brooks, A. L., and McClellan, R. O., 1970. Tissue distribution and dosimetry of ^{144}Ce in Chinese hamsters, *Radiat. Res.* 44:359–367.

Sugimae, A., 1980. Atmospheric concentrations and sources of rare earth elements in the Osaka area, Japan, *Atmos. Environ.* 14:1171–1175.

Swanson, A. A., and Truesdale, A. W., 1971. Elemental analysis in normal and cataractous human lens tissue, *Biochem. Biophys. Res. Commun.* 45:1488–1496.

Takada, K., 1978. Comparison of the metabolic behavior of ^{144}Ce injected intravenously with that absorbed from the wound site in rats, *Health Phys.* 35:537–543.

Takada, K., and Fujita, M., 1979. Effects of frequency of administration of DTPA on the excretion and tissue retention of ^{144}Ce from a contaminated wound in the rat, *Health Phys.* 37:401–405.

Talburt, D. E., and Johnson, G. T., 1967. Some effects of rare earth elements and yttrium on microbial growth, *Mycologia* 59:492–503.

Thomas, R. L., Scott, J. K., and Chiffelle, T. L., 1972. Metabolism and toxicity of inhaled ^{144}Ce in rats, *Radiat. Res.* 49:589–610.

Thomassen, P. R., and Leicester, H. M., 1964. Uptake of radioactive beryllium, vanadium, selenium, cerium, and yttrium in the tissues and teeth of rats, *J. Dent. Res.* 43:346–352.

Thomasset, M., Cuisinier-Gleizes, P., and Mathieu, H., 1976. Bone resorption measurement with unusual bone markers: critical evaluation of the method in phosphorus-deficient and calcium-deficient growing rats, *Calcif. Tissue Res.* 21:1–15.

Trnovec, T., Pleskova, A., and Chorvat, D., 1974. The effects of carbontetrachloride on radiocerium metabolism in rats, *Strahlentherapie* 147:521–530.

Tuchweber, B., Trost, R., Salas, M., and Sieck, W., 1976. Effect of praseodymium nitrate on hepatocytes and Kupffer cells in the rat, *Can. J. Physiol. Pharmacol.* 54:898–906.

Walker, L. A., 1950. Localization of radioactive colloids in lymph nodes, *J. Lab. Clin. Med.* 36:440–449.

Ward, N. I., 1988. Environmental analysis using ICP–MS, *in* Applications of ICP–MS (A. R. Date and A. L. Grey, eds.), Chapman and Hall, New York, pp. 189–219.

Wedow, H., 1967. The Morro do Ferro thorium and rare-earth deposit, Pocos de Caldas District, Brazil, U.S. Geological Survey Bulletin 1185-D, Washington, D.C.

Weinmann, H.-J., Brasch, R. C., Press, W. R., and Wesby, G. E., 1984a. Characteristics of gadolinium-DTPA complex: a potential NMR contrast agent, *Am. J. Radiol.* 142:619–624.

Weinmann, H. J., Laniado, M., and Mutzel, W., 1984b. Pharmacokinetics of GdDTPA/dimeglumine after intravenous injection into healthy volunteers, *Physiol. Chem. Phys. Med. NMR* 16:167–172.

Wester, P. O., 1965. Trace elements in human myocardial infarction determined by neutron activation analysis, *Akt. Atom. Stockholm* AE-188, 31 pp.

Wolf, G. L., and Fobben, E. S., 1984. The tissue proton T_1 and T_2 response to gadolinium DTPA injection in rabbits: a potential renal contrast agent for NMR imaging, *Invest. Radiol.* 19:324–328.

Wurm, M., 1951. The effect of lanthanum on growth and metabolism of *Streptococcus faecalis* R., *J. Biol. Chem.* 192:707–714.

Yamoto, A., 1984. Plutonium and radiocerium contents in shallow sea sediment collected at off shore Ibaraki prefecture, Japan, *Radioisotopes* 53:60–64.

Zilversmit, D. B., and Hood, S. L., 1953. Effect of buffer on paper electrophoretic studies on state of yttrium in blood, *Proc. Soc. Exp. Biol. Med.* 84:573–576.

Zylicz, E., Zablotna, R., Geisler, J., and Szot, Z., 1975. Effects of DTPA on the deposition of [144]Ce in pregnant rat and in foetoplacental unit, *Int. J. Radiat. Biol.* 28:125–136.

Toxicology and Pharmacology of the Lanthanides

8.1 Introduction

Broadly speaking, lanthanides interfere with normal metabolism in one of two general ways. The first is through specific biochemical interaction producing, as we shall see, a variety of toxic and pharmacologic sequelae. The second concerns the damage inflicted by radioactive lanthanides. Although the latter is an important matter related to such topical concerns as the safety of nuclear power plants and the dangers of nuclear weapons, it is not, in itself, a field of inquiry that is unique to the lanthanides. Rather, it belongs to the larger arena of radiation biology and biochemistry, a discussion of which lies outside the scope of this book. For this reason, the radiotoxicity of lanthanides is not given more than a passing mention in this chapter. Instead, we shall concentrate on the metabolic and physiological perturbations produced by nonradioactive lanthanides.

As discussed in the preceding chapter, only minute amounts of lanthanides occur naturally in living systems, where their presence is apparently of little physiological impact. Concerns about toxicity are raised by the increasing industrial and medical use of lanthanides. Their earliest industrial application was as a component of incandescent gaslight mantles (Section 1.2). Other uses are listed in Table 8-1. Possible future applications include their employment as catalysts in car exhaust systems (Pederson and Libby, 1972) and as "antiknock" agents in gasoline (Sievers and Sadlowski, 1978). In addition, the use of cobalt–samarium magnets in fixing overdentures has been suggested (Sarnat, 1983). At least one disease, "rare earth pneumoconiosis," has been attributed to industrial exposure to lanthanides (Heuck and Hoschek, 1968; Husain *et al.*, 1980; Section 8.5).

The pharmacologic properties of the lanthanides are not only an im-

Table 8-1. Some Industrial Uses of the Lanthanides

Use	Comment
Cored carbon arc lamps	Ln increase brightness and can provide an emission spectrum almost identical to sunlight. Used especially in lithographic and film industries.
Flints for lighters	
Phosphors in color television tubes	The red emisson of Eu is especially useful.
Lasers	
Polishes for lenses and mirrors	Most lenses produced in U.S. are polished with CeO_2.
Component of lenses	La_2O_3 used because of its high refractive index and low dispersion. Other Ln used to absorb specific wavelengths.
Ceramic glazes	
Opacifiers in certain enamels	
Petroleum cracking	
Steel industry	Nodulating agents for iron. Electrolysis of fused chlorides forms "mischmetall," a deoxidizer which improves mechanical qualities and refines grain size.
Alloying agent	Used with Al, Mg, Mn to increase resistance to creep and failure. Alloy of Mg, Ce, Zr used in jet engines.
Textiles	Prevents mildew; used in waterproofing.
Fireworks	
Mercury vapor and fluorescent lamps	
Luminous paint	^{147}Pm used as a low-energy β-emitter.
Nuclear power plants	Good neutron capture makes them useful in control rods. Dy is a burnable poison.

portant component of toxicological studies, but also an area of interest in their own right. Some of these properties are of potential medical application (Chapter 9). Various aspects of the toxicology and pharmacology of the lanthanides have been reviewed by Haley (1965, 1979), Venugopal and Luckey (1978), and Arvela (1979).

8.2 Effects of Lanthanides on Microorganisms

A surprising number of studies have been devoted to investigating the effect of lanthanides on microorganisms. While all agree that lanthanides inhibit the growth of bacteria, fungi, and yeast (Muroma, 1958), none have rigorously addressed the mode of inhibition. Rather high doses of 10^{-4}–$10^{-2} M$ Ln^{3+} are usually required for inhibition, while several authors have reported that low concentrations of around $10^{-5} M$ may stimulate bacterial growth (Muroma, 1958). In addition, suprainhibitory doses sometimes have reduced bacteriocidal potency (Muroma, 1958). Interestingly, the lethality curves of lanthanides in rodents share this phenomenon (e.g., Section 8.4, Fig. 8-1). Precipitation of lanthanides at high concentration under physiological conditions may be the cause in both cases.

In view of the complexities of many of the nutrient media used to culture microorganisms, it is quite impossible to gain much useful information from experiments in which free Ln^{3+} ions, or one of their complexes, are simply added directly to growth medium. Under such conditions, interactions between the lanthanide and the medium may predominate over those between the lanthanide and the cells. Indeed, Wurm (1951) has suggested that the inhibition of the growth of *Streptococcus faecalis* by La^{3+} largely reflects the precipitation of $LaPO_4$, with the concomitant depletion of soluble phosphate from the medium. Nevertheless, the phosphate-independent metabolism of *S. faecalis* is also inhibited by La^{3+}. Biochemical (Wurm, 1951) and histological (Cassone and Garaci, 1974) evidence confirms the binding of La^{3+} to the surface of bacteria. This reduces the surface charge and retards electrophoretic migration to the anode. When the surface charge is completely neutralized, flocculation occurs (Shearer, 1922). The older literature on the bactericidal and bacteriostatic properties of the lanthanides is reviewed by Muroma (1958). Far higher concentrations are required for bactericidal action than for bacteriostasis.

In general, bacteria seem to be more susceptible than fungi to lanthanides (Drossbach, 1897; Burkes and McCleskey,1947; Talburt and Johnson,1967), while the heavier lanthanides tend to be more toxic than the lighter ones (Dryfuss and Wolf, 1906; Eisenberg, 1919; Niccolini, 1930, 1931; Muroma, 1958, 1959). In a study of *E. coli, Staphylococcus aureus,* and five fungi, Talburt and Johnson (1967) found the following order of decreasing toxicity by 11 mM $Ln(NO_3)_3$: Lu, Y, Er, Ho, Dy, Eu, Nd, Ce. Exceptions to this sequence were the high toxicity of Nd^{3+} to *E. coli* and of Y^{3+} to *Saccharomyces cerevisiae*. While interesting, these results do not inform us whether the greater effectiveness of the smaller lanthanides reflects the chemistry of the microorganisms or that of the growth media and agar. In this work, three species of *Aspergillus* and one *Penicillium* detoxified the Ln^{3+} ions by forming insoluble crystals of lanthanide oxalate. Lanthanides also inhibit the formation (Sartory and Bailly, 1922) and germination (Miller and McCallan, 1957) of fungal spores.

Earlier in the century, lanthanides were tested as antibacterial drugs for the treatment of tuberculosis, leprosy, and cholera (Section 9.3.1), without lasting success. However, medical attention has recently returned to this theme, with successful clinical trials of $Ce(NO_3)_3$ as a topical antiseptic in burn wounds (Chapter 9.3.2). This, in turn, has resulted in the retesting of the antibacterial properties of Ce^{3+}. These properties have indeed been confirmed, both *in vitro* and *in vivo*, with gram-negative bacteria being the more sensitive, as Muroma (1958) had shown over 20 years earlier. Again, quite high concentrations are required for bacteriostasis (e.g., Saffer *et al.*, 1980).

Binding of Ln^{3+} ions to bacteria induces clumping (Shearer, 1922; Sobek and Talburt, 1968; Abaas, 1984), as occurs with erythrocytes and other types of eukaryotic cell (Chapter 6). With *Streptococcus mitus*, aggregation is not specific for Ln^{3+}, as Zn^{2+} and Al^{3+} have the same effect (Abaas, 1984). Likely explanations for clumping are the neutralization of surface charge and the formation of lanthanide bridges between anionic moieties on bacterial surfaces. Microbial nitrogen fixation is apparently inhibited by Ce^{3+} and Y^{3+} (Bahadur and Tripathi, 1978) but stimulated by La^{3+} in the presence or absence of molybdate ions (Bahadur *et al.*, 1978; Ranganayaki *et al.*, 1981). Carbon consumption was increased in each case. Lanthanides induce a number of morphological changes in microorganisms. With filamentous fungi, there is evidence of membrane destruction and cell death (Talburt and Johnson, 1967), and, on *E. coli*, the formation of small projections (Sobek and Talburt, 1968).

Bacterial respiration is strongly inhibited by lanthanides. Brooks (1921) found that concentrations of $La(NO_3)_3$ above $50 \mu M$ suppressed the production of CO_2 by *Bacillus subtilis*, although a concentration of $6 \mu M$ was slightly stimulatory. In agreement with this, $0.5 mM$ $Ce(NO_3)_3$ inhibited the endogenous and glucose-stimulated uptake of O_2 by *E. coli*. Release of $^{14}CO_2$ from [U-^{14}C]glucose and the incorporation of glucose into cellular constituents was also inhibited (Sobek and Talburt, 1968). In the experiments of Wurm (1951) mentioned earlier in this section, the basal, phosphate-independent metabolism of *S. faecalis* was also found to be inhibited by La^{3+}.

La^{3+} or Nd^{3+} ions can displace Ca^{2+} ions from the spores of *Bacillus cereus*, lowering their heat resistance (Bulman and Stretton, 1975). Silver ions increase the uptake of Ce^{3+} by *Neurospora sitophila* spores (Miller, 1959). There appears to be only one study of the interaction of lanthanides with viruses (Bjorkman and Horsfall, 1948). It records the strange observation that treatment of influenza virus with La^{3+}-acetate produces a heritable change in its rate of elution from red blood cells; the significance of this finding is completely obscure. Lanthanides increase the antiviral properties of interferon (Sedmak *et al.*, 1986). Dryfuss and Wolf (1906) have described the toxic effects of Nd^{3+} on protozoa.

In view of their antimicrobial properties, lanthanides have been added to the water used in fish farming in China (Anon, 1988). This apparently protects the fish from diseases of their scales, gills, and intestines. In addition, the growth of the fish is enhanced.

8.3 Effects of Lanthanides on Plants

Although a number of plants can assimilate lanthanides (Section 7.2; Tables 7-2, 7-3), little is known of the metabolic consequences. Evans (1913) reported that the carbonates of Ce and La increased the rate of cell division in hyacinth roots, while yttrium carbonate had the opposite effect. Lanthanum carbonate also increased the length of the flower stalk. At concentrations above $10^{-4}M$, La^{3+}, Pr^{3+}, or Nd^{3+} inhibited the growth of *Avena* coleoptiles either in the presence or absence of the plant hormone indoleacetic acid. This property is shared by Ca^{2+}. At higher concentrations, the lanthanides induced flaccidity (Pickard, 1970). Chinese scientists have recently investigated the effects of $CeCl_3$ upon the growth of corn seedlings. At low concentrations (0.5 ppm), growth was slightly stimulated, while a tenfold higher concentration inhibited growth of the shoots and, particularly, the roots. Lower numbers of electrophoretic bands corresponding to isozymes of peroxidase and esterase could be detected electrophoretically in inhibited roots, but the total enzymic activity was not reduced (Tang and Li, 1983).

Following their finding that lanthanides can enhance the growth of plants under certain conditions, the Chinese have developed a fertilizer known as "Nong-le" which contains lanthanides (Guo, 1985). When applied to crops at the appropriate stage of growth, at a concentration of 450–750 g Ln/ha, the yield is apparently increased. Excess amounts of the fertilizer are toxic to the crops, while application after the primary growth stage depresses the yield. "Nong-le" is reported to benefit the growth and quality of sugarcane, apples, wheat, rice, and various other crops. Its mode of action is unknown, but it has been suggested that lanthanides increase the uptake and transport of phosphates.

8.4 The Acute Toxicity of Lanthanides in Vertebrates

As might be expected from the discussion of their metabolism in Section 7.5, the toxicity of the lanthanides depends very much upon their chemical form and route of administration. Poor absorption from the gastrointestinal (G.I.) tract renders orally administered lanthanides benign (Table 8-2). Values for the LD_{50} of orally administered, poorly soluble

Table 8-2. Acute Oral Toxicity of Lanthanides

Administered substance	Animal	LD_{50} (mg/kg)	Reference[a]
La^{3+}-acetate	Rat	10,000	1
La^{3+}-ammonium nitrate	Rat	3,400	1
$LaCl_3$	Rat	4,200	1
$La(NO_3)_3$	Rat	4,500	1
$La_2(SO_4)_3$	Rat	>5,000	1
La_2O_3	Rat	>10,000	1
$Ce(NO_3)_3$	Rat ♀	4,200	2
$PrCl_3$	Mouse ♂	4,500	3
$Pr(NO_3)_3$	Rat ♀	3,500	2
$NdCl_3$	Mouse ♂	5,250	3
$Nd(NO_3)_3$	Rat ♀	2,750	2
$SmCl_3$	Mouse ♂	>2,000[b]	4
$Sm(NO_3)_3$	Rat ♀	2,900	2
$Eu(NO_3)_3$	Rat ♀	>5,000	2
$GdCl_3$	Mouse ♂	>2,000	4
$Gd(NO_3)_3$	Rat ♀	>5,000	2
$TbCl_3$	Mouse ♂	5,100	5
$Tb(NO_3)_3$	Rat ♀	>5,000	2
$DyCl_3$	Mouse ♂	7,650	6
$Dy(NO_3)_3$	Rat ♀	3,100	2
$HoCl_3$	Mouse ♂	7,200	6
$Ho(NO_3)_3$	Rat ♀	3,000	2
$ErCl_3$	Mouse ♂	6,200	6
$TmCl_3$	Mouse ♂	6,250	5
$YbCl_3$	Mouse ♂	6,700	5
$Yb(NO_3)_3$	Rat ♀	3,100	2
$LuCl_3$	Mouse ♂	7,100	3

[a] References: 1, Cochran et al., 1950; 2, Bruce et al., 1963; 3, Haley et al., 1964; 4, Haley et al., 1961; 5, Haley et al., 1963; 6, Haley, 1965.
[b] Solubility limitations restricted the use of higher doses.

salts, such as Ln_2O_3 and $Ln_2(SO_4)_3$, are so high as to have proved impossible to determine. Bruce et al. (1963), for instance, fed 10 different lanthanide oxides at doses of over 1 g/kg body weight to female rats without producing any deaths. Similarly, Cochran et al. (1950) could find no toxicity after feeding 10 g/kg of La_2O_3 or 5 g/kg $La_2(SO_4)_3$.

Subcutaneously injected lanthanides are slightly better absorbed, and

Table 8-3. Acute Toxicity of Subcutaneously
Injected Lanthanides

Injected substance	Animal	Minimum lethal dose (mg/kg)	Reference[a]
LaCl$_3$	Mouse	3,500	1
	Mouse	3,500	2
	Mouse	>500	3
	Mouse	>500	4
	Frog	~1,000	1
CeCl$_3$	Mouse	5,000–10,000	5
	Frog	~300	5
	Rat	2,000–4,000	5
	Mouse	569–843[b]	4
	Guinea pig	569–1,137[b]	4
	Guinea pig	2,130[b]	8
CeF$_3$	Guinea pig	5,000	8
PrCl$_3$	Mouse	2,500	2
	Mouse	900–1,500	6
	Rabbit	200–250	2
	Mouse	944	4
	Frog	~1,000–1,500	6
NdCl$_3$	Mouse	4,000	2
	Mouse	2,302	4
	Frog	250	9
SmCl$_3$	Rat	>2,000	6
	Frog	~150	5
	Guinea pig	750–1,000	6
	Guinea pig	500	8
	Guinea pig	1,000[b]	8
Sm(NO$_3$)$_3$	Guinea pig	500	7
	Frog	1,600	7
Y(NO$_3$)$_3$	Mouse	1,660	7
	Frog	350	7

[a] References: 1, Ajazzi-Mancini, 1926; 2, Vincke and Oelkers, 1938; 3, Laszlo *et al.*, 1952; 4, Kyker and Cress, 1957; 5, Hara, 1923; 6, Niccolini, 1930; 7, Steidle and Ding, 1929; 8, Venugopal and Luckey, 1978; 9, Guida, 1930.
[b] LD$_{50}$ (mg/kg).

consequently have greater toxicity. However, direct comparison with the oral toxicity data is difficult, as many of the values on subcutaneous toxicity are in the older literature where they are expressed as the "minimal lethal dose" (Table 8-3).

The only data on the lethality of lanthanides in the respiratory system are those of Arkhangelskaya and Spasskii (1967). They found that a single intratracheal injection of 50 mg of a mixture of lanthanide oxides killed all rats within a day from pulmonary edema. Two separate injections of 25 mg of the same material killed only 46% of animals. However, Tandon *et al.* (1977) injected 200 mg/kg of the lanthanide ore monazite (Table 7-1) intratracheally into mice without ill effects.

The LD_{50} values for intraperitoneally (i.p.) injected lanthanides (Table 8-4) are an order of magnitude or more lower than those for perorally (p.o.) administered salts. The scatter in the reported values is quite large and undoubtedly stems in part from the failure of many authors to pay attention to the degree of hydration of the salts used for injection. For this reason, it is difficult to distinguish patterns of variation in susceptibility between different species, lanthanides, or sexes. However, there appear to be no major differences between the toxicities of the different lanthanides. No consistent sex differences have been reported. However, Graca *et al.* (1957) found guinea pigs to be approximately twice as sensitive as mice to i.p. injections of Ln^{3+}-citrate complexes. According to Cochran *et al.* (1950), the i.p. toxicity of salts of La in rats decreases in the order sulfate > nitrate > ammonium nitrate > chloride > acetate. Mobilizing the lanthanides with a weak chelator, such as citrate (Table 2-8; Section 7.5.3.), increases their toxicity following i.p. injection (Table 8-5). When a stronger chelator, such as EDTA, is used, the effect on toxicity is more complicated (Table 8-5). The larger Ln^{3+} ions have a lower affinity than the smaller Ln^{3+} ions for EDTA (Section 2.3; Table 2-4). Thus, although EDTA prevents precipitation and transports all lanthanides from their injection site, it permits exchange of the larger members of the series with physiological ligands. In this case, the LD_{50} falls below those of both the $LnCl_3$ and, for the largest lanthanides, the citrate complexes. As the smaller Ln^{3+} ions have much greater affinity for EDTA, they are transported from the peritoneal cavity and excreted in the urine with much less exchange with physiological ligands (Section 7.5.3). Their EDTA complexes are consequently benign (Table 8-5).

Graca *et al.* (1957) reported that the toxicity of Ce^{3+}, Pr^{3+}, and Nd^{3+} complexes of EDTA showed biphasic dose–response curves. For Ce-EDTA, there was 60% mortality at 50 mg/kg but only 50% at 100 mg/kg and 33% at 200 mg/kg. However, at 250 mg/kg, toxicity was back up to 50%. This is interesting with regard to the bell-shaped dose–response mortality curves observed by Tuchweber *et al.* (1976a) for i.v.-injected $PrCl_3$ (Fig. 8-1). In both cases, aggregation or precipitation at high doses may explain the reduced toxicity. Evidence in favor of this explanation comes from the high LD_{50} values for i.v. injections of insoluble particles such as Gd_2O_3. According to Havron *et al.* (1980), the LD_{50} of i.v.-injected Gd_2O_3 in rabbits is 5 g/kg, although unpublished data cited in Burnett *et al.* (1985) suggest the much lower value of 125 mg/kg.

Table 8-4. Acute Toxicity of Intraperitoneally Injected Lanthanides

Injected substance	Animal	LD_{50} (mg/kg)	Reference[a]
La^{3+}-acetate	Rat	475	1
La^{3+}-citrate	Mouse	82	8
	Mouse	78	2
	Guinea pig	71	13
	Guinea pig	61	2
La^{3+}-ammonium nitrate	Rat	625	1
$LaCl_3$	Mouse	372	2
	Mouse	>160	3
	Mouse	211	13
	Mouse	362	8
	Mouse	121	13
	Rat	106	4
	Rat	197	13
	Rat	350	1
	Guinea pig	130	2
$La(NO_3)_3$	Rat	450	1
	Mouse ♀	410	5
$La_2(SO_4)_3$	Rat	275	1
Ce^{3+}-citrate	Guinea pig	56	2
	Guinea pig	104	13
	Mouse	50	8
	Mouse	147	2
	Rat	147	13
$CeCl_3$	Mouse	353	2,8
	Mouse	98	13
	Guinea pig	104	2
$Ce(NO_3)_3$	Mouse ♀	470	5
	Mouse	149	13
	Rat ♀	290	5
Pr^{3+}-citrate	Mouse	141	2
	Mouse	145	8
	Guinea pig	53	2
	Guinea pig	97	13
	Guinea pig	141	13
Pr^{3+}-propionate	Guinea pig	53	13
$PrCl_3$	Mouse	359	2,8
	Mouse	550[b]	13
	Mouse ♂	600	6
	Rat	<2000	5
	Guinea pig	190	13
	Guinea pig	125	2
	Rabbit	200[b]	13
$Pr(NO_3)_3$	Mouse ♀	290	5
	Rat ♀	245	5
Nd^{3+}-citrate	Mouse	138	2
	Mouse	140	9
	Guinea pig	41	2

(continued)

Table 8-4. *Continued*

Injected substance	Animal	LD$_{50}$ (mg/kg)	Reference[a]
NdCl$_3$	Mouse ♂	600	7
	Mouse	348	2
	Mouse	347	8
	Rat	375	13
	Rat	150–250[b]	9
	Guinea pig	202[b]	13
	Guinea pig	140	2
Nd$_2$O$_3$	Rat	800	13
Nd(NO$_3$)$_3$	Mouse ♀	270	5
	Mouse	138	13
	Rat ♀	270	5
	Guinea pig	41	13
Sm^{3+}-citrate	Mouse	164	8
	Guinea pig	75	13
SmCl$_3$	Mouse	365	5
	Mouse ♂	585	10
Sm(NO$_3$)$_3$	Mouse ♀	315	5
	Rat ♀	285	5
Eu^{3+}-citrate	Mouse	187	8
	Guinea pig	72	13
EuCl$_3$	Mouse ♂	535	12
	Mouse	387	8
	Guinea pig	156	13
Eu(NO$_3$)$_3$	Mouse ♀	320	5
	Rat ♀	210	5
Gd^{3+}-citrate	Mouse	153	8
	Guinea pig	60	13
GdCl$_3$	Mouse ♂	550	10
	Mouse	379	8
Gd(NO$_3$)$_3$	Mouse ♀	300	5
	Rat ♀	230	5
Gd$_2$O$_3$	Rat	1000	13
Tb^{3+}-citrate	Mouse	121	8
	Guinea pig	74	13
TbCl$_3$	Mouse	333	8
	Mouse ♂	550	11
Tb(NO$_3$)$_3$	Mouse ♀	480	5
	Rat ♀	260	5
Dy^{3+}-citrate	Mouse	113	8
	Guinea pig	54	13
DyCl$_3$	Mouse	343	8
	Mouse ♂	585	12
	Guinea pig	196	13
Dy(NO$_3$)$_3$	Rat ♀	295	5
	Mouse ♀	310	5
Ho^{3+}-citrate	Mouse	117	8
	Guinea pig	63	13
HoCl$_3$	Mouse	312	8
	Mouse ♂	560	12

Table 8-4. *Continued*

Injected substance	Animal	LD$_{50}$ (mg/kg)	Reference[a]
Ho(NO$_3$)$_3$	Mouse ♀	320	5
	Rat ♀	270	5
Er^{3+}-citrate	Mouse	122	8
	Guinea pig	63	13
ErCl$_3$	Mouse	227	8
	Mouse ♂	535	8
	Guinea pig	128	13
Er(NO$_3$)$_3$	Mouse ♀	225	5
	Rat ♀	230	5
Er$_2$O$_3$	Rat	600[b]	13
Tm^{3+}-citrate	Mouse	80	8
	Guinea pig	55	13
TmCl$_3$	Mouse	281	8
	Mouse	335	13
	Mouse ♂	485	11
	Guinea pig	144	13
Tm(NO$_3$)$_3$	Mouse ♀	255	5
	Rat ♀	285	5
Yb^{3+}-citrate	Mouse	143	8
	Guinea pig	69	13
YbCl$_3$	Mouse	300	8
	Mouse	285	13
	Mouse ♂	395	11
	Guinea pig	132	13
Yb(NO$_3$)$_3$	Mouse ♀	250	5
	Rat ♀	255	5
Lu^{3+}-citrate	Mouse	136	8
	Guinea pig	81	13
LuCl$_3$	Mouse ♂	315	6
	Guinea pig	161	13
Lu(NO$_3$)$_3$	Mouse ♀	290	5
	Mouse	350	13
	Rat ♀	325	5
Y^{3+}-citrate	Mouse	254	13
	Mouse	79	8
	Guinea pig	44	13
YCl$_3$	Mouse	88	4
	Rat	450	1
	Rat	45	4
	Guinea pig	85	13
Y(NO$_3$)$_3$	Rat	350	1
	Rabbit	515	13
	Mouse	1710	13
Y$_2$O$_3$	Rat	500	1

[a] References: 1, Cochran *et al.*, 1950; 2, Graca *et al.*, 1957; 3, MacDonald *et al.*, 1952; 4, Kyker and Cress, 1957; 5, Bruce *et al.*, 1963; 6, Haley *et al.*, 1964; 7, Niccolini, 1930; 8, Graca *et al.*, 1962; 9, Guida, 1930; 10, Haley *et al.*, 1961; 11, Haley *et al.*, 1963; 12, Haley, 1965; 13, Venugopal and Luckey, 1978.
[b] Minimum lethal dose.

Table 8-5. Comparison of Toxicities of
Intraperitoneally Injected Solutions of LnCl₃,
Ln³⁺-citrate, and Ln-EDTA in Mice[a]

	LD$_{50}$ (mg/kg)		
	LnCl$_3$	Ln^{3+}-citrate	Ln-EDTA
La	362	82	37
Ce	352	149	37
Pr	358	145	N.G.[b]
Nd	346	140	126
Sm	365	164	311
Eu	387	186	240
Gd	378	153	274
Tb	332	120	—[c]
Dy	342	113	—[c]
Ho	311	117	—[c]
Er	226	122	—[c]
Tm	381	79	—[c]
Yb	300	142	—[c]
Lu	N.G.	135	—[c]
Y	N.G.	79	—[c]

[a] The data used to construct this table were taken from Graca
  et al. (1962).
[b] N.G.: Not given.
[c] No deaths at 500 mg/kg.

Intravenously injected lanthanide salt solutions are the most toxic, with LD$_{50}$ values in the range of ten to hundreds of mg/kg body weight (Table 8-6). The wide range in reported values may reflect the biophasic nature of the lethal response which was just mentioned. Female rats and mice are far more susceptible than male animals to the lethal effects of i.v. injections of the lighter lanthanides (Bruce *et al.*, 1963; Tuchweber *et al.*, 1976a; Fig. 8-1). This may be linked to their far greater propensity toward the formation of a fatty liver (Section 8.7.2). With quail, however, males are more susceptible than females. When these birds were injected intravenously with 150 μmol (10.9 mg) La/kg body weight, there were no deaths among 34 laying females, while 3 of 13 males died within 24 h. Females may be protected from the toxic effects of lanthanides by their high plasma levels of Ca^{2+}. Other lanthanides from Ce to Dy were no more toxic than La. However, Yb^{3+} produced 2 deaths among 13 females during injection. At 18 h post injection, there had been one more death, while one further bird was moribund (Robinson *et al.*, 1978, 1980).

When complexed lanthanides are injected intravenously, the toxicity

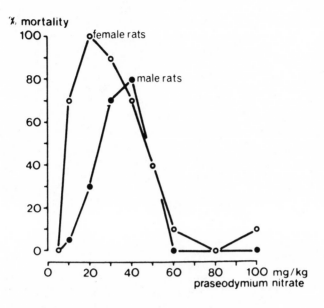

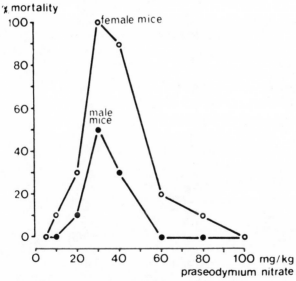

Figure 8-1. Mortality 14 days after intravenous injection of $Pr(NO_3)_3$ into male and female rats and mice. From Tuchweber *et al.* (1976a), with permission.

Table 8-6. Acute Toxicity of Intravenously Injected Lanthanides

Injected substance	Animal	LD_{50} (mg/kg)	Reference[a]
$LaCl_3$	Mouse	10	1
	Rat	91[b]	10
	Rat	3.5	1
	Rabbit	200–250[b]	2
	Rabbit	84[b]	10
$CeCl_3$	Rat	3.5[b]	1
	Rat	50–60[b]	3
	Rabbit	35	10
$Ce(NO_3)_3$	Rat ♀	4.3	4
	Rat ♂	49.6	4
Ce^{3+}-citrate	Rat	4.5	10
$PrCl_3$	Rat	3.5[b]	1
	Rat	9.5[b]	10
	Rabbit	75–94	1
$Pr(NO_3)_3$	Rat ♂	23	11
	Rat ♀	10	11
	Rat	3.5[b]	1
	Rat ♀	7.4	4
	Rat ♂	77.2	4
	Rat	10.8–13.9[b]	3
	Mouse ♂	30	11
	Mouse ♀	22	11
Pr^{3+}-citrate	Mouse	3500	10
$NdCl_3$	Guinea pig	70[b]	5
	Guinea pig	100[b]	10
	Rabbit	115–144	1
	Rabbit	200–250[b]	6
	Rat	8.7	10
$Nd(NO_3)_3$	Rat ♀	6.4	4
	Rat ♂	66.8	4
Nd^{3+}-citrate	Rabbit	50[b]	10
$Sm(NO_3)_3$	Rat ♀	8.9	4
	Rat ♂	59.1	4
Sm^{3+}-acetate	Mouse	1000	10
$GdCl_3$	Rat	79.1	7
	Mouse	100–200[c]	8
Gd_2O_3	Rabbit	5000	12
	Rabbit	125	13
$Er(NO_3)_3$	Rat	82.8–96.6[b]	3
	Rat ♀	35.8	4
	Rat ♂	52.4	4
	Rat	30	10
$Y(NO_3)_3$	Rat	20–30[b]	3
	Rat	75[b]	10
	Rabbit	500	9

[a] References: 1, Kyker and Cress, 1957; 2, Cochran *et al.*, 1950; 3, Maxwell *et al.*, 1931; 4, Bruce *et al.*, 1963; 5, Guida; 1930, 6, Vincke and Oelkers, 1938; 7, Weinmann *et al.*, 1984; 8, Caille *et al.*, 1983; 9, Steidle and Ding, 1929; 10, Venugopal and Luckey, 1978; 11, Tuchweber *et al.*, 1976a; 12, Havron *et al.*, 1980; 13, cited in Burnett *et al.*, 1985.
[b] Minimal lethal dose.
[c] Mean lethal dose.

reflects the stability constant. For example, whereas the LD_{50} of Gd-EDTA does not differ greatly from that of $GdCl_3$, Gd-DTPA has extremely low toxicity (Weinmann *et al.*, 1984). The toxicities of $GdCl_3$, Gd-EDTA, and Gd-DTPA are even lower when administered intracisternally. According to Zimakov (1973), the toxicity of Nd salts and complexes administered intravenously into mice increases in the order chloride < propionate < acetate < 3-sulfoisonicotinate < sulfate < nitrate.

A number of substances reduce the lethality of i.v.-administered lanthanides. Among these are agents which increase the activities of hepatic, microsomal drug-metabolizing agents, including pregnenolone-16α-carbonitrile, phenobarbital, spironolactone, and estradiol (Salas and Tuchweber, 1976). Adenine, adenosine, and tryptophan are also effective in this respect. Lethality is also reduced by conditioning the animals with small, regular injections of lanthanides (Oberdisse *et al.*, 1979) or by starving the animals for 24 h prior to injection (Schurig and Oberdisse, 1972). Dexamethasone (Salas and Tuchweber, 1976), glucocorticoids (Selye *et al.*, 1972), and triaminolone enhance the lethal effects of lanthanides. The mechanism through which the lethality of the lanthanides is modified by these substances is unknown. However, the observation that prednisone induces La^{3+} binding by HeLa cells (Boyd *et al.*, 1976; Section 6.2.2) may be relevant. Phenobarbital increases fecal excretion of cerium (Bjondahl *et al.*, 1974), but not by amounts which explain its protective effect. Although dexamethasone potentiates the lethal effects of Ce, it paradoxically protects animals from developing the rare earth fatty liver (Section 8.7.2). Rats sensitized to Ce with dexamethasone were found to have severe hemorrhagic necroses in the heart and liver 48 h after injection (Salas and Tuchweber, 1976). Glucose, methionine, and choline are without effect on the acute toxicity of lanthanides.

8.5 General Toxicological and Pharmacologic Effects of the Lanthanides

Aqueous solutions of $LnCl_3$ have been said to have "a markedly astringent taste" (Dryfuss and Wolf, 1906) but otherwise constitute an unremarkable drink. Haley *et al.* (1961, 1963, 1964) administered the chlorides of Lu, Yb, Tm, Tb, Sm, or Gd as 0.01, 0.1, or 1% of the diets of rats. There were no effects on hematologic parameters, and only the highest concentrations of Tm and Tb suppressed growth. The only changes found upon histological evaluation after 12 weeks occurred with Tb, Tm, Yb, and the highest dose of $GdCl_3$, where perinuclear vacuolization and cytoplasmic granularity were observed in the parenchymal cells of the liver. The severity of the changes, which were absent from female rats, appeared to be greater for the heavier lanthanides.

Suppression of growth has also been recorded for guinea pigs (Mogilevskaya and Roshchina, 1967) and mice (Haley, 1979) after prolonged oral administration of lanthanides. However, feeding several Ln_2O_3 at a dietary level of 0.5% to mice for three generations had no effect on growth, survival, development, or hematologic parameters. Nevertheless, dietary levels of 1% did retard growth (Hutcheson *et al.*, 1975). As well as slightly suppressing growth, the inclusion of 5 ppm of $Y(NO_3)_3$ in the drinking water of mice increased the rate of appearance of malignant tumors (Schroeder and Mitchener, 1971).

Large oral doses of solid $Ce(NO_3)_3$ have been shown to produce nausea, vomiting, and diarrhea with tenesmus in dogs (Baehr and Wessler, 1909). Equally large consumption of lanthanide oxalates was, however, symptomless. Gastric hemorrhage following oral ingestion of lanthanides has been reported (Haley *et al.*, 1963). The data of Sihvonen (1972) infer that the inclusion of lanthanides in the diets of rats increases their alcohol consumption.

Although Tb salts are apparently very irritating to intact skin (Haley *et al.*, 1963), direct contact of other lanthanides with the intact skin produces, at worst, depigmentization (Rapaport, 1982). Subcutaneous (s.c.) or intradermal injection produces calcification and the formation of granulomas at the injection site. Haley *et al.* (1961, 1964) have noted a severe reaction to crystals of $GdCl_3$, $SmCl_3$, or $LuCl_3$ by the abraded skin of rabbits. Within 7 days of application of $GdCl_3$ or $SmCl_3$, perforating ulcers formed, with penetration through the skin to the underlying muscle. By 14 days, there was no sign of healing. However, abraded skin exposed to $LuCl_3$ crystals healed by 35 days with scar formation.

Stineman *et al.* (1978) noted that following s.c. injection of Ce^{3+}-citrate, mice became hypoactive. Their livers had focal midzone necroses, although regenerative changes were indicated by mitotic figures in the hepatocytes. Hypertrophy, reticuloendothelial hyperplasia, and hyperactive lymphoid follicles were noted in the spleen.

Two groups have independently gathered data suggesting that subcutaneously injected pellets of metallic lanthanides or yttrium produce neoplasms. According to Ball *et al.* (1970), metallic Gd or Yb provoked the formation of sarcomas in adjacent mouse skin, with those caused by Yb metastasizing to the lung. In the studies of Talbot *et al.* (1965a), about one-half of the implanted mice which did not develop sarcomas at the site of the metal had histological evidence of a foreign body granulomatous reaction. According to the data presented in this paper, the incidence of spontaneous neoplasms unrelated to the metal pellets was nearly halved in implanted animals. This matter was not addressed by the authors. In view of the ability of foreign bodies, such as plastic, to induce sarcomas in rodents, it is unclear whether or not the tumors were induced specifically by the lanthanides. However, there is evidence that Ln^{3+} promote malignant transformation *in vitro* (Smith *et al.*, 1986). Blood coagulation times were prolonged 6 months after implantation of the metal (Talbot *et al.*, 1965a).

Large hypodermic injections of $Ce(NO_3)_3$ in dogs produced loss of appetite, diarrhea, gastric and intestinal peristalsis, and congestion of the intestinal mucosa (Baehr and Wessler, 1909). In addition, there were areas of apparently liquefied necrotic tissue at the site of injection. After intradermal injection of much smaller amounts of lanthanides into human volunteers, Shelly *et al.* (1958) noticed a delayed, noninflammatory, "whitish, fibrotic change" at the sites of injections.

Lanthanides also produce ocular irritation. The introduction of 1 mg of $GdCl_3$ or $SmCl_3$ crystals into the conjunctival sac of rabbits increased the rate of blinking and resulted in redness of the palpebral conjunctiva within an hour. These symptoms disappeared by 24 h (Haley *et al.*, 1961). However, solutions of $LuCl_3$ were more toxic (Haley *et al.*, 1964), producing redness, a profuse discharge, and swelling of the lids within 1 h. The swelling was severe enough to close the eyes. By 24 h, minute ulcers appeared. Complete healing required 2 weeks, but there was no evidence of residual damage.

The corneal epithelium serves to protect the underlying stroma from environmental insult. If this epithelium is removed, the cornea takes up large amounts of lanthanides (Grant and Kern, 1956). Much of the sequestered lanthanide is bound to collagen, a connective tissue protein with a high affinity for Ln^{3+} ions (Drouven and Evans, 1986; Section 4.12). As a result of exposure to lanthanides, the cornea undergoes a delayed opacification due to calcification. This is another example of the calcergenic properties of the lanthanides (Section 8.6). Interestingly, cataractous human lenses contain more La than do normal lenses (Swanson and Truesdale, 1971; Section 7.2; Table 7-6).

According to Gehrcke *et al.* (1939), inhalation of lanthanide fumes produces itching, insensitivity to heat, and a sharper sense of taste and smell. Although inhalation or intratracheal injection of mixtures of LnF_3 and Ln_2O_3 particles initially failed to produce pneumoconiotic fibrosis or any other gross pathological changes in guinea pigs (Schepers *et al.*, 1955), further work (Schepers, 1955a) led to the discovery of chemical hyperemia and cellular eosinophilia. Cellular, vascular granulomas occurred in animals surviving one year. When the guinea pigs were exposed to a mixture with a high fluoride content, a number of changes were produced (Schepers, 1955b), including pneumonitis, bronchitis, bronchiolitis, and emphysema.

Intratracheal administration of Ln_2O_3 induces delayed pulmonary fibrosis (Mezentseva, 1967), granulomas, and other proliferative reactions in rats and mice (Mogilevskaya and Roschchina, 1963). Additional pathological changes include erythrocythemia, neutrophilia, hyperhemoglobinemia, hypercoagulability of blood, pneumonia, abscesses, pneumosclerosis, ganglion hepatitis, and hepatic dystrophy (Arkhangelskaya and Spasskii, 1967). Apart from one report of increased incidence of pulmonary reticulosarcoma (Kikhachev *et al.*, 1972), neoplastic changes have not been observed.

Mice exposed to aerosols of Gd_2O_3 had an increased tendency to develop pneumonia 2–5 weeks later, but the incidence of lung tumors was not altered (Ball and Van Gelder, 1966). Particles of Gd_2O_3 (0.3 μm) did not reduce weight gain or change various hematologic parameters. Pulmonary macropages engulfed these particles, causing some cells to lyse. Under *in vitro* conditions, particles of Nd_2O_3 were toxic to pulmonary macrophages, with an LC_{50} value of 101 μM. Solutions of $LaCl_3$ and $CeCl_3$ had LC_{50} values of 52 μM and 29 μM, respectively, in this system. However, the toxicities of La_2O_3, Ce_2O_3, and $NdCl_3$ were very low (Palmer *et al.*, 1987). Around the macrophages exposed to Gd_2O_3 *in vivo* was an area of interstitial thickening, but little fibrosis (Ball and Van Gelder, 1966). One interesting finding was pulmonary calcification, which provides further evidence for the calcergenic properties of the lanthanides (Section 8.6). In similar experiments with guinea pigs, Gd_2O_3 lowered pulmonary elasticity, produced alveolar cell hypertrophy, septal cell thickening, lymphoid hyperplasia, and macrophage infiltration (Abel and Talbot, 1967).

Chronic exposure of industrial workers to dusts which contain lanthanides leads to a condition known as "rare earth pneumoconiosis" (Heuck and Hoschek, 1968; Nappee *et al.*, 1972; Husain *et al.*, 1980). This is seen most commonly with those employed in lithography, a process entailing the use of carbon arc lamps which emit Ln_2O_3 fumes. One case study of two such patients (Vocaturo *et al.*, 1983) reported enormously high levels of La, Ce, Nd, and Sm in the lung tissues and considerably elevated amounts in the prescalenic lymph nodes (Section 7.2; Table 7-8). Elevated concentrations of La occurred in the lungs of workers in a Swedish copper smelter (Section 7.2), but it was not possible to determine whether this was linked to any specific pathologies. Levels of La were particularly high in the subgroup of workers who died from causes other than malignancies or cardiovascular diseases (Gerhardsson *et al.*, 1984).

Death from i.p. injection of toxic doses of $LnCl_3$ usually occurs in the first few days. Haley *et al.* (1964) have described the preceding symptoms in mice as "writhing, ataxia, labored respiration and walking on toes with arched back." For i.p.-administered solutions of LuCl$_3$, the first death occurred at 24 h, with the peak number at 48 h (Haley *et al.*, 1964). With i.p. injections of SmCl$_3$ or GdCl$_3$, peak mortality occurred at the 4th and 5th day (Haley *et al.*, 1961). Most authors draw attention to the anorexia, dyspnea, and diarrhea that follow i.p. administration of LnCl$_3$ solutions. Death follows much more quickly after injection of Ln^{3+}-citrate complexes (Graca *et al.*, 1957), most deaths occurring within 6 h.

Necropsy reports on rodents receiving toxic i.p. doses of lanthanides agree on the prominent peritonitis, adhesions, and ascites. Graca *et al.* (1957) have given a detailed account of this for mice and guinea pigs. They found a loose, caseous precipitate of lanthanides in the peritoneal cavity, with generalized peritonitis, and hemorrhagic or, occasionally, serous ascites. In animals living beyond 24 h, large, irregular gray patches were found over the parietal and visceral surfaces of the abdominal cavity. The liver was firmer than normal, with fixation to the diaphragm and stomach. Apart from a surface coating of the precipitate, the spleen was normal, as were the lungs. There was evidence of thickening of the renal capsule with cortical adhesions. Sometimes a subserous intestinal hemorrhage was seen. Steffee (1959) reported similar findings for rats. The earliest peritoneal lesion was a surface deposit of fibrin and accumulation of neutrophils. The inflammatory exudate contained many macrophages and lymphocytes, while discrete granulomas formed. Some animals suffered focal necrosis of the kidneys. According to Steffee (1959), the granulomatous inflammatory response is triggered by the precipitation of insoluble salts following the i.p. injection of LnCl$_3$ solutions.

In animals dying from i.p. injection of Ln^{3+}-citrate complexes, lung edema and pleural effusion were constant findings, probably related to the dyspnea mentioned earlier (Graca *et al.*, 1957). For animals surviving for 24 h, pulmonary hyperemia, generalized visceral hyperemia, and marked hepatic congestion and edema were noted.

Chronic exposure to sublethal i.p. doses of lanthanides does not produce adverse symptoms, even though the cumulative dose may be exceedingly high. MacDonald *et al.* (1952) injected a 60-mg/kg dose of YCl$_3$ intraperitoneally into rats every other day for 5 months without toxic sequelae, although many of the rats showed abdominal trauma and intestinal adhesions. No precipitated yttrium salts were detected. A similar conditioning effect is found with i.v.-administered lanthanides (q.v.).

Intravenously injected lanthanides provoke a number of interesting physiological changes. As noted for i.p.-injected lanthanides, there is a marked loss of appetite. In a variety of experimental animals, there is a fall in blood pressure after i.v. injection of lanthanides (e.g., Mori, 1931;

Graca *et al.*, 1964). In cats, a 20–25-mg/kg dose of $GdCl_3$ or $SmCl_3$ produced transient hypotension of 15–60 mm Hg, with decreased femoral blood flow. Death of cats following i.v. injection of lanthanides usually occurred from cardiovascular collapse coupled with respiratory paralysis. In dogs, i.v. injections of lanthanide chlorides had an immediate hypotensive effect, with death from circulatory failure. Decreases in blood pressure varied from dog to dog, with some animals showing no response and others undergoing a marked, fatal drop in pressure. Inhibition of contraction of the vascular smooth muscle cells (Section 6.4.3) may cause the reduction in blood pressure. Whereas the first injection of $LnCl_3$ produced the biggest drop in blood pressure, the reverse was true of their citrate complexes. EDTA complexes had no immediate effect on blood pressure. The effects of yttrium were much smaller than those of the lanthanides. Concomitant with the drop in blood pressure was a decrease in heart rate without electrocardiogram abnormalities. One week after a sublethal i.v. dose of $LnCl_3$, the dogs became emaciated with extensive subseral hemorrhage and toxic hepatitis (Graca *et al.*, 1964). Epinephrine or atropine were unable to prevent death following i.v. injection of lanthanides (Haley *et al.*, 1964). Indeed, Haley (1979) comments that local injection of epinephrine after i.v. injection of $LnCl_3$ causes hemorrhagic lesions at the local injection site and in the kidneys.

Haley *et al.* (1961) have described the electrocardiographic changes that occur after the i.v. injection of toxic doses of $GdCl_3$ or $SmCl_3$ in cats. Among these were fluctuations in the height of the P-wave, T-wave, and QRS-complex, leading to ventricular fibrillation. Ojemann *et al.* (1961) have noted electroencephalographic abnormalities in cats after i.v. administration of Dy-EDTA.

In humans, i.v. injections of $LaCl_3$ solutions produced chills, fever, muscle pain, abdominal cramps, hemoglobinemia, and hemoglobinuria (Beaser *et al.*, 1942). Hemoglobinemia may result from increased erythrocyte fragility (Sarkadi *et al.*, 1977; Godin and Frohlich, 1981).

Intravenous injections of suspensions of Gd_2O_3 produce a persistent, subacute, multifocal hepatitis in rats (unpublished data cited in Burnett *et al.*, 1985). However, no long-term side effects appeared in rabbits (Seltzer *et al.* 1981).

Offspring from gravid mice which had received Ce^{3+}-citrate by s.c. injection were slightly lighter than normal, but behavioral parameters were normal (D'Agostino *et al.*, 1982). However, skeletal deformities have been produced in embryonic hamsters following i.v. injection of $YbCl_3$ into the mothers (Gale, 1975). Injections of Y^{3+} into young rats altered the morphology and pigmentation of their teeth (Castillo and Bibby, 1973). Introduction of $YbCl_3$ into chicken eggs produced malformations of the beak, eyes and legs (Oga, 1971). When injected into the yolk sacs of 8-day-old chick embryos, Eu^{3+} produced deformities of the legs and joints, inhibited feather formation, caused edema, and provoked subcutaneous blistering (Zanni, 1965). Intraperitoneal injection of $LaCl_3$ into pregnant mice reduced the number of successful pregnancies and the average litter size (Abramczuk, 1985). Preimplantation development of the embryos was unaffected. Indeed, *in vitro* studies showed that La^{3+} improved the proportion of embryos developing into blastocysts. Similar effects have been demonstrated for *Xenopus* oocytes (Schorderet-Slatkine *et al.*, 1976) and for mouse oocytes (Whittingham, 1980). On the other hand, the peri-implantation and near-term periods of pregnancy were particularly sensitive to La^{3+}. Intratesticular injections of lanthanides in goats produce calcification and sterility (Sharma *et al.*, 1973).

8.6 Lanthanide-Induced Calcification

Subcutaneously or intradermally injected aqueous solutions of $LnCl_3$ induce calcification at the site of injection (Selye, 1962; Haley and Upham, 1963). This process is known as calcergy. All of 14 Ln^{3+} ions and Y^{3+} tested by Garrett and McClure (1981) were calcergens in mice. This property is shared by a variety of other metal ions, including Cd^{3+}, Pb^{2+}, and Zn^{2+} (Gabbiani and Tuchweber, 1970). Histological examination of the calcification site by Garrett and McClure (1981) revealed close association of mineral with the collagen fibers of the dorsal fascia. Deposits were surrounded by mild fibrosis, with a peripheral accumulation of multinucleated giant cells. The granulomas formed following intradermal injection of La^{3+} also contained histiocytes and giant cells surrounding the mineral crystals (Haley and Upham, 1963).

McClure (1980) has provided a chronology of calcification following the injection of 200 μg of $CeCl_3$ or $LaCl_3$ into mouse skin. Within 1 h of injection, there was fluid accumulation in the dorsal fascia with mild inflammatory changes. This was superseded by a phase of acute inflammation which persisted for 2 days. By 18 h, mast cells in the dorsal fascia had degranulated, and the local capillaries had dilated. The first signs of calcification occurred at 18–24 h and were maximal after 3 days. At this time, there were numerous giant cells at the site of reactions.

Lanthanides also promote calcification at other sites of topical application. Pulmonary calcification follows inhalation of particles of Gd_2O_3 by mice (Ball and Van Gelder, 1966), while contact of lanthanide solutions with the denuded corneas of rabbits leads to their calcification and opacification (Grant and Kern, 1956). Opacification of the rabbit cornea began several hours after contact with lanthanides and proceeded gradually over a period of 2–3 weeks. This did not occur in intact corneas. Denuded, excised corneas which had been exposed to lanthanides had also reduced capacity to absorb water *in vitro*. Intratesticular injection of lanthanides causes calcification of the seminiferous tubules and the interstitium (Sharma *et al.*, 1973).

Intravenous injection of 8–15 mg of $LnCl_3$ into rats produced splenic calcification. All lanthanides, apart from $LaCl_3$, were active in this respect (Gabbiani *et al.*, 1966). Calcified material was restricted to the marginal zone and red pulp of the spleen, with a moderate fibrous reaction. Deposits of calcified material within hypertrophied Kupffer cells were also noted in the liver. In addition to promoting calcification of the spleen and liver, i.v.-injected lanthanides potentiate calcification by orthophosphate, but, unlike Pb^{2+}, they are not histamine liberators (Gabbiani and Tuchweber, 1965). The dose–response relationship is not linear. While small doses of $LnCl_3$ promote indirect calcification, high doses do not.

Sodium pyrophosphate inhibits both cutaneous and splenic calcification produced by lanthanides (Gabbiani *et al.*, 1966). However, mortality due to lanthanide injections is not affected by pyrophosphate. The mechanism through which pyrophosphate inhibits calcification is unknown. Of relevance are the antagonism that pyrophosphate exerts on the growth of apatite crystals *in vitro* (Fleisch and Neuman, 1961) and the affinity of phosphate groups for lanthanides (Section 2.3). In addition, Tuchweber and Gabbiani (1967) have shown that sodium pyrophosphate induces transient hypocalcemia without affecting excretion of calcium through the kidney.

Intravenous injections of solutions of $LnCl_3$ produce hypercalcemia and hyperphosphatemia in rats (Johansson *et al.*, 1968; Table 8-7). Again, the dose–response relationship is not linear, with high doses being ineffective. Hypercalciuria and hyperphosphaturia are not observed. Injections of $LnCl_3$ reduce the hypercalcemia and hyperphosphatemia of thyroparathyroidectomized rats. Johansson *et al.* (1968) screened a variety of lanthanides and other metal ions but found no correlation between their effects on serum calcium and phosphate levels and their direct calcifying activity. These authors also speculated that displacement of calcium and phosphorus from the bone mineral by bone-seeking lanthanides might explain the observed changes in serum concentration. However, the serum concentrations of calcium and phosphate increase within 1–3 h of lanthanide injection and then decline, whereas the skeletal accumulation of lanthanides is slower and progressive (Section 7.5). Furthermore, as the heavier lanthanides have the greater bone-seeking activity, they should be better inducers of hypercalcemia and hyperphosphatemia. This is not

Table 8-7. Effects of Intravenously Injected CeCl$_3$ on
Serum Calcium and Phosphate in Rats[a]

Approx. dose (mg/kg)	Serum calcium (mg %)	Serum phosphate (mg %)
0	10.0 ± 0.09	8.4 ± 0.12
5	11.3 ± 0.60	9.9 ± 0.20[c]
10	12.2 ± 0.30[b]	11.9 ± 0.30[b]
30	16.1 ± 0.40[b]	14.8 ± 0.60[b]
60	18.1 ± 0.60[b]	16.9 ± 0.90[b]
110	11.6 ± 0.90[c]	10.8 ± 0.60[c]
150	9.7 ± 0.20	9.8 ± 1.40

[a] Adapted from Johansson et al. (1968).
[b] $P < 0.01$.
[c] $P < 0.001$.

seen. Stimulation of the reticuloendothelial system with zymosan, BCG (Bacillus Calmette Guérin), or triolein protects against splenic calcification by GdCl$_3$. Suppression of the reticuloendothelial system with methyl palmitate has a weaker anticalcergenic effect (Lazar, 1973a).

The mechanism of both direct and indirect calcergy in response to lanthanides remains unknown.

8.7 Lanthanides and the Liver

8.7.1 General Considerations

From the discussion of Section 7.5, it is clear that the liver sequesters a considerable proportion of the lanthanides that become systemically available. This is especially so for the lighter lanthanides. Evidence of hepatotoxicity is provided by the increase in the serum levels of ornithine carbamoyl transferase (OCT), glutamic-pyruvic transaminase (GPT), sorbitol dehydrogenase (SDH), and glutamic-oxalacetic transaminase (GOT) (Magnusson, 1963; Tuchweber et al., 1976a,b; Marciniak and Baltrukiewicz, 1977; Strubelt et al., 1980). Serum levels of OCT begin to rise within a few hours of the injection of lanthanides into rats, peak at 1–2 days, and return to normal by 6–10 days (Magnusson, 1963). In this respect, they reflect the time course for the development and regression of the fatty liver, hypoglycemia, and the inhibition of microsomal enzyme activity (Sections 8.7.2, 8.7.3, and 8.7.4). Within the liver, the activities of nonspecific esterase, acid phosphatase, phosphorylase, lactate dehydrogenase, malate dehydrogenase, and succinate dehydrogenase all decrease during the first two hours of lanthanum treatment (Kadas et al., 1974a). One of the mysteries of Ln^{3+}-induced hepatotoxicity is the way in which liver function spontaneously returns to normal, despite a continued high hepatic concentration of lanthanides. Part of the reason presumably resides with the behavior of extrahepatic tissues. The dose–response curve for GPT and GOT resembles the mortality curves (Fig. 8-1) for rats injected with $Pr(NO_3)_3$ (Tuchweber et al., 1976a). Hepatotoxicity is also indicated by measurement of bromosulfthaleine retention in the serum (Strubelt et al., 1980). ATP levels are strongly depressed in the livers of lanthanide-treated animals (Salas et al., 1976).

There has been much speculation that the primary mode of action of lanthanides is through inhibition of transcription (Oberdisse et al., 1974; Sarkander and Brade, 1976; Grajewski et al., 1977; Arvela, 1979). Intravenous injection of the lighter lanthanides reduces the activities of RNA polymerases I and II in "run-on" assays with isolated hepatic nuclei (Sarkander and Brade, 1976). The effect decreases as the atomic number of the lanthanides increases in the order $Pr > Nd > Sm$. The heavier lanthanides Gd, Dy, and Er have no effect on the activity of polymerase I but enhance polymerase II activity. This dichotomy between heavy and light lanthanides mirrors their respective abilities to provoke the accumulation of lipid in the liver (Section 8.7.2). With the lighter lanthanides, there is a slight stimulation in both polymerase activities, peaking 3 h after injection of $Pr(NO_3)_3$, with a subsequent pronounced inhibitory phase, which was maximal after 12 h. Treatment with Pr^{3+} also reduces the activity of "free," nuclear RNA polymerase unattached to the chromatin and reduces the template activity of isolated chromatin. This may be coupled to a decreased rate of histone acetylation.

While provocative, these results do not indicate whether the effects on transcription were primary or secondary. Lanthanides do indeed bind strongly to DNA, RNA, and nucleotides in vitro (Chapter 5) and, at the quite high dose of 1 mM, modestly inhibit transcription within isolated nuclei (Novello and Stirpe, 1969). This does not necessarily occur in vivo. In particular, a large body of evidence suggests that lanthanides cannot normally penetrate the plasmalemmae of intact, living cells (Section 6.2.3). It is possible that lanthanides could enter hepatocytes after damage to the plasmalemma, in association with a carrier ligand or, as small colloidal particles, enter through pinocytosis. However, the lanthanides would still require transport to the nucleus in rather large quantities to fulfill the requirements of this hypothesis.

The hepatotoxic effects of lanthanides can be prevented by a number of substances. Phenobarbital, β-naphthoflavone (Arvela and Karki, 1972;

Arvela *et al.*, 1981), pregnenolone-16α-carbonitrile, and spironolactone do so by promoting the activities of microsomal, drug-metabolizing enzymes (Section 8.7.3). Othes, such as silybin (Tuchweber *et al.*, 1976b), tryptophan, adenine, and ATP (Salas *et al.*, 1976) have unknown mechanisms. Dexamethasone also prevents the accumulation of fats in the liver but paradoxically enhances the lethality of lanthanides (Salas and Tuchweber, 1976). Starvation and pretreatment with smaller doses of lanthanides are other ways to prevent liver damge (Oberdisse *et al.*, 1979). The preconditioning effect is presumably not mediated through metallothionein, as this protein sequesters metals through sulfhydryl groups for which lanthanides have little affinity (Section 2.3). In agreement with this conclusion, yeast strains which express metallothionein constitutively do not have increased resistance to La^{3+} (Ecker *et al.*, 1986). Suppression of the reticuloendothelial system with methylpalmitate prevents the development of the fatty liver, while promoters of reticuloendothelial function have no effect (Lazar, 1973a).

Specific aspects of lanthanide-induced hepatotoxicity are described in the remainder of this section.

8.7.2 The Rare Earth Fatty Liver

The most obvious hepatic lesion that follows i.v. injection of one of the lighter lanthanides is the formation of a fatty liver. This was first recorded by Kyker *et al.* (1957), who observed blanching of the livers of rats injected with $CeCl_3$. Further research revealed that a fatty infiltrate had caused the color change. It is now known that fatty infiltration may be induced in experimental animals by a wide variety of toxic substances or nutritional deficiencies (Lombardi and Recknagel, 1962). Despite the heterogeneity of these factors, the lipids accumulating in the liver are always predominantly triglycerides. However, in contrast to other such agents, lanthanides also produce increases in the triglyceride and phospholipid content of mitochondria (Neubert and Hoffmeister, 1960; Von Lehmann *et al.*, 1975). The phospholipid content of the cytoplasm is also elevated (Von Lehmann *et al.*, 1975).

Snyder *et al.* (1960a) found that the lipid content of fresh liver rose from about 6% in control animals to nearly 16%, 48 h after a single i.v. injection of a 2-mg/kg dose of $CeCl_3$ into female rats (Fig. 8-2). This response occurred after injecting the chlorides of La^{3+}, Ce^{3+}, Pr^{3+}, Nd^{3+}, or Sm^{3+} but not Gd^{3+}, Dy^{3+}, Ho^{3+}, Lu^{3+}, or Y^{3+}. The response of males was much weaker, a phenomenon possibly related to their lower hepatic sequestration of lanthanides (Schmantz, 1964). Rats, mice, hamsters, and

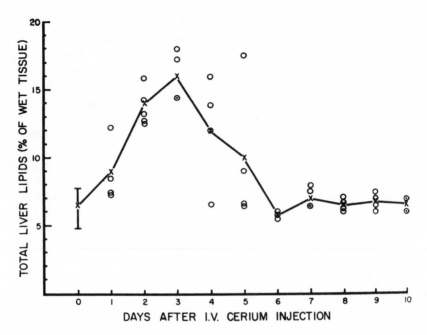

Figure 8-2. Total liver lipids as a function of time after intravenous Ce^{3+} (2 mg/kg) injection into rats. $\bigcirc$, Individual values; $\times$, mean values. From Snyder *et al.* (1960a), with permission.

some rabbits developed a fatty liver in response to $CeCl_3$, whereas guinea pigs, chickens, and dogs did not (Snyder *et al.*, 1960a). Neither choline nor methionine provided protection from the fatty infiltrate (Snyder *et al.*, 1959). The enormity of this hepatic effect, its rapid onset, and the ease with which the larger lanthanides induce it have led to the use of the rare earth fatty liver as an experimental model system of liver damage.

The majority of the lipids which accumulate in the rare earth liver are triglycerides. No differences in the concentration of liver cholesterol or phospholipid per unit weight have been found (Snyder *et al.*, 1959). However, when expressed on the basis of liver protein, there is a 50% increase in the concentration of hepatic phospholipids (Von Lehmann *et al.*, 1975). This increase is almost entirely due to increases in phosphatidylcholine and phosphatidylethanolamine. Subcellular fractionation revealed that although the phospholipid concentration of the smooth endoplasmic reticulum increased, that of the rough endoplasmic reticulum fell. This drop could be entirely accounted for by a marked decrease in lecithin content.

Although male rats were much poorer responders than female rats,

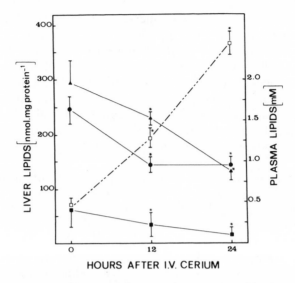

Figure 8-3. Changes in the concentrations of liver triglycerides (□), plasma triglycerides (■), plasma phospholipids (●), and cholesterol (▲) after intravenous injection of $CeCl_3$ into rats. From Renaud *et al.* (1980a), with permission.

castration increased their response to the female level, while testosterone reduced the extent of fatty infiltration in females. Hypophysectomy protected both females and males from developing a fatty liver. In the strain of rat used by Snyder *et al.* (1959), adrenalectomy abolished the fatty hepatic infiltrate in males only. However, in the rats examined by Renaud *et al.* (1980b), adrenalectomy also protected females. Furthermore, in the rats used by Arvela *et al.* (1977), males also developed a fatty liver in response to Ce^{3+}.

The plasma concentrations of free fatty acids begin to increase at about the same time as the fatty liver develops. Free fatty acids are not released if Ce^{3+} is injected in association with a strong chelating agent, such as EDTA or ATP, or when particles of $Ce(OH)_3$ are used. Under these conditions, the development of the fatty liver is also suppressed (Snyder and Stephens, 1961). As the levels of free fatty acids increase, there is a concomitant decrease in the plasma concentration of triglycerides, cholesterol, and phospholipids (Renaud *et al.*, 1980a; Grajewski *et al.*, 1977) (Fig. 8-3). Most of the phospholipid changes can be accounted for by phosphatidylcholine, while the decrease in cholesterol concentration largely reflects the levels of esterified cholesterol. According to Grajewski *et al.* (1977) and Arvela *et al.* (1981), the reduction in cholesterol is transient, with cholesterol levels increasing threefold by day 7. Coupled to these changes is a fourfold increase in the lipase activity in adipose tissue (Renaud *et al.*, 1980b), which presumably releases the free fatty acids into the bloodstream. However, La^{3+} has no effect on the rate of basal or hormone-stimulated lipolysis in isolated rat adipocytes (Schimmel, 1978). As Ce^{3+} has no direct effect on lipase extracted from adipose tissue, it presumably affects lipid mobilization indirectly. The uptake of free fatty acids by liver is directly related to their serum concentrations, indicating one possible component in the development of the fatty infiltrate.

Normal liver responds to an increased influx of fatty acids from the plasma by secreting low-density lipoproteins. However, in Ce^{3+}-treated rats, the plasma lipid concentrations did not increase (Renaud *et al.*, 1980a), suggesting a defect in their synthesis or secretion, or both. To test this, [^{3}H]oleate and [^{14}C]leucine were injected into rats as precursors of lipoproteins. Incorporation of oleate into liver triglyceride increased threefold within 12 h of Ce^{3+} injection, whereas phospholipid synthesis remained constant. Santos *et al.* (1982) have studied lipid biosynthesis *in vitro* by slices of liver from rats injected intramuscularly with Eu^{3+}, the heaviest lanthanide to produce a fatty liver. With [^{14}C]acetate as a precursor, synthesis of phospholipids and ketone bodies was unaltered. Although the cholesterol content of the livers was normal, slices of liver incorporated slightly greater amounts of acetate into cholesterol after injection of Eu^{3+}. Synthesis of triacylglycerols was reduced, but the acylglycerol content of the livers was higher.

Incorporation of oleate into serum very-low-density lipoprotein (VLDL) and low-density lipoprotein (LDL) triglycerides was reduced by Ce^{3+}, indicating reduced secretion by the liver (Renaud *et al.*, 1980a). However, incorporation into serum VLDL and LDL phospholipids was unchanged, although labeling of serum high-density lipoprotein (HDL) phospholipids was reduced. In addition, incorporation of [^{14}C]leucine into the protein moieties of the VLDL, LDL, and HDL in serum was inhibited by Ce^{3+}. This may reflect a general depression of protein synthesis by treated liver, as labeling of liver and plasma TCA (trichloroacetic acid)-precipitable proteins as a whole was also reduced. This may explain why, according to Arvela *et al.* (1977), the concentration of protein in the liver falls after Ce^{3+} injection. However, Von Lehmann *et al.* (1975) reported constant hepatic protein levels.

Removal of triglycerides from the plasma is inhibited by i.v. injection of the detergent triton. This permits the hepatic secretion of triglycerides to be estimated experimentally. In two such studies (Lombardi and Recknagel, 1962; Grajewski *et al.*, 1977), but not a third (Snyder and Kyker,

1964), the findings demonstrated that Ce^{3+} inhibited secretion of triglycerides by the liver. According to Grajewski *et al.* (1977), the alterations in the lipid pattern could be largely explained by diminished lecithin-cholesterol acetyltransferase activity, a prediction confirmed by subsequent studies (Godin and Frohlich, 1981).

Livers of Ce^{3+}-treated rats also degrade lipids more slowly. Studies of the respiratory quotient (Renaud *et al.*, 1980a) suggested that 20% less lipid was oxidized within 8 h of i.v. injection of Ce^{3+}. This may well be the earliest alteration in liver metabolism after Ce^{3+} injection. The results of Snyder *et al.* (1960b) and Glenn *et al.* (1962) suggested that inhibition of the fatty acid oxidase system was the reason. Both groups showed almost complete loss of ability to oxidize octanoic acid by mitochondria isolated from Ce^{3+} livers. Oxidative phosphorylation was also severely inhibited, although ATPase activity was actually greater in the liver mitochondria of Ce^{3+}-treated rats (Glenn *et al.*, 1962). These metabolic changes presumably account for the drastic fall in hepatic ATP concentrations after lanthanide injection (Salas *et al.*, 1976).

Thus, the production of the rare earth fatty liver involves increased uptake of fatty acids from the plasma, decreased oxidation of lipids by the mitochondria, and decreased synthesis and secretion of lipoproteins. However, the underlying biochemical disturbances are not well understood. Several lines of evidence suggest that extrahepatic organs are involved. Endocrine glands possibly play a key role, as castration, adrenalectomy, testosterone, and hypophysectomy all strongly influence the development of the fatty infiltrate. The adrenals are worthy of closer scrutiny, as the results of Renaud *et al.* (1980b) indicate that the liberation of free fatty acids from adipose tissue after Ce^{3+} injection is a response to adrenal hormones. In this regard, the finding of Borowitz (1972) that La^{3+} stimulates catecholamine release from isolated bovine adrenal glands (Section 6.4.8) is highly provocative. The notion that this also occurs *in vivo* is supported by a substantial decrease in the catecholamine concentration of rat adrenals following i.v. injection of Ce^{3+} (Arvela and Karki, 1972). Certain tracer studies have confirmed the accumulation of lanthanides by the adrenals (Section 7.5.2). Lanthanide stimulation of catecholamine release by isolated adrenal glands only occurred once (Borowitz, 1972). Further rounds of catecholamine release were inhibited by La^{3+} (Section 6.4.8). Thus, a hypothesis based upon adrenal stimulation also helps explain a puzzling feature of the rare earth fatty liver: the condition spontaneously reverts to normal within 6–12 days, even though the hepatic concentration of lanthanides remains high throughout. In further support of this idea are the autoradiographic studies of Ewaldsson and Magnusson (1964a,b) indicating that the lighter lanthanides, which produce a fatty liver, have a greater tendency to accumulate in the adrenals than the heavier lanthanides, which do not (Section 7.5.2).

However, an explanation resting upon adrenal stimulation alone does not explain all aspects of the fatty liver. It does not, for instance, readily explain the sex difference nor the effects of testosterone, castration, and hypophysectomy, nor does it easily account for the differences in the responses of different strains of rat to adrenalectomy. Another noteworthy, but unexplained, finding is that the rare earth fatty liver does not develop in alloxan diabetic rats (Snyder and Stephens, 1961). Graul and Hundeshagen (1959) have reported an unusually high retention of ^{90}Y by the pancreas of guinea pigs. However, guinea pigs have not been found to develop a fatty liver in response to lanthanides. Furthermore, Y^{3+} behaves as a heavy lanthanide in rats and does not induce a fatty infiltrate in the liver.

Presumably other agencies are at work to decrease hepatic lipid oxidation, synthesis, and secretion. These could well have a direct toxic effect on the liver itself.

8.7.3 Drug Metabolism and Lipid Peroxidation

An important function of the liver is to detoxify foreign compounds. It does this through a battery of microsomal enzymes which catalyze oxidation, reduction, hydrolysis, and conjugation reactions. Arvela and Karki (1971) found that i.v. injection of Ce^{3+} reduced the drug-metabolizing activities of the liver. This decline began on the first day of injection, was maximal at about 2–3 days, and spontaneously recovered by 6 days. In this respect, the time course resembled that of the appearance of liver-specific enzymes in the serum (Section 8.7.1), the development of the fatty liver (Section 8.7.2), and hypoglycemia (Section 8.7.4). The activity of cytochrome P-450, in particular, was extremely low in the livers of Ce^{3+}-treated animals. This is an important finding, as this cytochrome is thought to catalyze the rate-limiting step of drug metabolism by liver microsomes. Reduction in the levels of cytochromes P-450 and b_5 occurred in both the rough and smooth endoplasmic reticulum, although activities in the rough endoplasmic reticulum were the more strongly depressed (Arvela et al., 1980a). As cytochrome P-450 activity is dependent upon phospholipids, the selective loss of lecithin from the rough endoplasmic reticulum may account for the particularly low activity in this organelle (Von Lehmann et al., 1975). Similar mechanisms may account for the loss of activity of benzpyrene hydroxylase, aniline hydroxylase, and NADPH cytochrome c reductase (Von Lehmann et al., 1976). Ce^{3+}, but not Er^{3+}, was shown to inhibit the binding of cytochrome P-450 to both hexobarbital and aniline (Arvela et al., 1977).

Certain substances are able to induce the hepatic enzyme system responsible for the metabolism of drugs. Pretreatment of rats with one such compound, phenobarbital, strongly inhibited the effects of Ce^{3+} (Arvela and Karki, 1971). Another such drug, spironolactone, reduced the mortality and hepatotoxicity that followed i.v. injection of Ce^{3+} (Bjondahl et al., 1973). Silybin, a hepatoprotectant drug of unknown mechanism, also reduced the lanthanide-induced decline in drug-metabolizing enzyme activity (Strubelt et al., 1980). Hepatic drug-metabolizing enzymes convert CCl_4 into an active, hepatotoxic derivative. Paradoxically, treatment with Ce^{3+} protects the liver from CCl_4 (Sanna et al., 1976). Conversely, liver damage induced by CCl_4 decreases the uptake of $^{144}Ce^{3+}$ by the liver, with a corresponding increase in its uptake by the spleen (Trnovec et al., 1974). A similar phenomenon may occur in alcoholics (Section 7.2; Table 7-5).

Lanthanides inhibit lipid peroxidation by liver slices in vitro (Arvela, 1973). In vivo, however, a marked stimulation of liver lipid peroxidation occurs 1–6 h after injection of mice with Ce^{3+} (Arvela, 1979). An inhibitory phase follows 24–72 h later. The authors suggested that the early enhancement of lipid peroxidation was a result of membrane damage caused by Ce^{3+}. Oxidative damage to cells can be minimized in a number of ways. One makes use of glutathione, and another relies upon the activities of superoxide dismutase and catalase. Levels of both glutathione and oxidized glutathione were reduced in the livers of chicks injected 4–8 h earlier with $LaCl_3$ solution (Basu et al., 1984). Glutathione reductase activity was also reduced. Administration of cysteine normalized the concentration of oxidized glutathione and increased the activity of glutathione reductase, but it had no effect on the amount of hepatic glutathione. As found for Ce^{3+}-injected rats, the livers of La^{3+}-injected chicks showed stimulated rates of lipid peroxidation. Glutathione peroxidase activity was reduced, catalase activity enhanced, and superoxide dismutase activity slightly lower in these livers. In view of the possibility that peroxidation may increase soon after injection of lanthanides, it is interesting to note that α-tocopherol, an antioxidant, protected rats against certain aspects of Ce^{3+}-induced hepatotoxicity (Arvela, 1974).

8.7.4 Hepatic Gluconeogenesis

The pronounced hypoglycemia that follows i.v. injection of lanthanides was first described by Fischler and Roeckl (1938). All other reports

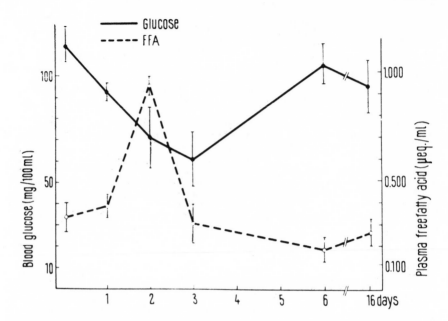

Figure 8-4. Effect of Ce^{3+} (2 mg/kg) on blood glucose and plasma free fatty acid levels in rats. From Arvela and Karki (1971), with permission.

except one (Mori, 1931) confirm this. Species differences may exist, as Mori (1931) used rabbits, while others have used rats or mice. The decrease in blood glucose begins within 6 h, is maximal at 2–3 days, and then, like several other lanthanide-induced changes, recovers by 6 days (Fig. 8-4).

In overall terms, hypoglycemia could result from decreased gluconeogenesis or elevated glucose catabolism. Rohling (1974) showed that intravenously administered glucose was removed from the blood of Pr^{3+}-treated animals at a slower rate than normal. One hour after injection of a 3-mg/kg dose of Pr^{3+}, the insulin concentration of the portal vein was doubled. However, as normal insulin values had been restored by 6 h, hypoglycemia could not be explained on this basis. Indeed, by 48 h, the pancreas was no longer secreting insulin in response to glucose (Oberdisse et al., 1973). Further evidence that insulin does not mediate the observed hypoglycemia is the observation that lanthanides increase the concentration of free fatty acids in the blood (Snyder and Stephens, 1961; Fig. 8-4). These results thus turned attention towards inhibition of gluconeogenesis as an explanation of lanthanide-induced hypoglycemia.

Schurig and Oberdisse (1972) measured the rates of gluconeogenesis

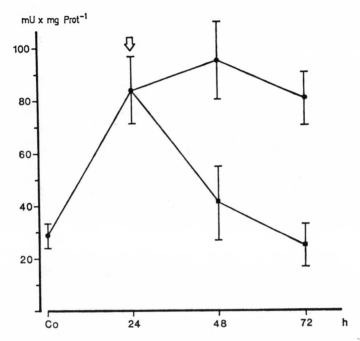

Figure 8-5. Effect of Pr^{3+} (3 mg/kg) on the activity of phosphoenolpyruvate carboxykinase. PrCl$_3$ was injected at the time indicated by the arrow. ●, Control; ■, Pr^{3+} treated. From Schurig and Oberdisse (1972), with permission.

from pyruvate by slices of liver. Livers from rats which had been injected with Pr^{3+} (3 mg/kg) 24 h earlier synthesized only 17% of the control levels of glucose. This was presumably a secondary effect, as direct addition of $10^{-4} M$ Pr(NO$_3$)$_3$ to control liver slices had no effect on net glucose production. Gluconeogenesis by slices of kidney was not inhibited. This presumably reflects the poor uptake of lanthanides by kidney (Section 7.5). When specific, key enzymes involved in gluconeogenesis were measured, the activities of pyruvate carboxylase and phosphoenolpyruvate carboxykinase were strongly reduced (Fig. 8-5). The activities of glucose-6-phosphatase and fructose-1,6-diphosphatase were only slightly affected, while glycerophosphate dehydrogenase, fructose-1,6-disphosphatase, aldolase lactate dehydrogenase, glyceraldehydephosphate dehydrogenase, and phosphofructokinase were unchanged. Pyruvate kinase and glucose-6-phosphate dehydrogenase increased. Starvation for 24 h before injection of Pr(NO$_3$)$_3$ inhibited the effect on blood glucose levels. This did not occur if food was withdrawn at the time of injection. Prestarved rats not only resisted lanthanide-enhanced hypoglycemia but had lower mortality rates and developed less fatty infiltration of the liver.

8.7.5 Morphological Changes

Detailed histological and electron microscopic studies have been made on the structural hepatic changes produced by lanthanides. All agree that the major structural changes are seen in the endoplasmic reticulum of the hepatocytes, which undergoes marked hypertrophy.

According to Magnusson (1963), fatty degeneration could be seen within a day following injection of a light lanthanide. By this time, small areas of intralobular necrosis were visible, but all the glycogen granules had disappeared. Kadas *et al.* (1974a) and Salas *et al.* (1976) also reported the disappearance of glycogen. With the heavy lanthanides, which do not produce a fatty liver (Section 8.7.2), few lipid droplets were seen. However, there were areas of necrosis with histiocytic and granulocytic infiltration from 4 days onwards. At 6–10 days, multinucleated giant cells were visible.

Under the electron microscope, the cytoplasm was seen to contain swollen mitochondria, lipid droplets, and vacuoles. According to Tuchweber *et al.* (1976a), the vacuoles contained a low-density, homogeneous material and were limited by a single membrane. Occasional small vacuoles containing a very dense inclusion were seen. Mitochondria were swollen, and their cristae had partly disappeared (Magnusson, 1963; Kadas *et al.,* 1974b). This may be related to their inability to oxidize fatty acids (Section 8.7.2). Nuclei became irregular in shape and showed degenerative changes, including granularity of the chromatin. The most striking morphological changes occurred in the endoplasmic reticulum. Magnusson (1963) noted that during the first day, there was extensive dissociation of ribosomes from the rough endoplasmic reticulum in large areas of the cells. Abundant free ribosomes were seen in the cytoplasm (Salas *et al.,* 1976). The smooth endoplasmic reticulum meanwhile underwent marked hypertrophy. According to Tuchweber *et al.* (1976a), the Golgi contained enlarged saccules, although Kakas *et al.* (1974b) reported it to be normal. The marked morphological changes of the endoplasmic reticulum are important with regard to the prominent changes in the activities of the microsomal enzymes that lanthanides induce (Section 8.7.4).

Hepatic lesions produced by $CeCl_3$ regressed spontaneously in 5–8 days despite the continued presence of Ce in the liver. Salas *et al.* (1976) suggested that at this time the Ce is no longer in the hepatocytes but in the Kupffer cells.

At high doses, i.v.-injected Pr^{3+} loses its toxicity (Tuchweber *et al.*, 1976a; Fig. 8-1). In accordance with the hypothesis that precipitates form at these higher doses, the morphology of the hepatocytes remains quite normal, while dense material accumulates in the Kupffer cells. According to unpublished data cited by Tuchweber *et al.* (1976a), these deposits contain Pr.

8.8 Blood: The Anticoagulant Properties of Lanthanides

Several of the biochemical changes that affect the blood of lanthanide-injected animals reflect alterations in the metabolism of the liver and other organs. Thus, reduced hepatic gluconeogenesis produces hypoglycemia, while alterations in lipid metabolism increase serum levels of free fatty acids and reduce levels of cholesterol, phospholipids, and triglycerides. In addition, hepatotoxicity liberates various cytoplasmic hepatocyte enzymes into the blood (Section 8.7.1). The biochemical changes in the serum which follow i.v. administration of lanthanides are listed in Table 8-8.

The most marked intrinsic change that lanthanides produce in blood is prolongation of the clotting time. Discovered at the Pharmacological Institute of the University of Florence in the 1920s, this occurs both *in vitro* and *in vivo*. Indeed, lanthanides have been used clinically as antithrombotic agents (Section 9.4). In rabbits, increased clotting times are seen after i.v. injection of $Nd(NO_3)_3$ at doses greater than 10 mg/kg. Blood withdrawn from rabbits 1 or 2 h after injection of a 60-mg/kg dose of Nd^{3+} failed to clot at all (Beaser *et al.*, 1942). The anticoagulant effect lasted for at least 4 h after a single i.v. injection of Nd^{3+}. Similar results were obtained with humans. A number of complexes of the lanthanides, such as nicotinate, 3-sulfoisonicotinate, and 3-acetylpropionate, also exhibit anticoagulant activity. However, Dorovini-Zis *et al.* (1983) stated that colloidal lanthanum rapidly clots blood; no experimental evidence was presented.

The anticoagulant mechanism remains largely unknown. However, it probably reflects the ability of lanthanides to inhibit certain Ca^{2+}-requiring enzymic reactions involved in blood clotting. A simplified scheme of the blood clotting cascade is shown in Fig. 8-6. Ca^{2+} is required for at least four reactions, namely, the activation of factors IX, and X, the conversion of prothrombin to thrombin, and the cross-linking of the fibrin clot. Biochemical studies have confirmed the ability of lanthanides to inhibit the activation of factor X and the conversion of prothrombin to

Table 8-8. Changes in Blood Biochemistry Produced by Lanthanides

Substance[a]	Change[b]	Comment	Reference
Free hemoglobin	+	Probably reflects erythrocyte fragility	Beaser et al., 1942
Ca^{2+}	+	See Table 8-7	Johansson et al., 1968
Phosphate	+	See Table 8-7	
Bilirubin	+		Salas et al., 1976
Bilirubin	0		Arvela et al., 1981
Albumin	0		Arvela et al., 1981
Urea	0		Arvela et al., 1981
Urate	0		Arvela et al., 1981
K	0		Arvela et al., 1981
Alkaline phosphatase	0		Arvela et al., 1981
γ-Glutamyl transpeptidase	0		Arvela et al., 1981
OCT	+	See Section 8.7.1	Salas et al., 1976
GPT	+		Magnusson, 1963
SDH	+		Tuchweber et al., 1976b
GOT	+		Marciniak and Baltrukiewicz, 1977
			Strubelt et al., 1980
Glucose	−	See Section 8.7.4	Fischler and Roeckl, 1938
Free fatty acids	+	See Section 8.7.2,	Arvela and Karki, 1971
Triglycerides	−	Fig. 8-4	
Cholesterol	−	See Section 8.7.2	Renaud et al., 1980a,b
Phospholipids	−		
Fibrinogen	+	See Section 8.8	Nagy et al., 1976
Insulin	Early + Late −	See Section 8.7.4	Oberdisse et al., 1973

[a] OCT, ornithine carbamyl transferase; GPT, glutamic-pyruvic transaminase; SDH, sorbitol dehydrogenase; GOT, glutamic-oxaloacetic transaminase.
[b] +, Concentration increased; −, concentration decreased; 0, no change in concentration.

thrombin (Section 4.5). In addition, lanthanides could interfere with blood clotting through inhibition of platelet aggregation (Section 6.4.6).

Whether any of these steps are inhibited *in vivo* is unknown. Vincke (1942) attributed the anticoagulant effect to decreased prothrombin production. However, others (Hunter and Walker, 1956; Beller and Mammen, 1956) suggested impaired activity of factors IX, X, and VII. In ascribing the effects of lanthanides to specific antagonism of Ca^{2+} in blood, Hunter and Walker (1956) became the first to articulate what has subsequently become the cornerstone of lanthanide biochemistry (Section 2.5). More

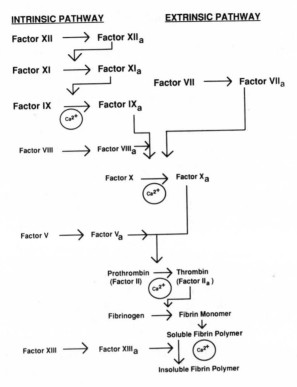

Figure 8-6. Simplified version of the intrinsic and extrinsic systems of blood coagulation.

recently, it has been shown that a single large dose of La^{3+} interferes with prothrombin activity. This blood has a high fibrinogen content and will clot if thrombin is added. Repeated, smaller doses of La^{3+} inhibit coagulation through a different mechanism, possibly involving a disturbance in liver function (Nagy *et al.*, 1976).

8.9 Neurological Effects of Lanthanides

Lanthanides strongly influence the physiology of nervous tissues *in vitro* by interfering with transmembrane Ca^{2+} fluxes (Section 6.3.2). However, marked neurological disturbances do not follow systemic administration of lanthanides because of their poor transport to the nervous system (Section 7.5). For example, subcutaneously or intragastrically injected cerium citrate had little or no effect on various behavioral parameters in mice (Stineman *et al.*, 1978; Morganti *et al.*, 1980; D'Agostino *et al.*, 1982). Nevertheless, i.p.-administered $LaCl_3$ was able to prevent, but not reverse, flaccid paralysis produced by ruthenium red in mice (Tapia, 1982). Presumably, this response was indirect.

However, if lanthanides are injected directly into the central nervous system, their effects are considerable. Best studied is the way in which lanthanides mimic the analgetic effects of opiates (Harris *et al.*, 1975). The effects of La^{3+}, like those of opiates, are antagonized by Ca^{2+} and naloxone. In addition, both La^{3+} and opiates produce differential analgesia for the upper and lower body, and for peripheral and central stimulation (Keresztes-Nagy and Rosenfeld, 1981; Rosenfeld and Hammer, 1983). Furthermore, rats tolerant to morphine are cross-tolerant to lanthanum (Harris *et al.*, 1976). On a molar basis, La^{3+} is about a tenth as effective as morphine in producing antinociceptive responses. Not all the affects of La^{3+} and morphine are identical, however. For instance, La^{3+} decreases motor activity, while morphine increases it. In addition, the effects of morphine are influenced by pretreatment with pargyline or cAMP, while those of La^{3+} are not.

Microinjection studies have shown that the periaqueductal gray region of the brain is the main site of action of both La^{3+} and morphine (Iwamoto *et al.*, 1978). It is thought that both these analgesics work by inhibiting cellular influx of Ca^{2+}. Tracer studies with $^{144}CeCl_3$ have shown that Ce^{3+} remains in the brain long after its analgetic effects have subsided.

Lanthanides also produce antinociceptive responses when injected intrathecally into the lumbar subarachnoid space. Nd and Eu are superior to La in this respect (Reddy and Yaksh, 1980). However, the authors concluded that, in the spinal cord, lanthanides induce analgesia in a manner unrelated to specific action at the opiate receptor. At concentrations above $0.3\,\mu$mol per rat, the lanthanides produced motor weakness in the hind limbs. Subarachnoid injection of $GdCl_3$, Gd-EDTA, or Gd-DTPA also produces lack of motor coordination and epileptoid fits in rats (Weinmann *et al.*, 1984).

8.10 Lanthanides and Reticuloendothelial Function

Lanthanides have a remarkable capacity to depress the reticuloendothelial system (RES) (Lazar, 1973a,b; Husztik *et al.*, 1980; Lazar *et al.*, 1985). Strong inhibition of phagocytosis by the Kupffer cells of the liver occurs with i.v. doses of 2 mg/kg in rats, a concentration well below the LD_{50} value (Section 8.4; Table 8-6). From their ultrastructural studies,

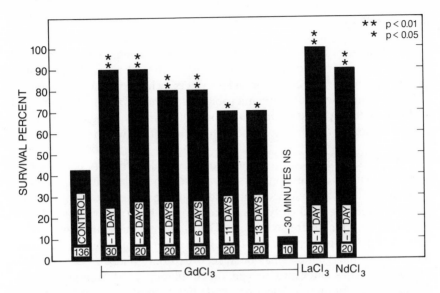

Figure 8-7. Effect of lanthanides on anaphylactic death in mice. Lanthanide chloride solutions (10 mg/kg) were injected at the indicated times before anaphylactic challenge. Redrawn from Lazar *et al.* (1985), with permission.

Husztik *et al.* (1980) suggested that Ln^{3+} ions displace Ca^{2+} from the surface of Kupffer cells, impairing attachment of particles and thus preventing phagocytosis. In view of the perceived importance of RES cells in the metabolism and pharmacology of the lanthanides, it is paradoxical that they should be particularly vulnerable to lanthanides.

Lanthanides suppress RES function not only in control animals, but also in those stimulated with zymosan, triolein, or BCG. This blockage has several consequences. Red cells are now sequestered in the spleen and lungs instead of the liver (Husztik *et al.*, 1980). Animals become much more sensitive to bacterial and yeast infection (Farkas and Karacsonyi, 1985) but are remarkably protected from anaphylactic death. Lazar *et al.* (1985) sensitized mice to ovalbumin before inducing anaphylaxis with an i.v. injection of 100 μg of ovalbumin 12 days later. Whereas more than half the control mice died, over 90% of those pretreated with $LnCl_3$ survived (Fig. 8-7).

Because of the easy and reproducible way in which lanthanides block RES function, they have found experimental use in studies related to RES physiology *in vivo*.

8.11 Summary

Lanthanides inhibit the growth of microorganisms, but the concentrations required to do this are in the 10^{-4}–$10^{-2} M$ range. Even higher concentrations are needed for cidal effects. The toxic mechanisms involved are unknown, although microbial respiration is strongly suppressed by lanthanides. Almost nothing is known of their physiological effects on higher plants, both inhibition and stimulation of growth having been reported.

Orally administered lanthanides are of very low toxicity in mammals, with LD_{50} values in the range of several grams per kilogram body weight. At high, chronic, p.o. doses, lanthanides suppress the growth of rats and mice, but feeding of slightly lower doses over three generations of mice produced no adverse consequences. Gastric hemorrhage and slight structural changes in the hepatocytes have been reported. However, increased incidence of malignancies has been recorded only once.

Lanthanides enter into the general circulation a little more effectively following subcutaneous injection, and their acute toxicities are consequently slightly higher, with minimum lethal doses typically in the range of 100–1000 mg/kg. Contact of lanthanides with intact skin is generally of little consequence. However, abraded skin reacts severely, with the formation of ulcers which heal only slowly. Intradermally injected lanthanides produce granulomas and induce calcification. Sarcomas form at sites of subcutaneously located pellets of metallic lanthanides and yttrium. Lanthanides are ocular irritants which produce swelling, conjunctivitis, and, in extreme cases, ulceration. The ulcers slowly heal without residual damage.

Intraperitoneally injected lanthanides are more toxic, with LD_{50} values typically 50–500 mg/kg. Guinea pigs are more sensitive than rats and mice. Weak chelators, such as citrate, increase i.p. lethality by increasing systemic availability of the lanthanides; strong chelators diminish lethality by facilitating excretion. Death from i.p.-injected lanthanides usually occurs within a few days, being preceded by symptoms which include loss of appetite, diarrhea, and respiratory distress. These are accompanied internally by peritonitis, adhesions, and ascites. Chronic exposure to sublethal i.p. doses of lanthanides does not produce adverse effects.

Inhalation of lanthanides fumes produces few immediate effects. Chronic exposure results in emphysema, pneumonitis, bronchitis, and delayed pulmonary fibrosis. There is one report of increased incidence of pulmonary reticulosarcoma. Rare earth pneumoconiosis has been diagnosed in humans chronically exposed to fumes of lanthanide oxides. Lanthanides show embryotoxicity, producing skeletal deformities and reducing litter size. Intratesticularly injected lanthanides provoke calcification and sterility.

Lanthanides are most toxic when injected intravenously. LD_{50} values for rats and mice range from 3 to 100 mg/kg. Rabbits are a little more resistant, while female rats and mice are much more susceptible than males. However, female quail are much more resistant than male birds. Death following i.v. injection occurs through cardiovascular collapse and respiratory failure. The dose–response curve is bell-shaped, with high doses being of low toxicity. This perhaps reflects precipitation of the lanthanides at high dose. At toxic doses, the major structural changes are observed in the hepatocytes, whose endoplasmic reticulum is particularly disturbed by lanthanides.

With both i.p. and i.v. injections, there is evidence that small, frequent injections of a lanthanide are much better tolerated than single, large doses, even if the cumulative dose eventually greatly exceeds the LD_{50}. Toxicity is also reduced by starvation, by substances which enhance the activities of microsomal, drug-metabolizing enzymes, and by ATP, adenosine, or tryptophan. Alloxan diabetic animals also resist lanthanides better. However, dexamethasone and glucocorticoids enhance the lethality of lanthanides. Glucose, choline, and methionine have no effect. Damage to the liver following injection of lanthanides is signaled by the increase in the serum concentrations of several hepatic enzymes.

Pharmacologic responses to i.v.-injected lanthanides include calcification (calcergy), hypotension, hypoglycemia, and increased clotting times. Lanthanides exhibit both direct calcergy, where calcification occurs at the injection site, and indirect calcergy, where it occurs at distal sites, such as the spleen. Calcification of the eye, skin, lungs, spleen, and testes has been recorded.

The best-studied pharmacologic response to i.v.-injected lanthanides is the rare earth fatty liver. Produced only by the lighter lanthanides, it reflects a massive hepatic accumulation of triglycerides. Plasma concentrations of free fatty acids concomitantly increase, while levels of cholesterol and phospholipids decline. The precise mechanisms which produce the fatty infiltrate are unclear, but they seem to entail increased hepatic sequestration of fatty acids from the plasma, decreased oxidation of lipids by hepatic mitochondria, and decreased hepatic synthesis and secretion of lipoproteins. Decreased hepatic gluconeogenesis is responsible for the hypoglycemic response and the disappearance of glycogen from these livers, whose ATP levels become very low. Lanthanides also reduce the activities of the hepatic microsomal enzymes responsible for metabolizing and detoxifying drugs. One curious feature of these changes in hepatic function is their time course. Each effect begins within a few hours of injection, is maximal at 2–3 days, and then spontaneously resolves by 6–8 days, even though the hepatic concentration of lanthanides remains high.

Lanthanides have marked anticoagulant properties which may result from antagonism of certain clotting reactions which require Ca^{2+}. Much effort was once expended in attempting to put this discovery to clinical use (Section 9.4). Intracranially injected lanthanides share the analgetic properties of opiates. These are blocked by the opiate antagonist naloxone, while rats tolerant to morphine are also cross-tolerant to lanthanides. Both are thought to modify Ca^{2+} fluxes across cellular membranes in specific regions of the brain. At higher doses, lanthanides cause loss of motor function and epileptoid fits. Phagocytic activity by cells of the reticuloendothelial system is markedly depressed following i.v. administration of quite low doses of lanthanides. Animals treated in this way are more susceptible to infection but are protected from anaphylactic death.

References

Abaas, S., 1984. Induction of aggregation in *Streptococcus mitis* by certain ions, *Acta Pathol. Microbiol. Immunol. Scand.* 92:253–259.

Abel, M., and Talbot, R. B., 1967. Gadolinium oxide inhalation by guinea pigs: a correlative functional and histopathologic study, *J. Pharmacol. Exp. Ther.* 157:207–213.

Abramczuk, J. W., 1985. The effects of lanthanum chloride on pregnancy in mice and on preimplantation mouse embryos *in vitro, Toxicology* 34:315–320.

Ajazzi-Mancini, M., 1926. Cited in Haley (1965).

Anon, 1988. RE in fresh-water fishery, *China Rare Earth Info.* 9:3.

Arkhangelskaya, L. N., and Spasskii, S. S., 1967. Cited in Haley (1979).

Arvela, P., 1973. The *in vitro* effects of lanthanons on lipid peroxidation and drug metabolism in rat, *Acta Physiol. Scand. Suppl.* 396:234.

Arvela, P., 1974. The effect of α-tocopherol on cerium induced changes in drug and lipid metabolism of rat, *Experientia* 30:1061–1062.

Arvela, P., 1979. Toxicity of rare-earths, *Prog. Pharmacol.* 2:71–114.

Arvela, P., and Karki, N. T., 1971. Effect of cerium on drug metabolizing activity in rat liver, *Experientia* 27:1189–1190.

Arvela, P., and Karki, N. T., 1972. Effect of phenobarbital pretreatment on cerium induced impairment of drug metabolism in rat liver, *Acta Pharmacol. Toxicol.* 31:380–386.

Arvela, P., Grajewski, O., Von Lehmann, B., and Oberdisse, E., 1977. Effect of lanthanons on substrate-induced difference spectra in rat liver microsomes, *Experientia* 32:491–493.

Arvela, P., Von Lehmann, B., Grajewski, O., and Oberdisse, E., 1980a. Effect of praseodymium on drug metabolism in rat liver smooth and rough endoplasmic reticulum, *Experientia* 36:860–861.

Arvela, P., Oberdisse, E., Merker, H. J., Grajewski, O., and Von Lehmann, B., 1980b. The influence of lanthanons on morphology and function of rat exocrine pancreas *in vivo*, *Pathol. Res. Pract.* 169:330–340.

Arvela, P., Reinila, M., and Pelkonen, O., 1981. Effects of phenobarbital and β-naphthoflavone on cerium induced biochemical changes in rat serum, *Toxicol. Lett.* 8:213–216.

Baehr, G., and Wessler, H., 1909. The use of cerium oxalate for the relief of vomiting: an experimental study of the effects of some salts of cerium, lanthanum, praseodymium, neodymium and thorium, *Arch. Intern. Med.* 2:517–531.

Bahadur, K., and Tripathi, P., 1978. Investigation of cerium and yttrium ions' effects on microbial nitrogen fixation and determination of inhibition, *Zentralbl. Bakteriol. II Abt.* 133:628–631.

Bahadur, K., Prakash, S., and Jyotishmati, U., 1978. Microbial fixation of nitrogen in presence of lanthanum sulphate with sodium molybdate, *Zentralbl. Bakteriol. II Abt.* 133:632–637.

Ball, R. A., and Van Gelder, G., 1966. Chronic toxicity of gadolinium oxide for mice following exposure by inhalation, *Arch. Environ. Health* 13:601–608.

Ball, R. A., Van Gelder, G., and Green, J. W., 1970. Neoplastic sequelae following subcutaneous implantation of mice with rare earth metals, *Proc. Soc. Exp. Biol. Med.* 135: 426–430.

Basu, A., Haldar, S., Chakrabarty, K., Santra, M., and Chatterje, G. C., 1984. Effect of cysteine supplementation on lanthanum chloride induced alterations in the antioxidant defense system of chick liver, *Indian J. Exp. Biol.* 22:432–434.

Beaser, S. B., Segel, A., and Vandam, L., 1942. The anticoagulant effects in rabbits and man of the intravenous injection of salts of the rare earths, *J. Clin. Invest.* 21:447–453.

Beller, F. K., and Mammen, E., 1956. Der angriffspunkt der seltenen erden neodym in Gerinnungssystem, *Arch. Gynaek.* 187:319–336.

Bjondahl, K., Mottonen, M., and Nieminen, L., 1973. The effect of spironolactone on acute toxicity and liver damage in mice induced by cerium chloride, *Experientia* 29:685–687.

Bjondahl, K., Isomaa, B., and Nieminen, L., 1974. Effect of pretreatment with phenobarbital and spironolactone on [144]Ce levels in blood, liver, urine and faeces at various time periods, *Biochem. Pharmacol.* 23:1509–1518.

Bjorkman, S. E., and Horsfall, F. L., 1948. The production of a persistent alteration in influenza virus by lanthanum or ultraviolet irradiation, *J. Exp. Med.* 88:445–461.

Borowitz, J. L., 1972. Effect of lanthanum on catecholamine release from adrenal medulla, *Life Sci.* 11:959–964.

Boyd, K. S., Melnykovych, G., and Fiskin, A. M., 1976. A replica technique for visualizing the distribution of lanthanum-binding sites over HeLa surfaces, *Cytobiology* 14:91–101.

Brooks, M. M., 1921. Comparative studies on respiration. XIV. Antagonistic action of lanthanum as related to respiration, *J. Gen. Physiol.* 3:337–342.

Bruce, D. W., Hietbrink, B. E., and DuBois, K. P., 1963. The acute mammalian toxicity of rare earth nitrates and oxides, *Toxicol. Appl. Pharmacol.* 5:750–759.

Bulman, R. A., and Stretton, R. J., 1975. Effect of the lanthanides lanthanum and neodymium on the heat resistance of *Bacillus cereus* spores, *Microbios* 12:167–174.

Burkes, S., and McCleskey, C. S., 1947. The bacteriostatic activity of cerium, lanthanum and thallium, *J. Bacteriol.* 54:417–424.

Burnett, K. R., Wolf, G. L., Schumacher, H. R., and Goldstein, E. J., 1985. Gadolinium oxide: a prototype agent for contrast enhanced imaging of the liver and spleen with magnetic resonance, *Magn. Reson. Imaging* 3:65–71.

Caillé, J. M., Lemanceau, B., and Bonnemari, B., 1983. Gadolinium as a contrast agent for NMR, *Am. J. Nucl. Med.* 4:1041–1042.

Cassone, A., and Garaci, E., 1974. Lanthanum staining of the intermediate region of the cell wall in *Escherichia coli, Experientia* 30:1230–1232.

Castillo, M. R., and Bibby, B. G., 1973. Trace element effects on enamel pigmentation, incisor growth and molar morphology in rats, *Arch. Oral Biol.* 18:629–635.

Cochran, K. W., Doull, J., Mazur, M., and DuBois, K. P., 1950. Acute toxicity of zirconium, columbium, strontium, lanthanum, cesium, tantalum and yttrium, *Arch. Ind. Hyg. Occup. Med.* 1:637–650.

D'Agostino, R. B., Lown, B. A., Morganti, J. B., and Massaro, E. J., 1982. Effects of in utero or suckling exposure to cerium (citrate) on the postnatal development of the mouse, *J. Toxicol. Environ. Health* 1:449–458.

Davison, F. C., and Ramsey, F. K., 1965. Cited in Haley (1979).

Dorovini-Zis, K., Sato, M., Goping, G., Rapoport, S., and Brightman, M., 1983. Ionic lanthanum passage across cerebral endothelium exposed to hyperosmotic arabinose, *Acta Neuropathol.* 60:49–60.

Drossbach, G. P., 1897. Über den Einfluss der Elemente der Cerund Zircon gruppe auf das Wachstum von Bakterien, *Zentralbl. Bakteriol. Parasitenkd. Abt. 1 Orig.* 21:57–58.

Drouven, B. J., and Evans, C. H., 1986. Collagen fibrillogenesis in the presence of lanthanides, *J. Biol. Chem.* 261:11792–11797.

Dryfuss, B. J., and Wolf, C. G. L., 1906. The physiological action of lanthanum, praseodymium and neodymium, *Am. J. Physiol.* 16:314–323.

Ecker, D. J., Butt, T. R., Sternberg, E. J., Neeper, M. P., Debouck, C., Gorman, J. A., and Crooke, S. T., 1986. Yeast metallothionein function in metal ion detoxification, *J. Biol. Chem.* 261:16895–16900.

Eisenberg, P., 1919. Untersuchungen über spezifische Desinfektionsvorgange. II. Über die Wirkung von Salzen und Ionen auf Bakterien, *Zentralbl. Bakteriol. Parasitenkd. Abt. 1 Orig.* 82:69–208.

Evans, W. H., 1913. The influence of the carbonates of the rare earths (cerium, lanthanum, yttrium) on growth and cell-division in hyacinths, *Biochem. J.* 7:349–355.

Ewaldsson, E., and Magnusson, G., 1964a. Distribution of radiocerium and radiopromethium in mice, *Acta Radiol. Ther. Phys. Biol.* 2:65–72.

Ewaldsson, E., and Magnusson, G., 1964b. Distribution of radioterbium, radioholmium and radioyttrium in mice, *Acta Radiol. Ther. Phys. Biol.* 2:121–128.

Farkas, B., and Karacsonyi, G., 1985. The effect of ketoconazole on *Candida albicans* infection following depression of the reticuloendothelial activity with gadolinium chloride, *Mykosen* 28:338–341.

Fischler, F., and Roeckl, K. W., 1938. Über experimentelle Beeinflussung er Leberfunktionen und der anatomischen Leberstruktur durch Einwirkung seltaer Erden, *Naunyn-Schmiedeberg's Arch. Exp. Pathol. Pharmacol.* 189:4–21.

Fleisch, H., and Neuman, W. F., 1961. Mechanisms of calcification: role of collagen, polyphosphates and phosphatase, *Am. J. Physiol.* 200:1296–1300.

Gabbiani, G., and Tuchweber, B., 1965. Studies on the mechanism of experimental soft tissue calcification, *Can. J. Physiol. Pharmacol.* 43:177–183.

Gabbiani, G., and Tuchweber, B., 1970. Studies on the mechanism of calcergy, *Clin. Orthop.* 69:66–74.

Gabbiani, G., Jacqmin, M. L., and Richard, R. M., 1966. Soft-tissue calcification induced by rare earth metals and its prevention by sodium pyrophosphate, *Br. J. Pharmacol.* 27:1–9.

Gale, T. F., 1975. The embryotoxicity of ytterbium chloride in golden hamsters, *Teratology* 11:289–296.

Garrett, I. R., and McClure, J., 1981. Lanthanide-induced calcergy, *J. Pathol.* 135:267–275.

Gehrcke, E., Lau, E., and Meinhart, O., 1939. The rare earths and the nervous system, *Z. Ges. Naturwiss.* 5:106–107.

Gerhardsson, L., Wester, P. O., Nordberg, G. F., and Brune, D., 1984. Chromium cobalt and lanthanum in lung, liver and kidney tissue from deceased smelter workers, *Sci. Total Environ.* 37:233–246.

Glenn, J. L., Tischer, K., and Stein, A., 1962. Rare-earth fatty liver. I. Octanoate oxidation and energy production, *Biochim. Biophys. Acta* 62:35–40.

Godin, D. V., and Frohlich, J., 1981. Erythrocyte alterations in praseodymium-induced lecithin:cholesterol acyltransferase (LCAT) deficiency in the rat: comparison with familial LCAT deficiency in man, *Res. Commun. Chem. Pathol. Pharmacol.* 31:555–556.

Graca, J. G., Garst, E. L., and Lowry, W. E., 1957. Comparative toxicity studies of stable rare earth compounds. I. Effect of citrate complexing on stable rare earth chloride toxicity, *A.M.A. Arch. Ind. Health* 15:9–14.

Graca, J. G., Davison, F. C., and Feavel, J. B., 1962. Comparative toxicity of stable rare earth compounds. II. Effect of citrate and edetate complexing on acute toxicity in mice and guinea pigs, *Arch. Environ. Health* 5:437–444.

Graca, J. G., Davison, F. C., and Feavel, J. B., 1964. Comparative toxicity of stable rare earth compounds. III. Acute toxicity of intravenous injections of chlorides and chelates in dogs, *Arch. Environ. Health* 8:555–564.

Grajewski, O., Von Lehmann, B., Arntz, H. R., Arvela, P., and Oberdisse, E., 1977. Alterations in rat serum lipoproteins and lecithin-cholesterol-acyltransferase activity in praseodymium-induced liver damage, *Naunyn-Schmiedeberg's Arch. Pharmacol.* 301: 65–73.

Grant, W. M., and Kern, H. L., 1956. Cations and the cornea. Toxicity of metals to the stroma, *Am. J. Ophthalmol.* 42:167–181.

Graul, E. H., and Hundeshagen, H., 1959. Investigations of radioyttrium (Y^{90}) metabolism. Studies on organ-distribution with special regard to the radioautographic method of demonstration and to paper electrophoresis under various experimental conditions, *Int. J. Appl. Radiat. Isotop.* 5:243–252.

Guida, G., 1930. Cited in Haley (1965).

Guo, B., 1985. Present and future of rare earth research in Chinese agriculture, *J. Chin. Rare Earth Soc.* 3:89–94.

Haley, T. J., 1965. Pharmacology and toxicology of the rare earth elements, *J. Pharm. Sci.* 54:663–670.

Haley, T. J., 1979. Toxicity, in *Handbook on the Physics and Chemistry of Rare Earths* (K. A. Gschneidner and L. R. Eyring, eds.), Vol. 4, North-Holland, Amsterdam, pp. 553–585.

Haley, T. J., and Upham, H. C., 1963. Skin reaction to intradermal injection of rare earths, *Nature* 200:271.

Haley, T. J., Raymond, K., Komesu, N., and Upham, H. C., 1961. Toxicological and pharmacological effects of gadolinium and samarium chlorides, *Br. J. Pharmacol.* 17:526–532.

Haley, T. J., Komesu, N., Flesher, A. M., Mavis, L., Cawthorne, J., and Upham, H. C., 1963. Pharmacology and toxicology of terbium, thulium and ytterbium chlorides, *Toxicol. Appl. Pharmacol.* 5:427–436.

Haley, T. J., Komesu, N., Efros, M., Koste, L., and Upham, H. C., 1964. Pharmacology and toxicology of lutetium chloride, *J. Pharm. Sci.* 53:1186–1188.

Hara, S., 1923. Beitrage zur Pharmakologie der Seltenen Erdmetalle. I. Mitteilung: Über das cerium, *Naunyn-Schmiedeberg's Arch. Exp. Pathol. Pharmacol.* 100:217–253.

Harris, R. A., Iwamoto, E. T., Loh, H. H., and Way, E. L., 1975. Analgetic effects of lanthanum: cross-tolerance with morphine, *Brain Res.* 100:221–225.

Harris, R. A., Loh, H. H., and Way, E. L., 1976. Antinociceptive effects of lanthanum and cerium in nontolerant and morphine tolerant-dependent animals, *J. Pharmacol. Exp. Ther.* 196:288–297.

Havron, A., Davis, M. A., Selter, S. E., Paskins-Hurlburt, A. J., and Hessel, S. J., 1980. Heavy metal particulate contrast materials for computed tomography of the liver, *J. Comput. Assist. Tomogr.* 4:642–648.

Heuck, F., and Hoschek, R., 1968. Cer-pneumoconiosis, *Am. J. Radiol.* 104:777–783.

Hunter, R. B., and Walker, W., 1956. Anticoagulant action of neodymium 3-sulfo-isonicotinate, *Nature* 178:47.

Husain, M. H., Dick, J. A., and Kaplan, Y. S., 1980. Rare earth pneumoconiosis, *J. Soc. Occup. Med.* 30:15–19.

Husztik, E., Lazar, G., and Parducz, A., 1980. Electron microscopic study of Kupffer-cell phagocytosis blockage induced by gadolinium chloride, *Br. J. Exp. Pathol.* 61:624–630.

Hutcheson, D. P., Gray, D. H., Verugopal, B., and Luckey, T. D., 1975. Nutritional safety of heavy metals in mice, *J. Nutr.* 105:670–675.

Iwamoto, E. T., Harris, R. A., Loh, H. H., and Way, E. L., 1978. Antinociceptive responses after microinjection of morphine or lanthanum in discrete rat brain sites, *J. Pharmacol. Exp. Ther.* 206:46–55.

Johansson, O., Perrault, G., Savoie, L., and Tuchweber, B., 1968. Action of various metallic chlorides on calcaemia and phosphataemia, *Br. J. Pharmacol. Chemother.* 33:91–97.

Kadas, I., Tanka, D., Keller, M., and Jobst, K., 1974a. Enzyme-histochemical and biochemical study of liver injury induced by lanthanum trichloride, *Acta Morphol. Acad. Sci. Hung.* 22:35–45.

Kadas, I., Lapis, K., and Jobst, K., 1974b. Electron microscopic studies of liver changes induced by lanthanum trichloride, *Acta Morphol. Acad. Sci. Hung.* 22:343–357.

Keresztes-Nagy, P., and Rosenfeld, J. P., 1981. Naloxone-reversible duplication by lanthanum of differential opiate analgesic effects on orofacial versus lower body versus central nociception, *Brain Res.* 208:234–239.

Kikhachev, Yu. P., Lyarskii, P. P., and Elovskaya, L. T., 1972. Cited in Haley (1979).

Kyker, G. C., and Cress, E. A., 1957. Acute toxicity of yttrium, lanthanum and other rare earths, *A.M.A. Arch. Ind. Health* 16:475–479.

Kyker, G. C., Cress, E. A., Sivaramakrishnan, V. M., Steffee, C. H., and Stewart, M., 1957. Fatty infiltration due to rare earths, *Fed. Proc.* 16:207.

Laszlo, D., Ekstein, D. M., Lewin, R., and Stern, K. G., 1952. Biological studies on stable and radioactive rare earth compounds. I. On the distribution of lanthanum in the mammalian organism, *J. Natl. Cancer Inst.* 13:559–573.

Lazar, G., 1973a. Effect of reticuloendothelial stimulation and depression on rare earth metal chloride-induced splenic calcification and fatty degeneration of the liver, *Experientia* 29:818–819.

Lazar, G., 1973b. The reticuloendothelial-blocking effect of rare earth metals in rats, *J. Reticuloendothel. Soc.* 13:231–237.

Lazar, G., Husztik, E., and Ribarszki, S., 1985. Reticuloendothelial blockade induced by gadolinium chloride. Effect on humoral immune response and anaphylaxis, in *Macrophage Biology* (S. Reichard and M. Kojima, eds.), Alan Liss, New York, pp. 571–582.

Lombardi, B., and Recknagel, R. O., 1962. Interference with secretion of triglycerides by the liver as a common factor in toxic liver injury, *Am. J. Pathol.* 40:571–586.

MacDonald, N. W., Nusbaum, R. E., Alexander, G. V., Ezmirnan, F., Spain, P., and Rounds, D. E., 1952. The skeletal deposition of yttrium, *J. Biol. Chem.* 195:837–842.

Magnusson, G., 1963. The behavior of certain lanthanons in rats, *Acta Pharmacol. Toxicol., Suppl. 3* 20:1–95.

Marciniak, M., and Baltrukiewicz, Z., 1977. Serum ornithine carbonoyltransferase (OCT) in rats poisoned with lanthanum, cerium and praseodymium, *Acta Physiol. Pol.* 28: 589–594.

Maxwell, L. C., Bischoff, F., and Ottery, E. M., 1931. Studies in cancer chemotherapy. X. The effect of thorium, cerium, erbium, yttrium, didymium, praseodymium, manganese and lead upon transplantable rat tumors, *J. Pharmacol. Exp. Ther.* 43:61–70.

McClure, J., 1980. The production of heterotropic calcification by certain chemical salts, *J. Pathol.* 131:21–33.

Mezentseva, N. V., 1967. Cited in Haley (1979).

Miller, L. P., 1959. Factors influencing the uptake and toxicity of fungicides, *Trans. N.Y. Acad. Sci.* 21:442–445.

Miller, L. P., and McCallan, S. E. A., 1957. Toxic action of metal ions to fungus spores, *J. Agric. Food Chem.* 5:116–122.

Mogilevskaya, O. Ya., and Roshchina, T. A., 1967. Cited in Haley (1979).

Morganti, J. B., Lown, B. A., Chapin, E., D'Agostino, R. B., and Monteverde, E. J., 1980. Effects of acute exposure to cerium nitrate on selected measures of activity, learning and social behavior of the mouse, *Gen. Pharmacol.* 11:369–373.

Mori, I., 1931. Pharmakologische Untersuchungen über die seltenen Erdmetalla Lanthan, Cerium, Praseodym und Neodym, *Jpn. J. Med. Sci.* 5:13–14.

Muroma, A., 1958. Studies on the bactericidal action of salts of certain rare earth metals, *Ann. Med. Exp. Biol. Fenn.* 36(Suppl. 6):1–54.

Muroma, A., 1959. The bactericidal action of the rare earth metals (further studies), *Ann. Med. Exp. Biol. Fenn.* 37(Suppl. 1–7):336–340.

Nagy, I., Kadas, I., and Jobst, K., 1976. Lanthanum trichloride induced blood coagulation defect and liver injury, *Haematologia* 10:353–359.

Nappee, J., Bobrie, J., and Lombard, D., 1972. Pneumoconiosis au cérium, *Arch. Mal. Prof. Med. Trav. Sec. Soc.* 33:13–18.

Neubert, D., and Hoffmeister, I., 1960. Intracellulare Lokalisation von Fettsubstanzen bei experimenteller Leberverfettong, *Naunyn-Schmiedeberg's Arch. Exp. Pathol. Pharmacol.* 237:519–537.

Niccolini, P. M., 1930. Contributo allo studio farmacologico delle terre rare—il praseodimio, *Arch. Int. Pharmacodyn. Ther.* 37:199–240.

Niccolini, P. M., 1931. Contributo allo studio farmacologico delle terre rare—il samario, *Arch. Int. Pharmacodyn. Ther.* 40:247–291.

Novello, F., and Stirpe, F., 1969. The effects of copper and other ions on the ribonucleic acid polymerase activity of rat liver nuclei, *Biochem. J.* 111:115–119.

Oberdisse, E., Rohling, G., Losert, W., Schurig, R., and Oberdisse, U., 1973. Influence of rare earths on insulin secretion in rats, *Naunyn-Schmiedeberg's Arch. Pharmacol.* 280: 217–221.

Oberdisse, E., Winkler, R., Grajewski, O., Von Lehmann, B., and Arntz, H. R., 1974. Pharmakologische Untersuchungen zum Mechanismus der Praseodym—ausgelosten Leberschadigung, *Verh. Dtsch. Ges. Inn. Med.* 80:1556–1558.

Oberdisse, E., Arvela, P., and Gross, U., 1979. Lanthanon-induced hepatotoxicity and its prevention by pretreatment with the same lanthanon, *Arch. Toxicol.* 43:105–114.

Oga, S., 1971. Actividade teratogenica do cloreto de iterbioem embrião de galinha, *Rev. Farm. Bioquim. Univ. São Paulo* 9:327–341.

Ojemann, R. G., Brownell, G. I., and Sweet, W. H., 1961. Possible radiation therapy of cephalic neoplasms by perfusion of short-lived isotopes. II. Dysprosium-165 metabolism in mouse and cat, *Neurochirurgica* 4:41–57.

Palmer, R. J., Buterhoff, J. L., and Stevens, J. B., 1987. Cytotoxicity of the rare earth metals cerium, lanthanum, and neodymium *in vitro:* comparisons with cadmium in a pulmonary macrophage primary culture system, *Environ. Res.* 43:142–156.

Pederson, L. A., and Libby, W. F., 1972. Unseparated rare earth-cobalt oxides as auto exhaust catalysts, *Science* 176:1355–1356.

Pickard, B. G., 1970. Comparison of calcium and lanthanum ions in the *Avena*-coleoptile growth test, *Planta* 90:314–320.

Ranganayaki, S., Bahadur, K., and Mohan, C., 1981. Microbial fixation of nitrogen in the presence of lanthanum sulphate, *Z. All. Mik.* 21:329–332.

Rapaport, M. J., 1982. Depigmentation with cerium oxide, *Contact Dermatitis* 8:282–283.

Reddy, S. V. R., and Yaksh, T. L., 1980. Antinociceptive effects of lanthanum, neodymium and europium following intrathecal administration, *Neuropharmacology* 19:181–185.

Renaud, G., Soler-Argilaga, C., and Infante, R., 1980a. Effect of cerium on liver lipids metabolism and plasma lipoproteins synthesis in the rat, *Biochem. Biophys. Res. Commun.* 95:220–227.

Renaud, G., Soler-Argilaga, C., Rey, C., and Infante, R., 1980b. Free fatty acid mobilization in the development of cerium-induced fatty liver, *Biochem. Biophys. Res. Commun.* 92:374–380.

Robinson, G. A., Wasnidge, D. C., and Floto, F., 1978. Distribution of ^{140}La and ^{47}Ca in female Japanese quail and in the eggs laid, *Poultry Sci.* 57:190–196.

Robinson, G. A., Wasnidge, D. C., and Floto, F., 1980. Radiolanthanides as markers for vitellogenin-derived proteins in the growing oocytes of Japanese quail, *Poultry Sci.* 59:2312–2321.

Rohling, G., 1974. Enfluss von Praseodym auf den Kohlenhydratstoffwechsel und die Insulinsekretion, Thesis, Freie Universität Berlin, cited in Arvela (1979).

Rosenfeld, J. P., and Hammer, M., 1983. Antagonism of opiate-like, lanthanum-induced analgesia by naloxone, 12 mg/kg, in rats, *Brain Res.* 268:189–191.

Saffer, L. D., Rodeheaver, G. T., Hiebert, J. M., and Edlick, R. F., 1980. *In vivo* and *in vitro* antimicrobial activity of silver sulfadiazine and cerium nitrate, *Surg. Gynecol. Obstet.* 151:232–236.

Salas, M., and Tuchweber, B., 1976. Prevention by steroids of cerium hepatotoxicity, *Arch. Toxicol.* 35:115–125.

Salas, M., Tuchweber, B., Kovacs, K., and Garg, B. D., 1976. Effect of cerium on the rat liver. An ultrastructural and biochemical study, *Beitr. Pathol. Biol.* 157:23–44.

Sanna, A., Mascia, A., Pani, P., and Congiu, L., 1976. Protection by cerium chloride on CCl_4-induced hepatotoxicity, *Experientia* 32:91–92.

Santos, M. C., Azevedo, M. D., and Jacobsohn, K., 1982. Effect of Eu^{3+} on lipid biosynthesis in rat liver, *Biochimie* 64:305–308.

Sarkadi, B., Szasz, I., Gerloczy, A., and Grardos, G., 1977. Transport parameters and stoichiometry of active calcium ion extrusion in intact human red cells, *Biochim. Biophys. Acta* 464:93–107.

Sarkander, H. I., and Brade, W. P., 1976. On the mechanism of lanthanide-induced liver toxicity, *Arch. Toxicol.* 36:1–17.

Sarnat, A. E., 1983. The efficiency of cobalt samarium (Co 5 Sm) magnets as retention units for overdentures, *J. Dent.* 11:324–333.

Sartory, A., and Bailly, P., 1922. Influence des sels de terres rares sur la structure du mycélium de l'*Aspergillus fumigatus* Fr. et sur la formation de l'appareil conidier, *C. R. Soc. Biol.* 86:601–604.

Schepers, G. W. H., 1955a. The biological action of rare earths: I. The experimental pulmonary histopathology produced by a blend having a relatively high oxide content, *A.M.A. Arch. Ind. Health* 12:301–305.

Schepers, G. W. H., 1955b. The biological action of rare earths: II. The experimental pulmonary histopathology produced by a blend having a relatively high fluoride content, *A.M.A. Arch. Ind. Health* 12:306–316.

Schepers, G. W. H., Delahart, A. B., and Redlin, A. J., 1955. An experimental study of the effects of rare earths on animal lungs, *Arch. Ind. Health* 12:297–316.

Schimmel, R. J., 1978. Calcium antagonists and lipolysis in isolated rat epididymal adipocytes: effects of tetracaine, manganese, cobaltous and lanthanum ions and D600, *Horm. Metab. Res.* 10:128–134.

Schmantz, E., 1964. Geschlechtsabhangigkeit der Verteilung von Radiocer kei der Ratte, *Strahlentherapie* 123:267–278.

Schorderet-Slatkine, S., Schorderet, M., and Bailieu, E.-E., 1976. Initiation of meiotic maturation in *Xenopus laevis* oocytes by lanthanum, *Nature* 262:289.

Schroeder, H. A., and Mitchener, M., 1971. Scandium, chromium(VI), gallium, yttrium, rhodium, palladium, indium in mice: effects on growth and lifespan, *J. Nutr.* 101: 1431–1438.

Schurig, R., and Oberdisse, E., 1972. The influence of rare earths on hepatic gluconeogenesis, *Naunyn-Schmiedeberg's Arch. Pharmacol.* 275:419–433.

Sedmak, J. J., MacDonald, H. S., and Kushnaryov, V. M., 1986. Lanthanide ion enhancement of interferon binding to cells, *Biochem. Biophys. Res. Commun.* 137:480–485.

Seltzer, S. E., Adams, D. F., Davis, M. A., Hessel, S. J., Hevron, A., Judy, P. F., Hurlburt, A. J., and Hollenbert, N. K., 1981. Hepatic contrast agents for CT: high atomic number contrast material, *J. Comput. Assist. Tomogr.* 5:370–374.

Selye, H., 1962. *Calciphylaxis,* University of Chicago Press, Chicago, p. 311.

Selye, H., Szabo, S., Tuchweber, B., and Lefebvre, F., 1972. Acute miliary necrosis resembling reliosis hepatis produced by the combination of glucocorticoid and gadolinium, *Proc. Soc. Exp. Biol. Med.* 139:887–889.

Sharma, S. N., Chatterjee, S. M., and Kamboj, V. P., 1973. Sterilization of male goats by formaldehyde and some metallic salts, *Indian J. Exp. Biol.* 11:143–148.

Shearer, C., 1922. Studies of the action of electrolytes on bacteria, *J. Hyg.* 21:77–86.

Shelley, W. B., Hurley, H. J., Mayock, R. L., Close, H. P., and Cathcart, R. T., 1958. Intradermal tests with metals and other inorganic elements in saracoidosis and anthracosilicosis, *J. Invest. Dermatol.* 31:301–303.

Sievers, R. E., and Sadlowski, J. E., 1978. Volatile metal complexes. Certain chelates are useful as fuel additives, as metal vapor sources and in trace metal analysis, *Science* 201:217–223.

Sihvonen, M. L., 1972. Accumulation of yttrium and lanthanoids in human and rat tissues as shown by mass spectrometric analysis and some experiments with rats, *Ann. Acad. Sci. Fenn. Ser. A* 168:1–62.

Smith, B. M., Gindhart, T. D., and Colburn, N. H., 1986. Possible involvement of a lanthanide-sensitive protein kinase C substrate in lanthanide promotion of neoplastic transformation, *Carcinogenesis* 7:1949–1956.

Snyder, F., and Kyker, G. C., 1964. Triglyceride accumulation and release in the rare-earth fatty liver, *Proc. Soc. Exp. Biol. Med.* 116:890–893.

Snyder, F., and Stephens, N., 1961. Plasma free fatty acids and the rare-earth fatty liver, *Proc. Soc. Exp. Biol. Med.* 106:202–204.

Snyder, F., Cress, E. A., and Kyker, G. C., 1959. Liver lipid response to intravenous rare earths in rats, *J. Lipid Res.* 1:125–131.

Snyder, F., Cress, E. A., and Kyker, G. C., 1960a. Rare-earth fatty liver, *Nature* 185: 480–481.

Snyder, F., Baker, F., Rafter, J., and Kyker, G. C., 1960b. Octanoate oxidation in the cerium-induced fatty liver, *Biochim. Biophys. Acta* 43:554–555.

Sobek, J. M., and Talburt, D. E., 1968. Effect of the rare earth cerium on *Escherichia coli*, *J. Bacteriol.* 95:47–51.

Steffee, C. H., 1959. Histopathologic effects of rare earths administered intraperitioneally to rats. A preliminary report, *A.M.A. Arch. Ind. Health* 20:414–419.

Steidle, H., and Ding, M., 1929. Beitrage zur Pharmakologie der Seltenen Erdmetalle. II. Mittelung Uber das Yttrium, *Naunyn-Schmiedeberg's Arch. Exp. Pathol. Pharmacol.* 141:273–279.

Stineman, C. H., Massaro, E. J., Lown, B. A., Morganti, J. B., and Al-Hakeeb, S., 1978. Cerium tissue/organ distribution and alterations in open field and exploratory behavior following acute exposure of the mouse to cerium (citrate), *J. Environ. Pathol. Toxicol.* 2:553–570.

Strubelt, O., Siegers, C. P., and Younes, M., 1980. The influence of silybin on the hepatotoxic and hypoglycemic effects of praseodymium and other lanthanides, *Arzneim.-Forsch.* 30:1690–1694.

Swanson, A. A., and Truesdale, A. W., 1971. Elemental analysis in normal and cataractous human lens tissue, *Biochem. Biophys. Res. Commun.* 45:1488–1496.

Talbot, R. B., Davison, F. C., Green, J. W., Reece, W. O., and Vangelder, G., 1965a. Effects of subcutaneous injection of rare earth metals, United States Atomic Energy Department Report 1170.

Talbot, R. B., Davison, F. C., and Reece, W. O., 1965b. Inhalation exposure of mice to an aerosol of gadolinium oxide, U.S. Atomic Energy Commission Report COO-1170-6.

Talburt, D. E., and Johnson, G. T., 1967. Some effects of rare earth elements and yttrium on microbial growth, *Mycologia* 59:492–503.

Tandon, S. K., Gaur, J. S., Behari, J., Mathur, A. K., and Singh, G. B., 1977. Effects of monazite on body organs of rats, *Environ. Res.* 13:347–357.

Tang, X., and Li, G., 1983. Effect of cerium on growth of corn seedling and its enzyme pattern, *J. Chin. Rare Earth Soc.* 1:56–59.

Tapia, R., 1982. Antagonism of the ruthenium red-induced paralysis in mice by 4-aminopyridine, guanidine and lanthanum, *Neurosci. Lett.* 30:73–77.

Trnovec, T., Pleskova, A., and Chorvat, D., 1974. The effect of carbon tetrachloride on radiocerium metabolism in rats, *Strahlentherapie* 147:521–530.

Tuchweber, B., and Gabbiani, G., 1967. Effect of sodium pyrophosphate on experimental soft tissue calcification and hypercalcemia, *Can. J. Physiol. Pharmacol.* 45:957–964.

Tuchweber, B., and Savoie, L., 1968. Rare earth metals and soft-tissue calcification, *Proc. Soc. Exp. Biol. Med.* 128:473–476.

Tuchweber, B., Trost, R., Salas, M., and Sieck, W., 1976a. Effect of praseodymium nitrate on hepatocytes and Kupffer cells in the rat, *Can. J. Physiol. Pharmacol.* 54:898–906.

Tuchweber, B., Trost, W., Salas, M., and Sieck, R., 1976b. Prevention of praseodymium-induced hepatotoxicity by silybin, *Toxicol. Appl. Pharmacol.* 38:559–570.

Venugopal, B., and Luckey, T. D., 1978. *Metal Toxicity in Mammals, Vol. 2, Chemical Toxicity of Metals and Metalloids*, Plenum Press, New York, Chapter 3, pp. 101–173.

Vincke, E., and Oelkers, H. A., 1938. Zur Pharmakologie der Selterer Erder: Toxizitat und Wirkung auf Stoffwechselvorgange, *Arch. Exp. Path. Pharmakol.* 188:465–476.

Vincke, E., 1942. Die blutgerinnungshemmende Wirkung der seltenen Erden, *Hoppe-Seyler's Z. Physiol. Chem.* 272:65–80.

Vocaturo, G., Colombo, F., Zanoni, M., Rodi, F., Sabbioni, E., and Pietra, R., 1983. Human exposure to heavy metals. Rare earth pneumoconiosis in occupational workers, *Chest* 83:780–783.

Von Lehmann, B., Oberdisse, E., Grajewski, O., and Arntz, H. R., 1975. Subcellular distribution of phospholipids during liver damage induced by rare earths, *Arch. Toxicol.* 34:89–101.

Von Lehmann, B., Arvela, P., Grajewski, O., and Oberdisse, E., 1976. Drug metabolizing enzymes in the rough and smooth endoplasmic reticulum during liver damage induced by rare earths, *Naunyn-Schmiedeberg's Arch. Pharmacol.* 293(Suppl. 1):R59.

Weinmann, H.-J., Brasch, R. C., Press, W. R., and Wesbey, G. E., 1984. Characteristics of gadolinium-DTPA complex: a potential NMR contrast agent, *Am. J. Radiol.* 142:619–624.

Whittingham, D. B., 1980. Parthenogenesis in mammals, in *Oxford Review of Reproductive Biology* (C. E. Finn, ed.), Vol. 2, Clarendon Press, Oxford, pp. 205–210.

Wurm, M., 1951. The effect of lanthanum on growth and metabolism of *Streptococcus faecalis* R., *J. Biol. Chem.* 192:707–714.

Zanni, A. C., 1965. Contribution to the pharmacology of europium, *Rev. Fac. Farm. Bioquim.* 3:199–240.

Zimakov, Yu. A., 1973. Cited in Haley (1979).

Past, Present, and Possible Future Clinical Applications of the Lanthanides

9.1 Introduction

Many of the properties of the lanthanides appear to lend themselves to clinical application. The lanthanides are antimicrobial and anticoagulant substances which suppress many of the types of Ca^{2+}-dependent cellular activation processes (Section 6.4) that occur in diseases. Lanthanides are of relatively low toxicity (Chapter 8), while their metabolism can be manipulated by the presence of specific chelators, by varying the site of injection, or both (Chapter 7). In addition, lanthanides appear to accumulate in tumors or at sites of inflammation. They provide a range of radioisotopes, with various half-lives, which emit α, β, or γ radiation, while certain members are strongly paramagnetic. In addition, most lanthanides are cheap, readily available, and straightforward to work with. As alluded to in the introductory chapter (Section 1.3), the evidence suggests that lanthanides are worth investigating as agents with which to attack several of the major diseases of the Western world.

Discussion of the medical potential of the lanthanides began with their prescription as antiemetics in the last century. In approximately chronological order, this chapter reviews the past and present medical uses of the lanthanides, with some speculations about possible future applications. Ellis (1977) has previously reviewed this field.

9.2 General Historical Medical Uses of Lanthanides

Cerium oxalate was introduced in the middle of the last century as an antiemetic, finding its greatest use in alleviating the reflex vomiting of early pregnancy. It subsequently came to be prescribed for all sorts of gastrointestinal (G.I.) disorders and even for coughs. However, success was mixed. From studies with dogs, Baehr and Wessler (1909) concluded that cerium oxalate did not inhibit vomiting of central origin but prevented vomiting due to local irritation of the gastric mucosa. They suggested that cerium oxalate served to form a protective lining to the wall of the G.I. tract. It ceased to be used for digestive disorders during the early twentieth century.

Cerium carbonate and salicylate have been prescribed as mild sedatives of the central nervous system. Its chloride, nitrate, sulfate, and acetate have been employed as astringents (Browning, 1961).

9.3 Applications Based on Antimicrobial Properties

9.3.1 Historical Introduction

As discussed in Section 8.2, plenty of evidence attests to the antimicrobial actions of lanthanides *in vitro*. However, quite high concentrations, usually in the millimolar to hundred millimolar range, are required to achieve this, with bacteria appearing to be more sensitive than fungi. The antimicrobial mechanisms are unknown.

The relatively high doses required for inhibition of microbial growth seem to preclude the systemic, clinical use of simple lanthanide salts. However, solutions of lanthanide salts have been injected intravenously into patients as a treatment for chronic pulmonary tuberculosis. Citing the unpublished observations of Frouin (1912) that lanthanides inhibit the growth of cultured tubercle bacilli, Grenet and Drouin (1920) intravenously (i.v.) injected $Sm_2(SO_4)_3$, $Nd_2(SO_4)_3$, or $Pr_2(SO_4)_3$ into several tuberculosis patients. Each patient received 4–5 ml of a 2% (ca. 34 mM) solution of the lanthanide every day or every other day for a series of 20 injections. Following a 20-day rest, a second and then a third series of injections was given. Despite the surprisingly encouraging results that Grenet and Drouin (1920) reported, there appears to have been no further development of this treatment. Attempts to use lanthanides to treat leprosy (Arvela, 1979), cholera (Frouin and Roudsky, 1914), and puerperal infections (Browning, 1961) seem also to have been without lasting success. In view of the restricted bodily distribution of i.v.-injected lanthanides (Section 7.5.2) and their ability to shut down the reticuloendothelial system (Lazar, 1973; Section 8.10), such lack of success is not surprising.

Topical application would seem a much more promising way to exploit medically the antimicrobial activities of the lanthanides. Indeed, back in 1906, potassium ceric sulfate was on the market under the trade name "Ceriform." This generated Ce(IV) ions which, like many oxidizing agents, have strong bactericidal action. It proved a useful antiseptic for external application to wounds. Various trivalent cerium salts have also been used as antiseptics (Muroma, 1958).

However, topical use of cerium salts does not prevent tooth decay (Regolati et al., 1975), nor does LaCl$_3$ (0.3%) mouthwash reduce plaque formation (Beazley et al., 1980). Indeed, La^{3+} ions reduce the effectiveness of chlorhexidine in this respect (Waler and Rolla, 1983). Nevertheless, administration of Y(NO$_3$)$_3$ either in the drinking water or by i.p. injection reduced the incidence of caries in rats (Mercado and Ludwig, 1973). However, the authors attributed this not so much to the antimicrobial activity of Y^{3+} ions as to their ability to enter into the chemical structure of the teeth. In particular, they drew attention to the ability of Y^{3+} to reduce the solubility of dental enamel (Manly and Bibby, 1949).

Striking, albeit disputed, success has been reported for the treatment of burns.

9.3.2 Cerium Salts in the Treatment of Burns

Many burn patients die through sepsis originating from infection of the open wound. Topical antiseptics are used to combat wound infection, silver sulfadiazine having been the agent of choice for nearly 20 years. However, it is not the final answer to topical antimicrobial therapy. It is expensive and does not reliably suppress the growth of gram-negative bacteria in large wounds. Furthermore, resistant strains arise from time to time.

Domotor (1969) first reported the topical use of a lanthanide in burn therapy. An ointment containing "phlogosam," the Sm^{3+} complex of pyrocatechol disulfonate (Fig. 9-5, Section 9.9), was successfully used to treat first- and second-degree burns. Monafo et al. (1976) introduced cerium nitrate for topical application to burns. Used alone, or in combination with silver sulfadiazine, the results were spectacular. None of the patients treated with Ce^{3+} ions developed a necrotizing wound infection. Furthermore, the use of cerium was associated with a decrease of about 50% in the anticipated death rate. Ce(NO$_3$)$_3$ was particularly active in vivo against fungi and gram-negative bacteria, including Pseudomonas aeruginosa, a bacterium which is a troublesome colonizer of burns. In agreement with the results of in vitro studies (Section 8.2), the bacteriology of Ce^{3+}-treated wounds suggested that gram-positive bacteria were more resistant. However, as silver salts are more active against gram-positive bacteria, the bactericidal spectra of the two antiseptics are complementary. A combination of Ce(NO$_3$)$_3$ and silver sulfadiazine indeed showed

Table 9-1. Mortality in Burned Patients Tested with
Cerium Nitrate–Silver Sulfadiazine[a]

Burned surface area (%)	No. of patients	Deaths observed	Deaths predicted
0–9	213	2	4.1
10–19	110	2	5.2
20–29	63	3	10.6
30–39	44	8	13.7
40–49	32	11	16.0
50–59	19	7	11.7
60–69	18	7	13.5
70–79	10	2	8.2
80–89	10	7	9.9
90–100	11	11	11.0
Total	530	60 (11%)	103.9 (20%)

[a] From Monafo (1983), with permission.

wide bactericidal activity and was suggested as a superior topical antiseptic for use in burns. As Ag^+ ions increase the uptake of Ce^{3+} ions by *Neurospora* spores (Miller, 1959), the two agents have the potential to act synergistically against fungi.

In these studies, Ce^{3+} was supplied in one of two ways. In some cases, bandages were saturated with a 40 mM solution of $Ce(NO_3)_3$ in saline. In most cases, 50 mM $Ce(NO_3)_3$ was incorporated into a water-soluble cream. Such creams are now commercially available in Europe. Allergy was not a problem, and no toxicity was experienced. Despite the adverse dermatologic effects that lanthanides have in experimental animals (Section 8.5), spontaneous healing was not impaired by Ce^{3+} ions, and skin grafts were readily accepted. Neutron activation analysis of blood and urine failed to detect cerium, confirming the poor systemic absorption of lanthanides from sites of topical application (Section 7.5.1). In a subsequent larger study involving 530 burn victims, halving of the mortality was confirmed (Monafo, 1983) (Table 9-1). Supportive results were obtained by Fox *et al.* (1977), whose data suggested that cerium sulfadiazine would be superior to cerium nitrate as an antibacterial agent. Unfortunately, not all subsequent investigators have been able to repeat these successes. Helvig *et al.* (1979), Munster *et al.* (1980), Saffer *et al.* (1980), and Bowser *et al.* (1981) all found that the presence or absence of Ce^{3+} ions made little difference to wound bacteriology or clinical outcome.

A number of factors complicate comparative studies between institutions. One is that each burn center is home to its own strains of microbial flora. Not only do these differ between centers, but they also change with time. Several studies have compared the effects of Ce^{3+} on rates of morbidity and mortality *expected* from previous experience. In view of temporal changes in the flora and recent improvements in other aspects of the treatment of burn victims, such comparisons may be invalid.

As an ingenious way to limit some of these variables, Hermans (1984) allowed each patient to serve as his or her own control. In 16 patients with symmetrical burns, one side of each patient was treated with silver sulfadiazine alone and the other with a mixture of $Ce(NO_3)_3$ and silver sulfadiazine. The mixture proved far superior in reducing the concentration of bacteria on the wound surface.

Reports of studies of the *in vitro* sensitivities of strains of bacteria isolated from burn wounds to $Ce(NO_3)_3$ and silver sulfadiazine are equally contradictory. Rosenkranz (1979) tested creams containing these antiseptics in an agar plate diffusion assay. He found that $Ce(NO_3)_3$ alone had no inhibitory effect. However, the inhibitory effect of silver sulfadiazine was as good, or better, in the presence of $Ce(NO_3)_3$. In broth, saline or water, silver sulfadiazine is much more toxic than $Ce(NO_3)_3$ to *P. aeruginosa* (Saffer *et al.*, 1980). Slight synergism was noted in water or saline, but not in broth. However, other studies (Heggers *et al.*, 1979; Holder, 1982) have demonstrated antagonism between the two agents. Of 37 strains of microorganism tested in an agar diffusion test, $Ce(NO_3)_3$ reduced the zone of growth inhibition by silver sulfadiazine in 30 strains; 5 strains were unaffected and only 2 were enlarged (Holder, 1982).

Much of the difficulty in reproducibly evaluating the bactericidal properties of Ce^{3+} ions resides in their readiness to precipitate or to form strong complexes with multidentate ligands with oxygen donor atoms (Section 2.3). In several of the test systems, the concentration of free Ce^{3+} ions would seem to have been very low. This helps to explain why the bulk of the recently published work suggests that silver sulfadiazine is the more active antibacterial agent and that its potency is not improved by adding $Ce(NO_3)_3$. It is likely that this state of affairs also obtains *in vivo*. If Ce^{3+} ions are really not such good antiseptic agents, we are left in need of an explanation for the apparent efficacy of $Ce(NO_3)_3$ in several impressive clinical trials. The answer may lie with burn toxins and immunomodulation.

Allgower's group (Allgower *et al.*, 1968) first identified a burn toxin which appears to be a lipoprotein of M.W. 3×10^6 (Allgower *et al.*, 1973). As Ce^{3+} ions bind to this toxin *in vitro* (Kremer *et al.*, 1981), it has been suggested that the improvement seen in burn patients is due to inactivation of the toxin by Ce^{3+} ions.

A related observation is that topically applied Ce^{3+} ions prevent postburn immunosuppression in mice (Hansbrough *et al.,* 1984; Peterson *et al.,* 1985). Depressed cell-mediated immunity after burning is thought to result from a factor released from burned tissue. Its relationship to Allgower's burn toxin is unclear, although Hansbrough *et al.* (1984) consider them to be different factors. Cerium's anti-immunosuppressive effect is presumably unrelated to any direct bactericidal activity, as silver sulfadiazine fails to inhibit postburn immunosuppression.

Given the importance of cell-mediated immunity in fighting infection, postburn immunosuppression should promote infection, and Ce^{3+} ions should counteract this. Such was the case with mice in which cecal ligation and puncture was used as a septic challenge (Zapata-Sirvent *et al.,* 1986). Unburned mice showed a 63.7% survival following this challenge. In burned mice, the survival rate fell to 20%, but daily topical application of $Ce(NO_3)_3$ increased survival to 54.1%.

As an alternative to direct inactivation of burn toxins or immunosuppressive factors by Ce^{3+} ions, one could speculate that these ions act directly upon the cells which synthesize these molecules. Many examples exist where lanthanides prevent the cellular secretion of specific products in response to stimuli (Section 6.4). Inhibition of the secretion of toxins or immunoregulators in response to burning could constitute one more example of this.

9.4 Lanthanides as Anticoagulants

As discussed in Section 8.8, intravenously injected lanthanides prolong the clotting time of mammalian blood. When administered orally or by intraperitoneal (i.p.) injection, they are ineffective. The anticoagulant mechanism is incompletely understood but may involve antagonism of Ca^{2+}-dependent clotting reactions.

Guidi's (1930) early discovery of the lanthanides' anticoagulant activities sparked over two decades of research aiming to harness these properties for clinical use (e.g., Vincke and Oelkers, 1937; Dyckerhoff and Goossens, 1939; Beaser *et al.,* 1942; Hunter and Walker, 1956). Despite claims to the contrary (Dyckerhoff and Goossens, 1939), i.v. injections of anticoagulant doses (ca. 10 mg/kg) of simple lanthanide salt solutions proved too toxic for clinical use. Beaser *et al.* (1942), for example, while confirming their potent and long-lasting anticoagulant properties, reported several undesirable side effects. These included chills, fever, muscle pain, abdominal cramps, hemoglobinemia, and hemoglobinuria. For this reason, much effort was directed toward developing lanthanide derivatives with fewer side effects.

Vincke and Sucker (1950) described neodymium 3-sulfoisonicotinate as an improved anticoagulant lacking toxic effects when administered as a 2.5% solution (5 mg/kg) by i.v. injection. Although Hunter and Walker (1956) confirmed its efficacy, this line of research eventually petered out as heparin became freely and more cheaply available. However, according to Divald and Joullie (1970), some complexes, such as those with 3-sulfoiso-nicotinic acid and β-acetylpropionic acid, were still in fairly recent clinical use.

9.5 Antitumor Therapy

Attempts to use lanthanides in cancer chemotherapy date back to the work of Maxwell *et al.* (1931), who were unable to inhibit the growth of transplantable sarcomas in rats. Aqueous solutions of lanthanide salts were injected intraperitoneally. Given the poor absorption of lanthanides from i.p. locations (Section 7.5.1), their lack of effect would not seem surprising. However, Anghileri (1979) claimed that daily i.p. injection of 2.5 mg of $LaCl_3$ retarded the growth of sarcoma tumors in rats. Inhibition of tumor growth was based on weight and was actually a modest 23% decrease. In the same series of experiments, lanthanum aspartate increased the weight of the tumors by nearly 20%, although this increase was said not to be statistically significant. Although rather high concentrations of $LaCl_3$ inhibit the respiration of ascites tumor cells *in vitro*, $LaCl_3$ has no *in vivo* therapeutic activity in mice (Lewin *et al.*, 1953). However, Ln^{3+} have been found to promote the malignant transformation of cells *in vitro* (Smith *et al.*, 1986).

In subsequent work, Anghileri and co-workers have used La^{3+} as an adjunct to the killing of tumors by other methods. Whereas hyperthermia or intratumor injection of $1 mM$ $LaCl_3$ alone had no effect on the growth of sarcoma cells in mice, a combination of the two was said to give remarkable inhibition of tumor growth and increased survival (Anghileri *et al.*, 1983). This work has recently been reviewed by Anghileri (1988). Other studies have shown that La^{3+} enhances the uptake of a derivative of the anticancer drug hematoporphyrin by Ehrlich ascites cells (Crone-Escanye *et al.*, 1985). Whether any eventual clinical use will result from such studies is unclear.

A more promising approach to tumor therapy has been to localize suitable radioactive lanthanides at the site of the malignancy. Most commonly used has been ^{90}Y, as this has a favorable half-life of 60 h and emits purely β-radiation which penetrates tissue to an approximate depth of only 1 cm. Two methods may be employed to localize the lanthanide in the tumors. One takes advantages of the apparent metabolic affinity that certain lanthanide complexes have for tumors. Although successfully used in tumor scanning (Section 9.6.2), this approach has found little application to tumor therapy. Major restrictions are unacceptably high radiation doses to normal tissues and the relatively rapid clearing of radioactivity from the tumor. More widely used has been the second method, whereby the lanthanide is introduced directly at the site of the malignancy. To do this, lanthanides may be administered by intra- or peritumor injections or by surgical implantation.

Metabolic studies quickly revealed that i.p.-injected lanthanides remained largely within the peritoneal cavity (Laszlo $et\,al.$, 1952), especially if large amounts of "carrier" were added (Section 7.5.1). With this information, the possible therapeutic effects of i.p.-administered lanthanides on ascites tumors were investigated. Initial studies showed that the tumors accumulated 15–22% of an i.p. dose of $^{140}LaCl_3$ in mice (Laszlo $et\,al.$, 1952). Radioactive, but not stable, La^{3+} inhibited the growth of Ehrlich ascites tumors, reduced the numbers of viable tumor cells, and prolonged the survival of mice (Lewin $et\,al.$, 1953). Intrapleural, intraprostatic, and subcutaneous use of radioactive lanthanide and yttrium ions have also been evaluated in animals (Lewin $et\,al.$, 1954; Cooper $et\,al.$, 1956; Goldie and West, 1956).

When administered in the form of insoluble particles, lanthanides are particularly well retained at the injection site (Section 7.5.4). On this basis, direct intratumor or intracavity application of colloidal suspensions of radioactive lanthanides has been tested experimentally. In mice, $^{90}YF_3$ caused extensive necrosis of sarcoma tumors (Mayer and Morton, 1956). Although i.p. injections of colloidal suspensions of ^{165}Dy and ^{90}Y extended the survival time of mice with experimental ascites tumors, they were not curative (Bloomer $et\,al.$, 1984). Selective irradiation of the lymph nodes with colloidal $^{177}Lu^{3+}$ has also been investigated (Kyker $et\,al.$, 1956).

Nearly all radiotherapeutic use of lanthanides in human tumors has involved surgical implantation of ^{90}Y. This has found successful employment in the treatment of Nelson's syndrome (Cassar $et\,al.$, 1976), Cushing's syndrome (Nadjmi, 1970; Burke $et\,al.$, 1973) and breast cancer (Fasching $et\,al.$, 1974).

One possible additional use of lanthanides in tumor therapy, which, as far as I know, has not been explored, would exploit their ability to inhibit reticuloendothelial function (Section 8.10). The full anticancer potential of monoclonal antibodies and liposome-encoated drugs has not been realized because of their rapid sequestration by the reticuloendothelial system (RES). One wonders, therefore, whether they might not be more effective if used in conjunction with lanthanides.

Tissue concentrations of lanthanides may alter during cancer, suggesting that their measurement could provide diagnostic and prognostic information (Esposito *et al.*, 1986b). Other examples of diseases in which lanthanide abundances change are given in Section 7-2, Tables 7-6, 7-7, 7-8, and 9-5.

9.6 Imaging and Tracer Studies

9.6.1 Introduction

Presently, there is much clinical interest in developing ways to image specific parts of the body for diagnostic purposes. Traditionally, imaging has been limited to the taking of X-ray photographs. Newer methods include scintigraphy, computed tomography (CT), and nuclear magnetic resonance (NMR) imaging. Lanthanides have been employed in each of these three techniques.

Scintigraphy is a technique which autoradiographically visualizes specific parts of the body. It involves the introduction of a suitable radionuclide and the subsequent generation of images which reflect the bodily distribution of the radioactive material. This process is often referred to as "scanning." In general terms, the radionuclide should either selectively accumulate in the specific organ, tissue, or lesion that requires visualization or remain in the extracellular compartment to delineate cavities and fluid flow. The latter technique is known as cisternography.

A number of strict requirements limit the selection of radioactive substances which are suitable for scintigraphy. Only the radiation from γ-emitting radioisotopes provides sufficient penetration for autoradiographic detection by scintigraphy. As excessively energetic emissions are too toxic, energies in the 63–198 keV range are usually sought. The isotope should ideally have no β emissions, as these interact most readily with adjacent biological material, causing tissue damage. Its half-life should be such as to permit scintigraphy but to limit exposure time.

Biochemically, the radionuclide or its derivative should either show high and selective affinity for the parts to be imaged or have no such affinities at all but act as an inert filler of spaces. In the latter case, it would remain in, and permit visualization of, extracellular cavities. As the agent may remain in the body long after its radioactivity has disappeared, its decay products should have no toxic effects. Lastly, the isotope should be readily available through local production or be amenable to easy shipping and storage.

Based on these criteria, the lanthanides have much to offer scintigraphy. A number of their isotopes are γ emitters with suitable energies and half-lives. As discussed in Chapter 8, nonradioactive lanthanides are of relatively low toxicity. Particularly useful is the way in which their metabolic distribution can be altered by the route of application and by the inclusion of suitable complexing agents (Section 7.5). Indeed, experimental clinical use has been made of radioactive lanthanides both as selective, tumor-scanning and bone-scanning reagents and as inert space-fillers for cisternography.

The same general principles hold for NMR and CT scanning, except that the image is based upon magnetic disturbances in the former method and upon subtle differences in radiolucency in the latter. With NMR (Section 3.2), images result from the responses of magnetically susceptible centers, usually protons, in biological materials. Here, the paramagnetic lanthanides (Section 2.2; Table 2-2) have obvious potential application. As heavy metals which attenuate X-rays, lanthanides also have potential application in CT imaging.

As an alternative to imaging the distribution of radioactive, magnetic, or electron-dense substances in the body, suitable biological materials can be directly sampled and their content of these probes measured. Tracing the passage of such materials through various bodily compartments in this way serves to indicate functional integrity. Obvious examples are blood and urine, where the urinary clearance of such markers helps monitor kidney function. There are several reports of the use of lanthanides in tracer studies of this type.

9.6.2 Scintigraphic Imaging

A growing number of reports draw attention to the affinity that lanthanides have for tumors *in vivo*. Hisada and Ando (1973) tested $^{140}LaCl_3$ and the citrate complexes of ^{141}Ce, ^{153}Sm, ^{153}Gd, ^{160}Tb, ^{170}Tm, ^{169}Yb, and ^{177}Lu. As $^{67}Ga^{3+}$-citrate has been widely used in tumor scanning, it was included for comparison. Solutions of these reagents were intravenously injected into the tail veins of rats bearing the Yoshida sarcoma. Rats were killed and the specific radioactivities of the tumors and various organs compared to provide an index of specificity (Table 9-2).

Inspection of Table 9-2 reveals that in every case the specific radio-activity of the tumor exceeded that of blood or muscle. As expected (Section 7.5.2), the lighter lanthanides showed greater affinity for the liver. The highest overall tumor specificity, both in terms of the number of organs for which the tumor:organ specific activity exceeded unity and in the magnitude of these ratios, was exhibited by $^{170}Tm^{3+}$. Lanthanides

Table 9-2. Organ Distribution of Lanthanides and Gallium in Rats Bearing
Yoshida Sarcoma[a]

Nuclide	Tumor/Organ ratio at 24 h					
	Blood	Muscle	Liver	Kidney	Spleen	Bone
^{140}LaCl$_3$	10.29	40.00	0.05	0.56	0.54	N.G.[b]
^{141}Ce^{3+}-citrate	23.75	15.20	0.04	0.30	0.70	0.69
^{153}Sm^{3+}-citrate	20.00	10.23	0.05	0.19	1.10	0.56
^{153}Gd^{3+}-citrate	20.56	18.50	0.18	0.29	2.64	0.12
^{160}Tb^{3+}-citrate	19.52	7.32	0.29	0.19	1.95	0.14
^{170}Tm^{3+}-citrate	95.71	53.60	2.53	1.89	7.44	0.44
^{169}Yb^{3+}-citrate	51.43	37.90	1.90	0.86	2.67	0.38
^{177}Lu^{3+}-citrate	26.82	23.60	1.26	0.92	1.84	N.G.
^{67}Ga^{3+}-citrate	6.00	22.80	8.09	1.52	8.84	N.G.

[a] Data recalculated from Hisada and Ando (1973), with permission.
[b] N.G.: Not given.

showed several advantages over ^{67}Ga as a scanning agent, including lower
retention by the blood and muscle. However, there was less accumulation
of ^{67}Ga in the liver, kidney, and spleen. Inspection of the raw data (not
shown) reveals that although greater amounts of radioactivity accumulated in the tumor when ^{67}Ga was used as the tracer, other soft tissues
also retained more radioactivity, thus lowering the tumor:organ ratio.

Subsequent work (Yano and Chu, 1975; Sullivan *et al.*, 1975; Beyer
et al., 1978; Woolfenden *et al.*, 1983; Schomacker *et al.*, 1986), using a
variety of rodents and different tumors, has generally confirmed the conclusions of Hisada and Ando (1973). Lanthanide chlorides are poor tumor-
seeking agents (Higasi *et al.*, 1973; Woolfenden *et al.*, 1983), but bleomycin complexes (Section 4.2) show promise (Sullivan *et al.*, 1975).

Although Tm shows the greatest tumor specificity, its usefulness is
limited by practical considerations. Of the various gamma-emitting isotopes of Tm, ^{167}Tm has the most favorable half-life and energy of emission,
but it can only be produced with the aid of a cyclotron. However, ^{169}Yb^{3+}-
citrate (Table 9-2), which is more readily available, has been tested clinically in 15 patients with primary lung and liver cancer or lymphosarcoma
(Hisada *et al.*, 1974). Successful delineation of the tumors was achieved
in 13 of the 15 patients. An example is shown in Fig. 9-1. In the other
two patients, accumulation of Yb^{3+} ions in the spinal column and sternum
was thought to have obscured the lesion. However, Hisada *et al.* (1974)
remarked that skeletal deposition of ^{169}Yb^{3+} ions helped to provide "land-
marks" for accurately localizing soft tissue tumors. They found that back-

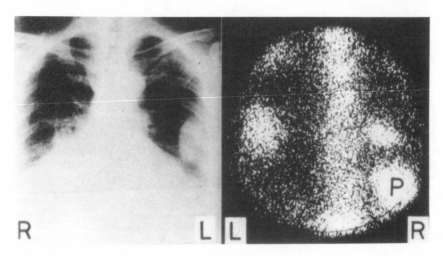

Figure 9-1. Tumor scanning with $^{169}Yb^{3+}$-citrate. Patient with right pulmonary cancer. Chest X-ray (left panel) shows primary focus in right lower lobe with metastasis to left fourth rib. Scintigraph (right panel) shows marked accumulation of ^{169}Yb in primary focus (P), in metastasis in fourth rib, and in a hilar metastasis not seen on the X-ray. From Hisada *et al.* (1974), with permission.

ground radioactivity in the soft tissues was extremely low, permitting clear images to be obtained. According to Tarjan *et al.* (1975), malignant lung tumors accumulate $^{169}Yb^{3+}$-citrate, while benign tumors do not. However, a large clinical trial involving 10 hospitals in Japan showed a 65% detection rate for malignant tumors and 29% false positives for benign tumors. Of all tumors, squamous cell carcinoma was best diagnosed, with a 77% detection rate. Worst visualized were adenocarcinomas, which had a 55% detection rate. The efficiency depended on the part of the body being screened. Whereas tumors of the extremities were detected on every occasion, the success rate fell to 78.5% for tumors of the head and neck, 77.8% for lungs, and only 48.3% for the abdomen (Hisada *et al.*, 1975). Studies on humans by Chatal *et al.* (1975) with $^{169}Yb^{3+}$-citrate confirmed good imaging of lung cancers, primary and secondary brain tumors, liver tumors, and bone metastases of various origins. Unfortunately, $^{169}Yb^{3+}$-citrate can no longer be used in Japan due to concerns over waste disposal (Hisada, personal communication).

The mechanism of the transport of lanthanides to, and accumulation by, the tumors is unknown. Once in the blood, the Ln^{3+} ion is presumably displaced from its citrate complex by competing carrier ligands, as discussed in Section 7.5. Autoradiographic investigation of the distribution of ^{153}Sm (Friedman *et al.*, 1976) or ^{169}Yb (Ando *et al.*, 1977) in tumors reveals that the radioactivity is concentrated in the peripheral regions of the tumor where the blood supply is greatest. The largest amounts of ^{167}Tm were found in connective tissues surrounding the tumor at sites of inflammation. None was detected in necrotic tumor tissue, but areas of viable tumor were radioactive (Ando *et al.*, 1983).

Ando *et al.* (1981, 1982) have homogenized tumors in attempts to determine the putative subcellular location of ^{169}Yb following administration of ^{169}Yb^{3+}-citrate. Although they recovered radioactivity from acidic carbohydrates, these sorts of studies failed to take into account the redistribution of label that occurs on homogenization of whole tissue. Thus, as discussed in Section 6.2.3, they provide little information on the localization or nature of lanthanide ligands *in vivo*.

The ability of lanthanides to accumulate in tumors may have relevance to cancer therapy, as discussed in Section 9.5.

Although the bone-seeking nature of the heavier lanthanides interferes with the imaging of tumors close to bone, these properties suggest their use in bone scanning. Chandra *et al.* (1971) have claimed good results with intravenously injected ^{170}TmCl$_3$ and ^{170}Tm^{3+}-citrate in rabbits. The citrate complexes of ^{171}Er^{3+}, ^{157}Dy^{3+}, and ^{170}Tm^{3+} have been evaluated in the detection of bone lesions in patients with multiple myeloma and solitary plasmacytoma (Hubner *et al.*, 1977). Although neither ^{171}Er^{3+}-citrate nor ^{170}Tm^{3+}-citrate was found to be satisfactory, ^{157}Dy^{3+}-citrate showed promise. Among its advantages were those of not concentrating in the kidneys, bladder, or gastrointestinal tract and thus not obscuring the abdomen or pelvis.

Possible future applications of lanthanides in scintigraphy include the localization of abscesses (Woolfenden *et al.*, 1983) and sites of inflammation (Chatal *et al.*, 1975; Ando *et al.*, 1983) and the detection of cartilaginous lesions in arthritic joints. Lippiello *et al.* (1984) found that the binding of ^{169}Yb^{3+} ions to human articular cartilage *in vitro* increased with the degree of osteoarthritic degeneration. Although their *in vivo* studies with arthritic rabbit knees failed to show a similar correlation, changes from the normal pattern were seen.

As discussed in Section 7.5, many lanthanides tend to accumulate in the liver after intravenous injection. This is especially so if the lanthanides are administered as particles. Agha *et al.* (1982) have reported that ^{169}Yb^{3+}-phytate forms colloidal suspensions which are rapidly cleared by reticuloendothelial cells in the liver. In patients with hepatomegaly, good scintigraphs of the liver were obtained.

9.6.3 Cisternography and Tracer Studies with Radioactive Lanthanides

For purposes of cisternography, lanthanides are usually administered as a complex with diethylenetriaminepentaacetic acid (DTPA). Such complexes are remarkably stable (Section 2.3; Table 2-4) with formation constants of the order of 10^{22}. This ensures that the Ln-DTPA complex remains intact, diffusing into accessible extracellular spaces, with rapid excretion through the kidneys (Section 7.5.3).

Among the first cisternographic uses of these complexes was the measurement of spinal fluid kinetics with ^{169}Yb-DTPA (Deland et al., 1971; Andrews and David, 1974). Following injection into the lumbar intrathecal space, scintography of the brain, spinal region, and abdomen was achieved. Reporting on a series of 125 cisternograms, Deland et al. (1971) claimed excellent-quality images for as long as 108 h after injection, with normal individuals as well as patients suffering from a variety of disorders, including hydrocephalus, obstructed cerebrospinal fluid (CSF) flow, and CSF leaks.

As a diffusible, noninteractive tracer, ^{169}Yb-DTPA has been used to detect CSF rhinorrhea (Wagner et al., 1970; Doge and Johannsen, 1977). The tracer was injected intrathecally or intraventricularly, after which leakage of the CSF became apparent as increased radioactivity in the gastric juice. The agent had a biological half-life in the CSF of about 10 h and could distinguish rhinorrhea from hydrocephalus (Wagner et al., 1970). However, concerns have been raised over the long-term retention of small amounts of ^{169}Yb in the CNS if the Yb-DTPA preparation is improperly formulated (Bolles, 1977). The 32-day half-life of ^{169}Yb complicates this and other concerns (Pauwels and Van Damme, 1974).

As mentioned earlier, Ln-DTPA complexes are rapidly excreted through the kidneys. They thus serve as good markers of the efficiency of urinary excretion. In this capacity, ^{169}Yb-DTPA has been used to monitor glomerular filtration rates (Hosain et al., 1969; Russell et al., 1985).

Lanthanides have also been used as digesta markers in nutritional studies on experimental animals (e.g., Crooker et al., 1982; Dixon et al., 1983; Siddons et al., 1985) and humans (Hutcheson et al., 1979). Microspheres labeled with ^{144}Ce have been used to measure blood flow in dogs (Delgado et al., 1983). Intra-articular injections of ^{169}Yb-DTPA permit scintographic detection of synovial cysts (Adiseshan et al., 1975).

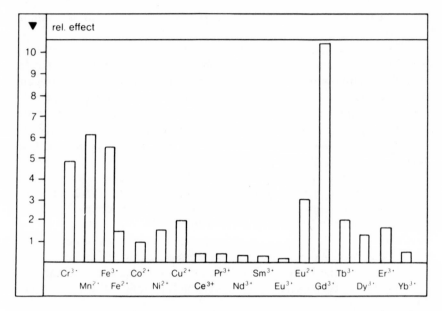

Figure 9-2. Influence of paramagnetic ions on proton spin–lattice relaxation time. From Weinmann *et al.* (1984), with permission.

9.6.4 Nuclear Magnetic Resonance Imaging

Differentiation of tissues by NMR imaging is possible when there exist differences in proton spin–lattice or spin–spin relaxation times, or both. Paramagnetic lanthanides have been introduced as reagents which increase such differences and thereby enhance the contrast obtained. To do this, it is necessary to administer a paramagnetic lanthanide such that its bodily distribution is nonuniform. The Ln^{3+} ion then acts as a local, paramagnetic center which influences the response of susceptible nuclei in its environment to the applied magnetic field. The resulting images are thus affected by the degree to which the lanthanide is present in the target tissue. Contrast enhancement is obtained when the concentration of the lanthanide differs in two adjacent structures.

Examination of the effect of various lanthanides and transition metal ions on proton spin–lattice relaxation times (Fig. 9-2) reveals the superiority of Gd^{3+} ions in this respect. The precise physical and chemical form in which Gd^{3+} is administered depends upon the organ or lesion to be imaged.

Taking advantage of the rapid clearing of particulate lanthanides by the RES, Burnett *et al.* (1985) intravenously injected suspensions of Gd_2O_3 into rabbits and obtained enhanced NMR images of the liver and spleen. Improved pictures of the plasma, kidney, liver, and brain have been obtained at different times after i.v. injection of $GdCl_3$ into rats and mice (Caillé *et al.*, 1983). As orally administered lanthanides are poorly absorbed, Gd^{3+}-oxalate has been fed to dogs to visualize better the G.I. tract (Runge *et al.*, 1984).

However, the greatest potential for clinical application lies with the freely diffusible, nonexchanging DTPA complex of Gd^{3+} (Wolf and Fobben, 1984; Weinmann *et al.*, 1984). As discussed in Sections 2.3 (Table 2-4), 7.5.3, and 9.6.3, Ln-DTPA complexes are remarkably stable, are nontoxic, and are quickly cleared from the body via the urine. Animal studies with rats (Weinmann *et al.*, 1984) have confirmed that the half-life of Gd-DTPA in the blood is about 20 min, and by 3 h after i.v. administration 80% has been excreted unchanged in the urine (Section 7.5.3; Fig. 7-10).

Further experiments with animals have produced good image enhancement in most cases (Table 9-3). Unger *et al.* (1985) attempted to modify the technique by covalently attaching DTPA to antibodies directed against the carcinoembryonic antigen. Addition of Gd^{3+} then provided Gd-DTPA-labeled antibodies for tumor imaging. However, this reagent failed to provide better images of human colon carcinoma tumors in hamsters. Apparently the accumulation of Gd^{3+} in the tumors was only 0.1 μM, three orders of magnitude below the minimum necessary to produce proton relaxation enhancement at 0.35 T. Based upon the affinity of Ln^{3+}-citrate complexes for tumors (Section 9.6.2), it should be possible to detect malignancies by NMR imaging using these chelates.

Gd-DTPA alone permits improved imaging of tumors in humans. Clinical studies with human volunteers began in 1983 (Carr *et al.*, 1984b). In the first published clinical trials, Gd-DTPA was used as a contrast-enhancing agent in NMR imaging of cerebral tumors (Fig. 9-3), hepatic tumors, and transition cell carcinoma of the bladder (Carr *et al.*, 1984a,b; Schorner *et al.*, 1984). However, no improvement was obtained in one patient with hepatic cysts. In most cases, the degree of image enhancement seen with the tumors permitted better imaging than with CT. Gd-DTPA also delineated the margin between the cerebral tumor and peritumoral edema as clearly as is possible with CT. Of interest is the observation that Gd-DTPA forms a circular region of image enhancement around the rim of the tumor (Schorner *et al.*, 1984). This has also been observed by

Table 9-3. Use of Gd-DTPA in NMR Imaging: Animal Studies

Animal	Organ lesion or other location	Contrast enhancement[a]	Reference
Dog	Urine within renal pelvis	+	Brasch et al., 1984
	Soft tissue abscesses	+	
	Brain lesion	+	
	Normal brain	−	
Rabbit	Normal kidney	+	Carr et al., 1984a
	Cerebral infarct	+	
Rabbit	Normal kidney	+	Wolf and Fobben, 1984
Dog	Splenic and renal infarct	+	Runge et al., 1984
Dog	Early myocardial ischemia	+	McNamara et al., 1984
Monkey	CSF cavities	+	Di Chiro et al., 1985
Rat	Pulmonary edema	*	Schmidt et al., 1985
Cat	Cerebral ischemia	−	McNamara et al., 1986a
Dog	Reversible vs. irreversible myocardial injury	+	McNamara et al., 1986b

[a] +, Contrast enhancement improved; −, contrast enhancement not improved; *, contrast enhancement unaltered, but Gd-DTPA permitted quicker imaging.

scintigraphy (Section 9.6.2) and CT scanning (Section 9.6.5). In each case, it seems that Ln-DTPA complexes associate with viable tumor cells and peritumoral inflammation but not with the central necrotic cells. At about the same time, Laniado et al. (1984) reported signal enhancement of normal kidney and bladder in human volunteers.

Good results have subsequently been reported for imaging meningiomas (Bydder, 1985), intracranial tumors (Claussen et al., 1985), and mammary tumors (Heywang et al., 1986). Preliminary data also suggest the intraarticular use of Gd-DTPA in detecting small lesions in cartilage (Gylys–Mosin et al., 1987).

From the results so far obtained, Gd-DTPA seems likely to become routinely used as an adjunct to NMR imaging.

9.6.5 Computed Tomography

Havron et al. (1980) injected suspensions of CeO_2, Dy_2O_3, and Gd_2O_3 into rabbits. About 95% of the injected dose accumulated in the liver, greatly increasing its opacity (Fig. 9-4), thus sharply brightening the image.

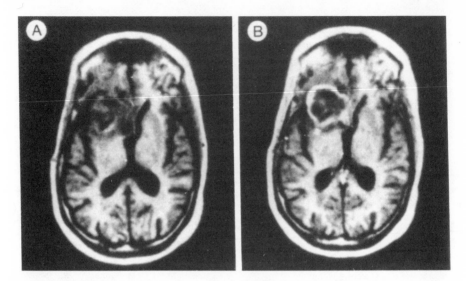

Figure 9-3. NMR imaging of tumors with Gd-DTPA: cerebral tumor before (a) and after (b) injection of Gd-DTPA. From Carr *et al.* (1984b), with permission.

As liver tumors do not phagocytose the particles, they stand out clearly from the surrounding liver. Tumors as small as 5 mm have been detected in this way. One disadvantage of the technique is the slow clearance of the lanthanide from the liver.

Claussen *et al.* (1985) obtained good contrast enhancement of cranial tumors with Gd-DTPA by CT.

9.7 Radiosynovectomy

The synovium is a tissue which lines all diarthrodial joints. In many arthritic conditions, this tissue becomes inflamed, causing articular destruction, pain, swelling, and loss of motion. As drugs are not always successful in controlling this condition, surgical removal of the synovium has become a common orthopedic procedure. In many patients, synovectomy provides symptomatic relief for several years.

An alternative approach to surgical synovectomy is the nonsurgical or medical destruction of the inflamed synovium. Radioisotopes have been widely used to accomplish this. Targeting is made easier by the accessibility of many joints to direct, intra-articular injection and the phagocytic properties of synoviocytes and many of the white blood cells which col-

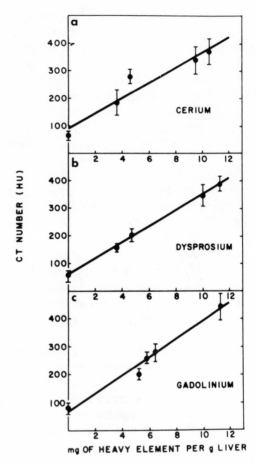

Figure 9-4. Relationship between CT number and concentration of lanthanide in the liver. From Havron *et al.* (1980), with permission.

onize the synovium during inflammation. Thus, the method of choice has been to inject a colloidal suspension of the radionuclide into the joint. As with other forms of radiation therapy, the ideal isotope should provide β-irradiation with a half-life of hours or a few days. Maximum penetration should match the thickness of the inflamed synovium it is to destroy.

Colloidal [198]Au was the first radionuclide to be used (Ansell *et al.*, 1963). While proving an effective agent with which to destroy synovium, unacceptably high leakage of radioactivity from the joint and evidence of chromosome damage to circulating lymphocytes (Stevenson *et al.*, 1971) led to its being supplanted by [90]Y. Colloidal ^{90}Y has been used in several clinical trials with reasonable success (e.g., Spooren *et al.*, 1985; Gumpel *et al.*, 1975; Oka, 1975). However, it also leaked from the joint and caused chromosome damage (De La Chapelle *et al.*, 1972; Doyle *et al.*, 1977). The relatively long half-lives of 2.7 days for ^{198}Au and 60 h for ^{90}Y compounded the effects of their leakage from the joint.

Erbium-169 has found use in nonsurgical synovectomy (Menkes *et al.*, 1977), especially for small joints, such as the finger, where its limited maximum penetration of 0.9 mm is a big advantage. Despite being a purely β-emitter, ^{169}Er has a comparatively long half-life of 9.4 days.

Sledge *et al.* (1977, 1986) have investigated ^{165}Dy as a possible new agent for nonsurgical synovectomy. Among its advantages are a short half-life of 140 min and a maximum tissue penetration of 5.7 mm, which is about the thickness of an inflamed synovium in a human knee. Leakage of ^{165}Dy from the joint was reduced by increasing the particle size to 1–5 μm. Laboratory studies (Sledge *et al.*, 1977) demonstrated that particles comprising mixed aggregates of ^{165}Dy and ferric hydroxide destroyed the synovium in experimentally arthritic rabbits with low leakage from the joint and minimum radiation exposure to other tissues. Clinical trials involving 93 patients and 108 knees (Sledge *et al.*, 1986) confirmed these results. The clinical improvement matched that produced by ^{90}Y, ^{198}Au, or surgical synovectomy, while leakage of radioactivity from the joint was minimal. The main disadvantage of ^{165}Dy is a half-life which requires a local means of generating the isotope.

As an alternative, Bard *et al.* (1985) have experimented with ^{177}Lu, a pure β-emitter, trapped in chelator liposomes. Intra-articular injection of these particles inhibited the progress of experimental arthritis in rabbits at doses at 87, 175, and 350 μCi. Only at the highest dose was there damage to the cartilage, while leakage rates were below 1% per day.

9.8 Atherosclerosis

During the development of atherosclerosis, plaques form on the internal surfaces of certain arteries. These increase in size, partially or totally obstructing blood flow, eventually leading to a myocardial infarction. Among the constituents of atherosclerotic plaques are collagen, elastin, cholesterol, and calcium.

Animal studies have shown that the atherosclerotic process can be inhibited by agents which prevent the deposition of calcium onto artery walls (Kramsch and Chan, 1978). On the basis that La^{3+} ions often antagonize Ca^{2+}-dependent biological processes (Section 2.5; Chapters 4, 5, and 6), Kramsch *et al.* (1980) investigated the effect of orally administered $LaCl_3$ on experimental atherosclerosis in rabbits. The results were surprising and dramatic.

At all concentrations tested (20–40 mg $LaCl_3$/kg body weight), $LaCl_3$ strongly reduced coronary atherosclerosis. The aortic surface area involved with atherosclerotic lesions decreased from 64% in untreated atherosclerotic animals to 16% with 20-mg/kg doses of $LaCl_3$ and to 3% with 40-mg/kg doses. Microscopic examination of the thoracic aortae showed a similar degree of protection by $LaCl_3$. Aortas of atherosclerotic rabbits showed increased cellularity of the intima, accumulation of collagen, damaged elastic fibers, and deposits of calcium, lipid, and glycosaminoglycans. $LaCl_3$ provided a dose-dependent protection from all these changes. Similar protection was found in the pulmonary arteries and the major coronary arteries. Biochemical analyses (Table 9-4) were compatible with the macroscopic and histological findings. According to the authors, oral $LaCl_3$ affords similar protection to monkeys fed an atherosclerotic diet. In both animals, protection occurred without evidence of detrimental effects on other tissues or on such physiological parameters as cardiac function, muscular function, and coordination, and, in monkeys, treadmill performance. Later work by Kramsch has shown that $LaCl_3$ not only inhibits the development of atherosclerotic plaques but actually causes established plaques to regress.

Given these remarkable results and abundant evidence that orally administered La^{3+} salts are of extremely low toxicity (Section 8.4; Table 8-2), one wonders whether any clinical application will evolve from these studies. To ignore them would surely be irresponsible. The most puzzling aspect of this work is the mechanism of action of $LaCl_3$. Numerous studies agree that lanthanide salts are very poorly absorbed from the G.I. tract (Section 7.5.1). Kramsch et al. (1980) reported that 65% of the daily ingested La was eliminated in the feces. Very small amounts were detected in urine, while 4 h after a test meal, only 17.1 μg of La^{3+} were detected per 100 ml of whole blood. It is presumably to these tiny amounts of absorbed La^{3+} that the antiatherosclerotic effects must be ascribed. Whether the effects of lanthanides on hepatic lipid metabolism (Section 8.7.2) are involved remains unknown. However, $LaCl_3$ did not affect the rise in serum cholesterol that accompanied the atherosclerotic diet and had only minor effects on serum calcium. Examination of autopsy material has revealed that infarcted myocardial tissue contains greater amounts of lanthanides than normal (Sihvonen, 1972).

Its specificity of action is quite remarkable, as no other tissues appear to have been affected. Furthermore, it is paradoxical that one of the effects of La^{3+}, a recognized calcergen (Section 8.6), was to inhibit calcification of the aorta. In view of the low circulating levels of La^{3+} and the high

Table 9-4. Components of Aortic Intima Media from Control Rabbits and Rabbits on the Atherogenic Diet with and without Lanthanum[a,b]

Experimental group	Collagen	Elastin	Total cholesterol	Percent ester cholesterol	Total calcium	Nonlipid phosphorus
Control diet	2.8 ± 0.4	10.3 ± 2.8	0.3 ± 0.2	7 ± 2	0.03 ± 0.007	0.04 ± 0.02
Atherogenic diet without LaCl$_3$	6.5 ± 1.3[c]	22.1 ± 1.7[c]	3.5 ± 1.2[c]	50 ± 6[c]	0.06 ± 0.012[c]	0.09 ± 0.03[c]
Atherogenic diet + 20 mg LaCl$_3$/kg body wt.	4.2 ± 0.6[c]	17.3 ± 1.9[c]	1.9 ± 1.0[c]	55 ± 12[c]	0.04 ± 0.010[d]	0.06 ± 0.01[d]
Atherogenic diet + 30 mg LaCl$_3$/kg body wt.	3.6 ± 0.6[d]	14.1 ± 2.6	1.4 ± 0.7[c]	42 ± 5[c]	0.03 ± 0.010	0.04 ± 0.03
Atherogenic diet + 40 mg LaCl$_3$/kg body wt.	2.9 ± 0.8	13.6 ± 4.7	0.6 ± 0.1[d]	19 ± 6[c]	0.03 ± 0.008	0.05 ± 0.02
Control diet + 40 mg LaCl$_3$/kg body wt.	2.4 ± 0.9	10.4 ± 2.0	0.1 ± 0.1	N.D.[e]	0.02 ± 0.011	0.03 ± 0.03

[a] From Kramsch et al. (1980), with permission.
[b] Absolute amounts in milligram per whole aorta per kilogram body weight; mean ± SD.
[c] $P < 0.01$ compared to untreated control.
[d] $P < 0.05$ compared to untreated control.
[e] N.D.: Not detectable.

Ca^{2+} content of bodily fluids, it is hard to envisage how La^{3+} could mediate its effects through competitive antagonism of Ca^{2+}-dependent processes. However, verapamil (Rouleau *et al.*, 1982) and nifedipine (Henry and Bentley, 1981), which, like La^{3+}, block voltage-operated Ca^{2+} channels on cells (Section 6.3), also suppress experimental atherosclerosis. Like La^{3+}, these blockers inhibited atherogenesis without reducing hypercholesterolemia.

Ginsburg *et al.* (1983) confirmed that orally administered $LaCl_3$ (40 mg/kg per day) suppressed atherogenesis in the rabbit by 37.4% in the thoracic aorta and 59.5% in the abdominal aorta. However, $LaCl_3$ did not affect the extent of atherosclerotic disease in the coronary artery. Two organic calcium blockers, diltiazem and flunarizine, had qualitatively similar effects to $LaCl_3$. However, they were much weaker suppressors of atherosclerotic changes in the abdominal aorta.

With cardiovascular disease remaining the major killer in many societies, further investigation of the antiatherosclerotic properties of lanthanides seems mandated.

9.9 Inflammation and Arthritis

The cardinal signs of inflammation are heat, swelling (edema), redness, and pain. The underlying biochemistry and physiology are very complex. Early changes in the permeability of the blood vessels permit the efflux of plasma, leading to edema. Certain types of white blood cells migrate to the site of inflammation under the influence of chemotactic signals. Stimulus-coupled activation of both blood cells and the cells normally resident at the inflammatory site leads to the production of cytokines and proteinases causing local tissue destruction and pain.

A variety of lanthanides and their complexes inhibit one or more of these processes and possess anti-inflammatory properties. Among the first to demonstrate this was Jancso (1962). Working on the hypothesis that inflammation and blood coagulation were closely linked processes and knowing the anticoagulant properties of the lanthanides (Section 8.8), he examined their possible anti-inflammatory properties. Inorganic salts of La^{3+}, Ce^{3+}, Nd^{3+}, Pr^{3+}, and Sm^{3+} inhibited angiotaxis and edema following the increase in vascular permeability caused by inflammatory agents. Certain complexes of Ln^{3+}, including the Dy^{3+} salt of β-acetylpropionic acid (Helodym 88) and the Nd^{3+} salt of sulfoisonicotinic acid (Thrombodym), were also active. Because of the excessive toxicity of Helodym 88 and Thrombodym, Jancso (1962) prepared the rare earth complexes of pyrocatechol sodium disulfonate as less harmful anticoagulant and anti-

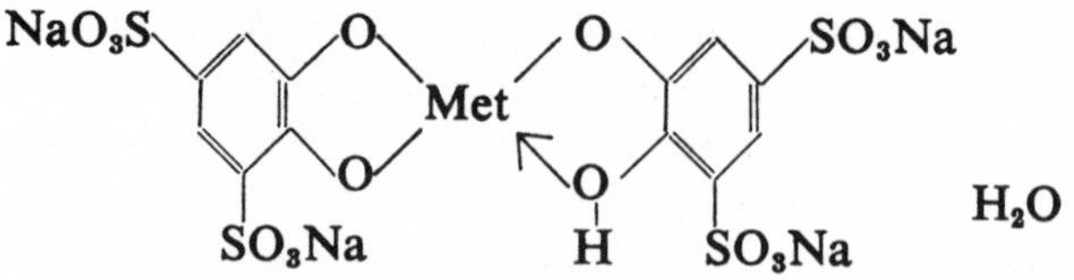

NaO₃S, O, O, SO₃Na — Met — O, O, SO₃Na — SO₃Na — H — SO₃Na — H₂O

Figure 9-5. Structure of lanthanide complex of pyrocatechol sodium disulfonate. From Jancso (1962), with permission. Met = metal ion.

inflammatory agents (Fig. 9-5). Depending on the lanthanide incorporated, these complexes go under various names prefixed by "phlogo" from the Greek word for flame; for example, the Sm^{3+}-substituted compound is called phlogosam. The La^{3+}, Nd^{3+}, Pr^{3+}, and Sm^{3+} complexes are well tolerated by rats and rabbits. They inhibit the inflammatory reactions produced in rats by the subplantar injection of bee venom, cobra venom, compound 40/80, or dextran. At doses of 250–350 mg/kg, inhibition of edema by greater than 80% was claimed.

The anti-inflammatory properties of lanthanides have been confirmed by Basile and Hanada (1979). In rats, increases in capillary permeability induced by histamine or serotonin were inhibited by i.p. injections of 50-mg/kg doses of $PrCl_3$, $GdCl_3$, or $YbCl_3$. When injected i.p. at 20 mg/kg for 6 days, these salts prevented granuloma induction by implanted cotton pellets. In later work (Basile *et al.*, 1984), it was further shown that i.p. injections of 15–75-mg/kg doses of these salts also inhibited paw edema induced by carrageenan, nystatin, or myobacterial adjuvant. These $LnCl_3$ were ineffective when given orally.

Whitehouse and Ellis (cited in Ellis, 1977) have reported similar findings, with Nd^{3+} being the most effective Ln^{3+} ion. The Nd-ATP complex showed greater potency than $NdCl_3$, phlogodym, Nd-catechol, or Nd-AMP in suppressing carrageenan-induced paw inflammation in rats.

Reports in the Hungarian literature claim that suitable preparations of lanthanide complexes can be used as topical anti-inflammatory agents in eczema. According to Wozniak (1970), phlogosam was equally as effective as a 0.25% prednisolone ointment in treating eczema, including refractory eczema cruris madidans. Unlike steroids, anti-inflammatory agents based on lanthanides do not increase the likelihood of infection (Section 9.3). Phlogosam has also been used to treat inflammation of the gums and oral mucosa (Balogh, 1974), skin lesions caused by radiation, and other sites of inflammation amenable to topical application (Wozniak, 1970). However, Lazar and Karady (1965) have reported that phlogodym renders rats more sensitive to shock.

The anti-inflammatory mechanism of lanthanides is unclear. Although

Jancso (1962) favors an explanation based on their anticoagulant properties, his own data show that heparin has no anti-inflammatory activity. A possible alternative explanation (Evans, 1988) makes reference to the ability of Ln^{3+} ions to antagonize certain types of stimulus-coupled cellular activation which depend upon the influx of Ca^{2+} ions (Section 6.4). Northover and Northover (1985) have independently proposed that drugs which modify the cellular handling of Ca^{2+} might be useful anti-inflammatory agents. In addition, they propose that the activities of presently available nonsteroidal anti-inflammatory drugs may involve the antagonism of Ca^{2+}-mediated cellular responses.

Examples of important cellular responses in inflammation include the activation of lymphocytes and macrophages, and polymorphonuclear (PMN) leukocyte chemotaxis and degranulation. In addition, the cells which are normally resident in the uninflamed tissue often become activated during inflammatory episodes. Several of these types of activation are inhibited by Ln^{3+} ions. Examples include antigen and mitogen activation of lymphocytes (Yamage and Evans, 1989), PMN leukocyte chemotaxis (Boucek and Snyderman, 1976), and phagocytosis (Mircevova et al., 1984). PMN degranulation is less strongly inhibited (O'Flaherty et al., 1978). Furthermore, extensive studies by Lazar and his colleagues (Lazar, 1973; Husztik et al., 1980) have shown that a variety of $LnCl_3$ depress the RES in rats (Section 8.10). This occurs at doses as low as 2 mg/kg, well below the LD_{50}. From their studies of the Kupffer cells in the liver, Husztik et al. (1980) suggested that La^{3+} displaces Ca^{2+} from the cell surface, impairing attachment of particles and thus preventing phagocytosis.

It has been speculated that lanthanides or their complexes may find eventual use in the treatment of certain forms of arthritis (Evans, 1979, 1988). There are two aspects to this. One involves the inhibition of certain neutral metalloproteinases such as collagenase (Evans, 1979; Evans and Ridella, 1985; Section 4.11) which are thought to erode connective tissue matrices, including cartilage, in arthritic joints. In this capacity, lanthanides would act as antierosive agents. The other aspect involves the suppression of stimulus-coupled cellular activation in diseased joints. This would not only have a general anti-inflammatory effect, as discussed above, but would provide an additional antierosive mechanism by inhibiting the release of neutral proteinases from synoviocytes, macrophages, and PMN leukocytes.

Following this suggestion, Esposito et al. (1986) measured the concentrations of lanthanides in the plasma and synovial fluid of normal subjects and patients with rheumatoid arthritis. Huge differences were found. In normal synovial fluid, the lanthanide concentration was below the limit of detection. In the synovial fluid of rheumatoid patients, the

Table 9-5. Lanthanide Concentrations in Normal and Rheumatoid
Plasma and Synovial Fluid[a]

| | Concentration (mean ± SD; μg/liter) | | |
| | Healthy subjects[b] | RA patients | |
Element	Plasma	Plasma	Synovial fluid
La	4.49 ± 0.51	6.30 ± 0.86	11.73 ± 2.32
Ce	45.10 ± 4.98	59.60 ± 5.44	94.40 ± 7.41
Nd	27.98 ± 1.02	42.50 ± 6.50	63.75 ± 19.43
Eu	1.94 ± 0.42	3.26 ± 0.51	5.19 ± 1.69
Yb	2.04 ± 0.11	3.36 ± 0.62	5.86 ± 1.61
Lu	0.16 ± 0.06	0.28 ± 0.06	0.53 ± 0.19

[a] From Esposito et al. (1986), with permission.
[b] No lanthanides could be detected in the synovial fluid of healthy subjects.

concentration of the various lanthanide ions totaled nearly 200 μg/liter. Smaller increases were found in the plasma concentrations of lanthanides in rheumatoid patients (Table 9-5). These results remain unexplained but are extremely provocative.

Targeting of potential antiarthritic drugs, such as lanthanide complexes, is facilitated by the ease with which intra-articular injection is possible in many joints. Furthermore, imaging studies (Section 9.6) have confirmed the affinity of lanthanides for sites of inflammation. Radioactivity from intra-articularly injected $^{69}YbCl_3$ solutions is retained in rabbit knee joints through interaction with the synovium and cartilage (Section 7.5.1; Table 7-8). Twenty-eight days after injection, accumulation of $^{169}Yb^{3+}$ in such extra-articular locations as the kidney, liver, bone, and ear cartilage was small, and no ^{169}Yb was detected in blood at this time. The rate of clearance of ^{169}Yb following intra-articular injection into human knees was different in each of four patients tested (Fig. 7-1). It is of interest that clearance was slowest from the most severely diseased knees. This may be related to the possibly greater affinity of $^{169}Yb^{3+}$ ions for damaged cartilage (Lipiello et al., 1984) or to the greater mass of synovium in diseased joints. In each patient, more than 99% of the nuclide cleared from the joint appeared in the urine within 72 h (McCarty et al., 1979).

Thus, if Ln^{3+} ions do have antiarthritic properties, their use will be aided by their high degree of intra-articular retention and limited accumulation in other parts of the body. High concentrations of La^{3+} extract proteoglycans from cartilage (Section 3.8.1), leaving it more deformable (Sokoloff, 1963). However, these concentrations are well above those envisaged for therapeutic use. Furthermore, a suitable lanthanide complex would presumably lack this property.

9.10 Summary

Although not fulfilling their early promise as medically useful anti-emetics, anticoagulants, or systemic anti-infectious agents, lanthanides remain of potential clinical value. Very promising clinical trials have been reported with topical cerium-based ointments for burns, with colloidal radioactive lanthanides for radiosynovectomy, with various radioactive Ln^{3+} ions for tumor scanning, and with Gd-DTPA for enhanced NMR imaging. At the time of writing, the last of these is becoming something of a growth industry and looks set to become a standard procedure. In addition, some success has been obtained with radioactive lanthanides in the radiotherapy of tumors. Lanthanides possess remarkable antiathero-sclerotic properties in experimental animals, a finding with great clinical implications.

More speculatively, there are suggestions of possible future employ-ment of lanthanides in the treatment of inflammation and arthritis. Whether any of these will eventually become clinically useful is presently unknown.

References

Adiseshan, N., Johnson, F. L., and Buttfield, I. H., 1975. Detection of synovial cysts by transmission–emission scintigrams following intraarticular [169]Yb-DTPA, *Aust. N.Z.J. Med.* 5:256–260.

Agha, N. H., Al-Hilli, A. M., Hassan, H. A., Al-Hissoni, M. H., and Miran, K. M., 1982. Ytterbium-169-phytate: a potential new radiopharmaceutical for functional scintigraphy of the liver, *Int. J. Appl. Radiat. Isotop.* 33:673–677.

Allgower, M., Burri, C., Cueri, L., Engley, M., Gruber, U. F., Harder, F., and Russel, R. G. G., 1968. Study of burn toxins, *Ann. N.Y. Acad. Sci.* 150:808–815.

Allgower, M., Cueri, L. B., and Stadtler, K., 1973. Burn toxin in mouse skin, *J. Trauma* 13:95–111.

Ando, A., Doishita, K., Ando, I., Sanada, S., Hiraki, T., Midsukami, M., and Hisada, K., 1977. Study of distribution of [169]Yb, ^{67}Ga and ^{111}In in tumor tissue by macroautora-diography: comparison between viable tumor tissue and inflammatory infiltration around tumor, *Radioisotopes (Tokyo)* 26:421–422.

Ando, A., Ando, I., Takeshita, M., Hiraki, T., and Hisada, K., 1981. Subcellular distribution of ^{111}In and [169]Yb in tumor and liver, *Eur. J. Nucl. Med.* 6:221–226.

Ando, A., Ando, I., Hiraki, T., Takeshita, M., and Hisada, K., 1982. Mechanism of tumor and liver concentration of ^{111}In and [169]Yb: [111]In and [169]Yb binding substances in tumor tissues and liver, *Eur. J. Nucl. Med.* 7:298–303.

Ando, A., Ando, I., Sakamoto, K., Hiraki, T., Hisada, K., and Takeshita, M., 1983. Affinity of [167]Tm-citrate for tumor and liver tissue, *Eur. J. Nucl. Med.* 8:440–446.

Andrews, R. P., and David E., 1974. Cisternography with chelated ytterbium 169, *J. Maine Med. Assoc.* 65:313–332.

Anghileri, L. J., 1979. Effects of gallium and lanthanum on experimental tumor growth, *Eur. J. Cancer* 15:1459–1462.

Anghileri, L., 1988. Potentiation of hyperthermia by lanthanum, *Recent Results Cancer Res.* 109:126–135.

Anghileri, L. J., Marchal, C., Crone, M. C., and Robert, J., 1983. Enhancement of hyperthermia lethality by lanthanum, *Arch. Geschwulstforsch.* 53:335–339.

Ansell, B. M., Crook, A., Mallard, J. R., and Bywaters, E. G. L., 1963. Evaluation of intra-articular colloidal gold [198]Au in the treatment of knee effusions, *Ann. Rheum. Dis.* 22:435–439.

Arvela, P., 1979. Toxicity of rare-earths, *Prog. Pharmacol.* 2:71–114.

Baehr, G., and Wessler, H., 1909. The use of cerium oxalate for the relief of vomiting: an experimental study of the effects of some salts of cerium, lanthanum, praseodymium, neodymium and thorium, *Arch. Intern. Med.* 2:517–531.

Balogh, G., 1974. Phlogosol therapy in inflammations of the oral mucosa, *Ther. Hung.* 22:83–89.

Bard, D. R., Knight, C. G., and Page-Thomas, D. P., 1985. Effect of the intra-articular injection of lutetium-177 in chelator liposomes on the progress of an experimental arthritis in rabbits, *Clin. Exp. Rheumatol.* 3:237–242.

Basile, A. C., and Hanada, S., 1979. Inhibitory effects of praseodymium, gadolinium and ytterbium chlorides on the increase of vascular permeability and on granuloma tissue formation in rats, *An. Farm. Quim. São Paulo* 19:3–26.

Basile, A. C., Hanada, S., Sertie, J. A. A., and Oga, S., 1984. Anti-inflammatory effects of praseodymium, gadolinium and ytterbium chlorides, *J. Pharmacobio-Dyn.* 7:94–100.

Beaser, S. B., Segel, A., and Vandam, L., 1942. The anticoagulant effects in rabbits and man of the intravenous injection of salts of the rare earths, *J. Clin. Invest.* 21:447–453.

Beazley, V. C, Thrane, P., and Rolla, G., 1980. Effect of mouthrinses with SnF_2, $LaCl_3$, NaF and chlorhexidine on the amount of lipoteichoic acid formed in plaque, *Scand. J. Dent. Res.* 88:193–200.

Beyer, G. J., Franke, W. G., Hennig, K., Johannsen, B. A., Khalkin, V. A., Kretzschmar, M., Lebedev, N. A., Munze, R., Novgorodov, A. F., and Thieme, K., 1978. Comparative kinetic studies of simultaneously injected [167]Tm and ^{67}Ga-citrate in normal and tumour bearing mice, *Int. J. Appl. Radiat. Isotop.* 29:673–681, 1978.

Bloomer, W. D., McLaughlin, W. H., Lambrecht, R. M., Atcher, R. W., Mirzadeh, S., Madera, J. L., Milius, R. A., Zalutsky, M. R., Adelstein, S. J., and Wolf, A. P., 1984. [211]At radiocolloid therapy: further observations and comparisons with radiocolloids of ^{32}P, [165]Dy and ^{90}Y, *Int. J. Radiat. Oncol. Biol. Phys.* 10:341–348.

Bolles, T. F., 1977. Suitability of Yb 169 DTPA for cisternography, *Semi. Nucl. Med.* 7:201.

Boucek, M. M., and Snyderman, R., 1976. Calcium influx requirements for human neutrophil chemotaxis: inhibition by lanthanum chloride, *Science* 193:905–907.

Bowser, B. H., Caldwell, F. T., Cone, J. B., Eisenach, K. D., and Thompson, C. H., 1981. A prospective analysis of silver sulfadiazine with and without cerium nitrate as a topical agent in the treatment of severely burned children, *J. Trauma* 21:558–563.

Brasch, R. C., Weinmann, H. J., and Wesbey, G. E., 1984. Contrast-enhanced NMR imaging: animal studies using gadolinium-DTPA complex, *Am. J. Radiol.* 142:625–630.

Browning, E., 1961. *Toxicity of Industrial Metals,* Butterworth, London.

Burke, C. W., Doyle, F. H., Joplin, G. F., Arndt, R. N., Macerlean, D. P., and Fraser,

T. R., 1973. Cushing's disease: treatment by pituitary implantation of radioactive gold or yttrium seeds, *Quart. J. Med.* 42:693–714.

Burnett, K. R., Wolf, G. L., Schumacher, H. R., and Goldstein, E. J., 1985. Gadolinium oxide: a prototype agent for contrast enhanced imaging of the liver and spleen with magnetic resonance, *Magn. Reson. Imaging* 3:65–71.

Bydder, G. M., Kingsley, D. P., Brown, J., Niendorf, H. P., and Young, I. R., 1985. MR imaging of meningiomas, including studies with and without gadolinium-DTPA, *J. Comput. Assist. Tomogr.* 9:690–697.

Caillé, J. M., Lemenceau, B., and Bonnemain, B., 1983. Gadolinium as a contrast agent for NMR, *Am. J. Nucl. Med.* 4:1041–1042.

Carr, D. H., Brown, J., Leung, A. W., and Pennock, J. M., 1984a. Iron and gadolinium chelates as contrast agents in NMR imaging: preliminary studies, *J. Comput. Assist. Tomogr.* 8:385–389.

Carr, D. H., Brown, J., Bydder, G. M., Steiner, R. E., Weinmann, H. J., Speck, U., Hall, A. S., and Young, I. R., 1984b. Gadolinium-DTPA as a contrast agent in MRI: initial clinical experience in 20 patients, *Am. J. Radiol.* 143:215–224.

Cassar, J., Doyle, F. H., Lewis, P. D., Mashiter, K., Van Noorden, S., and Joplin, G. F., 1976. Treatment of Nelson's syndrome by pituitary implantation of yttrium-90 or gold-108, *Br. Med. J.* 2:269–272.

Chandra, R., Hernberg, J., and Brauenstein, P., 1971. ^{167}Tm: a new bone scanning agent, *Radiology* 100:687–689.

Chatal, J. F., LeMevel, B. P., Guihard, R., Guihard, E., and Moigneteau, C., 1975. Intérêt diagnostique du citrate d'ytterbium 169 en cancérologie, *J. Radiol. Electrol.* 56:401–409.

Claussen, C., Laniado, M., Kazner, E., Schorner, W., and Felix, R., 1985. Application of contrast agents in CT and MRI (NMR): their potential in imaging of brain tumors, *Neuroradiology* 27:164–171.

Cooper, J. A. D., Bulkley, G. J., and O'Conor, V. J., 1956. Intraprostatic injection of radioactive yttrium chloride in the dog, in *Rare Earths in Biochemical and Medical Research* (G. C. Kyker and E. B. Anderson, eds.), U.S. Atomic Energy Commission, Report ORINS-12, pp. 323–331.

Crone-Escanye, M. C., Anghileri, L. J., and Robert, J., 1985. Enhancement of hematoporphyrin derivative uptake *in vitro* and *in vivo* by tumor cells in the presence of lanthanum, *Tumor* 71:39–43.

Crooker, B. A., Clark, J. H., and Shanks, R. D., 1982. Rare earth elements as markers for rate of passage measurements of individual feedstuffs through the digestive tract of ruminants, *J. Nutr.* 112:1353–1361.

De La Chapelle, A., Oka, M., Rekonen, A., and Ruotsi, A., 1972. Chromosomal damage after intraarticular injection of radioactive yttrium. Effect of immobilization on the biological dose, *Ann. Rheum. Dis.* 31:508–512.

Deland, F. H., James, A. E., Wagner, H. N., and Hosain, F., 1971. Cisternography with ^{169}Yb-DTPA, *J. Nucl. Med.* 12:683–689.

Delgado, G., Butterfield, A. B., Dritschilo, A., Hummel, S., Harbert, J., Petrilli, E. S., and Kot, P. A., 1983. Measure of blood flow by the multiple radioactive microsphere technique in radiated gastrointestinal tissue, *Am. J. Clin. Oncol.* 6:463–467.

Di Chiro, G., Knop, R. H., Girton, M. E., Dwyer, A. J., Doppman, J. L., Patronas, N. J., Gansow, O. A., Brechbiel, M. W., and Brooks, R. A., 1985. MR cisternography and myelography with Gd-DTPA in monkeys, *Radiology* 157:373–377.

Divald, S., and Joullie, M. M., 1970. Coagulants and anticoagulants, in *Medicinal Chemistry* (A. Burger, ed.), 3rd ed., Part II, Wiley Interscience, New York, pp. 1092–1122.

Dixon, R. M., Kennelly, J. J., and Milligan, L. P., 1983. Kinetics of [103]Ru phenanthroline and dysprosium particulate markers in the rumen of steers, *Br. J. Nutr.* 49:463–473.

Doge, H., and Johannsen, B. A., 1977. Radioactivity in gastric juice—a simple adjunct to the Yb-169 DTPA cisternographic diagnosis of CSF rhinorrhea: concise communication, *J. Nucl. Med.* 18:1202–1204.

Domotor, E., 1969. Treatment of first and second degree burns with phlogosam ointment, *Ther. Hung.* 17:40–43.

Doyle, D. V., Glass, J. J., Grow, P. J., Daker, M., and Grahame, R., 1977. A clinical and prospective chromosomal study of yttrium-90 synovectomy, *Rheumatol. Rehab.* 16: 217–222.

Dyckerhoff, H., and Goossens, N., 1939. Über die thromboseverhutende Wirkung des Neodyms (Neodympraparat "Amer 144"), *Z. Exp. Med.* 106:181–192.

Ellis, K. J., 1977. The lanthanide elements in biochemistry, biology and medicine, *Inorg. Perspect. Biol. Med.* 1:101–135.

Esposito, M., Oddone, M., Accardo, S., and Cutolo, M., 1986a. Concentrations of lanthanides in plasma and synovial fluid in rheumatoid arthritis, *Clin. Chem.* 32:1598.

Esposito, M., Collecchi, P., Brera, S., Mora, E., Mazzucotelli, A., Cutolo, M., and Oddone, M., 1986b. Plasma and tissue levels of lanthanide elements in malignant and nonmalignant human tissues, *Sci. Total Environ.* 50:55–63.

Evans, C. H., 1979. Enzyme inhibitors as possible therapeutic agents for arthritis, *Orthop. Surv.* 3:63–69.

Evans, C. H., 1987. Alkaline earths, transition metals and lanthanides, in *Calcium in Drug Actions* (P. F. Baker, ed.), Springer-Verlag, Heidelberg, pp. 527–546.

Evans, C. H., and Ridella, J. D., 1985. Inhibition, by lanthanides, of neutral proteinases secreted by human, rheumatoid synovium, *Eur. J. Biochem.* 151:29–32.

Fasching, W., Wense, G., and Zangl, A., 1974. Résultats du traitement des cancers avancés du sein par l'implantation intrahypophysaire d'yttrium radioactif, *Bull. Soc. Int. Chir.* 2:81–84.

Fox, C. L., Monafo, W. W., Ayvazian, V. N., Skinner, A. M., Modak, S., Stanford, J., and Condict, C., 1977. Topical chemotherapy for burns using cerium salts and silver sulfadiazine, *Surg. Gynecol. Obstet.* 144:668–672.

Friedman, A. M., Sullivan, J. C., Ruby, S. L., Lindenbaum, A., Russell, J. J., Zabransky, B. J., and Rayudu, G. U., 1976. Studies of tumor metabolism—I: By use of Mössbauer spectroscopy and autoradiography of [153]Sm, *Int. J. Nucl. Med. Biol.* 3:37–40.

Frouin, A., and Roudsky, D., 1914. Action bactéricide et antitoxique des sels de lanthanum et de thorium sur le vibrion cholérique. Action thérapeutique de ces sels dans le choléra experimental, *C. R. Acad. Sci.* 159:410–413.

Ginsburg, R., Davis, K., Bristow, M. R., McKennett, K., Kodsi, S. R., Billingham, M. E., and Schroeder, J. S., 1983. Calcium antagonists suppress atherogenesis in aorta but not in the intramural coronary arteries of cholesterol-fed rabbits, *Lab. Invest.* 49:154–158.

Goldie, H., and West, H. D., 1956. Effect of peritumoral tissue infiltration with radioactive yttrium on growth and spread of malignant cells, *Cancer Res.* 16:484–489.

Grenet, H., and Drouin, H., 1920. Les sels de terres de la série du cérium dans le traitement de la tuberculose pulmonaire chronique, *Gaz. Hop.* 93:789–791.

Guidi, G., 1930. Contributo alla farmacologia delle terre rare; il neodimio, *Arch. Int. Pharmacodyn. Ther.* 37:305–348.

Gumpel, J. M., Beer, T. C., Crawley, J. C., and Farran, H. E., 1975. Yttrium 90 in persistent synovitis of the knees: a single center comparison of four radiocolloids, *Br. J. Radiol.* 48:377–381.

Gylys–Morin, V. M., Hajek, P. C., Sartoris, D. J., and Resnick, D., 1987. Articular cartilage defects: detectability in cadaver knees with MR, *Am. J. Radiol.* 148:1153–1157.

Hansbrough, J. F., Zapata-Sirvent, R., Peterson, V., Wang, X., Bender, E., Claman, H., and Boswick, J., 1984. Characterization of the immunosuppressive effect of burned tissue in an animal model, *J. Surg. Res.* 37:383–393.

Havron, A., Davis, M. A., Selter, S. E., Paskins-Hurlburt, A. J., and Hessel, S. J., 1980. Heavy metal particulate contrast materials for computed tomography of the liver, *J. Comput. Assist. Tomogr.* 4:642–648.

Heggers, J. P., Ko, F., and Robson, M. C., 1979. Cerium nitrate silver sulphadiazine: synergism or antagonism as determined by minimum inhibitory concentrations, *Burns* 5: 308–311.

Helvig, E. I., Munster, A. M., Su, C. T., and Oppel, M., 1979. Cerium nitrate–silver sulfadiazine cream in the treatment of burns: a prospective, randomized study, *Am. Surg.* 45:270–272.

Henry, P. D., and Bentley, K. I., 1981. Suppression of atherogenesis in cholesterol-fed rabbit treated with nifedipine, *J. Clin. Invest.* 68:1366–1396.

Hermans, R. P., 1984. Topical treatment of serious infections with special reference to the use of a mixture of silver sulphadiazine and cerium nitrates: two clinical studies, *Burns Incl. Therm. Inj.* 11:59–62.

Heywang, S. H., Hahn, D., Schmidt, H., Krischke, I., Eiermann, W., Bassermann, R., and Lissner, J., 1986. MR imaging of the breast using gadolinium-DTPA, *J. Comput. Assist. Tomogr.* 10:199–204.

Higasi, T., Ito, K., Tobari, H., and Tomura, K., 1973. On the accumulation of rare earth elements in animal tumor, *Int. J. Nucl. Med. Biol.* 1:98–101.

Hisada, K., and Ando, A., 1973. Radiolanthanides as promising tumor scanning agents, *J. Nucl. Med.* 14:615–617.

Hisada, K., Tonami, N., Hiraki, T., and Ando, A., 1974. Tumor scanning with [169]Yb-citrate, *J. Nucl. Med.* 15:210–212.

Hisada, K., Suzuki, Y., Hiraki, T., Sanos, H., and Suzaki, K., 1975. Clinical evaluation of tumor scanning with [169]Yb-citrate, *Radiology* 116:389–393.

Holder, I. A., 1982. *In vitro* inactivation of silver sulphadiazine by the addition of cerium salts, *Burns Incl. Therm. Inj.* 8:274–277.

Hosain, F., Reba, R. C., and Wagner, H. N., 1969. Measurement of glomerular filtration rate using chelated ytterbium 169, *Int. J. Appl. Radiat.* 20:517–521.

Hubner, K. F., Andrews, G. A., Hayes, R. L., Poggenburg, J. K., and Solomon, A., 1977. The use of rare-earth radionuclides and other bone seekers in the evaluation of bone lesions in patients with multiple myeloma or solitary plasmacytoma, *Radiology* 125: 171–176.

Hunter, R. B., and Walker, W., 1956. Anticoagulant action of neodymium 3-sulpho-isonicotinate, *Nature* 178:47.

Husztik, E., Lazar, G., and Parducz, A., 1980. Electron microscopic study of Kupffer-cell phagocytosis blockade induced by gadolinium chloride, *Br. J. Exp. Pathol.* 61:624:630.

Hutcheson, D. P., Venugopal, B., Gray, D. H., and Luckey, T., 1979. Lanthanide markers in a single sample for nutrient studies in humans, *J. Nutr.* 109:702–707.

Jancso, H., 1962. Inflammation and the inflammatory mechanisms, *J. Pharmacol.* 13:577–594.

Kramsch, D. M., and Chan, C. T., 1978. The effect of agents interfering with soft tissue calcification and cell proliferation on calcific fibrous-fatty plaques in rabbits, *Circ. Res.* 42:562–571.

Kramsch, D., Aspen, A. J., and Apstein, C. S., 1980. Suppression of experimental atherosclerosis by the Ca^{2+}-antagonist lanthanum. Possible role of calcium in atherogenesis, *J. Clin. Invest.* 65:967–981.

Kremer, B., Allgower, M., Graf, M., Schmidt, K. H., Schoetmerich, J., and Schoenenberger, G. A., 1981. The present status of research in burn toxins, *Intensive Care Med.* 7:77–87.

Kyker, G. C., Christopherson, W. M., Berg, H. F., and Brucer, M., 1956. Selective irradiation of lymph nodes by radiolutecium (Lu[177]), *Cancer* 9:489–498.

Laniado, M., Weinmann, H. J., Schorner, W., Felix, R., and Speck, U., 1984. First use of GdDTPA/dimeglumine in man, *Physiol. Chem. Phys. Med. N.M.R.* 16:257–265.

Laszlo, D., Ekstein, D. M., Lewin, R., and Stern, K. G., 1952. Biological studies on stable and radioactive rare earth compounds. 1. On the distribution of lanthanum in the mammalian organism, *J. Natl. Cancer Inst.* 13:559–571.

Lazar, G., 1973. The reticuloendothelial-blocking effect of rare earth metals in rats, *J. Reticuloendothel. Soc.* 13:231–237.

Lazar, G., and Karady, S., 1965. Traumatic and endotoxin shock in rats, *J. Pharm. Pharmacol.* 17:517–518.

Lewin, R., Stern, K. G., Ekstein, D. M., Woidowsky, L., and Laszlo, D., 1953. Biological studies on stable and radioactive rare earth compounds. II. The effect of lanthanum on mice bearing Ehrlich ascites tumor, *J. Natl. Cancer Inst.* 14:45–56.

Lewin, R., Hart, H. E., Greenberg, J., Spencer, H., Stern, K. G., and Laszlo, D., 1954. Biological studies on stable and radioactive rare earth compounds. III. Distribution of radioactive yttrium in normal and ascites-bearing mice and in cancer patients with serous effusions, *J. Natl. Cancer Inst.* 15:131–143.

Lippiello, L., Prellwitz, J., Schmetter, R., Connolly, J., and Quaife, M., 1984. Investigation of ytterbium-169 as a predictor of joint degeneration in osteoarthritis, *Trans. Orthop. Res. Soc.* 9:68.

Manly, R. S., and Bibby, B. G., 1949. Substances capable of reducing the acid solubility of tooth enamel, *J. Dent. Res.* 43:346–352.

Maxwell, L. C., Bischoff, F., and Ottery, E. M., 1931. Studies in cancer chemotherapy. X. The effect of thorium, cerium, erbium, yttrium, didymium, praseodymium, manganese and lead upon transplantable rat tumors, *J. Pharmacol. Exp. Ther.* 43:61–70.

Mayer, S. W., and Morton, M. E., 1956. Preparation and distribution of yttrium 90 fluoride, in *Rare Earths in Biochemical and Medical Research* (G. C. Kyker and E. B. Anderson, eds.), U.S. Atomic Energy Commission, Report ORINS-12, pp. 263–279.

McCarty, D. J., Palmer, D. W., and Halverson, P. B., 1979. Clearance of calcium pyrophosphate dihydrate crystals *in vivo*. I. Studies using [169]Yb labeled triclinic crystals, *Arthritis Rheum.* 22:718–727.

McNamara, M. T., Higgins, C. B., Ehman, R. L., Revel, D., Sievers, R., and Brasch, R. C., 1984. Acute myocardial ischemia: magnetic resonance contrast enhancement with gadolinium-DTPA, *Radiology* 153:157–163.

McNamara, M. T., Brant-Zawadzki, M., Berry, I., Pereira, B., Weinstein, P., Derugin, N., Moore, S., Kucharczyk, W., and Brasch, R. C., 1986a. Acute experimental cerebral ischemia: MR enhancement using Gd-DTPA, *Radiology* 158:701–705.

McNamara, M. T., Tscholakoff, D., Revel, D., Soulen, R., Schechtmann, N., Botvinick, E., and Higgins, C. B., 1986b. Differentiation of reversible and irreversible myocardial injury by MR imaging with and without gadolinium-DTPA, *Radiology* 158:765–769.

Menkes, C. J., LeGo, A., Verrier, P., Aignan, M., and Delbarre, F., 1977. Double-blind study of erbium-169 injection in rheumatoid digital joints, *Ann. Rheum. Dis.* 36:254–256.

Mercado, R. C., and Ludwig, T. G., 1973. Effect of yttrium on dental caries in rats, *Arch. Oral Biol.* 18:637–640.

Miller, L. P., 1959. Factors influencing the uptake and toxicity of fungicides, *Trans. N. Y. Acad. Sci.* 21:442–445.

Mircevova, L., Viktora, L., and Hermanova, E., 1984. Inhibition of phagocytosis of poly-morphonuclear leucocytes by adenosine and $HoCl_3$ *in vitro, Med. Biol.* 62:326–330.

Monafo, L., 1983. The use of topical cerium nitrate–silver sulfadiazine in major burn injuries, *Panminerva Med.* 25:151–156.

Monafo, W. W., Tandon, S. N., Ayvazian, V. H., Tuchschmidt, J., Skinner, A. M., and Deitz, F., 1976. Cerium nitrate: a new topical antiseptic for extensive burns, *Surgery* 80:465–473.

Munster, A. M., Helvig, E., and Rowland, S., 1980. Cerium nitrate–silver sulfadiazine cream in the treatment of burns: a prospective evaluation, *Surgery* 88:658–660.

Muroma, A., 1958. Studies on the bactericidal action of salts of certain rare earth metals, *Ann. Med. Exp. Biol. Fenn.* 36(Suppl. 6):1–54.

Nadjmi, M., 1970. The long-term results after the stereotactic implantation of yttrium 90 into the pituitary gland as a treatment for Cushing-syndrome, *Confin. Neurol.* 32: 203–206.

Northover, A. M, and Northover, B. J., 1985. Calcium ions in acute inflammation: a possible site for anti-inflammatory drug action, in *Handbook of Inflammation, Vol. 5, The Pharmacology of Inflammation* (I. L. Banta, M. A. Bray, and M. J. Parnham, eds.), Elsevier, New York, pp. 235–254.

O'Flaherty, J. T., Showell, H. J., Becker, E. L., and Ward, P. A., 1978. Substances which aggregate neutrophils. Mechanism of action. *Am. J. Pathol.* 92:155–166.

Oka, M., 1975. Radiation synovectomy of the rheumatoid knee with yttrium 90, *Ann. Clin. Res.* 7:205–210.

Pauwels, E. K. J., and Van Damme, K. J., 1974. An unexpected complication in the use of ¹⁶⁷Yb-DTPA with respect to hospital health physics, *Int. J. Nucl. Med. Biol.* 1:228–229.

Peterson, V. M., Hansbrough, J. F., Wang, X. W., Zapata-Sirvent, R., and Boswick, J. A., 1985. Topical cerium nitrate prevents postburn immunosuppression, *J. Trauma* 25: 1039–1044.

Regolati, B., Schait, A., Schmid, R., and Muhlemann, H. R., 1975. Effect of enamel sol-ubility reducing agents on erosion in the rat, *Helv. Odontol. Acta* 19:31–36.

Rosenkranz, H. S., 1979. A synergism effect between cerium nitrate and silver sulphadiazine, *Burns* 5:278–281.

Rouleau, J. L., Parmley, W. W., Stevens, J., Wilkman-Coffelt, J., Sievers, R., Mahley, R., and Havel, R. J., 1982. Verapamil suppresses atherosclerosis in cholesterol-fed rabbits, *Am. J. Cardiol.* 49:889.

Runge, V. M., Clanton, J. A., Foster, M. A., Smith, F. W., Lukehart, C. M., Jones, M. M., Partain, C. L., and James, A. E., 1984. Paramagnetic NMR contrast agents. Develop-ment and evaluation, *Invest. Radiol.* 19:408–415.

Russell, C. D., Bischoff, P. G., Katzen, F. N., Rowell, K. L., Yester, M. U., Lloyd, L. K., Tauxe, W. N., and Dubovsky, E. V., 1985. Measurement of the glomerular filtration rate: single injection plasma clearance method without urine collection, *J. Nucl. Med.* 26:1243–1247.

Saffer, L. D., Rodeheaver, G. T., Hiebert, J. M., and Edlick, R. F., 1980. *In vivo* and *in vitro* antimicrobial activity of silver sulfadiazine and cerium nitrate, *Surg. Gynecol. Obstet.* 151:232–236.

Schmidt, H. C., McNamara, M. T., Brasch, R. C., and Higgins, C. B., 1985. Assessment of severity of experimental pulmonary edema with magnetic resonance imaging. Effect of relaxation enhancement by Gd-DTPA, *Invest. Radiol.* 20:687–692.

Schomacker, K., Franke, W. G., Henke, E., Fromm, W. D., Maka, G., and Beyer, G. J.,

1986. The influence of isotopic and nonisotopic carriers on the biodistribution and biokinetics of M^{3+}-citrate complexes, *Eur. J. Nucl. Med.* 11:345–349.

Schorner, W., Kazner, E., Laniado, M., Sprung, C., and Felix, R., 1984. Magnetic resonance tomography (MRT) of intracranial tumors: initial experience with the use of the contrast medium gadolinium-DTPA, *Neurosurg. Rev.* 7:303–312.

Siddons, R. C., Paradine, J., Beever, D. E., and Cornell, P. R., 1985. Ytterbium acetate as a particulate-phase digesta flow marker, *Br. J. Nutr.* 54:509–519.

Sihvonen, M. L., 1972. Accumulation of yttrium and lanthanoids in human and rat tissues as shown by mass spectrometric analysis and some experiments with rats, *Ann. Acad. Sci. Fenn. Ser. A* 168:1–62.

Sledge, C. B., Nolde, J., Hnatowich, D. J., Kramer, R., and Shortkroff, S., 1977. Experimental radiation synovectomy by ^{165}Dy ferric hydroxide macroaggregate, *Arthritis Rheum.* 20:1334–1342.

Sledge, C. B., Zuckerman, J. D., Zalutsky, M. R., Atcher, R. W., Shortkroff, S., Lionberger, D. R., Rose, H. A., Hurson, B. J., Lankenner, P. A., Anderson, R. J., and Bloomer, W. A., 1986. Treatment of rheumatoid synovitis of the knee with intraarticular injection of dysprosium 165-ferric hydroxide macroaggregates, *Arthritis Rheum.* 29:153–159.

Smith, B. M., Gindhart, T. D., and Colburn, N. H., 1986. Possible involvement of a lanthanide-sensitive protein kinase C substrate in lanthanide promotion of neoplastic transformation, *Carcinogenesis* 7:1949–1956.

Sokoloff, L., 1963. Elasticity of articular cartilage: effects of ions and viscous solutions, *Science* 141:1055–1057.

Spooren, P. F., Rasker, J. J., and Arens, R. P., 1985. Synovectomy of the knee with ^{90}Y, *Eur. J. Nucl. Med.* 10:441–445.

Stevenson, A. C., Hill, A. G. S., and Hills, H., 1971. Chromosome damage in patients who have had intraarticular injections of radioactive gold, *Lancet* i:837–839.

Sullivan, J. C., Friedman, A. M., Rayudu, G. V., Fordham, E. W., and Ramachandran, P. C., 1975. Tumor localization studies with radioactive lanthanide and actinide complexes, *Int. J. Nucl. Med. Biol.* 2:44–45.

Tarjan, G., Karika, Z., Pal, I., and Schweiger, O., 1975. The role of ^{169}Yb-citrate in the diagnosis of lung tumors, *Nucl. Med.* 13:267–271.

Unger, E. C., Totty, W. G., Neufeld, D. M., Otsuka, F. L., Murphy, W. A., Welch, M. S., Connett, J. M., and Philpott, G. W., 1985. Magnetic resonance imaging using gadolinium labeled monoclonal antibody, *Invest. Radiol.* 20:693–700.

Vincke, E., and Oelkers, H. A., 1937. Zur pharmacologie der seltenen Erden: Workung auf die Blutgerinnung, *Arch. Exp. Pathol.* 187:594–603.

Vincke, E., and Sucker, E., 1950. Über ein neues antithromboticum das neodymsulfoisonicotinat, *Klin. Wochenschr.* 28:74–75.

Wagner, H. N., Hosain, F., DeLand, F. H., and Som, P., 1970. A new radiopharmaceutical for cisternography: chelated ytterbium 169, *Radiology* 95:121–125.

Waler, S. M., and Rolla, G., 1983. Effect of chlorhexidine and lanthanum on plaque formation, *Scand. J. Dent. Res.* 91:260–262.

Weinmann, H.-J., Brasch, R. C., Press, W. R., and Wesbey, G. E., 1984. Characteristics of gadolinium-DTPA complex: a potential NMR contrast agent, *Am. J. Radiol.* 142:619–624.

Wolf, G. L., and Fobben, E. S., 1984. The tissue proton T_1 and T_2 response to gadolinium DTPA injection in rabbits. A potential renal contrast agent for NMR imaging, *Invest. Radiol.* 19:324–328.

Woolfenden, J. M., Hall, J. N., Barber, H. B., and Wacks, M. E., 1983. [^{153}Sm] citrate for tumor and abscess localization, *Int. J. Nucl. Med. Biol.* 10:251–256.

Wozniak, R. D., 1970. Efficiency of phlogosam in the treatment of eczema, *Ther. Hung.* 18:111–112.

Yano, Y., and Chu, P., 1975. Cyclotron-produced thulium-169 for bone and tumor scanning, *Int. J. Nucl. Med. Biol.* 2:135–139.

Zapata-Sirvent, R. L., Hansbrough, J. F., Bendo, E. M., Bartle, E. J., Mansour, M. A., and Carter, W. H., 1986. Postburn immunosuppression in an animal model. IV. Improved resistance to septic challenge with immunomodulating drugs, *Surgery* 99:53–58.

Appendix: Review Articles on Lanthanide Biochemistry

Anghileri, L., 1988. Potentiation of hyperthermia by lanthanum, *Recent Results Cancer Res.* 109:126–135.

Arvela, P., 1979. Toxicity of rare-earths, *Prog. Pharmacol.* 2:71–114.

Dos Remedios, C., 1981. Lanthanide ion probes of calcium-binding sites on cellular membranes, *Cell Calcium* 2:29–51.

Ellis, K. J., 1977. The lanthanide elements in biochemistry, biology and medicine, *Inorg. Perspect. Biol. Med.* 1:101–135.

Evans, C. H., 1983. Interesting and useful properties of lanthanides, *Trends Biochem. Sci.* 8:445–449.

Evans, C. H., 1988. Alkaline earths, transition metals and lanthanides, in *Calcium in Drug Actions* (P. F. Baker, ed.), Springer-Verlag, Heidelberg, pp. 527–546.

Glasel, J. A., 1973. Lanthanide ions as nuclear magnetic resonance chemical shift probes in biological systems, in *Current Research Topics in Bioinorganic Chemistry* (S. J. Lippard, ed.), Vol. 18, Wiley, New York, pp. 383–413.

Gschneidner, K. A., and Capellen, J. (eds.), 1987. 1787–1987: Two hundred years of rare earths, North-Holland, Amsterdam.

Haley, T. J., 1965. Pharmacology and toxicology of the rare earth elements, *J. Pharm. Sci.* 54:663–670.

Haley, T. J., 1979. Toxicity, in *Handbook on the Physics and Chemistry of Rare Earths* (K. A. Gschneidner and L. Eyring, eds.), Vol. 4, North-Holland, Amsterdam, pp. 553–585.

Horrocks, W. DeW., 1982. Lanthanide ion probes of biomolecular structure, in *Advances in Inorganic Biochemistry* (G. L. Eichhorn and L. G. Marzilli, eds.), Vol. 4, Elsevier Biomedical, New York, pp. 201–261.

Horrocks, W. DeW., and Sudnick, D. R., 1981. Lanthanide ion luminescence probes of the structure of biological macromolecules, *Acc. Chem. Res.* 14:383–392.

Karraker, D. G., 1970. Coordination of trivalent lanthanide ions, *J. Chem. Educ.* 47:424–430.

Kyker, G. C., 1956. The distribution of interstitial and intracavitary injections of certain radiochemical preparations of medical interest, *Bol. Assoc. Med. Puerto Rico* 46:362–374.

Kyker, G. C., 1962. Rare earths, in *Mineral Metabolism. An Advanced Treatise* (C. L. Comar and F. Bronner, eds.), Vol. 2, Part B, Academic Press, New York, pp. 499–541.

Kyker, G. C., and Anderson, E. B. (eds.), 1956. Rare earths in biochemical and medical research, U.S. Atomic Energy Commission, Report ORINS-12.

Martin, R. B., 1983. Structural chemistry of calcium: lanthanides as probes, in *Calcium in Biology* (T. G. Spiro, ed.), Wiley, New York, pp. 237–270.

Martin, R. B., and Richardson, F. S., 1979. Lanthanides as probes for calcium in biological systems, *Quart. Rev. Biophys.* 12:181–209.

Mikkelsen, R. B., 1976. Lanthanides as calcium probes in biomembranes, in *Biological Membranes* (D. Chapman and D. F. H. Wallach, eds.), Academic Press, New York, pp. 153–190.

Nieboer, E., 1975. The lanthanide ions as structural probes in biological and model systems, *Struct. Bonding* 22:1–47.

O'Hara, P. B., 1987. Lanthanide ions as luminescent probes of biomolecular structure, *Photochem. Photobiol.* 46:1067–1070.

Ramsden, E. N., 1961. A review of experimental work on radio-yttrium comprising 1. the tissue distribution, 2. the mechanism of deposition in bone, and 3. the state in the blood, *Int. J. Radiat. Biol.* 3:399–410.

Reuben, J., 1975. The lanthanides as spectroscopic and magnetic resonance probes in biological systems, *Naturwissenschaften* 62:172–178.

Reuben, J., 1979. Bioinorganic chemistry: lanthanides as probes in systems of biological interest, in *Handbook on the Physics and Chemistry of Rare Earths* (K. A. Gschneidner and L. Eyring, eds.), Vol. 4, North-Holland, Amsterdam, pp. 515–552.

Reuben, J., and Elgavish, G. A., 1979. Shift reagents and NMR of paramagnetic lanthanide complexes, in *Handbook on the Physics and Chemistry of Rare Earths* (K. A. Gschneidner and L. Eyring, eds.), Vol. 4, North-Holland, Amsterdam, pp. 483–514.

Richardson, F. S., 1982. Terbium(III) and europium(III) ions as luminescent probes and stains for biomolecular systems, *Chem. Rev.* 82:541–552.

Shaklai, M., and Tavassoli, M., 1982. Lanthanum as an electron microscopic stain, *J. Histochem. Cytochem.* 30:1325–1330.

Steidle, H., 1935. Seltene Erdmetalle, in *Handbuch der Experimentellen Pharmakologie* (A. Heffter and W. Heubner, eds.), Vol. 3, Part 4, Springer-Verlag, Berlin, pp. 2189–2213.

Switzer, M. E., 1978. The lanthanide ions as probes of calcium ion binding sites in biological systems, *Sci. Prog. Oxf.* 65:19–30.

Trapman, H., 1959. Neuartige metallionerkatalytische Vorgänge, in Sonderheit in Bereich der selterer Erdmetalle, und ihre Auswirkunger auf das Zellgescheher. *Arzneim-Forsch.* 9:341–346, 403–410.

Venugopal, B., and Luckey, T. D., 1978. Toxicity of group III metals, in *Metal Toxicity in Mammals. Vol. 2. Chemical Toxicity of Metals and Metalloids*, Plenum Press, New York, pp. 101–173.

Weiss, G. B., 1974. Cellular pharmacology of lanthanum, *Annu. Rev. Pharmacol.* 14:343–354.

J. Chinese Rare Earth Soc., September 1985, Special issue on hygiene and toxicology.

Index